T0271308

Foundations of Artificial Intelligence and Robotics

Artificial intelligence (AI) is a complicated science that combines philosophy, cognitive psychology, neuroscience, mathematics and logic (logicism), economics, computer science, computability, and software. Meanwhile, robotics is an engineering field that compliments AI. There can be situations where AI can function without a robot (e.g., Turing Test) and robotics without AI (e.g., teleoperation), but in many cases, each technology requires each other to exhibit a complete system: having "smart" robots and AI being able to control its interactions (i.e., effectors) with its environment. This book provides a complete history of computing, AI, and robotics from its early development to state-of-the-art technology, providing a roadmap of these complicated and constantly evolving subjects.

Divided into two volumes covering the progress of symbolic logic and the explosion in learning/deep learning in natural language and perception, this first volume investigates the coming together of AI (the mind) and robotics (the body), and discusses the state of AI today.

Key Features:

- Provides a complete overview of the topic of AI, starting with philosophy, psychology, neuroscience, and logicism, and extending to the action of the robots and AI needed for a futuristic society.
- Provides a holistic view of AI, and touches on all the misconceptions and tangents to the technologies through taking a systematic approach.
- Provides a glossary of terms, list of notable people, and extensive references.
- Provides the interconnections and history of the progress of technology for over 100 years as both the hardware (Moore's Law, GPUs) and software, i.e., generative AI, have advanced.

Intended as a complete reference, this book is useful to undergraduate and postgraduate students of computing, as well as the general reader. It can also be used as a textbook by course convenors. If you only had one book on AI and robotics, this set would be the first reference to acquire and learn about the theory and practice.

Wendell H. Chun has 30-plus years of experience designing and building robots with autonomous capability at the former Martin Marietta Space Systems Company. He is currently a Director in Systems Engineering and an Assistant Professor at the University of Denver. Professor Chun bridges industry, academia, and government in the areas of robotics and AI, serving as a consultant, subject matter expert, and reviewer in this field since the early 1980s.

Foundations of Artificial Intelligence and Robotics

Volume 1 A Holistic View

Wendell H. Chun
University of Denver

CRC Press
Taylor & Francis Group
Boca Raton London New York

CRC Press is an imprint of the
Taylor & Francis Group, an **informa** business
A CHAPMAN & HALL BOOK

First edition published 2025
by CRC Press
2385 NW Executive Center Drive, Suite 320, Boca Raton FL 33431

and by CRC Press
4 Park Square, Milton Park, Abingdon, Oxon, OX14 4RN

CRC Press is an imprint of Taylor & Francis Group, LLC

ISBN: 9781032673110 (hbk)
ISBN: 9781032673127 (pbk)
ISBN: 9781032673134 (ebk)

DOI: 10.1201/9781032673134

Typeset in Palatino
by codeMantra

Contents

Preface

Artificial intelligence (AI) is a complicated science that combines philosophy, cognitive psychology, neuroscience, mathematics and logic (logicism), economics, computer science, computability, and software. Linguistics is also important so that its "AI" can communicate with human operators at its user interface. AI has a long history, starting way before the formalization of the term artificial intelligence at the famous Dartmouth Conference in 1956. Robotics is an engineering field that complements AI. There can be situations where AI can function without a robot (e.g., Turing Test, Robot Process Automation) and robotics without AI (e.g., teleoperation), but in many cases, each technology requires each other to exhibit a complete system: having "smart" robots and AI being able to control its interactions (i.e., effectors) with its environment. At one point in time, robotics and AI were parallel disciplines, but because they are so interwoven, some researchers would consider robotics a subset of AI, while others view robotics separate from AI as a supporting discipline.

AI is viewed as a science, while robotics is an engineering endeavor that focuses on a solution. This is based on the critical mind and brain problem. The human brain has been used as a model, but there are many things that we do not understand about the human brain and the mind itself. This issue has been a topic of discussion for centuries by philosophers. Neuroscience tries to understand the brain, and cognitive psychology sees the mind as an information-processing system, with memory a major component. Included in this research is the notion of perception and the ability to communicate with natural language and the military's need for language translation (based on the Cold War at the time). Similar to AI's human model, the robot requires AI techniques to navigate, search, and manipulate objects. Robotics' "man-in-the-loop" is a mature technology, but AI enables an unmanned vehicle or robot to operate autonomously and ultimately intelligently. Central to almost all AI systems is the need for machine learning. The ultimate robot is a humanoid that was intelligent as envisioned by Karl Capek and later Isaac Asimov. It is important to remember that many of AI's promises and claims in the 20th century were wildly over-optimistic. Finally, the typical test for human intelligence is the Turing Test, with its strengths and weaknesses.

This is a history book on the subject, as well as a description of the subject which provides a roadmap of this complicated topic that is still evolving. This book encompasses the entire subject and provides a status of the "state-of-the-art" from an academic and industry perspective. This book was not intended to be a textbook (can be used as such) but as a complete source of reference. It tries to bring together multiple disciplines in a wide and in-depth subject, however, which is still early in its development. This book has two volumes: Background and History. The Background Volume provides a primer on AI and robotics, starting with the mind-body problem from philosophy. This volume focuses on cognition and logicism, and introduces cybernetics, robotics, mechatronics, and system theory. This is followed by human–robot theory, autonomy (architectures and decision making), and autonomous robots. There is an extensive discussion of modern AI and all of its main pillars. The first half of Volume 1 investigates the coming together of AI (the mind) and the robot (the body). The second half of Volume 1 discusses where AI is today including science, engineering, and math. There was a considerable amount of time used on the symbolic approach to AI (1956 – 1990s) and the current approach is based on connectionism and the neural network. This volume ends with a potential solution with neuro-symbolic AI, and an example where traditional computer vision is enhanced with deep learning.

The second volume goes more in depth into the history of AI, starting with key people and important events that bring us to today. This includes the early pioneers in the field, AI winters, military funding, and the DARPA Strategic Computing Initiative. There are groundbreaking events that chronicle robot systems in factories, underwater, vacuum cleaners, self-driving cars, playing games, rovers, and drones. AI technologies and robots are constantly being improved in the areas of entertainment, warehouses, surgery, exoskeletons, quadrotors, and walking robots. There is considerable effort placed on machine learning, conversational AI, and the newer Generative AI with Generative Adversarial Networks (GANs). Volume 2 ends with looking toward the future with miscellaneous topics such as the singularity, ethics, and AI sentience. We look at AI and robotic efforts all over the world, which will impact jobs and employment. At the end of Volume 2, an epilogue and an incomplete list of contributors to the AI and robotics technology effort are provided. Each volume has a glossary of terms, and includes relevant figures and references.

The two volumes complement each other and are made available as a set. A substantial bibliography at the end of each volume allows readers to explore and expand their knowledge based on their individual interests.

About the Author

Wendell H. Chun is the Director of Systems Engineering and an Assistant Professor in Electrical Engineering at the University of Denver. He has had a long career in robotics and artificial intelligence. Both robotics and systems engineering are multi-disciplinary engineering disciplines. His classes are based on his work in academia, industry (small, medium, and large businesses), and government as a SETA (Support Engineering and Technical Assistance) contractor to the Department of Energy. Volume 1 is based on the author's graduate courses in:

- Robotics
- Intelligent Machines
- Computational AI
- Advanced Robotics
- Mechatronics
- Systems Engineering
- Advanced Ground System Robotics
- Unmanned Aerial Vehicles/Drones
- Fault Diagnosis and Prognosis for Engineering Systems
- Advanced Systems Design
- Senior Design
- Fundamentals of System Electrical, Mechanical, and Software Design

The keys to any multi-disciplinary design project are requirements development, architecture design, hardware–software integration, and verification and validation (V&V) through testing. Professor Chun's academic teaching has been complemented by professional teaching/workshops on:

- Space Robotics (Launchspace Ltd.)
- Mobile Robots (SPIE)
- Fundamentals of Drones: UAV Concepts, Designs, and Technologies (AIAA)
- Artificial Intelligence (Launchspace Ltd.)
- Introduction to Robotics (Department of Energy)
- Unmanned Systems 101 (AUVSI)
- Space Discovery Graduate Course (US Space Foundation)

1

Artificial Intelligence Background

The expectations of smart machines (robots and artificial intelligence [AI]) have been high. The promise of machines that "think" intrigues everyone, including the public in general. There is an expectation that machines someday will display human intelligence, and it is speculated that expert human knowledge should be innate. Inherent knowledge is those skills that are beneficial to robots in completing planned goals using robotic behaviors. With the ever-increasing speed and computing power of modern computers, we may be able to construct smart machines for specific problems (e.g., autonomous vehicle control and credit card fraud detection), and to be sure, the complexity of the problems for which smart machines are deployed is increasing as we progress, but will we ever construct machines that can learn for themselves from scratch, that is, machines that can truly reason? Similar promises are expected from robots, for example, remember that amazing robot you were promised when you were a kid? The one that would: fold the laundry, clean the toilets, or do your homework? On TV and in the movies, such robotic machines have been with us for quite a while, from Gort in The Day the Earth Stood Still to R2-D2 in Star Wars and T-800 in The Terminator. Along with flying cars and teleportation, it seemed just a matter of time before we all would have our own reliable robot. So where are we today? This is a story about AI and robotics, but neither technology can be developed in isolation. In past history, these technologies were developed in parallel paths where a robot without intelligence would be in the same category as a brain without a body. Roboticists assumed that AI would be developed in parallel, while AI researchers assumed that robotics would be a sub-element of their technology. The reality is that AI without a body is hard to demonstrate to the public. In essence, we require the mind and body together (Figure 1.1).

It is difficult to talk about the whole without talking about its pieces and to talk about the pieces without referencing the whole. This is especially true when describing AI where the whole is "strong" AI or artificial general intelligence (AGI), and the pieces are "weak" AI such as expert systems, search algorithms, machine learning, planning, machine vision, natural language processing, speech, reasoning, inference, and problem-solving.

So, what is AI?

- AI is *psychology*, where a complete model of human thought is put on a machine.
- AI is *mathematics*, such that knowledge representation conforms to established formalisms and logics.
- AI is *linguistics* by building grammars of English (and other languages) and putting these grammars on a machine. This provides the means to communicate with a human.

FIGURE 1.1
Robot toys.

DOI: 10.1201/9781032673134-1

- And finally, AI is *software* programs on the machine or robot that provide a means to add domain knowledge. Software science also includes "sets of algorithms".

The fact that intelligence, defined as the knowledge of a certain set of associations appropriate to a domain, can always be accounted for in terms of relations among a number of highly abstract features of a skill domain does not, however, preserve the rationalist intuition that these explanatory features must capture the essential structure of the domain, that is, that one could base a theory on them. Intelligence is the ability to learn and understand, to solve problems, and to make decisions. It is important to remember that there is no AI without a learning component. AI has two goals: build an intelligent machine and then explore the nature of intelligence. One way to attack the problem is to list some features that we would expect an intelligent entity to have. For example, critical features include communication (internal and external to the machine), internal knowledge, world knowledge, intentionality (goal-driven), and creativity (having the ability to learn and adapt).

As proposed by Roger Schank, AI can be defined by its issues:

1. representation (of knowledge),
2. decoding (translating the real world into representation, i.e., natural language and vision),
3. inference (extracting knowledge from messages),
4. the control of combinatorial explosion,
5. indexing (organizing knowledge),
6. prediction and recovery (understanding its own working to predict what will happen next),
7. dynamic modification (learning by assimilating information and changing the nature of its program),
8. generalization (connect experiences that are not obviously connectable),
9. curiosity (ability to wonder why to explain a hypothesis), and
10. creativity (e.g., consciousness).

Dietrich Dörner makes an underlying assumption that humans can adapt to uncertainty. Humans can form hypotheses about situations marked by uncertainty and can anticipate their actions by planning. They can expect the unexpected and take precautions against it. And AI is not just about robots. It is also about understanding the nature of intelligent thought and action using computers as experimental devices. By 1944, for example, Herb Simon had laid the basis for the information processing, symbol-manipulation theory of psychology:

> Any rational decision may be viewed as a conclusion reached from certain premises.... The behavior of a rational person can be controlled, therefore, if the value and factual premises upon which he bases his decisions are specified for him.
>
> **Newell & Simon [1972]**

AI in its formative years was influenced by ideas from many disciplines. These came from people working in engineering (such as Norbert Wiener's work on cybernetics, which includes feedback and control), biology (e.g., W. Ross Ashby, Warren McCulloch, and Walter Pitts's work on neural networks in simple organisms), experimental psychology (see Newell and Simon), communication theory (e.g., Claude Shannon's theoretical work), game theory (notably by John Von Neumann and Oskar Morgenstern), mathematics and statistics (e.g., Irving J. Good), logic and philosophy (e.g., Alan Turing, Alonzo Church, and Carl Hempel), and linguistics (such as Noam Chomsky's work on grammar). Early programs were necessarily limited in scope by the size and speed of memory, and processors by the relative clumsiness of the early operating systems and languages. Memory management, for example, was the programmer's problem until the invention of garbage collection.

Philosophy of Mind-Body

Aristotle (384–322 BC) was the first to formulate a precise set of laws governing the rational part of the mind. He developed an informal system of syllogisms for proper reasoning, allowing one to generate conclusions mechanically given initial premises. In 1308, Catalan poet *Ramon Llull* (1232–1316) realized that useful reasoning could actually be carried out by a mechanical artifact. Llull is one of the most intriguing figures of the High Middle Ages, and his thoughts are expressed in a huge corpus of works, which was important not only in his own day, but throughout the Renaissance and Early Modern period. His manuscripts include the *Ars brevis,* which has been called "Llull's single most influential work", and a text on preaching, the *Ars abbreviate praedicandi*, a Latin translation of a work Llull wrote in Catalan. He also completed his *Ars generalis ultima* (The Ultimate General

Art), a method of using a paper-based mechanical means to create new knowledge from combinations of concepts. Llull devised a system of thought that he wanted to impart to others to assist them in theological debates, among other intellectual pursuits. He wanted to create a universal language using a logical combination of terms. Called a Concept Wheel, it contains nine principles, namely, goodness, greatness, etc., and nine letters, namely, B, C, D, E, etc. This figure is circular because the subject is transformed into the predicate and vice versa, as in saying "goodness is great", "greatness is good", and so forth. In this figure, the artist seeks out the naturally proportionate connection standing between the subject and the predicate, in order to find media for drawing conclusions. The principles of this figure implicitly contain everything in existence, given that everything that exists is good, great, etc. For instance, God and angels are good, etc. Hence, all existing things can be reduced to the said principles. The nine subjects of the wheel are God, Angels, Heaven, Man, Imaginative Power, Sensitive Power, Vegetative Power, Elementative, and Instrumentative. To apply this wheel, there are three parts: apply the implicit to the explicit, apply the abstract to the concrete, and apply questions to the loci. The next step, the Hundred Forms with its definition will put the subjects closer to the intellect's reach. Llull's system was based on the belief that only a limited number of undeniable truths exist in all fields of knowledge and by studying all combinations of these elementary truths, humankind could attain the ultimate truth. His art could be used to "banish all erroneous opinions" and to arrive at "true intellectual certitude removed from any doubt".

It's easy to forget that our quest to create computers capable of thinking like humans first began with humans learning to think like computers. One of the key individuals who taught us the basis for the kind of computational thinking that underpins today's most robust AI systems was Ramon Llull. He was passionate about translating the message of Christ and had an ardent faith in the Catholic doctrine of God's perfection. With those two pillars set in place, Llull went on to produce a prolific output of over 200 books on a diverse array of topics and developed what he called the Ars Magna, or the Great Art – an incredibly complex memory system and symbolic science that used tables and cipher-wheels to codify the common language with which all the Abrahamic faiths use to describe God and God's perfect creation. When this codification is turned outward to the world and applied, the same systematic architecture that worked to create a calculation of our perceptual world becomes the root dynamics of our current attempts to reproduce those perceptual states through mechanical calculations. Thus, Llull's inner art when turned outwards becomes the inner workings of our current information systems. Just the fact that in seeking to create a mechanism of logic and reason, he opened a space to begin thinking about mechanisms that are capable of logic and reason.

Thomas Hobbes (1588–1679) proposed that reasoning was like numerical computation in that "we add and subtract in our silent thoughts". Around 1500, *Leonardo da Vinci* (1452–1519) designed but did not build a *mechanical calculator.* The first known *calculating machine* was constructed around 1623 by the German scientist *Wilhelm Schickard* (1592–1635), and in 1942, *Blaise Pascal* (1623–1662) built the Pascaline (or arithmetic machine). Schickard's "Calculating Clock" is composed of a multiplying device, a mechanism for recording intermediate results, and a six-digit decimal-adding device (also described as a rotating clock). *Gottfried Wilhelm Leibniz* built a mechanical device intended to carry out operations on concepts rather than numbers. Now that we have the idea of a set of rules that can describe the formal, rational part of the mind, the next step is to consider *the mind as a physical system.* Next, *René Descartes* gave the first clear discussion of the distinction between mind and matter, and of the problems that arise. Unfortunately, the truth of the matter is that the mind itself is unobservable.

Galen of Pergamum (129–216 AD) was *a prominent Greek physician and philosopher.* He believed, however, that there is no sharp distinction between the mental and the physical, but this position was a controversial argument at the time. Galen agreed with some Greek philosophical schools (especially Platonism, a metaphysical view but not attributable to Plato) in believing that the mind and body were not separate faculties, and he believed that this could be scientifically shown, but his results/discoveries made his position less clear. This paradigm is known as monism. On the other hand, the Newtonian Paradigm is built on the theory of Cartesian Reductionism. The Newtonian Paradigm, also known as the mechanistic paradigm, assumes that things in the environment around humans are more like machines than like life itself (circa 19th century). In this *Machine Metaphor,* along with Cartesian Dualism (as described by Descartes): "the Body is a biological machine and the mind as something apart from the body". The intuitive concept of a machine that is built up from distinct parts and can be reduced to those parts without losing its machine-like character, thus, *Cartesian Reductionism.* The *Newtonian Paradigm* and the three General Laws of Motion are used as the foundation of the modern scientific method where *dynamics* is the center of the framework and leads to trajectory.

But the real world is made up of complex things and a world of simple mechanisms is fictitious and created by science. Our approach to experiments involves *reducing the "system" to its parts* (the essence of systems theory) and then studying those parts in a context formulated according to dynamics. AI Science is (1) *Sensing* (observing the world) + (2) *Mental activity* (making sense out of that sensory information using the mind/computer). We *encode natural systems* (NSs) into *formal systems* (FSs); and manipulate FS to mimic the causal change in the NS. From the FS, we derive an *implication* that corresponds to the causal event in the FS. We then *decode* the FS and check its success in representing the causal event in the NS. The FS uses representations, typically in the form of symbols and models.

We begin with the mind-body problem, which is a debate concerning the relationship between thought and consciousness in the human mind and the brain (as part of the physical, human body). In philosophy, dualism is a set of views about the relationship between mind and matter, which begins with the claim that mental phenomena are in some respects, nonphysical. Minds and bodies are often supposed to be very *different kinds of things*: bodies are physical or material and minds are mental or immaterial. Minds and bodies are supposed to have very different properties. For example, bodies occupy space and have weight, while minds do not. Minds may have ideas, feelings, and possibly dreams, but not have bodies. The question is whether mind and body are considered distinct because they are different in nature as postulated by René Descartes, so can we address each distinction separately? This is the dualism of mind and body. In 1950, Alan Turing started asking the question – can machines think? If yes, then are we machines? Thomas Hobbes believes thinking is the manipulation of symbols and reasoning is computation. "By reasoning", he says,

> I understand computation. And to compute Is to collect the sum of many things added together at the same time, or to know the remainder when one thing has been taken from another. To reason therefore is the same as to add or to subtract.

Accordingly, he divided human acts up into two distinct kinds, mechanical and rational. Mechanical acts were those that could be imitated by automata: walking, eating, and playing the flute. The rational, however, could not be imitated: judgment, will, choice. Descartes argued that in the pineal gland, buried deep within the brain, and soul, the director of rational behavior, met and interacted with the material body. Though what could or could not be imitated has changed, this scheme bears some surprising structural similarities to descriptions of human cognition offered by late-20th-century workers. And as modern researchers have been deeply influenced by the computer model, so was Descartes influenced by the automata (Figure 1.2).

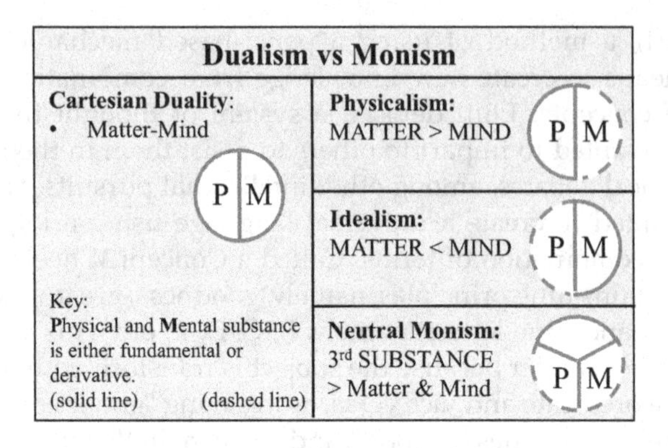

FIGURE 1.2
René Descartes' dualism vs. monism.

The symbols are mental symbols. These symbols are defined as a perceptible something that stands for something else, for example, alphabet symbol, numeral, road sign, and music symbol. For example, a symbol such as an "apple" symbolizes something which is edible and red in color. In some other languages, we might have some other symbol that symbolizes the same edible object. René Descartes regarded "thoughts" themselves as symbolic representations and "perception" as an internal process. He also believes a symbol and what it symbolizes are two different things. The mind is a nonphysical, and therefore a non-spatial substance. Descartes clearly identified the mind with consciousness and self-awareness and distinguished this from the brain as the seat of intelligence. Thus thinking, the algebraic manipulation of symbols, is related to the mind. Now how thought and matter interact is the mind-body problem. Descartes also believed that the universe is written in the language of mathematics and pointed out that all of reality is mathematical. Thus, geometry can be expressed as algebra, which is the study of mathematical symbols and the rules for manipulating these symbols. A different way to create AI was to build machines that have a mind of their own.

Wilhelm Wundt opened the first laboratory dedicated to psychology, and its opening is usually thought of as the beginning of modern psychology. He is often regarded as the father of psychology. Wundt was important because he separated psychology from philosophy by analyzing the workings of the mind in a more structured way, with the emphasis being on objective measurement and control. Wundt's aim was to record thoughts and sensations and to analyze them into their constituent elements, in much the same way as a chemist analyses chemical compounds, in order to get at the underlying structure. Wundt wanted to study the structure of the human mind (using introspection which is experimental self-observation). Wundt believed in reductionism. That is, he believed consciousness could be broken down (or reduced) to its basic elements without sacrificing any of the properties of the whole. Wundt concentrated on three areas of mental functioning: thoughts, images, and feelings.

Ken Craik was a pioneer in the use of computation as a theoretical model for human thought. He wrote *The Nature of Explanation*. In this book, he first laid the foundation for the concept of mental models that the mind forms models of reality and uses them to predict similar future events. He was thus one of the earliest practitioners of cognitive science. One of the most fundamental properties of thought is its power of predicting events. In all of his observations on the process of thought, he reduced thought to its simplest terms, is as follows: a man observes some external event or process and arrives at some "conclusion" or "prediction" expressed in words or numbers that "mean" or refer to or describe some external event or process which comes to pass if the man's reasoning was correct. He suggested that the human mind is a kind of machine that constructs small-scale models of reality that it uses to anticipate events. During the process of reasoning, Craik may also have availed himself of words or numbers. Here there are the three essential processes required:

1. "Translation" of external process into words, numbers, or other symbols,
2. Arrival at other symbols by a process of "reasoning", deduction, inference, etc., and
3. "Retranslation" of these symbols into external processes (as in building a bridge to a design) or at least recognition of the correspondence between these symbols and external events (as in realizing that a prediction is fulfilled).

Here we have a very close parallel to the three stages of reasoning – the translation' of the external processes into their representatives (positions of gears, etc.) in the model; the arrival at other positions of gears, etc., by mechanical processes in the instrument; and finally, the retranslation of these into physical processes of the original type. Roboticists have translated Craik's process of reasoning to a Sense–Plan–Act paradigm where Sense–Plan–Act (SPA) is the predominant robot control methodology. Sense (gather information using the sensors), Plan (create a world model using all the information, and plan the next move), and Act or action. Action is the basis of robotics. SPA is used in iterations: After the acting phase, the sensing phase, and the entire cycle, is repeated. In the case for robotics, the robot is the body (including the brain), and AI represents the mind. The SPA is also known as the "deliberate paradigm" and the function of "planning" in the SPA paradigm is equated to the operations of the "mind". However, this approach has significant drawbacks. First, the planning component has been a computational bottleneck that held up the controller subsystem. Second, since the controller did not have direct access to the sensor data, the overall system was not very *reactive* (Figure 1.3).

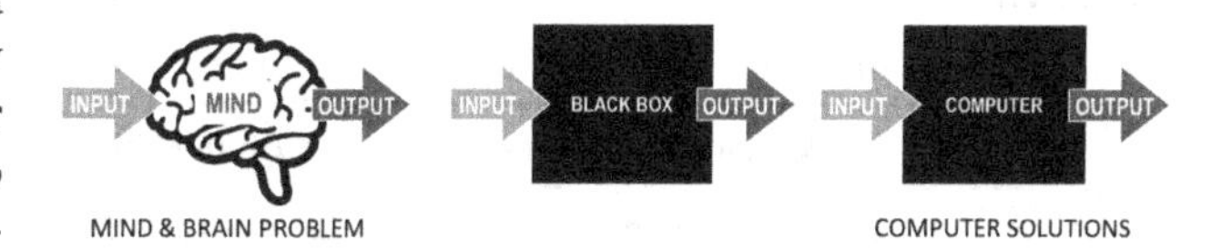

FIGURE 1.3
Mind and brain diagram.

There are around 86 billion neurons (or nerve cells) in the human brain. Each neuron is the basis for information processing and is connected to at least 10,000 others, giving well over 1,000 trillion connections (1 quadrillion connections). They all connect at a junction called a synapse, which can be electrical or a higher percentage of them are chemical. It is estimated that the human brain is capable of performing about 10,000 trillion calculations per second to as much as 1 exaFLOP (a billion calculations per second), that is, a lot. The cycle time for the brain is 1 ms. Computers today have about 10^9 or more transistors per CPU with tens of thousands of cores. Cycle times for modern CPUs are in the order of 10^{-9} seconds. The numbers between the brain and CPU are getting close (refer to Moore's Law) and the CPU will surpass the human brain within the next 20 years. As a comparison, the brain is parallel and slow, but is continuous (neurons in the brain), and the computer is sequential and rapid, but discrete. In understanding the workings of the brain, its inner working is demonstrated using a process model. A process model describes the flow of activities, usually in a graphic format, that contribute to accomplishing a specific goal. Process models are typically used to represent and analyze a series of activities that occur repeatedly and on a regular basis. For example, process models can be used to model the flow of activities in a computer system or application. Process models have a clear beginning and end, intended outcome, order of activities, and different results based on the decisions that are made through the course of the process. The intent of the model is

TABLE 1.1

Personal Computer versus Human Brain Table

	Personal Computer	Human Brain
Processing units	1 CPU, 2–10 cores 10^{10} transistors 1–2 graphics cards/GPUs 10^3 cores/shaders 10^{10} transistors	10^{11} neurons
Storage capacity	10^{10} bytes main memory (RAM) 10^{12} bytes external memory	10^{11} neurons 10^{14} synapses
Processing speed	10^{-9} seconds 10^9 operations per second	$>10^{-3}$ seconds $>10^{-3}$ seconds
Bandwidth	10^{12} bits/second	10^{14} bits/second
Neural updates	10^6 per second	10^{14} per second

to provide a representation of the process that allows understanding, analysis, and decisions (Table 1.1).

The key assumption in the mind-brain issue is that if "human thought" can be mechanized. If this is true, *logical reasoning* and *rational thinking* can be reproduced. This implies that an analogous "physical system" is required. This requires taking physical patterns like our thoughts, being able to combine the thoughts with itself such as applying the association rules and being able to manipulate a set of expressions. This sounds a lot like mathematics where mathematics is a language. For example, formal reasoning is characterized by the rules of logic and mathematics, and one conclusion reached was that "reasoning" can be viewed as a "search" problem. The brain is a predicting machine.

Understanding is the simplest procedure of all human beings. "Understanding" means to have the ability to determine some new knowledge from a given knowledge. For each action of a problem, the mapping of some new actions is very necessary. Mapping the knowledge means transferring the knowledge from one representation to another representation. For example, if you will say "I need to go to New Delhi" for which you will book the tickets. If the computer is the brain, then the software is the mind in the mind-body problem. Historically, the human mind has been compared to the highest levels of contemporary technology. The brain started with the water clock, followed by the clockworks, than industrial organizations, telephone switching circuits, and now digital computers. This is an assumption since the human brain is biological, while the computer and software are digital or digital/analog. No one can refute a computer's ability to process logic. But to many it is unknown if a machine could *think*. Most people think computers will never be able to think. That is, really think. Not now or ever. To be sure, most people also agree that computers can do many things that a person would have to be thinking to do. Then how could a machine seem to think but not actually think? Well, setting aside the question of what thinking actually is, I think that most of us would answer that by saying that in these cases, what the computer is doing is merely a superficial imitation of human intelligence. It has been designed to obey certain simple commands, and then it has been provided with programs composed of those commands. Because of this, the computer has to obey those commands, but without any idea of what's happening.

Turing argued that if the brain was a machine, then a machine could be a brain: If it is accepted that real brains, as found in animals, and in particular in men, are a sort of machine it will follow that our digital computer suitably programmed, will behave like a brain. Analog computers, unlike digital computers, represented real-world systems physically as specially configured electrical, mechanical, or hydraulic systems that were expected to behave analogously to the systems under investigation. A slide rule is a very basic form of an analog computer. Analog computers arrived at answers to problems by emulation rather than calculation. The question was postulated that "What did Turing mean by 'machine' as applied to a brain?" Turing's was not completely fixed. The principal meaning of "mechanical" is "algorithmic", that is, operating according to a finite set of instructions which, in their context of use, are explicit, unambiguous, and capable of being performed. The new digital computers were algorithmic, but they were also discrete-state machines engaged in computable operations. In discrete-state machines, only certain states are possible (or, at any rate, only certain states matter as the intermediate states during transitions have no significance). The extent to which the working of human and animal brains could be likened to that of digital computers was, but Turing held that brains must be governed by physical laws. Between computability and law-governed behavior, there was no obvious connection. For instance, might a brain, though governed by physical laws, be capable of processes that were not possible in principle for an algorithmic machine? It is postulated that an abstract model of a digital computer is rich enough to capture any computational process. Inspired by Turing's model, data and programs are stored in the computer's memory (Figure 1.4).

The basic idea of the reducibility of thought to these ultimately very simple, formalized logical, mathematical or mechanical rules and processes lies at the heart of AI. The bottleneck in the SPA is the speed and capacity of the brain in terms of the manipulation of symbols and its reasoning is

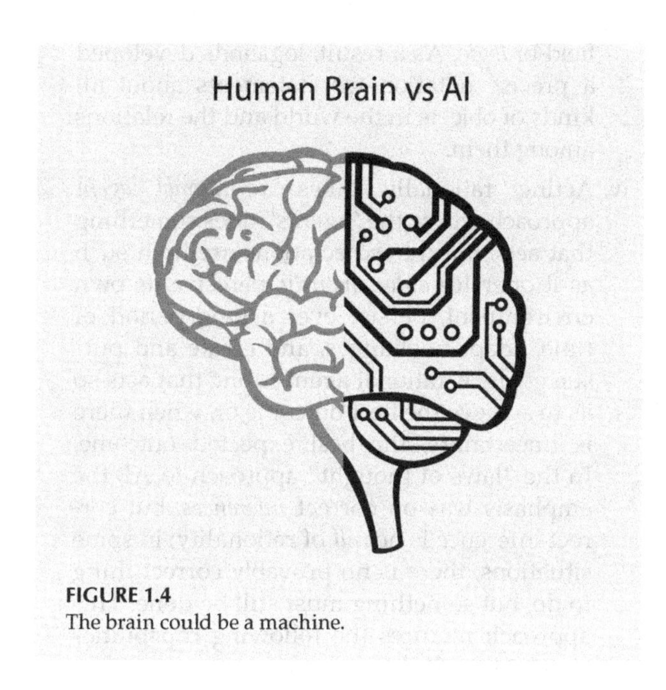

FIGURE 1.4
The brain could be a machine.

Thinking Humanly	Thinking Rationally
Acting Humanly	Acting Rationally

FIGURE 1.5
Four categories of AI.

computationally limited. The notion of "planning" in robotics equates to thinking (or thinking longer) as it limits the actions of the robot as it thinks. In designing robots, practitioners such as myself understand hardware always lead software in terms of maturity. So, we have two problems; the hardware/computer is not performing at human-level as realized using Moore's Law. And at the same time, software (especially AI software) is also a maturing technology that is still early in its developmental process. Through research of intelligent systems, we can try to understand how the human brain works and then model or simulate it on the computer. There are many interesting philosophical questions surrounding intelligence and AI. We humans have consciousness; that is, we can think about ourselves and even ponder that we are able to think about ourselves. Consciousness is an experience of reality. And how does consciousness come to be? Many philosophers and neurologists now believe that the mind and consciousness are linked with matter, that is, with the brain. The question of whether machines could 1 day have a mind or consciousness could at some point in the future become relevant.

So, what is AI? In 1955, John McCarthy, one of the pioneers of AI, was the first to define the term *AI*, roughly as follows: "The goal of AI is to develop machines that behave as though they were intelligent". Or more plainly, AI is the simulation of human intelligence processes by machines, especially computer systems. AI is a wide-ranging branch of computer science concerned with building smart machines capable of performing tasks that typically require human intelligence. At its core, AI is the branch of computer science that aims to answer Turing's question in the affirmative and is the endeavor to replicate or simulate human Intelligence In machines. In an attempt to make computers more "intelligent" and to better understand human intelligence, Peter Norvig and Stuart Russell go on to explore four different approaches that have historically defined the field of AI into four categories:

These four categories are an attempt to answer the questions: is it about thought .... Or action ... and oriented toward a human model (with all of its defects such as human error) .... Or normative (or simply how should a rational being think/act)? It is important to remember that a system may be able to act like a human without thinking like a human, thus it could easily "fool" us into thinking it was human (Figure 1.5).

1. Systems that *think* like humans: Method must not just exhibit behavior sufficient to fool a human judge but must do it in a way demonstrably. Requires detailed matching of computer behavior and timing to detailed measurements of human subjects gathered in psychological experiments. Based on cognitive modeling, an interdisciplinary field (AI, psychology, linguistics, philosophy, and anthropology) that tries to form computational theories of human cognition analogous to human cognition.
2. Systems that *think* rationally: or laws of thought, formalize "correct" reasoning using a mathematical model (e.g., of deductive reasoning). The Logicist Program encodes knowledge in formal logical statements and uses mathematical deduction to perform reasoning.
3. Systems that *Act* like humans: for example, if the response of a computer to an unrestricted textual natural-language conversation cannot be distinguished from that of a human being then it can be said to be intelligent, that is, passing the Turing test.
4. Systems that *Act* rationally: a rational agent, an entity that perceives its environment and is able to execute actions to change it, having

inherent goals that they want to achieve (e.g., survive, reproduce), and acts in a way to maximize the achievement of its goals.

Thinking humanly and acting humanly measure performance are based on the fidelity to human performance, while thinking rationally and acting rationally are measured against an ideal performance measure called rationality. Both thinking humanly and thinking rationally are concerned with the thought process and reasoning, while acting humanly and acting rationally address behavior.

i. Acting humanly is based on the Turing's Test, which is an operational definition of intelligence. However, this approach requires the following capabilities:
   - *Natural Language* processing to enable it to communicate successfully in English.
   - *Knowledge Representation* to store what it knows or hears.
   - *Automated Reasoning* to use the stored information to answer questions and to draw new conclusions.
   - *Machine Learning* to adapt to new circumstances and to detect and extrapolate patterns.
   - In addition to computer vision to perceive objects, a robot is needed to manipulate objects and move about.

ii. Thinking humanly takes a *cognitive modeling* approach, having some way of determining how humans think. There are three ways to do this: through introspection (trying to catch our own thoughts as they go by), through psychological experiments (observing a person in action), and through brain imaging (observing the brain in action). Once we have a sufficiently precise theory of the mind, it becomes possible to express the theory as a computer program. The interdisciplinary field of cognitive science brings together computer models from AI and experimental techniques from psychology to construct precise and testable theories of the human mind (refer back to Craik).

iii. Thinking rationally takes the "Law of Thought" approach which tries to codify "right thinking" which equates to an irrefutable reasoning process. Aristotle provided patterns for argument structures that always yielded correct conclusions when given correct premises that was supposed to govern the operation of the mind, thus initiating the field of *logic*. As a result, logicians developed a precise notation for statements about all kinds of objects in the world and the relations among them.

iv. Acting rationally takes a *rational agent* approach where the "agents" does something that acts, but the expectations are high such as it operates *autonomously*, perceive its own environment, persist over a long period of time, adapt to changes, and create and pursue goals. A rational agent is one that acts so as to achieve the best outcome or, when there is uncertainty, the best expected outcome. In the "laws of thought" approach to AI, the emphasis was on correct *inferences*, but correct inference is not *all* of rationality; in some situations, there is no provably correct thing to do, but something must still be done. This approach requires the following capabilities to act rationally:
   - *Natural Language* processing to enable it to communicate successfully in English.
   - *Knowledge Representation* to store what it knows or hears.
   - *Automated Reasoning* to use the stored information to answer questions and to draw new conclusions.
   - *Machine Learning* to adapt to new circumstances and to detect and extrapolate patterns, but also because it improves our ability to generate effective behavior.
   - *Inference.*

An agent denotes rather generally a system that processes information and produces an output from an input. Agents may be classified in many different ways. This rational-agent approach has two advantages over the other approaches: First, it is more general than the "laws *of* thought" approach because correct inference is just one of several possible mechanisms for achieving rationality and second, it is more amenable to scientific development than are approaches based on human behavior or human thought. The standard of rationality is mathematically well defined and completely general and can be "unpacked" to generate agent designs that integrate well with robots. Human behavior, on the other hand, is well adapted for one specific environment and is defined by the sum total of all the things that humans do. A running AI program is at most a simulation of a cognitive process but is not itself a cognitive process. The simplest agent program is a lookup table.

The "thinking rationally" approach to AI uses symbolic logic to capture the laws of rational thought as symbols that can be manipulated. Reasoning involves manipulating the symbols according to well-defined rules, kind of like algebra. *Rationality* is a behavior that is evaluated on its basis for reasoning and being able to explain itself in terms of ends, norms and values. Thus, rationality makes decisions by some set of rules that are grounds for truth. It is some correspondence for reality and a soundness of inference. For example, Aristotle noted that "Man is a rational animal". A rational machine is possible:

1. If the world obeys math equations (solvable step by step).
2. If a machine is able to simulate the world and make predictions about it.
3. To the extent that rational thought corresponds to logic rules.
4. If the robot can carry out rational thought.
5. To the extent that language can be captured by grammatical rules.
6. If it can produce grammatical sentences.
7. If thinking consists of applying well-specified rules, thus enabling a thinking machine to be built.

The *Gemini theorem* asserts that given certain reasonable assumptions, no physical system can be certainly aware of its own existence. The theorem can be proved algorithmically, but the proof of this theorem is somewhat obscure, and there exists very little literature on it. A key feature of the Gemini theorem is that it does not require any assumptions to be made about the nature of consciousness itself. It relies instead on an operational property possessed by all healthy conscious human beings. This is the property of being able to assert that they do, in some sense, exist. Human beings do not need to "prove" that they are conscious or that they exist – the fact that they are capable of asking the question "Am I conscious" already establishes that.

Weak AI or narrow AI focuses on performing a specific task, such as answering questions based on user input or playing chess. Weak AI is one intended to reproduce an observed behavior as accurately as possible. It can carry out a task for which they have been precision trained. Such AI systems can become extremely efficient in their own field but lack generalizational ability. In a report released by the Obama Administration in 2016, it states narrow AI is all around us and is easily the most successful realization of AI to date. With its focus on performing specific tasks, narrow AI has experienced numerous breakthroughs in the last decade that have had "significant societal benefits and have contributed to the economic vitality of the nation", according to "Preparing for the Future of Artificial Intelligence". Narrow AI is often focused on performing a single task extremely well and while these machines may seem intelligent, they are operating under far more constraints and limitations than even the most basic human intelligence. Even advanced chess programs are considered weak AI. This categorization seems to be rooted in the difference between supervised and unsupervised programming. However, there is a caution in being "weak" such that the technology may or may not scale to large or difficult problems. "Weak AI" is sometimes called "narrow AI", but the latter is usually interpreted as subfields within the former. It claims the digital computer is a useful tool for studying intelligence and developing useful technology (Figure 1.6).

The software can perform one type of task, but not both, whereas *Strong AI* can perform a variety of functions, eventually teaching itself to solve new problems. *AGI*, sometimes referred to as "Strong AI", is the kind of AI we see in the movies and is a machine with general intelligence and, much like a human being, it can apply that intelligence to solve any problem. "Strong" AI claims that a digital computer can in principle be programmed to actually be a mind, to be intelligent, to understand, perceive, have beliefs, and exhibit other cognitive states normally ascribed to human beings. It does not classify but uses clustering and association to process data. In short, it means there isn't a set answer to your keywords. The function will mimic the result, but in this case, we aren't certain of the result. From the perspectives of researchers, the more an AI system approaches the abilities of a human, with all its intelligence, emotion, and broad applicability of knowledge of humans, the "stronger" that AI is. The term "strong" AI can alternatively be understood as broad or general AI. AGI is focused on creating intelligent

FIGURE 1.6
What is inside the computer diagram.

machines that can successfully perform any intellectual task that a human being can.

> If you want the computer to have general intelligence, the outer structure has to be common sense knowledge and reasoning.
>
> – **John McCarthy**

"Meaning" is the central concept of AI because:

- For an agent to be "intelligent", it must be able to understand the meaning of information.
- Information is acquired/delivered/conveyed in messages which are phrased in a selected representation language.
- There are two sides in information exchange: the source (text, image, person, program, etc.) and the receiver (person or an AI agent). They must speak the same "language" for the information to be exchanged in a meaningful way.
- The receiver must have the ability to interpret the information correctly according to the intended by the source meaning or semantics of it.

Basically,

MEANING = SEMANTICS

In practice, AI comprises advanced algorithms that follow a mathematical function, which is able to handle higher processes similar to humans. Algorithms running on programmable computer does: sorting, rendering, search, and image classification. Some AI examples include visual perception, speech recognition, decision-making, and translating between languages. The creation of a robot with human-level intelligence that can be applied to any task is the Holy Grail for many AI researchers, but the quest for AGI has been fraught with difficulty. The search for a "universal algorithm for learning and acting in any environment", isn't new, but time hasn't eased the difficulty of essentially creating a machine with a full set of cognitive abilities. Intelligent robots and artificial beings first appeared in the ancient Greek myths of Antiquity. Aristotle's development of syllogism and its use of deductive reasoning was a key moment in mankind's quest to understand its own intelligence.

Narrow AI span a wide, but common range of disciplines:

- Understanding *Natural Language*, for example, speech recognition and translation between languages
- *Knowledge Representation*, for example, how should the robot represent itself, its task, and the world
- Automated *Reasoning* (including deduction, logic, and decision-making)
- Machine *Learning*
- *Inference*, for example, generating an answer when there isn't complete information
- Perception or *Computer Vision*
- *Planning* and *Problem-solving*, for example, mission, task, and path planning.

It is important to note that the "weak AI" view is not a mind, but good intelligent process. In contrast, the "strong AI" view is to build a mind in a computer.

Problems dealt with in AI generally use a common term called "state". A state represents a status of the solution at a given step of the problem-solving procedure. The solution of a problem, thus, is a collection of the problem states. The problem-solving procedure applies an operator to a state to get the next state. Then it applies another operator to the resulting state to derive a new state. Many of these disciplines are based on the basic AI fundamental methods of "search". *Search* in AI is the process of navigating from a starting state to a goal state by transitioning through intermediate states and almost any AI problem can be defined in these terms, for example,

- State: A potential outcome of a problem.
- Transition: The act of moving between states.
- Starting State: Where to start searching from.
- Intermediate State: The states between the starting state and the goal state that we need to transition to.
- Goal State: The state to stop searching.
- Search Space: A collection of states.

The fundamental state in robotics is called Configuration Space (or C-Space). C-Space is the state of everything in a robot's environment, including the robot itself. For a practical example, an expert system is a software program that incorporates AI techniques such as using an expert knowledge to offer advice or make decisions. As another example, search is finding answers in a knowledge base or finding objects in the world. Search engines are computer algorithms used to search data for specified information. The fundamental question is what is the difference between the mind and the human brain? Similarly, what is intelligence and is it computation? Is the mind software and does it follow the *Computational Theory of the Mind*? Is the brain hardware analogous to a neural network? Connectionism is a movement in cognitive science that hopes to explain intellectual abilities using artificial neural networks. Neural networks are simplified models of the brain composed of large numbers of units together with weights that measure the strength of connections between the units. Philosophers have become interested in connectionism because it promises to provide an alternative to the classical theory of the mind, the widely held view that the mind is something akin to a digital computer processing a symbolic language. Connectionisms are structured propositions with hidden layers.

The process of applying an operator to a state and its subsequent transition to the next state, thus, is continued until the goal (desired) state is derived. Such a method of solving a problem is generally referred to as state space approach. For example, in order to solve the problem, play a game, which is restricted to two-person table or board games, we require the rules of the game and the targets for winning as well as a means of representing positions in the game. The opening position can be defined as the initial state and a winning position as a goal state, and there can be more than one legal moves that allow for transfer from initial state to other states leading to the goal state. However, the rules are far too copious in most games especially chess where they exceed the number of particles in the universe. Thus, the rules cannot in general be supplied accurately and computer programs cannot easily handle them. The storage also presents another problem but searching can be achieved by hashing. The number of rules that are used must be minimized and the set can be produced by expressing each rule in as general a form as possible. The representation of games in this way leads to a state space representation and it is natural for well-organized games with some structure. This representation allows for the formal definition of a problem which necessitates the movement from a set of initial positions to one of a set of target positions. It means that the solution involves using known techniques and a systematic search. This is quite a common method in AI.

René Descartes believed that nonhuman animals could be reductively explained as automata, from his *De Homines in 1622.* The word automata comes from the Greek word αὐτόματος, which means "self-acting, self-willed, self-moving". An automaton (automata in plural) is an abstract self-propelled computing device that follows a predetermined sequence of operations automatically. *Automata Theory* is a theoretical branch of computer science. It established its roots during the 20th century as mathematicians began developing both theoretical and literal machines that imitated certain features of man and completed calculations more quickly and reliably. The word *automaton* itself, closely related to the word "automation", denotes automatic processes that carry out the production of specific processes. Simply stated, automata theory deals with the logic of computation with respect to simple machines, referred to as *automata.* Through automata, computer scientists are able to understand how machines compute functions and solve problems, and what it means for a function to be defined as *computable* or for a question to be described as *decidable.* The major objective of automata theory is to develop methods by which computer scientists can describe and analyze the dynamic behavior of discrete systems in which signals are sampled periodically. The behavior of these discrete systems is determined by the way that the system is constructed, from storage and combinational elements. Characteristics of such machines include:

- Inputs: assumed to be sequences of symbols selected from a finite set I of input signals. Namely, set I is the set $\{x_1, x_2, x_3 \dots x_k\}$ where k is the number of inputs.
- Outputs: sequences of symbols selected from a finite set Z. Namely, set Z is the set $\{y_1, y_2, y_3 \dots y_m\}$ where m is the number of outputs.
- States: finite set Q, whose definition depends on the type of automaton (Figure 1.7).

A process diagram, also known as a flowchart in engineering, is a visual representation of the steps required to complete a given task or achieve a desired outcome. The purpose of a process diagram is to show how a "process" works and the different stages it goes through. Ernest Nagel, in his 1961 book, *The Structure of Science*, developed some foundational work in the logic of scientific explanation. Nagel believed in the philosophy of mathematics (i.e., logical positivism) and that reductionists are those who take one theory or phenomenon to be reducible to some other theory or phenomenon. For example, most reductionists regarding mathematics might take any given mathematical theory to be reducible to logic or set theory. The types of reductionism that is of most interest are in metaphysics and in philosophy of mind that involves the claim

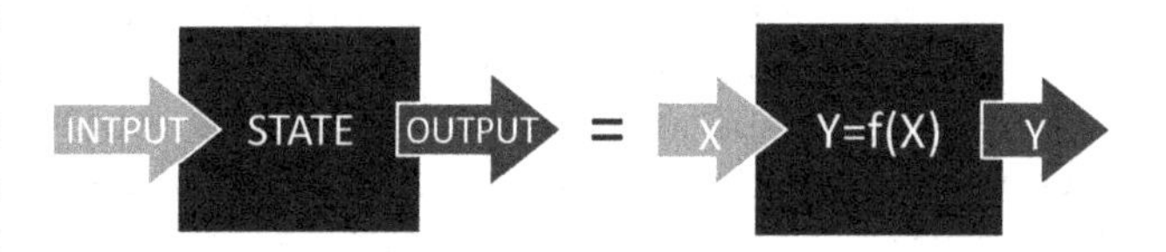

FIGURE 1.7
Automata theory diagram.

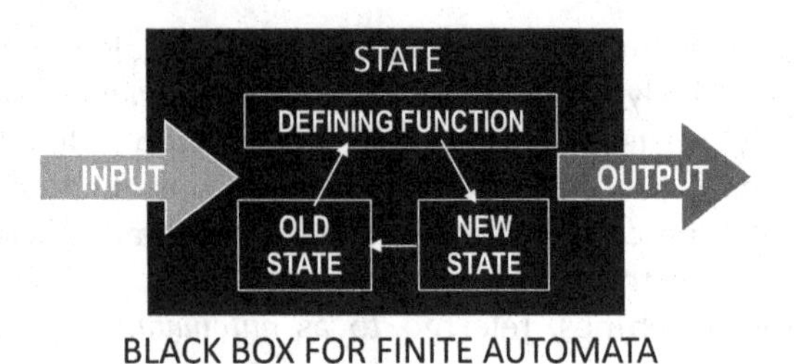

FIGURE 1.8
Black box for finite automata.

that all sciences are reducible to physics. This is usually taken to entail that all phenomena (including mental phenomena like consciousness) are identical to physical phenomena. The target science will contain terms that do not occur in the theory of the base discipline, so the reduction will be heterogeneous in nature. This does not necessarily mean that one must *translate* terms from the target science into the language of the base science. As an example, one interested in reducing psychology to physics will notice that psychological theories contain terms like "belief", "desire", and "pain", which do not occur in the base, physical theory. $Y=f(X)$ where $f(\ )$ is the model, Y is the model output, and X is input data. It is not GenAI when Y is a number, is discrete, a class, and a probability. It is GenAI when Y is natural language, an image, or a natural language (Figure 1.8).

In these cases, assumptions must be added to the laws of the base science (physics) stating relations between these (psychological) terms and the terms already present in the base science. These assumptions, often called "bridge laws", will then allow one to derive the laws and theorems of the target science from the theory of the base discipline. Bridge Laws are used to connect different theories by identifying terms (and things) across theories. All basic theory incorporates higher-level theories. As a result, we can deduce higher-level science from lower level, requiring connectability (via the bridge laws) and deducibility. Many philosophers of science agree that the reduction of one scientific theory to another requires "bridge laws" such that there are statements linking concepts of the reduced theory to concepts of the reducing theory. One area that uses reductionism extensively is computer modeling. For example, if a scientist designs a computer program to model and predict thinking patterns, they cannot possibly include every single permutation of such a vast and complicated system. It can be seen as a function that maps an ordered sequence of input events into a corresponding sequence, or set, of output events.

State transition function: $\mathbf{I} \rightarrow \mathbf{Z}$

The other convention used in this book is "→" or "moving forward", that is, Goal-Based Decision-Making moving forward to planning. As another example, a way of looking at this would be reasoning requires logic and logic requires search.

Search → Logic → Reasoning

(Note: "→" literally means "moving forward")

Finite-state machines are ideal computation models for a small amount of memory and do not maintain memory (except the Mealy Machine). This mathematical model of a machine can only reach a finite number of states and transitions between these states, and its main application is in mathematical problem analysis.

A *production system* is a model of computation that can be applied to implement search algorithms and model human problem-solving. Such problem-solving knowledge can be packed up in the form of little quanta called productions. A production is a rule consisting of a situation recognition part and an action part. It is a situation–action pair in which the left side is a list of things to watch for and the right side is a list of things to do so. When productions are used in deductive systems, the situation that triggers production is specified combination of facts. The actions are restricted to assertions of new facts deduced directly from the triggering combination. Production systems may be called premise conclusion pairs rather than situation–action pairs.

In philosophy of mind, the *Computational Theory of Mind* (CTM), also known as computationalism, is a family of views that hold that the human mind is an information-processing system and that cognition and consciousness together are a form of computation. A key task facing computationalists is to explain what one means when one says that the mind "computes" and a second task is to argue that the mind "computes" in the relevant sense. In a representational/computational theory of mind, computers show us how to connect semantic properties of a symbol with its causal properties via its syntax and thus offer an explanation of how there could be nonarbitrary content relations among causally related thoughts.

Alan Turing advocated that intelligence arises from information and information is the correlation between two things produced by a lawful process! One problem with a purely physical conception of the mind is that it seems to leave little ***room for free will***: if the mind is governed entirely by physical laws, then it has no more *free will* than a rock "deciding" to fall toward the center of the earth. Dualism is the division of something conceptually into two opposed or contrasted aspects (mind-body), or the state of being so divided. Alternative to dualism is *materialism,* which holds that the brain's operation according to the laws of physics constitutes the mind. Materialism is a form of philosophical monism that holds matter to be the fundamental substance in nature, and all things, including mental states and consciousness, are results of material interactions. *Free will* is simply the way that perception of available choices appears to the choice process. According to John McCarthy, Human *free will* is a product of evolution and contributes to the success of the human animal. Useful robots will also require *free will* of a similar kind, and we will have to design it into them. Some agents have more *free will,* or *free will* of different kinds, than others. He studied what aspects of *free will* can make robots more useful, and we will not try to console those who find determinism distressing. Thus, *Free will* does not require a very complex system. Consider a machine, for example, a computer program, that is entirely deterministic, that is, is completely specified and contains no random element. A major question for philosophers is whether a human is deterministic in the above sense. If the answer is yes, then we must either regard the human as having no *free will* or regard *free will* as compatible with determinism. An automobile has none, a chess program has a minimal kind of *free will,* and a human has a lot. Human-level AI systems, that is, those that match or exceed human intelligence will need a lot more than present chess programs, and most likely will need almost as much as a human possesses, even to be useful servants. One such requirement for intelligence is the ability to store and manipulate symbols. Human-level AI requires the ability of the agent to reason about its past, present, future, and hypothetical choices. Finally, what people can do and know about what they can do is similar to what robots can do and know (Figure 1.9).

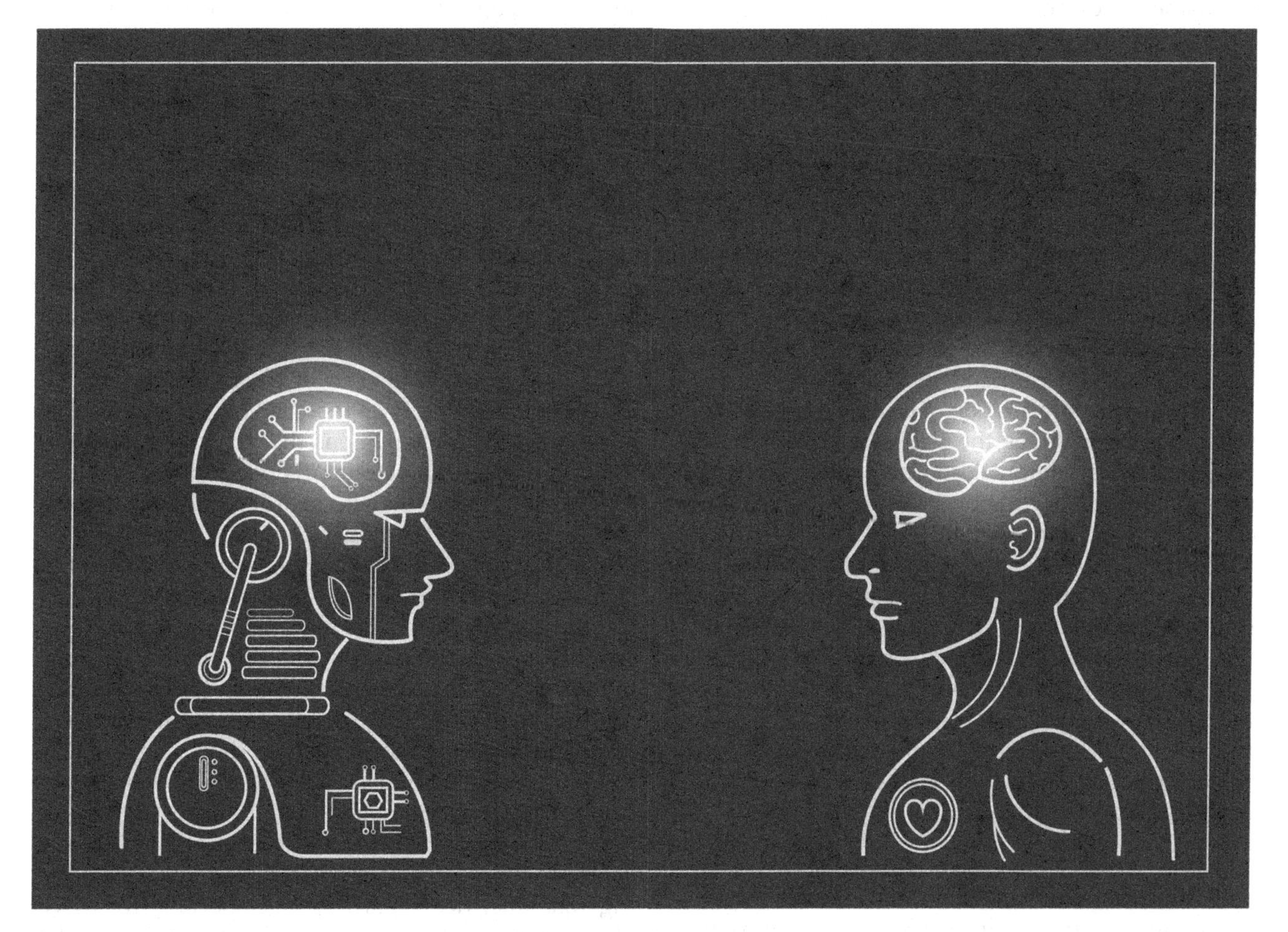

FIGURE 1.9
AI (left) vs. human (right).

Early AI development was conducted by mathematicians, philosophers, and psychologists until the field of computer science was formalized with the help of J.C.R. Licklider who became interested in information technology. While at MIT in the 1950s, Licklider worked on "SAGE" (Semi-Automatic Ground Environment), a Cold War project to create a computer-aided air defense system. The SAGE system included computers that collected and presented data to a human operator, who then chose the appropriate response. Licklider worked as a human factors expert, which helped convince him of the great potential for human/computer interfaces. His ideas foretold of graphical computing, point-and-click interfaces, digital libraries, e-commerce, online banking, and software that would exist on a network and migrate wherever it was needed. Much like Vannevar Bush's contribution, Licklider's contribution to the development of the internet consisted of ideas, not inventions. He foresaw the need for networked computers with easy user interfaces. Licklider as director of ARPA's Information-Processing Techniques Office (IPTO) from 1962 to 1964, he funded the Project MAC (Project on Mathematics and Computation) at MIT where a large mainframe computer was designed to be shared by up to 30 simultaneous users, each sitting at a separate "typewriter terminal". Project MAC sought to create a functional time-sharing system. He also funded similar projects at Stanford University, UCLA, U.C. Berkeley, and the System Development Corporation. This was the beginning of developing Computer Science Departments in American Universities such as at U.C. Berkeley, CMU, MIT, and Stanford. Today, computer science is the study of computers and computing, including their theoretical and algorithmic foundations, hardware and software, and their uses for processing information. The discipline of computer science includes the study of algorithms and data structures, computer and network design, modeling data and information processes, and AI. Thus, computer science began to be established as a distinct academic discipline in the 1950s and early 1960s. The world's first computer science degree program, the Cambridge Diploma in Computer Science, began at the University of Cambridge Computer Laboratory in 1953. Some problems used to be thought of as AI but are now considered not, for example, compiling Fortran in 1955, symbolic mathematics in 1965, and pattern recognition in 1970. There are two main lines of research:

- One is *biological,* based on the idea that since humans are intelligent, AI should study humans and imitate their psychology or physiology.
- The other is *phenomenal,* based on studying and formalizing commonsense facts about the world and the problems that the world presents to the achievement of goals.

Cognition

Neuroscience is the study on how the brain works relative to the body's nervous system, Psychology helps to explain how the mind thinks, and Perception is the organization, identification, and interpretation of sensory information in order to represent and understand the presented information or environment. All perception involves signals that go through the nervous system, which in turn result from physical or chemical stimulation of the sensory system. The scientific effort to come to terms with the nature of intelligence, perception, thought, and behavior is usually labeled cognitive science. Cognitive Science considers *memory* and *cognition* to be the mental events and knowledge that we use when we recognize an object, remember a name, have an idea, understand a sentence, or solve a problem. It is the scientific study of the mind, with contributions of disciplines such as linguistics, computer science, and the neurosciences. According to Francisco Varela, there have been four major stages in cognitive science development over the last 40 years: cybernetics, cognitivism, connectionism, and enaction.

The human brain is orders of magnitude more sophisticated than the most complex AI deep learning networks ever built, including those using activation functions. A single neuron in the brain is complex, and its behavior has been modeled by a complex set of partial differential equations that try to capture neuronal dynamics (Ref. Hodgkin and Huxley). We have roughly 100 billions of these neurons, and each neuron is connected to as many as a thousand other neurons in a highly structured but complex three-dimensional network. Unlike the smooth activation functions used in today's deep learning models, which bear no resemblance to the real ones in our brains, neurons in biology are largely based on spike trains, communicating with their neighbors with what looks like a complex version of Morse code. In short, simple ideas like backpropagation used everywhere in deep learning are profoundly wrong. The brain cannot work using back propagation and this fact has been known for more than three decades. However, as a processing machine, deep learning may be able to arrive at the answers that we seek without mimicking the human brain exactly.

BIO GRAND CHALLENGE (GC2): THE BRAIN

Unraveling the mystery of how the brain works is one of the greatest challenges of science. Neuroscience is challenging *scientists and engineers* to develop new research tools to investigate multi-scale, non-linear dynamical systems *as problems* of fundamental interest essential for constructing a comprehensive brain activity map. Fundamental neuroscience research includes species-comparative approaches that will provide the necessary foundation for understanding how the genomic architecture, nuclear organization, synaptic activity, and neural circuitry underlying the emergent properties of the brain.

and emergent properties of neurons, glia, and neural circuits. For example, cognitive neuroscience is the scientific study of the Influence of brain structures on mental processes, done through the use of brain scanning techniques such as fMRI. The understanding of the biology is the basis of learning, memory, behavior, perception, and consciousness as described by Eric Kandel (Kandel discovered the central role synapses play in memory and learning) as the "epic challenge" of the biological sciences. For example, unraveling the mystery of how the brain works is one of the greatest challenges of science. Cognitive neuroscience aims to find out how brain structures influence the way we process information and map mental cognitive functions to specific areas of the brain. Neuroscience research investigates multi-scale non-linear dynamical systems as problems of fundamental interest essential for constructing a comprehensive brain activity map.

Neuron Doctrine, circa 1894, is the idea that the neuron is the basic unit of the nervous system. Within a nervous system, a neuron is an electrically excitable cell that fires electric signals called action potentials across a neural network. Neurons communicate with other cells via synapses, specialized connections that commonly use minute amounts of chemical neurotransmitters to pass the electric signal from the presynaptic neuron to the target cell through the synaptic gap. The neuron doctrine is the now fundamental idea that neurons are the basic structural and functional units of the nervous system. It held that neurons are discrete cells (not connected in a meshwork), acting as metabolically distinct units. Thus, the nervous system consists of multiple individual cells called "neurons", which have an individual structure and function, working together in order to create a singular and refined machinery that controls the entire human body.

The hundred billion neurons in our brains are all chattering to each other in a mysterious language that we are still unable to decipher. At this point, we just don't have good models of spike train dynamics to be useful in deep learning research, yet this is probably the single biggest reason why the brain is so efficient in terms of power consumption. The brain has more computing power than the most sophisticated super computer ever built, yet the spike train design is probably the biggest reason why our brains are so efficient. The mysteries of the human brain are much deeper than the sheer number of neurons using complicated inter connectivity and spike train dynamics. The whole human body, including the brain with all its 10^{15} connections, has to be encoded in the human genome, which is about 30 billion base pairs of letters — A, T, C, and G for the four amino acids that encode hereditary information. The mystery of the workings of the brain only gets more puzzling and more insurmountable, the deeper you dig into its workings. It turns out that as the brain develops and grows, neurons are born and move physically to their designated final location. So, in short, the puzzle AI is trying to figure out is how to build the most sophisticated supercomputer there is, which runs on 10W of power, which grows from a single cell, uses complex spike train dynamics to convey information to effortlessly understand auditory, visual, language and motor control behavior than any computer ever designed by humans.

Neuroscience is the scientific study of the nervous system (the brain, spinal cord, and peripheral nervous system) and its functions. It is a multidisciplinary science that combines physiology, anatomy, molecular biology, developmental biology, cytology, psychology, physics, computer science, chemistry, medicine, statistics, and mathematical modeling to understand the fundamental

Neuroscience studies the nervous system and in particular the brain. The exact way in which the brain enables thought is one of the great mysteries of science. It has been appreciated for thousands of years that the brain is somehow involved in thought, because of the evidence that strong blows to the head can lead to mental incapacitation. The brain is recognized as the seat of consciousness. Before then, candidate locations included the heart, the spleen, and the pineal gland. Paul Broca's study of *aphasia* (speech deficit) in brain-damaged patients in 1861 reinvigorated the field and persuaded the medical establishment of the existence of localized areas of the brain responsible for specific cognitive functions. Despite these advances, we are still a long way from understanding how any of these cognitive processes actually work. The truly amazing conclusion is that *a collection of simple cells can lead to thought, action, and consciousness* or, in other words, that *brains cause minds* according to Searle. The only real alternative theory is mysticism: that there is some mystical realm in which minds operate that is beyond physical science. The brain and computers perform quite different tasks and have different properties. Moore's Law predicts that the CPU's gate count will equal the brain's neuron count sometime around the year 2020. Moore's Law says that the number of transistors per square inch doubles every 1 to 1.5 years, while the human brain capacity doubles roughly every 2–4 million years. Contrary to what we believe, the "computer is a million times faster in raw switching speed, but the brain ends up being 100,000 times faster at what it does".

Computational neuroscience can be divided into two parts. On the one hand, it builds computational models of neural phenomena, analogously to the way computational chemistry, climate science, computational economics, etc. build computational models of their respective phenomena. On the other hand, computational neuroscience also studies the ways nervous systems compute and process information. Computational neuroscientists often assume that the nervous system performs computations and process information. As to computation, there is a precise and powerful mathematical theory that defines which functions of a denumerable domain, such as the natural numbers or strings of letters from a finite alphabet, which can be computed by following an algorithm. The same theory shows how to build machines that can compute any function that is computable with algorithms, that is, universal computers. But the mathematical theory of computation does not tell us whether and how nervous systems perform computations, and in what sense. This is because the mathematical theory of computation was never intended to be and is not a theory of physical computation, namely, of physical computing systems such as the brain. Thus, there might be hypothetical physical systems that compute functions that are not Turing machine computable (Figure 1.10).

As to information, there is also a precise and powerful mathematical theory that defines information as the reduction of uncertainty about the state of a system. The same theory can be used to quantify the amount of information that can be transmitted over a communication channel. Again, the mathematical theory of information does not tell us whether and how the brain processes information, and in what sense. A related question concerns whether every physical object is a computer. Hilary Putnam argues that every physical system satisfying minimal conditions implements every finite-state automaton. Assuming that to compute is to satisfy Putnam's minimal condition, and this implies that every physical object computes practically everything! Many have argued that Putnam assumes a much too liberal notion of implementation. David Chalmers, for example, concedes that everything computes something, but insists that only few objects implement the kind of automata that suffice for minds.

There is no doubt that the human nervous system contains internal variables that correlate reliably with other variables, both internal and external to it. For instance, neuronal spike trains correlate reliably with other neuronal spike trains from other neurons and with aspects of the environment such

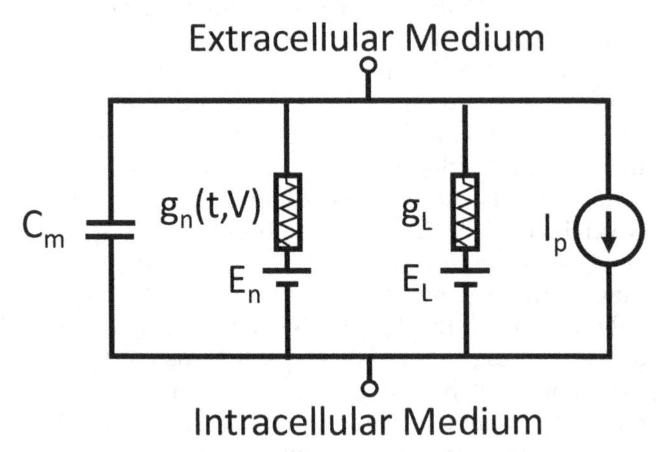

FIGURE 1.10
Hodgkin–Huxley model.

In the 1930s, Alan Hodgkin and Andrew Huxley started a series of experiments and modeling to elucidate the flow of electric current through an axonal membrane. This led to the formulation of the Hodgkin–Huxley model in 1952. Their work provided fundamental insights into nerve cell excitability. The *Hodgkin–Huxley* model, or conductance-based model, is a mathematical model that describes how action potentials in neurons are initiated and propagated. Their work not only establishes our understanding of how voltage-gated ion channels give rise to propagating action potentials but also the very framework for studying and analyzing ion channel kinetics.

as light, sound, pressure, and temperature. Nervous systems carry information in two senses. First, they carry information in Claude Shannon's sense: some of their variables reduce uncertainty about other variables. Information in Shannon's sense has to do with the uncertainty that characterizes the process as a whole, including all of the possible alternative messages at once. Shannon information, generated by the selection of a particular message, is a function of how many alternative messages may be selected instead and the probability with which any possible message is selected. Semantic information has to do with what a particular signal stands for or means. To capture the semantics of a signal, it is not enough to know which other signals might have been selected instead, and with what probabilities. We also need to know what a particular signal stands for. Different equiprobable messages carry the same amount of Shannon information, but they may well mean completely different things. We call "semantic information" the information a signal carries by reducing uncertainty about a specific state of affairs. Nervous systems carry semantic information in the sense that specific states of some of their variables make it likely that other variables (which they reliably correlate with) are in certain specific states.

Some philosophers have tried to explain what it takes for a physical system to perform computations by using notions found in logic and computability, or automata theory. They describe computation as program execution, syntactic operations, automatic FSs, or implementation of automata. These notions might apply to digital computers. But, as many have noted, the brain is very different from the familiar digital computer. In nervous systems, the functional relevance of neural signals depends on non-digital aspects of the signals such as firing rates and spike timing. Therefore, there is a strong case to be made that typical neural signals are not strings of digits, and neural computation is not, in the general case, digital computation. According to the modeling view of computation, physical computation is a special form of representation. It is a dynamic model in the sense that it represents a target domain in a way that preserves its high-order structures (i.e., the modeling view is stronger than the semantic view that merely identifies computation with information processing). Computation is a specific kind of mechanistic process; it has to do with the processing of variables to obtain certain relationships between inputs, internal states, and outputs independently of how the variables are physically implemented, and this is so regardless of whether the variables carry any information about the environment.

Brain in a Vat Argument – this thought experiment is most commonly used to illustrate global or Cartesian skepticism. You start by being told to imagine the possibility that at this very moment, you are actually a brain hooked up to a sophisticated computer program that can perfectly simulate experiences of the outside world. Here is the skeptical argument. If you cannot now be sure that you are not a brain in a vat, then you cannot rule out the possibility that all of your beliefs about the external world are false. Or, to put it in terms of knowledge claims, we can construct the following skeptical argument. Let "*P*" stand for any belief or claim about the external world, say, that snow is white.

1. If I know that *P*, then I know that I am not a brain in a vat
2. I do not know that I am not a brain in a vat
3. Thus, I do not know that *P*.

Putnam's argument is designed to attack the possibility of global skepticism that is implied by metaphysical realism. Putnam defines metaphysical realism as the view which holds that "... the world consists of some fixed totality of mind-independent objects". There is exactly one true and complete description of "the way the world is".

There is some consensus that nervous systems process information by performing computations and that computation must be characterized by abstracting away from certain aspects of a physical system (from the implementing media for the mechanistic view or from the mechanisms themselves for the modeling view). The next logical question is which kind of computations? The traditional answer is that nervous systems perform digital computations, like those performed by our artificial digital computers. But the theory that nervous systems are digital computers does face a serious problem, that is, the nervous system is primary computational vehicles, spike trains, are irreducibly graded in their functional properties. In other words, the functional relevance of neural signals depends on non-digital aspects of the signals such as firing rates and spike timing. Therefore, there is an argument to be made that typical neural signals are not strings of digits, and neural computation is not, in the general case, digital computation. This is not to say that neural computations are analog. Strictly speaking, analog computation employs continuous signals, whereas neural signals are made out of discontinuous functional units, neuronal spikes, or action potentials. As a result, neural computation appears to be neither digital nor analog; it appears to be a distinct kind of computation.

It is believed that the nervous systems, as well as artificial computational systems, have many levels of mechanistic organization. Computational neuroscience studies neural systems at all of these mechanistic levels, and then it attempts to discover how the properties exhibited by the components of a system at one level when they are suitably organized into a larger system, give rise to the properties exhibited by that larger system. If this process of linking explanations at different mechanistic levels is carried out, the hopeful result is an integrated, multi-level explanation of neural activity. David Marr's computational theory specifies the function computed and why it is computed, without saying what representations and procedures are used in computing it. Specifying the representations and procedures is the job of the "algorithmic theory". Finally, an "implementation theory" specifies the mechanisms by which the representations and algorithms are implemented. Some argue that the distinctive feature of Marr's computational level is anchoring the computed function in the individual's environment. Marr implicitly assumes that the brain models the environment, in the sense of preserving certain mathematical relationships between different environmental variables, for example, variables in the visual field. Others argue that computational explanation is fully mechanistic, whereas computational and algorithmic theories are sketches of mechanisms. They are just partial descriptions of a neurocomputational mechanism at one or more mechanistic levels. It is important to note that how we integrate computational explanations in psychology and neuroscience is still an on-going debate that is constantly affected by new advances in neuroscience. If thinking is information processing, in other words, then hypotheses about how we think can either be proven or falsified by being successfully "modeled", or by being proven intractable, through the development and testing of software systems.

If the brain is a "processing" system, then a *process* is work performed by that system in response to incoming data, information, or conditions. It is a synonym for *transform* (e.g., transforming data into useful information). All information systems include processes, and usually many of them. Modeling processes are used to help understand the interactions with the system's environment, other systems, and other processes. A system is itself a process. *A Policy* is a set of rules that govern a process and indicate when the process is completed. The *decision table* is a tabular form of presenting the specifics of a set of conditions and their corresponding actions for a state machine. Decision tables are required to implement a policy. A simple policy statement dictates the appropriate conditions and actions in order that the policy can lead to a decision. Models are the language that represents the system and allows it to reason about characteristics of the real system. Modeling languages consist of semantics (signs and symbols) and syntax (ways in which signs and symbols are put together). Finite-State Machines (FSMs) are used only when a finite number of inputs having a finite set of values that can lead to a finite number of outputs (or set of actions) having a finite set of values (typically systems that are not complex). The logic for an FSM can be represented by various decision methods: State Transition Diagrams, State Event Matrices, Decision Tables, and Process Activation Tables. Generally, a process is activities that transform data (Figure 1.11).

Statistics starts with the data. Think of the data as being generated by a black box in which a vector of input variables x (independent variables) go in one side, and on the other side, the response variables y come out. Inside the black box, nature functions to associate the predictor variables with the response variables, so the figure is y nature x. There are two goals in analyzing the data:

- Prediction: To be able to predict what the responses are going to be to future input variables; and

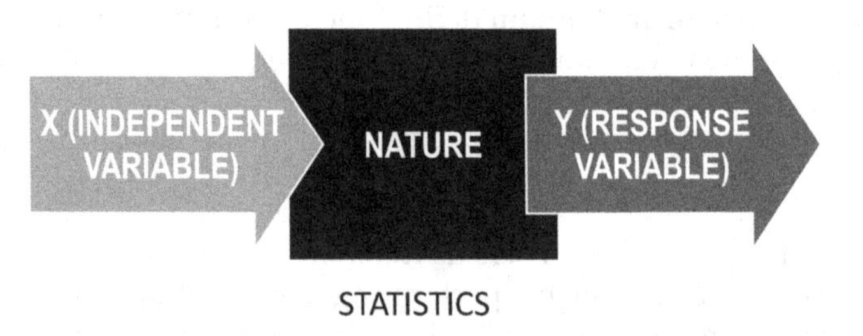

FIGURE 1.11
Statistics diagram.

- Information: To extract some information about how nature is associating the response variables to the input variables.

How do we represent or abstract or model the world? There are two different approaches toward these goals: (1) The analysis in the modeling culture starts with assuming a stochastic data model for the inside of the black box for "nature" or (2) The analysis in the algorithmic modeling culture considers the inside of the box complex and unknown as "nature". The data modeling group rely on techniques such as linear regression, logistic regression, and the Cox model. The *Cox model* provides hazard ratios for variables included in the model. Regression is also known as Method of Least Squares (also known as Ordinary Least Squares, abbreviated OLS). The problem with logistic regression is that it works only for two-layer perceptrons. A more general approach is to employ gradient descent. The general idea behind gradient descent is to approach the minimum of the error function in small steps. However, the necessary condition to use gradient descent is to have differentiable activation and output functions. The algorithmic modeling group still has the black box (nature) as unknown but uses decision trees and neural networks to connect x and y by fitting data. Their approach is to find a function $f(x)$ – an algorithm that operates on x to predict the responses y. The algorithmic approach serves as an introduction to machine learning.

Gerald M. Edelman defended the notion that the brain works by a process of random variation and selection similar to that characteristic of Darwinian evolution. He proposed that variation and selection within neural populations play key roles in the development and function of the brain. The theory of neuronal group selection was proposed to provide a framework to connect biology and psychology in a fashion consistent with developmental and evolutionary mechanisms. Although the real stimulus world certainly obeys the laws of physics, it is not uniquely partitioned into "objects" and "events". To survive in its econiche, an organism must either inherit or create criteria that enable it to partition the world into perceptual categories according to its adaptive needs. Even after that partition occurs as a result of experience, the world remains to some extent an unlabeled place that is full of novelty. Neuronal group selection argues that the ability of organisms to categorize an unlabeled world and behave in an adaptive fashion that arises not from instruction or information transfer, but from processes of selection upon variation. In this view, the otherwise puzzling variability of individual brains is a feature that is central to their function. And like the theories of natural selection and of clonal selection in immunity, the theory of neuronal group selection is a population theory. It is widely believed that in neuroscience, its memory is not absolute and there is emotion connected with a memory.

Edelman proposed neuronal group selection (NGS) as an organizing principle for higher brain functions, mainly a biological basis for perception, primarily applicable to the mammalian (and specifically, human) nervous system. The essential idea is that groups of cells, structurally varied as a result of developmental processes, comprise a population from which are selected those groups whose function leads to adaptive behavior of the system. Similar notions appear in immunology and, of course, evolutionary theory, although the effects of NGS are manifest in the lifetime of the organism. According to the theory of NGS, the world becomes "labeled" or perceptually categorized as a consequence of two interactive processes of selection upon variation. The first occurs largely in embryonic and postnatal development, during which adjacent neurons tend to be strongly interconnected in collectives of variable size and structure called neuronal groups. The second process consists of alterations in synaptic strengths during an animal's activity, selecting the correlated responses of those neuronal groups that yield adaptive behavior. If we shift from structure to the level of psychological function, two other remarkable observations deserve notice. First, even in species without language, animals are capable of a remarkable range of generalization in their perceptual categorizations. When presented with a few examples of particular shapes belonging to a common class (i.e., photographs of fish), pigeons can respond effectively to large numbers of different novel shapes in the same class. Second, objects and their properties are perceived to be unitary, despite the fact that a given perception results from parallel activity in the brain of many different maps, each with different degrees of functional specialization.

The theory of NGS proposes three mechanisms to account for the production of adaptive behavior by rich nervous systems: developmental selection, experiential selection, and reentrant signaling. Each mechanism acts within and among collectives consisting of hundreds to thousands of strongly interconnected neurons, called neuronal groups. Neurons within a group are highly interconnected, and changes in their synaptic strengths tend differentially to enhance the adaptive responses of the group as a whole. While the structures underlying neuronal groups arise from local anatomical connections, the groups themselves are dynamic entities whose borders and characteristics are affected by such synaptic changes and by the nature of the signals the groups receive. The theory of NGS, including aspects of memory, learning, and consciousness can account for key features of such higher-order functions of the brain. The purpose is to stress the explanatory and predictive usefulness of the theory of NGS by addressing several fundamental phenomena that must be explained before considering those subjects. It does not seem likely that neuroscientific research can lead to a deep view of how the brain functions, unless such global theories and the models that reflect them are available to bridge the experimental results obtained in a variety of disparate fields ranging from molecular biology to psychology. Human memory is the mental processes of acquiring and retaining information for later retrieval and the mental storage system that enables these processes. Cognition is the collection of mental processes and activities used in perceiving, remembering, thinking, and understanding, as well as the act of using those processes.

In 2005, the Blue Brain Project was launched by Ecole Polytechnique Fédérale de Lausanne (EPFL) and IBM at the Brain Mind Institute. The mission is to develop radically new strategy to understanding the human brain and its diseases and developing brain-like technologies. The goal is to answer questions about how we think, remember, learn, and feel. The cortical column is the basic unit of the cortex, a functional unit where each column seems to perform a simple function and is the essential first step toward achieving a complete virtual brain. IBM's Blue Gene/L supercomputer allows a quantum leap in the level of detail at which the brain can be modeled. Adaptation and learning algorithms have massively enhanced the power of these systems, but it could also be claimed that these approaches merely enable the system to automatically acquire more if–then rules. As modern computers approach petaFLOPS (one quadrillion FLOPS) speeds, it might now be possible to retrace these elementary steps in the emergence of biological intelligence using a detailed, biologically accurate model of the brain. The current state-of-the-art is a model of a thalamocortical column comprising 3,650 multi-compartment neurons (~100 compartments) representing diverse types, including superficial and deep pyramidal neurons, spiny stellates, fast-spiking interneurons, low-threshold spiking interneurons, thalamocortical relay neurons, and reticular nucleus neurons. These Blue Brain studies provide sound proof of principle that "multi-compartment, multi-neuron circuit" simulations are possible and give valuable insight into cortical network properties.

Sir David MacKay's learning architecture is based on the principles of neuronal organization found in biological neural networks that constitute animal brains. A *Self-Organizing Neural Network (SONN)* is an unsupervised learning model in artificial neural networks, termed as Self-Organizing Feature Maps or Kohonen Maps. These feature maps generate a two-dimensional discretized form of an input space during the model training (based on competitive learning) process. This phenomenon is very similar to biological systems and in the human cortex, sensory input spaces (e.g., auditory, motor, tactile, visual, and somatosensory) of multi-dimension are represented by two-dimensional maps. Such projection of higher dimensional inputs to reduced dimensional maps is termed topology conserving. MacKay's model is constructed around his notion of self-organizing systems, whereas Newell and Simon's model is based on high-level symbol manipulation.

In neuroscience, we want to highlight a surprising fact that is often denounced but seldom believed; namely, that most of current neuroscientists, contrary to often-heralded physicalist credo, embrace dualism. Second, we want to introduce an original explanation of such a fact: an explanation that casts a disturbing light on many notions of current usage in the field of neuroscience. We will claim that the implicit assumptions adopted by most neuroscientists invariably lead to some sort of dualistic framework. The ultimate aim of computational neuroscience is to explain how electrical and chemical signals are used in the brain to represent and process information. Neuroscience would be the foundation for the connectionist approach to AI.

Psychology: As the science of psychology began to develop in the 1800s, particularly experimental psychology, researchers began to search for specific characteristics that were common for the human mind. In pursuit of consistency and understanding, the scientific community adopted the idea of *behaviorism,* a point of view that treated the human mind as little more than an array of programmed behaviors, occurring entirely as biological reactions to stimuli. In other words, the behaviorists didn't view you much differently than they did a dog or a single-celled creature. They saw humans as just more advanced versions of "cause and effect", stimulus and response. In more recent times, beginning early in the 1900s, scientists began projecting the idea that there is much more to the human mind than merely programmed responses. As computer models were being built that simulated levels of human thought, scientists began to understand more about the reasoning process, becoming aware of the complexity of the operations that go on within the mind.

In psychology, how do humans and animals think and act? The origins of scientific psychology are usually traced to the work of the German physicist Hermann von *Helmholtz* and his student Wilhelm *Wundt*. Helmholtz applied the scientific method to the study of human vision, and his *Handbook of Physiological Optics* is even now described as "the single most important treatise on the physics and physiology of human vision". In 1879, Wundt opened the first laboratory of experimental psychology at the University of Leipzig. Wundt insisted on carefully controlled experiments in which his workers would perform a perceptual or associative task while introspecting on their thought processes. Biologists studying animal behavior on the other hand, lacked introspective data and developed an objective methodology. Applying this viewpoint to humans, the *behaviorism* movement, led by *John Watson*, rejected *any* theory involving mental processes on the grounds that *introspection could not provide reliable evidence*. Behaviorism was concerned primarily with the learning of associations, particularly in nonhuman species, and it constrained theorizing to stimulus–response notions. *Behaviorists emphasize* on studying only objective measures of the percepts given to an animal and its resulting actions (or *response*). The so-called mental constructs such as knowledge, beliefs, goals, and reasoning steps were dismissed as unscientific "folk psychology". *Behaviorism discovered a lot about rats and pigeons but had less success at understanding humans*. The view of the *brain as an information-processing device*, which is a principal characteristic of *cognitive psychology*, can be traced back at least to the works of William James. Cambridge's Applied Psychology Unit reestablished the legitimacy of "mental" terms as beliefs and goals, arguing that they are just as scientific as, say, using pressure and temperature to talk about gases, despite their being made of molecules that have neither. Kenneth Craik specified the three key steps of a knowledge-based agent as:

1. the stimulus must be translated into an internal representation,
2. the representation is manipulated by cognitive processes to derive new internal representations, and
3. these are in turn retranslated back into action.

Craik clearly explained why this was a good design for an agent: If the organism carries a "small-scale model" of external reality and of its own possible actions within its head, it is able to try out various alternatives, conclude which is the best of them, react to future situations before they arise, utilize the knowledge of past events in dealing with the present and future, and in every way to react in a much fuller, safer, and more competent manner to the emergencies which face it [86]. Meanwhile, in the United States, the development of computer modeling led to the creation of the field of *cognitive science*. The field can be said to have started at a workshop in September 1956 at MIT. This is just 2 months after the conference at which AI itself was "born". At the workshop, George *Miller* presented *The Magic Number Seven*, Noam *Chomsky* presented *Three Models of Language*, and Allen *Newell* and Herbert *Simon* presented *The Logic Theory Machine*. These three influential papers showed how computer models could be used to address the psychology of memory, language, and logical thinking, respectively. It is now a common view among psychologists that "a cognitive theory should be like a computer program" that is, it should describe a detailed information-processing mechanism whereby some cognitive function might be implemented.

Cognitive psychology is the branch of psychology that focuses on how a person acquires, processes, and stores information. Prior to the 1950s, the dominant school of thought had been behaviorism. For the next 20 years, the psychology world began to shift away from studying observable behaviors and moved toward studying internal mental processes, focusing on topics such as attention, memory, problem-solving, perception, intelligence, decision-making, and language processing. Cognitive psychology differed from psychoanalysis because it used scientific research methods to learn about mental processes, instead of simply relying on the subjective perceptions of a psychoanalyst. The 1950s through the 1970s is now commonly referred to as the "cognitive revolution" because it was during this time period that processing models and research methods were created. American psychologist Ulric Neisser first used the term in his 1967 book, *Cognitive Psychology*.

Epistemology from ancient Greek *ἐπιστήημη* (*epistēmē*) "knowledge", and -logy), or the *theory of knowledge*, is the branch of philosophy concerned with *knowledge*. Epistemology is considered a major subfield of philosophy, along with other major subfields such as ethics, logic, and metaphysics. Epistemologists study the nature, origin, and scope of knowledge, epistemic justification, the rationality of belief, and various related issues. Potential sources of knowledge are justified beliefs, such as perception, reason, memory, and testimony. Bertrand Russell brought a great deal of attention to the distinction between "knowledge by description" and "knowledge by acquaintance". One of the most important distinctions in epistemology is between what can be known *a priori* (independently of experience) and what can be known *a posteriori* (through experience). Views that emphasize the importance of *a priori* knowledge are generally classified as *rationalist* (refer back to Stewart's definition of AI). One of the core concepts in epistemology is *belief*. A belief is an attitude that a person holds regarding anything that they take to be true and truth is the property or state of being in accordance with facts or reality. There are many proposed sources of knowledge and justified belief, which we take to be actual sources of knowledge in our everyday lives. For example, Bayesian epistemology is a formal approach to various topics in epistemology that has its roots in Thomas Bayes' work in the field of probability theory. One advantage of its formal method in contrast to traditional epistemology is that its concepts and theorems can be defined with a high degree of precision. It is based on the idea that beliefs can be interpreted as subjective probabilities. As such, they are subject to the laws of probability theory, which act as the norms of rationality.

As a result, cognition is defined as the "mental action or process of acquiring knowledge and understanding through thought, experience, and the senses". Two examples of cognition are learning and reasoning through logic. Cognition is also known as "thinking" and refers to the ability to process information, hold attention, store and retrieve memories, and to select appropriate responses and actions. Cognitive science studies both "human thinking" and "machine thinking". Cognitive psychology is human thinking, while machine thinking is known as artificial intelligence (AI). The human's brain and mind is used as the model for thinking. There are three important cognitive psychology theories:

- Developmental Theory (Piaget)
- Socio-cultural Cognitive Theory (Vygotsky)
- Information Processing (Atkinson and Shiffrin)

Developmental psychology attempts to understand the nature and sources of growth in children's cognitive, language, and social skills. This tracks with Alan Turing's idea of "child programme" where the mind starts off simple and matures as the child grows up. There are four central themes that are unique to a developmental perspective. The first is *nature versus nurture* in shaping development. The second question focuses on whether child's mind growth proceeds in a continuous or more stage-like fashion such as with a fixed predetermined progression. The third question is if there is a critical or sensitive period, defined as a time of growth during which an organism is maximally responsive to certain environmental or biological events. The final theme concerns the importance of early "experience" in shaping later growth and development. Socio-Cognitive Theory tells us that portions of an individual's knowledge acquisition can be directly related to observing others within the context of social interactions, experiences, and outside media influences. Socio-Cognitive learning theory starts from evolving an assumption that humans can learn by observing others. The key processes during this type of learning are observation, imitation, and modeling, which as such involve attention, memory, and motivation. Thus, humans can learn through observing others' behaviors, attitudes, and the outcomes of those behaviors. Analogical thinking consists of dealing with new situations by adapting a similar familiar situation.

The Psi theory was developed by Dietrich Dörner and is a systemic psychological theory covering human action regulation, intention selection, and emotion. It models the human mind as an information-processing agent, controlled by a set of basic physiological, social and cognitive drives. Perceptual and cognitive processing are directed and modulated by these drives, which allow the autonomous establishment and pursuit of goals in an open environment. The Psi theory suggests a neuro-symbolic model of representation, which encodes semantic relationships in a hierarchical spreading activation network. The representations are grounded in sensors and actuators and are acquired by autonomous exploration. The Psi theory supplies a conceptual framework highlighting the interrelations between perception and memory, language and mental representation, reasoning and motivation, emotion and cognition, autonomy, and social behavior. One of the most interesting aspects of the theory is the suggestion of hierarchical neuro-symbolic representations, which are grounded in a dynamic environment. These representations employ spreading activation mechanisms to act as associative memory, and they propose the combination of neural learning with symbolic reasoning and planning. "MicroPsi" is a comprehensive software framework designed to implement and execute models of the Psi theory as multi-agent systems.

Information-processing theory is *an approach to cognitive development that aims to explain how information is encoded into memory*. It is based on the idea that humans do not merely respond to stimuli from the environment, but instead, humans process the information they receive (just like computers). Learning is what is happening when our brains receive information, record it, adapt it, and store it. This theory lists three stages of our memory that works together in this order: *sensory memory, short-term or working memory, and long-term memory.* Information-processing theory is used by psychologists to describe the stages of how individuals receive, decipher, and react to daily occurrences in their lives. As a result, information processing is a cognitive theory that uses computer processing as a metaphor for the workings of the human brain. It is theorized that the brain works in a set sequence, as does a computer. Proposed by George A. Miller and other psychologists in the 1950s, the theory describes how people focus on information and encode it into their memories. Data processing deals with extraction of information from raw material; information processing broadly applies to all information, but generally information implies data that has been processed.

Piaget's stage theory suggests that the cognitive limitations of the mind are based on what developmental stage the child is in; the information-processing theory suggests that limitations are due to a child's functional short-term memory capacity which is linked to age. Vygotsky's socio-cultural theory views human development as a socially mediated process in which children acquire their cultural values, beliefs, and problem-solving strategies through collaborative dialogues with more knowledgeable members of society.

In 1948, Shannon published his paper "A Mathematical Theory of Communication" in the *Bell Systems Technical Journal.* He showed how information could be quantified with absolute precision and demonstrated the essential unity of all information media. Telephone signals, text, radio waves, and pictures, essentially every mode of communication, could be encoded in bits. This hypothesis is as follows: if thinking is indeed a form of mechanical calculation, then development of calculating machines (i.e., computers) has the potential to unlock the "mathematical laws of human reasoning" posited in the West's rationalistic tradition. If thinking is information processing, in other words, then hypotheses about how we think can either be proven or falsified by being successfully "modeled", or by being proven intractable through the development and testing of software systems. In general, psychologists do not agree on a definition of intelligence or to what extent intelligence has to do with the related capacities of:

- Learning from experience, and
- Being able to adapt to one's environment.

Machine thinking builds the foundation for AI. Information-Processing Theory uses a computer model to describe human learning. Basically, information comes in, it gets processed, and then it gets stored and retrieved. This is akin to the Craik model. This process has four steps:

1. Step 1: Information is Sensed and Registered: In human terms, this means that we sense, or perceive, something in our environment and a decision is made as to whether or not to attend to it.
2. Step 2: Information is Momentarily Held in Short-Term, or Working, Memory: Fairly robust research indicates that we can hold approximately 7 "chunks" of information at any one time in working memory.
3. Step 3: Information is Encoded and Put in Long-Term Memory: Encoding occurs while information is in working memory, often by connecting it to existing knowledge (or schemas). Well-organized information is easier to encode because it will be "filed" in a more easily findable location.
4. Step 4: Information is Retrieved: Depending on how well it was encoded (which largely has to do with how much it was worked with in working memory), information is retrieved with the right environmental cues.

Throughout information processing, there is an executive function that enforces self-regulation: maintaining attention, planning ahead, organizing thoughts, completing tasks, adapting to changes such as avoiding obstacles, and emotion regulation. The *Atkinson–Shiffrin model* (also known as the *multi-store model* or *modal model*) is a human model of memory proposed in 1968 by Richard Atkinson and Richard Shiffrin. The model asserts that human memory has three separate components:

i. A *sensory register,* where sensory information enters memory,
ii. A *short-term store,* also called *working memory* or *short-term memory,* which receives and holds input from both the sensory register and the long-term store, and
iii. A *long-term store,* where information which has been rehearsed (explained below) in the short-term store is held indefinitely.

The levels of processing approach, also known as the depth of processing approach, were proposed by Fergus I.M. Craik and Robert Lockhart in 1972 [130]. This approach looks at how information is remembered depending on how it is processed. People can analyze stimuli at a number of different levels. Two types of processing are those of shallow processing and deep processing. The shallow level involves analysis in terms of physical or sensory characteristics, such as brightness or pitch. The deep level requires analysis in terms of meaning. When you analyze for meaning, you may think of other associations, images, and past experiences related to the stimulus. This means that the way that we process material influences how well we recall it. Deep processing (e.g., making judgments about a word's meaning) produces more permanent retention than shallow levels of processing (e.g., making judgments about a word's physical appearance). The by-product of all this analysis is a memory trace. If the stimulus is analyzed at a very shallow level (perhaps in terms of whether it had capital letters or whether it was printed in red or similar), then that memory trace will be fragile and may be quickly forgotten. However, if the stimulus is analyzed at a very deep level (perhaps in terms of its semantic appropriateness in a sentence or in terms of the meaning category to which it belongs), then that memory trace will be durable; it will be remembered. Like Atkinson–Shiffrin, Craik and Lockhart also focused on rehearsal, the process of cycling information through memory. They both proposed two different kinds of rehearsal – maintenance rehearsal and elaborate rehearsal. Maintenance rehearsal is where information is simply repeated and elaborative rehearsal refers to trying to rehearse the information in relation to its meaning and trying to associate images and past experiences in the rehearsal. In 1972, Donald Hebb presented a basic information-processing model in which information is received via receptors and transmitted through afferent and internuncial pathways to the brain stem and cerebrum. These pathways to the brain stem and cerebrum are believed to be "hard wired" to make a function a permanent feature in a computer by means of permanently connected circuits so that it cannot be altered by software (built to work in a particular way) (Figure 1.12).

Endel Tulving also developed a model in 1972, which distinguished between the short-term memory and other types of memories. He focused on the nature of the material that is stored in memory and distinguished between two different kinds of memory – episodic and semantic. In 1987, Tulving added another category called procedural memory. Episodic memory is where information is stored about events that have happened and are going to happen, and the relationships between these events, basically it refers to personal experiences. Semantic memory is made up of all the information one knows about the outside world. Information in this memory is very consistent compared to that of information in which episodic memory is forever changing. It includes knowledge about words and also information that may not be able to be explained through words. Tulving emphasized that short-term memory constitutes a separate memory system. He argued that procedural knowledge is the first system to develop during infancy, followed by semantic knowledge and last of all episodic memory.

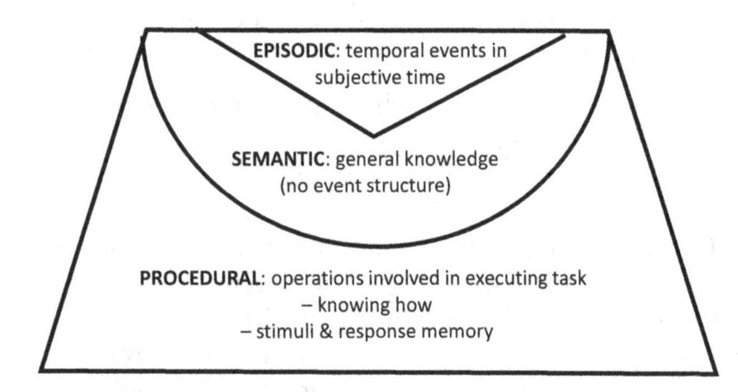

FIGURE 1.12
Endel Tulving model.

Donald Broadbent's model outlined the transfer of information from short- to long-term memory and the filter model became the salient point of the dual-memory models developed in the 1970s. His Filter Model of Attention proposed the existence of a theoretical filter device, located between the incoming sensory register and the short-term memory storage. This filter functions together with a buffer and enables the subject to handle two kinds of stimuli presented at the same time. One of the inputs is allowed through the filter, while the other waits in the buffer for later processing. The filter theory has difficulties explaining the famous cocktail party effect, which tries to explain how we are able to focus our attention on the stimuli that we find most interesting. The basic notions are that the episodic memory trace may be thought of as a rather automatic by-product of operations carried out by the cognitive system and that the durability of the trace is a positive function of "depth" of processing, where depth refers to greater degrees of semantic involvement. Computational demands continue to outpace readily available technology. In recent years, fiber optics technology has revolutionized the rate at which data can be carried between any two points. The theoretical communication bandwidth offered by fiber optic channels is of the order of 1,016 Mbytes/s. Satellite-generated data must be processed at a rate of 1,010 Hz. For example, before and during surgery, when a team of surgeons require a 3-D view of human body parts on a TV screen using tomography technology, information must be processed at speeds on the order of 1,015 Hz. Many more applications demand increased processing speeds: speech recognition, spatial and temporal pattern recognition, modeling fusion rectors, oil explorations, astronomy, robotics, and the solutions of large differential equations for numerical simulations of earthquakes and for atomic and nuclear physics with thousands of variables.

The earliest attempts to apply mathematical modeling to memory probably date back to the late 19th century when pioneers such as Ebbinghaus and Thorndike started to collect empirical data on learning and memory. Given the obvious regularities of learning and the forgetting curves, it is not surprising that the question was asked whether these regularities could be captured by mathematical functions. Ebbinghaus, for example, applied the following equation to his data on savings as a function of the retention interval:

$$S = \frac{100k}{(\log t)^c + k} \tag{1.1}$$

where S is the percentage saving, t is the retention interval in minutes, and k and c are constants. Since no mechanisms were described that would lead to such an equation, these early attempts at mathematical modeling can best be described by the terms curve fitting or data descriptions.

Modern memory models have their roots in the models developed in the 1950s by mathematical psychologists such as Estes, Bush, Mosteller, and Restle. Initially, these models were mainly models of learning, describing the changes in the probability of a particular response as a function of the event that occurs on a certain trial. Differences between recognition and recall performance may be attributed to differences in storage processes, retrieval processes, or some combination of both. Human memory is divided into a short-term working memory and a long-term permanent memory. Control processes act within the short-term working memory to make decisions and regulate information flow, thereby controlling learning and forgetting. Mental models are cognitive representations of external reality. The notion of a mental model was originally postulated by the psychologist Kenneth Craik who proposed that people carry, in their minds, a small-scale model of how the world works. These models are used to anticipate events, reason, and form explanations.

Parallel computing is a type of computation in which many calculations or processes are carried out simultaneously. Large problems can often be divided into smaller ones, which can then be solved at the same time. In cognitive psychology, parallel processing refers to our ability to deal with multiple stimuli simultaneously. The concept of parallel processing originated around the same time as the concept of information processing, which came about with the invention of computers in the mid-20th century. These models assume that information processing takes place through the interactions of a large number of simple processing elements called units, each sending excitatory and inhibitory signals to other units. In some cases, the units stand for possible hypotheses about such things as the letters in a particular display or the syntactic roles of the words in a particular sentence. In these cases, the activations stand roughly for the strengths associated with the different possible hypotheses, and the interconnections among the units stand for the constraints the system knows between the hypotheses. In other cases, the units stand for possible goals and actions, such as the goal of typing a particular letter, or the action of moving the left index finger, and the connections relate goals to subgoals, subgoals to actions, and actions to muscle movements. In still other cases, units stand not for particular hypotheses or goals, but for aspects of these things.

Psychologist Edward Throndike's early work on comparative psychology and the learning process has led to the theory of connectionism. His "Law of Exercise or Use or Frequency" states that all things being equal, the more often a situation connects with or evokes or leads to or is followed by a certain response, the stronger the tendency for it to do so in the future. Or simply, a behavior that is followed by a positive consequence is more likely to be repeated, while behavior that is followed by a negative consequence is less likely to be repeated. In Throndike's second law, his "Law of Effect" states that what happens as an effect or consequence or accompaniment or close sequel to a situation-response, works back upon the connection to strengthen or weaken it. And psychologist Clark L. Hull believes the process of learning is wholly automatic where it occurs as the result of the interaction of the organism with its environment. He sought to explain learning and motivation by the scientific laws of behavior.

A number of different rules for adjusting connection strengths have been proposed is due to Donald Hebb, but his actual proposal was not sufficiently quantitative to build into an explicit model. However, a number of different variants can trace their ancestry back to Hebb. Perhaps the simplest version is: When unit A and unit B are simultaneously excited, increase the strength of the connection between them. Hebb also introduced the concept of cell assemblies – a concrete example of a limited form of distributed processing – and discussed the idea of reverberation of activation within neural networks. Karl Lashley introduced the concept of cell assemblies – a concrete example of a limited form of distributed processing – and discussed the idea of reverberation of activation within neural networks. Later, Rosenblatt articulated the promise of a neurally inspired approach to computation, and he developed the perceptron convergence procedure, an important advancement over the Hebb rule for changing synaptic connections.

Much of the emphasis and success within the connectionist community to date has been on the distributed and bottom-up aspect of the connectionist approach associated with the so-called Parallel Distributed Processing (PDP) model. The PDP model of memory is based on the idea that the brain does not function in a series of activities but rather performs a range of activities at the same time, parallel to each other. PDP (also referred to as connectionism) differs from other models as it does not focus on distinctions among different kinds of memory. Instead, it proposes that cognitive processes can be represented by a model in which activation flows through networks that link together neuron-like units or nodes, that is, (1) Parallel – more than one process occurring at a time and (2) Distributed processing – processing occurring in a number of different locations. For example, information about a person, object, or event is stored in several interconnected units rather than in a single place. This approach to memory explains why people can still come to a correct conclusion even when incorrect information has been given. We are also able to make "generalizations" based on the links between information that were already known. One reason for the appeal of the PDP models is their obvious "physiological" flavor; they seem so much more closely tied to the physiology of the brain than are other kinds of information-processing models.

Learning involves gaining access to one or more of these units, which then activate the other units and re-create your knowledge of the object or event. Each unit is involved in the representation of several different individuals or objects. James McClelland is one of the major developers of the PDP approach. He described how our knowledge might be stored by connections that link these people with their personal characteristics. According to the PDP approach, memory consists of networks of units linked according to varying connection weights; when a unit reaches a critical level of activation, it may affect another unit either by exciting it (if the connection weight is positive) or by inhibiting it (if the connection weight is negative). Cognitive processes involve parallel operations; new events change the strength of the connections, but sometimes we have only partial memory for some information, rather than complete, perfect memory. The brain's ability to produce partial recollection is called graceful degradation.

There are many advantages to the PDP approach:

- system can often function reasonably well even if a unit is damaged or imperfect information is supplied to it
- "spontaneous generalizations"
- "default assignment"
- convincing account of how general info can emerge from specific information
- its account of amnesia.

Analogy is a mapping of knowledge from one domain to another supported by abstract representations. It has become a gateway to opening our eyes – to redefine problems (and commonplace tasks like writing require problem-solving) and find fresh solutions. There are two approaches to finding analogies: to use biomimetic principles to search, and to use functional modeling and comparison of functional similarity. An artificial neural network is only a rough analogy of how the brain works. For instance, it models synapses as numbers in a matrix, when in reality they are complex pieces of biological machinery that use both chemical and electrical activity to send or terminate signals, and that interact with their neighbors in dynamic patterns. Machine learning's main strength lies in recognizing patterns that might be too subtle or too buried in huge data sets for people to spot. *Reinforcement learning* (RL) is a technique that enables machines to learn by trial and error, much like how humans learn from experience. This approach has proven highly effective in training AI agents for complex tasks, such as playing games and controlling autonomous vehicles. RL enables a robot to autonomously discover an optimal behavior through trial-and-error interactions with its environment. Instead of explicitly detailing the solution to a problem, in RL the designer of a control task provides feedback in terms of a scalar objective function that measures the one-step performance of the robot. A RL problem is to find a policy that optimizes the long-term sum of rewards $R(s, a)$; a RL algorithm is one designed to find such a (near)-optimal policy. Both RL and optimal control address the problem of finding an optimal policy (often also called the controller or control policy) that optimizes an objective function (i.e., the accumulated cost or reward), and both rely on the notion of a system being described by an underlying set of states, controls and a plant or model that describes transitions between states. Problems in robotics are often best represented with high-dimensional, continuous states and actions (note that the 10–30-dimensional continuous actions common in robot RL are considered large).

The learning theory of Thorndike (Thorndike's Learning) represents the original *S–R* (Stimuli–Response) framework of behavioral psychology: *Learning is the result of associations forming between stimuli and responses.* Such associations or "habits" become strengthened or weakened by the nature and frequency of the *S–R* pairings. In the *Law of Effect*, a chance act becomes a learned behavior when a connection is formed between a stimulus (*S*) and a response (*R*) that is rewarded. As a *Law of Exercise*, the *S–R* connection is strengthened by use and weakened with disuse. In the law of readiness, motivation is needed to develop an association or display changed behavior. In associative shifting, a learned behavior (response) can be shifted from one stimulus to another. And once a behavior is learned, the stimulus is gradually changed. The puzzle box is used to study instrumental or operant conditions in cats. Thorndike found that the cats took less and less time to get out of the box the more trials of training had been given. Thus, trial-and-error learning describes an organism's attempt to learn/solve a problem by trying alternative possibilities until a correct solution or desirable outcome is achieved. The correct response is rewarded; learning by trial and error is a gradual process.

Cognitive computing is *an attempt to have computers mimic the way a human brain works.* To accomplish this, cognitive computing makes use of AI and other underlying technologies, including the following: expert systems and neural networks. The view that cognition can be understood as computation is ubiquitous in modern cognitive theorizing, even among those who do not use computer programs to express models of cognitive processes. One of the basic assumptions behind this approach, sometimes referred to as "information-processing psychology", is that cognitive processes can be understood in terms of formal operations carried out on symbol structures. It thus represents a formalist approach to theoretical explanation. In practice, tokens of symbol structures may be depicted as expressions written in some lexicographic notation (as is usual in linguistics or mathematics), or they may be physically instantiated in a computer as a data structure or an executable program.

In the most general sense, an algorithm is a series of instructions telling a computer how to transform a set of facts about the world into useful information. The facts are data, and the useful information is knowledge for people, instructions for machines, or input for yet another algorithm. To a computer, input is the information needed to make decisions. Next comes the heart of an algorithm – computation. Computations involve arithmetic, decision-making, and repetition. The last step of an algorithm is output for expressing the answer. To a computer, output is usually more data, just like input. It allows computers to string algorithms together in complex fashions to produce more algorithms. However, output can also involve presenting information. An algorithm is the process a computer uses to transform input data into output data. A special category of algorithms, machine learning algorithms, tries to "learn" based on a set of past decision-making examples. Thus, machine learning is commonplace for things like recommendations, predictions, and looking up information.

Algorithms: The name "algorithm" is derived from the Latin transliteration of Muhammad ibn Musa al-Khwarizmi's name. Al-Khwarizmi was a Persian scholar during the 9th century whose books introduced the western world to the decimal positional numeral system, as well as to the solutions of linear and quadratic equations. AI often revolves around the use of algorithms. An algorithm is a sequence of instructions telling a computer what to do and is a set of instructions that a mechanical computer can execute. An algorithm is thus a sequence of computational steps that transform the input into the output. A complex algorithm is often built on top of another, simpler, one and a common way to visualize it is with a tree design. Any one algorithm for a particular problem is not applicable over all types of problems in a variety of situations. So, there should be a general problem solving algorithms, which may work for different strategies of different problems.

According to the *information-processing model*, the human brain takes essentially meaningless information and turns it into meaningful patterns. Memory in cognitive psychology refers to the processes used in *acquiring, storing, retaining, and retrieving information.* It does this through three steps: (1) Encoding, (2) Storage, and (3) Retrieval. Encoding is the modification of information to fit the preferred format for the memory system. Storage is the retention of encoding material over time. There are three stages of memory: sensory, working (short-term), and long-term. Retrieval is the location and recovery of information from memory. Procedural memory is implicit and the part of long-term memory where we store memories of how things are done. Declarative memory or explicit is the part of long-term memory where we store specific information such as facts and events. To create a new memory, information must first go through encoding so it can be transformed into a usable form. Following encoding, the information is stored in our memory so that it can be used later.

There is a direct relationship between a system's inputs and a system's outputs.

- In linear systems, this relationship reduces to the system's transfer function.
- However, most systems aren't linear, but this general statement still holds for most of them.
- This relationship is therefore encoded in words, tables, graphs, mathematics, fuzzy statements, and other forms.

But it is still the formal relationship of inputs to outputs (Figure 1.13).

One of the defining characteristics of humans is our use of complex languages used in its ability to communicate. Language is the spoken, written, or gestured works and the way we combine them to communicate meaning. In language, semantics is the study of meaning and the set of rules by which we derive meaning from morphemes, words, and sentences in a given language. Syntax is the rule for combining words into grammatically sensible sentences in a given language. When thinking, a concept is a mental grouping of similar objects, events, ideas, or people. Inference is used to figure out an episode, and we have tools to put context and stimulus of a situation. The first tool is a schema, a general framework that provides expectations about topics, events, objects, people, and situations. The second tool is scripts, schemas about sequences of events and actions expected to occur in particular settings. So, when the brain is faced with a problem, there are a few options to figure out a solution: algorithms (problem-solving procedures or formulas that guarantee a correct outcome if correctly applied) and *heuristics, as simple and basic rules that serve as shortcuts to solve complex mental tasks*. However today, people are still far better at perceiving objects in natural scenes and noting their relations, at understanding language and retrieving contextually appropriate information from memory, at making plans and carrying out contextually appropriate actions, and at a wide range of other natural cognitive tasks. Yet people are also far better at learning to do these things more accurately and fluently through processing experiences. What is the basis for these differences between humans and machines? One answer that we might expect from AI is "software". The typical response heard is if we only had the right computer program, the argument goes, we might be able to capture the fluidity and adaptability of human information processing (Figure 1.14).

Similar to Craik's SPA paradigm, the US military's warfighter has adopted Air Force Colonel John Boyd's decision-making loop called the observe-orient-decide-act (OODA) as a model for the mind. Observations are the raw information on which decisions and actions are based, and analogous to the mind's concept of perception. The observed information must be processed to orient it for decision-making. The second "O", orientation, does filtering of the observed information through knowledge about our culture, genetics, ability to analyze and synthesize, and applicable previous experience. The observed information must be processed to orient it for decision-making and the OODA loop's decisions are based on observations of the evolving situation tempered with implicit filtering of the problem being addressed. The last step is the action. Decision-making by the pilot requires processing the cycle quickly, observing and reacting to unfolding events more rapidly than a human opponent, thereby "getting inside" the opponent's decision cycle and gaining the advantage. Boyd's emphasis on the "loop" is actually a set of interacting loops that are to be kept in continuous operation during combat.

The OODA loop emphasizes the importance of speed and agility in decision-making and action-taking. The goal is to complete the loop as quickly and efficiently as possible so that the warfighter can adapt to changing circumstances and take advantage of opportunities as they arise. By inspection of the above diagram, there are two critical considerations for leveraging AI to enable decision superiority: AI processing applies to every part of the process and minimizes latencies among the four steps (Figure 1.15).

Cognitive science is an enabler for understanding the human mind and can be used as a model for AI. There is a hypothesis that "a system using human-like representations and processes will enable better collaboration with people than a computational system that does not". Thus, similar representations and

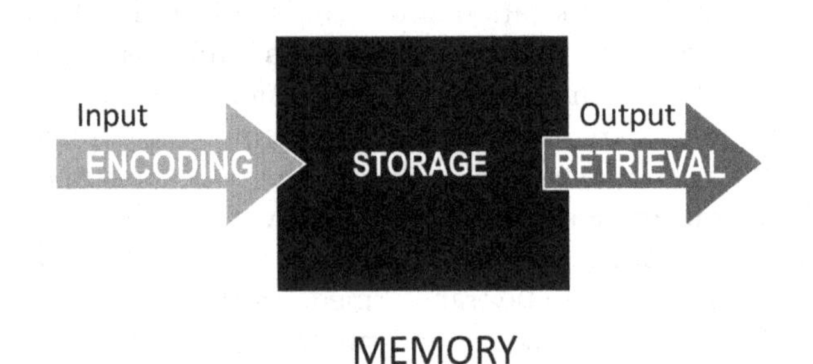

FIGURE 1.13
Memory diagram.

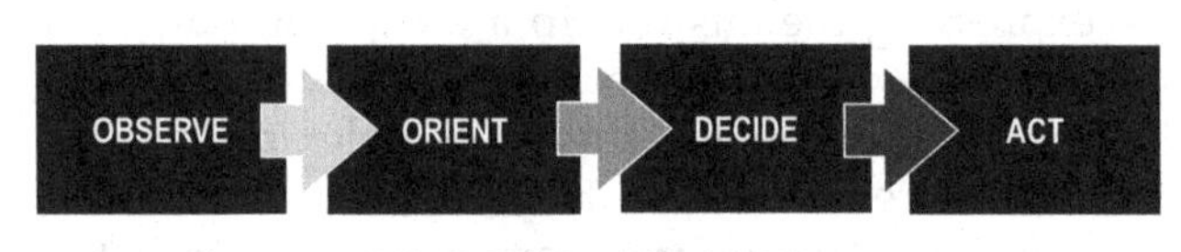

FIGURE 1.14
Warfighter OODA loop diagram.

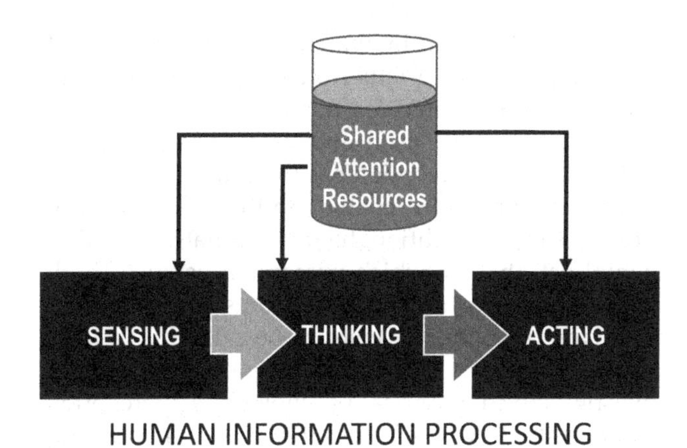

FIGURE 1.15
Human information-processing diagram.

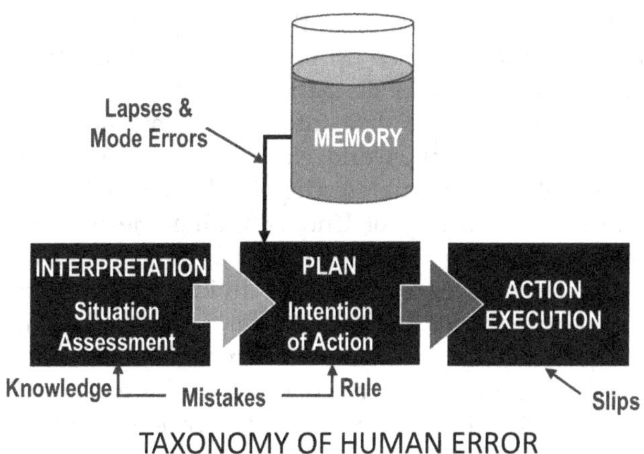

FIGURE 1.16
Taxonomy of human error diagram.

reasoning make it easier for a human to work with the robot, making conversation between the two systems compatible. For close collaboration between machine and human, the machine should accommodate the human. Cognitive skills required of a proposed AI system include appropriate knowledge representations, having the ability to solve problems, having the capability to learn, having the ability to see the human to recognize gestures and have object permanence and tracking, and being able to communicate with natural language. AI, from the cognitive perspective, should exhibit anticipation, temporal reasoning, spatial reasoning, and be able to take perspective, both spatial and social. A human error is an *unintentional* deviation from desired performance due to a mistake, oversight, misunderstanding, lack of awareness, lack of knowledge, miscommunication, lapse, or slip. Human error is a generic term used to describe all those occasions where a planned sequence of mental or physical activities fails to achieve its intended outcome, and when these failures cannot be attributed to outside intervention. E. Mach in 1905 observed that knowledge and error flow from the same mental source in memory; but only success can tell one from the other. As a theory of human error, there are three stages of cognitive processing for every task: (1) planning, (2) storage, and (3) execution (Figure 1.16).

During planning, a goal is identified and a sequence of actions is selected to reach the goal. For storage, the selected plan is stored in memory until it is appropriate to carry it out. And for execution, the plan is implemented by the process of carrying out the actions specified by the plan. Human errors in the cognitive architecture are due to lapses and mode errors, slips, and both knowledge mistakes and rule mistakes. *Mistakes* are a failure to come up with appropriate solution and take place at levels of perception, memory, or cognition. Knowledge-based mistakes are wrong solutions because the human did not accurately assess the situation, caused by poor heuristics/biases, insufficient info, and information overload. Rule-based mistakes occur when invoking the wrong rule for a given situation. *Slips* are having the right intentions but are incorrectly executed. These capture errors might involve having a similar situation (from past experience) that elicits a similar action, but the solution may be wrong for this particular situation. Slips happen when intended action is similar to routine behavior, either the stimulus or response is related to incorrect response, or the response is relatively automated and not monitored by consciousness. *Lapse* error is the failure to carry out an action, typically an error of omission (related to working memory). A *Mode* error is when the mind makes the right response, but while in the wrong mode of operation. For a robot or machine operator, human errors can contribute to events and unwanted outcomes. Error rates in humans can never be reduced to zero, but with proper design, the consequences of errors can be eliminated. Finally, there have been great breakthroughs in our understanding of cognition as a result of the development of expressive high-level computer languages and powerful algorithms.

One of the deepest problems in cognitive science is that of understanding how people make sense of the vast amounts of raw data constantly bombarding them from their environment. The essence of human perception lies in the ability of the mind to differentiate order from this chaos, whether this means simply detecting movement in the visual field, recognizing sadness in a tone of voice, perceiving a threat on a chessboard, or coming to understand the Iran – Contra affair in terms of Watergate. It has long been recognized that perception goes on at many levels. Immanuel Kant divided the perceptual work of the mind into two parts: the faculty of Sensibility, whose job it is to pick up raw sensory information, and the faculty of Understanding, which is devoted to organizing these data into a coherent, meaningful experience of the world. Kant found the faculty of Sensibility rather uninteresting, but he devoted much effort to the faculty of Understanding. He went so far as to propose a detailed model of the higher-level perceptual processes involved, dividing the faculty into 12 Categories of Understanding. Corresponding roughly to Kant's faculty of Sensibility, we have low-level perception, which involves the early processing of information from the various sensory modalities. High-level perception, on the other hand, involves taking a more global view of this information, extracting meaning from the raw material by accessing concepts, and making sense of situations at a conceptual level. This ranges from the recognition of objects to the grasping of abstract relations, and on to understanding entire situations as coherent wholes. Low-level perception is far from uninteresting, but it is high-level perception that is most relevant to the central problems of cognition. The study of high-level perception leads us directly to the problem of mental representation. Thus, representations are the fruits of perception (Figure 1.17).

The principal subject of Christian Wolff's empirical psychology, as well as his rational psychology, is mind (or soul) and its activities. The two disciplines of psychology, in his construction, are distinguished chiefly by their methods. Empirical psychology, while not neglecting observations of external behavior, has as its primary method the mind's direct introspection of its own activities, either by catching on the wing its normal operations or by contriving experiments in order to elicit particular acts. Wolff details two sets of assumptions, one regarding the structure of perceptual experience and the other its certitude, that make the introspective method of empirical psychology possible. According to Wolff, perception

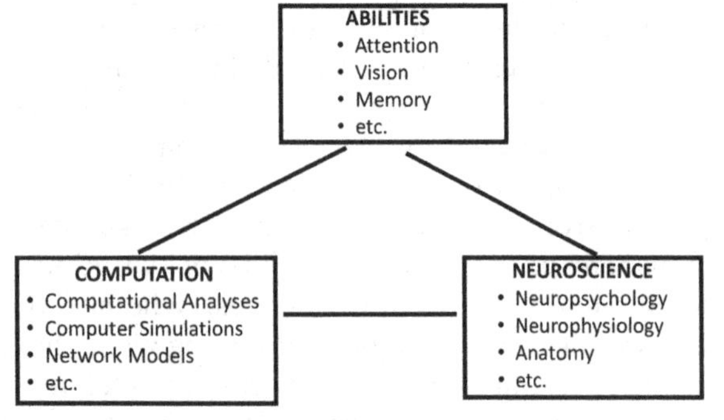

FIGURE 1.17
Christian Wolff's empirical psychology figure.

is an act of the mind by which it represents to itself something occurring either outside or within itself. Thus, the mind perceives not only colors, sounds, odors, etc. but also itself and its own activities. In Wolff's scheme, perception has two fundamental features: the represented content and the mind's act of representing. In addition, he discriminates a further mental act, that by which perception as such (both the act and the content) becomes consciously present to the mind. Following Leibniz, he calls this "apperception" and sees it as the principal instrument of investigation in empirical psychology. Apperception is a willful act by which one attends to mental operations and, through effort and perhaps extrinsic aid, brings perceptions from obscurity and vagueness to more luminous states of clarity and distinctness. Psychology would be the foundation for what the mind must exhibit in AI.

Perception: In cognition, the lowest level of perception occurs with the reception of raw sensory information by various sensed organs. For example, light impinges on the retina, sound waves cause the eardrum to vibrate, and so on. Other processes further along the information-processing chain may also be usefully designated as low-level. In the case of vision, for instance, after information has passed up the optic nerve, much basic processing occurs in the lateral geniculate nuclei and the primary visual cortex, as well as the superior colliculus. Included here is the processing of brightness contrasts, of light boundaries, and of edges and corners in the visual field, and perhaps also location processing. It is an important subject of study, and a complete theory of perception will necessarily include low-level perception as a fundamental component. The transition from low-level to high-level perception is of course quite blurry, but we may delineate it roughly as follows. High-level perception begins at that level of processing where concepts begin to play an important role. Processes of high-level perception may be subdivided into a spectrum from the concrete to the abstract. At the most concrete end of the spectrum, we have object recognition, exemplified by the ability to recognize an apple on a table, or to pick out a farmer in a wheat field. Then there is the ability to grasp relations. This allows us to determine the relationship between a blimp and the ground ("above"), or a swimmer and a swimming pool ("in"). As one moves further up the spectrum toward more abstract relations ("George Bush is in the Republican Party"), the issues become distant from particular sensory modalities. The most abstract kind of perception is the processing of entire complex situations, such as a love affair or a war.

For example, light enters the human eyes, but it is the brain that interprets color. Humans develop a mental model of the world based on what they are able to perceive with their limited senses. The decisions and actions we make are based on this internal model. Jay Wright Forrester, the father of system dynamics, described a mental model as:

> The image of the world around us, which we carry in our head, is just a model. Nobody in his head imagines all the world, government or country. He has only selected concepts, and relationships between them, and uses those to represent the real system.

To handle the vast amount of information that flows through our daily lives, our brain learns an abstract representation of both spatial and temporal aspects of this information. We are able to observe a scene and remember an abstract description thereof. Evidence also suggests that what we perceive at any given moment is governed by our brain's prediction of the future based on our internal model. The starting point for knowledge representation is the *knowledge representation hypothesis*, first formalized by Brian C. Smith in 1985. Our world model can be trained quickly in an unsupervised manner to learn a compressed spatial and temporal representation of its environment. By using features extracted from the world model as inputs to an agent, we can train a very compact and simple policy that can solve the required task. We can even train our agent entirely inside of its own hallucinated dream generated by its world model and transfer this policy back into the actual environment.

The eye and the video camera (the so-called "electronic eye") are often the subject of a superficial comparison. It is interesting to examine the limitation of this analogy as a means of emphasizing the actual role of the eye. The functional analogy probably has its roots in the discovery of the image-forming properties of lenses, and observations by natural philosophers such as Descartes of structural similarities with the image-forming optics of the single-chambered eye. The image-forming, single-chambered eye is, however, just one of an intriguing collection of adaptations that serve as the transducers for a visual sense. The compound eye, for example, has similar spatial discriminatory capacity to the single-chambered eye yet does not form any type of projected "image" in the conventional sense. Taking the collection of different "eye" types as a whole, the function of the eye becomes much clearer. It is to maximize the amount of information available from the changing optic array needed to guide the animal's or the person's actions.

Visual perception can be defined as:

i. that the fundamental operational unit or event of a perceptual system is itself a primitive type of observation, measurement, or classification event.
ii. that perception involves many of these primitive "observations" arranged as an interacting system or network – not as a single layer of transducers or detectors, but where the output of one "signal-to-symbol" transition can be used as the input of another.
iii. that these "observations" may not necessarily be implemented as discrete physical or physiological units but may result from the dynamic functioning of an underlying structure at a different level of description.
iv. that without an aspect of its operation that can be described in terms of these "observations", it does not make sense to ascribe perceptual functionality to a system.
v. that any particular percept involves the simultaneous activation of many primitive observations spread across the system or network. There is nothing else required to "look at" any output of the perceptual mechanism for perception to occur. The act of making sensory information explicit is perception.
vi. that the primitive observations involve inductive generalization, a classification process not unlike that in classical pattern recognition, which overcomes the ambiguity inherent in implicit information.
vii. The classification process is one of measuring a predicate: applying a universal property or general concept to the input stimulus data for, or relative to, each primitive observation. Each particular measurement output is one and only one of a mutually exclusive set of possible outputs, making an explicit decision that the input stimulus data satisfies a certain general property or concept.

One of the most important properties of high-level perception is that it is extremely flexible. A given set of input data may be perceived in a number of different ways, depending on the context and the state of the perceiver. Due to this flexibility, it is a mistake to regard perception as a process that associates a fixed representation with a particular situation. Both contextual factors and top-down cognitive influences make the process far less rigid than this. Some of the sources of this flexibility in perception are as follows: for example, perception may be influenced by belief. Numerous experiments by the "New Look" theorists in psychology in the 1950s (e.g., Bruner) showed that our expectations play an important role in determining what we perceive even at quite a low level. At a higher level, that of complete situations, such influence is ubiquitous. Take for instance the situation in which a husband walks in to find his wife sitting on the couch with a male stranger. If he has a prior belief that his wife has been unfaithful, he is likely to perceive the situation one way; if he believes that an insurance salesman was due to visit that day, he will probably perceive the situation quite differently.

In the 1980s, both Pylyshyn and Fodor have argued against the existence of top-down influences in perception, claiming that perceptual processes are "cognitively impenetrable" or "informationally encapsulated". These arguments are highly controversial, but in any case they apply mostly to relatively low-level sensory perception. Few would dispute that at the higher, conceptual level of perception, top-down and contextual influences play a large role. Perception may be influenced by goals. If we are trying to hike on a trail, we are likely to perceive a fallen log as an obstacle to be avoided. If we are trying to build a fire, we may perceive the same log as useful fuel for the fire. In another example, reading a given text may yield very different perceptions depending on whether we are reading it for content or proofreading it. Perception may be influenced by external context. Even in relatively low-level perception, it is well known that the surrounding context can significantly affect our perception of visual images. For example, an ambiguous figure halfway between an "A" and an "H" is perceived one way in the context of "C - T", and another in the context of "T - E". At a higher level, if we encounter somebody dressed in tuxedo and bowtie, our perception of them may differ depending on whether we encounter them at a formal ball or at the beach. Perceptions of a situation can be radically reshaped where necessary.

In Maier's well-known two-string experiment, subjects are provided with a chair and a pair of pliers and are told to tie together two strings hanging from the ceiling. The two strings are too far apart to be grasped simultaneously. Subjects have great difficulty initially, but after a number of minutes some of them hit upon the solution of tying the pliers to one of the strings and swinging the string like a pendulum. Initially, the subjects perceive the pliers first and foremost as a special tool; if the weight of the pliers is perceived at all, it is very much in the background. To solve this problem, subjects have to radically alter the emphasis of their perception of the pair of pliers. Its function as a tool is set aside, and its weightiness is brought into the foreground as the key feature in this situation. The distinguishing mark of high-level perception is that it is semantic; it involves drawing meaning out of situations. The more semantic the processing involved, the greater the role played by concepts in this processing, and thus the greater the scope for top-down influences. The most abstract of all types of perception, the understanding of complete situations, is also the most flexible.

Concepts of space and time, objects and depths, and 3-D representations are not innate; nor can they be proved to be really as we conceive them. They are simply our limited conception of what our perceptual system is doing based on interpretative mechanisms provided by our perceptual system. What is real, is the coordinated activity of layers of neurons, their development and adaption, the structure of their maps, their relationships to their neighbors – and more fundamentally – why things are this way. That is, not "why?" in the sense of "top-down" and imposed computational theories and representations, but "why?" in the sense of the only thing that perceptual mechanisms deal with – data signals transduced from the external world, and their own organization at any particular moment.

> Similarity Theory: Similarity is one of the central problems of psychology. It underlies object recognition and categorization, which are crucial to much of modern cognitive research, not to mention survival in the real world. It underlies the transfer of learning, errors of memory, perceptual organization, social bonding, and many other experimental problems one might choose almost at random from the psychological literature.

The commonsensical view of classes is that of a collection of similar objects. One might say that after many centuries of scholastic detour, the philosophers with and after Hume have come back to this simple-minded yet robust commonsensical view. The theory of general concepts based on similarity has regained some respectability, but we should note that the most important contribution by Hume, namely, that similarity is a product of association formed by mental habit, is left out of consideration by most of the proponents of the similarity theory. The pixel representation of an image is the standard interpretation of image data, where each pixel or variable represents the light intensity imaged at a point in the image plane from the original scene. In general, in the pixel representation or in an arbitrary unitary transform of it, the coefficients (pixel values or transform coefficients, respectively) are correlated. The third stage of the defect recognition process is to use the features or coefficients coded in the previous stage for classification. A multi-layer perceptron network structure with error-back propagation is used as the basis of the classification algorithm. This is augmented with a number of improvements to increase the rate and reliability of classification.

Researchers have (1) considered functional analysis, the computer model of the mind's approach to intelligence, (2) distinguished intelligence from intentionality, and (3) considered the idea of the brain as a syntactic engine. The idea of the brain as a syntactic engine explains how it is that symbol-crunching operations can result in a machine "making sense". A major rationale for accepting the language of thought has been one form or another of produdiuily argument, stemming from Chomsky's work. The idea is that people are capable of thinking vast numbers of thoughts that they have not thought before indeed that no one may ever have thought before. And among AI researchers, this idea of the computer and mind being equal in one or more fundamental aspects, as mirroring each other's operations and contents, has come to be known as the "strong AI" thesis. According to this way of thinking, the value of computer programs and models is not that they provide one predictive or falsifiable account of mental phenomena among others.

For the Standard Representational Approach, the basic assumption here is that the environment of our machine is sufficiently well defined or constrained, that we can capture it in a suitable type of representation, and that we can give our machine enough information to manipulate and update this representation as it requires. We first (possibly explicitly but more usually implicitly) determine a symbolic representation of our world. Then, based on the assumption that this representation or description is also an appropriate description of what we see as the system's environment, we "ground" the symbols in the system's formal processes, by providing the appropriate links or heuristics. In this context, information should be interpreted as fixed and instructive, and we are ultimately the creators of this information. The actual means of instructing the machine within this approach may be made increasingly flexible by the use of sophisticated interfaces, such as natural language interfaces or voice activation, or alternatively by using sophisticated programming languages like PROLOG and programming techniques like sub-symbolic computation as in artificial neural networks. Machine vision works, not because its representations are necessarily identical to what we see as our picture or representation of the world, but because the representations and processes used are geared to the problem, and the problem is relatively tightly constrained.

Mental Representations include:

- Visual Images: Template in a two-dimensional, picture-like mosaic
- Phonological Representation: A stretch of syllables
- Grammatical Representation: Nouns, verbs, phrases, phonemes, and syllables arranged into hierarchical trees
- Mentalese: The language of thought in which our conceptual knowledge is couched

In practice, representation techniques are primarily: (1) production rules (sets of if-then rules, similar to production rules used to specify a grammar) and (2) semantic networks that are sets of nodes and the links between them. Moreover, increasing evidence shows that vision, action, and language should not be regarded as a set of disembodied processes. Instead, they form a closely integrated and highly dynamic system that is attuned to the constraints of its bodily implementation as well as to the constraints coming from the world with which this body interacts. One consequence of such embodiment of cognition is that seeing an object, even when there is no intention to handle it, activates plans for actions directed toward it. This discussion emphasizes a point also made by Francis Crick about the brain as a whole. We must be very careful about interpreting what any part of the brain is doing because:

> the brain handles information in ways quite different from those we might have guessed at... we are deceived at every level of our introspection... our capacity for deceiving ourselves about the operation of our brain is almost limitless.

This is a point that Craik believes cannot be emphasized too much. It is interesting to try to interpret the function of a particular part of the eye or brain. The brain, however, is so vast and so complex that such interpretations must be tentative to say the least. But more influential than communications engineering was computer science since computer science developed a machine that reflected the essence of the human mind. Because those things are unseen when both computers and humans do them, there was good reason for drawing the *computer analogy* to cognition. Basically, this analogy said that human information processing may be similar to the steps and operations in a computer program, similar to the flow of information from input to output. If so, then thinking about how a computer does various tasks gives some insights into how people process information. In moving away from the simpler computer analogy, the cognitive science approach embraced the ideas that cognition needs to be understood with some reference to the brain. An important lesson we have learned from neuroscience is that the brain shows countless ways in which different cognitive components and processes operate simultaneously, in parallel. Furthermore, there is now ample neurological evidence that different regions of the brain are more specialized for different processing tasks, such as encoding, responding, memory retrieval, and controlling the stream of thought. The *Desiderata for Theory of Psychology* is:

- Should Predict: Complex representations for difficult task (compared to an easier task),
- Two similar things have more similar symbols, and
- Salient entities have different representations from their neighbors.

Linguistics: B.F. Skinner (1904–1990) studied how organisms learn and also how behavior could be controlled. His theories emphasize the effects of a response on the response itself. Skinner thought that most animal and human behavior is controlled by the events that precede the behavior (antecedents) and also those that follow (consequences) the behavior. In general, the antecedent tells a person what to do, and the consequence either strengthens or weakens the behavior. In 1957, B.F. Skinner published *Verbal Behavior,* a comprehensive, detailed account on the behaviorist approach to language learning. Behaviorism, also known as behavioral psychology, is a theory of learning that states all behaviors are learned through interaction with the environment through a process called conditioning. Thus, behavior is simply a response to environmental stimuli. A review of *Verbal Behavior* became as well known as the Skinner book itself and served to almost kill off interest in behaviorism. The author of the review is *Noam Chomsky,* who had just published a book on his own theory, called *Syntactic Structures.* Chomsky showed how the behaviorist theory did not address *the notion of creativity in language.* However, it did not explain how a child could understand and make up sentences that he or she had never heard before. Chomsky's theory is based on syntactic models that go back to the Indian linguist Panini (350 BC), who could explain this, and unlike previous theories, it was formal enough that it could in principle be programmed. Modem linguistics and AI, then, were "born" at about the same time, and grew up together, intersecting in a hybrid field called computational linguistics or natural language processing. The problem of understanding language soon turned out to be considerably more complex than it seemed in 1957. Understanding language requires an understanding of the subject matter and context, not just an understanding of the structure of sentences. This might seem obvious, but it was not widely appreciated until the 1960s. Much of the early work in *knowledge representation* (the study of how to put knowledge into a form that a computer can reason with) was tied to language and informed by research in linguistics, which was connected in turn to decades of work on the philosophical analysis of language.

Logic and Computation

The mind-body problem concerns the explanation of the relationship that exists between minds, or mental processes, and bodily states or processes. The most frequently used argument in favor of dualism appeals to the commonsense intuition that conscious experience is distinct from inanimate matter. If asked what the mind is, the average person would usually respond by identifying it with their self, their personality, their soul, or another related entity. They would almost certainly deny that the mind simply is the brain, or vice versa, and finding the idea that there is just one ontological entity at play to be too mechanistic or unintelligible.

In 1619, René Descartes began work on an unfinished treatise regarding the proper method for scientific and philosophical thinking. He did not believe that the information we receive through our senses is necessarily accurate. Descartes' dream argument began with the claim that *dreams and waking life can have the same content.* There is, Descartes alleges, a sufficient similarity between the two experiences for dreamers to be routinely deceived into believing that they are having waking experiences while they are actually asleep and dreaming. Although Descartes mistrusted the information received through the senses, he did believe that certain knowledge can be acquired by other means, arguing that the strict application of *reason* to all problems is the only way to achieve certainty in science. In his book, *Rules for the Direction of the Mind,* Descartes argues that all problems should be broken up into their simplest parts and that problems can be expressed as abstract equations. Descartes hoped to minimize or remove the role of the unreliable sense perception in the sciences. If all problems are reduced to their least sense-dependent and most abstract elements, then objective reason can be put to work to solve the problem. Descartes planned for 36 rules in total, although only 21 were actually written. The first 12 rules dealt with his proposed scientific methodology in general.

Descartes' work combining algebra and geometry is an application of this principle. By creating a two-dimensional graph on which problems could be plotted, he developed a visual vocabulary for arithmetic and algebraic ideas. In other words, he made it possible to express mathematics and algebra in geometric forms. He also developed a method for understanding the properties of objects in the real world by reducing their shapes to formulae and approaching them through reason rather than sense perception. In his formulation: "I think, therefore I am", Descartes took himself to have established not only his existence, but also his nature, and he is essentially a thing that thinks. Thought, that is to say, is the essence of mind. There are two aspects of thought that are of particular philosophical interest: its representation of things beyond itself, that is, its intentionality; and its movement from one representation to another in accordance with the laws of logic, that is, its rationality. Contemporary philosophers of mind typically take the problems of qualia and consciousness to pose the most serious challenge to a materialist concept of the mind, with intentionality and rationality being more readily explicable in naturalistic terms. There is a certain irony in this view; in so far as it effectively takes sensation and feeling, the capacities we seem to share with other (obviously material) animals, to be more mysterious than thought, which we (arguably) do not share with them. Many think that this conclusion is bolstered by an important set of arguments associated with John Searle, a critic of the notion that the human mind ought to be thought of as a kind of software and the brain as a kind of computer hardware. The first and most famous of these arguments involves a thought experiment that has come to be known as the "Chinese room" and is directed at the claim that the implementation of the right sort of program, whether in a computer, a sophisticated robot, or a human being; is sufficient for genuine intelligence. Searle claimed that computationalism is that the human mind is identical to a computer program, a piece of software implemented in the brain. The brain, that is to say, is on this view, literally a kind of computer. But by virtue of what, exactly, does something count as a computer in the first place?

Aristotle's logical works, collected in the *Organon* (meaning instrument), contain the earliest formal study of logic. However, in the last century, Aristotle's reputation as a logician has undergone two remarkable reversals such as having strong limitations following the rise of modern formal logic based on the works of Frege and Russell. Deductions are one of two species of arguments recognized by Aristotle. It is important to acknowledge that Aristotle's definition of deduction is not a precise match for the modern definition of validity. The other species is *induction* (*epagôgê*). Induction (or something very much like it) plays a crucial role in the theory of scientific knowledge in the *Posterior Analytics*: it is induction, or at any rate a cognitive process that moves from particulars to their generalizations, that is the basis of knowledge of the indemonstrable first principles of sciences. Aristotle sets out the conditions under which scientific arguments will provide true knowledge; where true conclusions are deduced from first principles and basic principles are used to explain more complex ones. Logic is derived from the word "logos" meaning *reason.* According to Aristotle, the word "logos" or "argument itself" consists of material such as data and narrative, as well as the cogent reasoning that allows us to make sense of our stories. Logos is commonly defined as a set of logical (and therefore inevitable) conclusions drawn from assertions or claims, such as syllogism (Figure 1.18).

The terms "logic", "physics", and "ethics" come from the Greek words λογική, φυσική, ἠθική, to which ἐπιστήμη is always to be added. Επιστήμη means roughly the same as the German term Wissenschaft, or "science". Wissenschaft, like the German word Landschaft, refers to a region, and in this case a specific self-enclosed whole comprised of a manifold of grounded knowledge, that is, of cognitions drawn exclusively and judiciously from the very things the science seeks to know. The history of logic deals with the study of the development of the science of valid inference (logic). Formal logics developed in ancient times in India, China, and Greece. Greek methods, particularly Aristotelian logic (or term logic) as found

FIGURE 1.18
Robot doing calculations.

in the *Sir Francis Bacon*'s Novum Organon of 1620, found wide application and acceptance in Western science and mathematics. Logic revived in the mid-19th century, at the beginning of a revolutionary period when the subject developed into a rigorous and formal discipline that took as its exemplar the exact method of proof used in mathematics, a hearkening back to the Greek tradition. The development of the modern "symbolic" or "mathematical" logic during this period by the likes of *Boole*, *Frege*, *Russell*, and *Peano* is the most significant in the history of logic. The title of Boole's 1854 book suggests an ambitious motivation: Because the rational human brain uses logic to think, if we were to find a way in which logic can be represented by mathematics, we would also have a mathematical description of how the brain works. Of course, nowadays this view of the mind seems to us quite naive. Boole invented a kind of algebra that looks and acts very much like conventional algebra. In conventional algebra, the operands (which are usually letters) stand for numbers, and the operators (most often + and x) indicate how these numbers are to be combined. Logic gates perform simple tasks in logic by blocking or letting through the flow of electrical current.

Mathematics: Philosophers staked out most of the important ideas of AI, but the leap to a formal science required a level of *mathematical formalization* in three fundamental areas: *logic, computation, and probability*. The idea of formal logic can be traced back to the philosophers of ancient Greece, but its mathematical development really began with the work of *George Boole*, who worked out the details of *propositional or Boolean logic*. In 1879, *Gottlob Frege* extended Boole's logic to include objects and relations, creating the *first-order logic* that is used today as the most basic knowledge representation system. *Alfred Tarski* introduced a theory of reference that shows *how to relate the objects in a logic to objects in the real world*. The next step was to determine the limits of what could be done with logic and computation. The first nontrivial *algorithm* is thought to be Euclid's algorithm for computing the greatest common denominators. The study of algorithms as objects in themselves goes back to Muhammad al-Khwarizmi, a Persian mathematician of the 9th century, whose writings also introduced Arabic numerals and algebra to Europe. Boole and others discussed algorithms for logical deduction and by the late 19th century, efforts were under way to formalize general mathematical reasoning as logical deduction. In 1900, David Hilbert presented a list of 23 problems that he correctly predicted would occupy mathematicians for the bulk of the century. The final problem asks whether there is an algorithm for deciding the truth of any logical proposition involving natural numbers – the famous "decision problem". Essentially, Hilbert was asking whether there were fundamental limits to the power of effective proof procedures.

Philosophy of Mind is a branch of philosophy that studies the ontology and nature of the mind and its relationship with the body. The mind-body problem is a paradigmatic issue in philosophy of mind, although a number of other issues are addressed, such as the hard problem of consciousness and the nature of particular mental states. Aspects of the mind that are studied include mental events, mental functions, mental properties, consciousness and its neural correlates, the ontology of the mind, the nature of cognition and thought, and the relationship of the mind to the body. In psychology, *Theory of Mind* refers to the capacity to understand other people by ascribing mental states to them. Such mental states may be different from one's own states and include beliefs, desires, intentions, emotions, and thoughts. There are two kinds of theory of mind representations: cognitive (concerning the mental states, beliefs, thoughts, and intentions of others) and affective (concerning the emotions of others). The litmus test for Theory of Mind has been the ability to attribute *false beliefs*, where prediction and explanation of action cannot be based simply on its own convictions or the state of the world.

In the second half of the 19th century a new branch of logic took shape: mathematical logic. Its aim was to link logic with the ideas of arithmetic and algebra, in order to make logic accessible to the algebraic techniques of formula manipulation. Deductive reasoning could be then reduced to algebraic formalisms. The author who made this possible was George Boole (1815–1864). By reducing Logic to Algebra, Boole made it possible to introduce mathematics into logical thinking through the Boolean algebra. At a certain point, Boole realized Leibniz's conception of the algebra of thought. With only numbers 0 and 1, Boole built up an entire algebra, turning logic into a fully symbolic practice that could be computed easily. So easy that its implementation into a computational binary framework made Boolean circuits possible. This late idea came from Claude Shannon, whose 1938 master's thesis in electrical engineering showed how to apply Boole's algebra of logic into electronic switching circuits. Although some of the first computers were not working on a binary approach, like ENIAC or Harvard's Marc I (still decimal), very soon all computers used Boolean algebra implemented with Boolean circuits (Figure 1.19).

Between 1890 and 1905, Ernst Schröder wrote his *Vorlesungen über die Algebra der Logik*, where he developed the ideas of Boole and De Morgan, including at the same time ideas of C.S. Peirce that showed that the algebraization of logic was possible and with a great impact for new logic studies. Independently, the mathematics of Charles Babbage and his co-researcher Augusta Ada Byron, Countess of Lovelace and a young Alan Turing (mathematician), developed the difference engine, first, and the Analytical Engine, second. To be accurate, Ada only collaborated on some aspects of the second machine; however, she expanded considerable effort on developing the first programs for the planned machine and on documenting its design and logic. In fact, Ada developed a method for calculating a sequence of Bernoulli numbers with the Analytical Engine, which would have run correctly whether the Analytical Engine had ever been built. Both machines were far beyond the capabilities of the technology available at the time. Nevertheless, as a theoretical concept, the idea of the Analytical Engine and its logical design are of enormous significance. This is the first realization that, by mechanical means, it might be possible to program complicated algorithms. The Analytic Engine had, in principle, all of the important components (Memory, Processor, and Input/Output protocol) that are present in modern-day computer systems. For this reason, Babbage has a strong claim to being the inventor (even if not the first builder) of the modern computer. Designed for military purposes (by calculating firing maritime ballistic tables), the difference engine was intended to evaluate polynomial functions, using a mathematical technique called the *Method of Differences*, on which Babbage had carried out important work. Future digital computers were made possible by applying propositional calculus to the two voltage levels, arbitrarily assigning the symbols 0 and 1 to these voltage levels.

FIGURE 1.19
Charles Babbage's difference engine.

With his *Begriffschrift* (1879), Gottlob Frege pioneered the first attempts to put into symbols natural languages, which were used to make science. He tried to achieve the old Leibniz's dream, using calculus as rational thinking (with the *characteristica universalis*), trying to develop a syntactic proof calculus that could be as good as mathematical proofs. Unfortunately, the system Frege eventually developed was shown to be inconsistent. It entails the existence of a concept *R* which holds of all and only those extensions that do not contain themselves. A contradiction known as "Russell's Paradox" followed. Very often, Zermelo and Russell discovered the set-theoretic paradoxes, which demolished the main logic proof building. Russell and Whitehead published their *Principia Mathematica* between 1910 and 1913, in which they reestablished the foundations of pure mathematics in logical terms.

The same bad news for idealistic approaches to mathematical entities were found in Hilbert's Program about the foundations of mathematics after Gödel's ideas on incompleteness theory. According to him, any logical system powerful enough to include natural numbers was also necessarily incomplete. Beyond this idea, Gödel also introduced the fundamental technique of arithmetization of syntax ("Gödel-numbering"), which led to the transformation of kinds of data useful for the purposes of computation. At the same time, Gödel introduced the first rigorous characterization of computability, in his definition of the class of the primitive recursive functions.

Gödel's incompleteness theorems on the formal axiomatic model of mathematical thought can be used to demonstrate that the mind is not mechanical, in opposition to a Formalist-Mechanist Thesis. Following the explanation of the incompleteness theorems and their relationship with Turing machines, we concentrate on the arguments of Gödel (with some caveats) and Lucas among others for such claims; in addition, Lucas brings out the relevance to the free will debate.

There is an argument in support of determinism that dates back to Aristotle, if not farther. It rests on acceptance of the *Law of Excluded Middle*, $\neg(p \wedge \neg p)$ according to which every proposition is either true or false, no matter whether the proposition is about the past, present, or future. In 1929, Alfred Tarski created the notion of independent semantics, a very important idea for the future of logics. The syntactic concepts comprised the notation for predicates and the rules governing inference forms needed to be supplemented by appropriate semantic ideas, and Tarski was the researcher who created the fundamentals of the semantics of first-order logic (FOL). With these conceptual tools, Tarski provided a rigorous mathematical concept of an interpretation. He was interested in defining truth as a property of sentences, but any definition of truth for sentences must be relativized to languages.

Most modern philosophers of mind adopt either a reductive physicalist or non-reductive physicalist position, maintaining in their different ways that the mind is not something separate from the body. With a deep insight for the work of the future, David Hilbert proposed the stimulant problems which from enthusiastic mathematicians generations are required for their solution. The *Entscheidungsproblem* (German for "decision problem") is a challenge posed by David Hilbert and Wilhelm Ackermann in 1928. The problem asks for an algorithm that considers, as input, a statement and answers "Yes" or "No" according to whether the statement is *universally valid*, that is, valid in every structure satisfying the axioms. The origin of the *Entscheidungsproblem* goes back to Gottfried Leibniz, who in the 17th century after having constructed a successful mechanical calculating machine, dreamt of building a machine that could manipulate symbols in order to determine the truth values of mathematical statements. He realized that the first step would have to be a clean formal language, and much of his subsequent work was directed toward that goal. In the 1930s, well before there were computers, various mathematicians from around the world invented precise, independent definitions of what it means to be computable. In 1936, Turing wrote a crucial paper, *On Computable Numbers*, with an application to the Entscheidungsproblem, in which he proposed the idea of a universal machine (the antecedent of the programmable processor). At the same time, Turing demonstrated that this problem was undecidable for FOL. The infallibility was off the agenda, because no fixed computation procedure by means of which every definite mathematical assertion could be decided (as being true or false). In the same year of 1936, Emil Post published a characterization of computability remarkably similar to that of Turing.

Nevertheless, 4 years before Turing's crucial paper, Alonzo Church introduced lambda calculus, as a new kind of formulation of logic (basing the foundation of mathematics upon functions rather than sets) and a way to go away from the Russell paradox. Nevertheless, the Kleene-Rosser paradox showed that the lambda calculus was unable to avoid set-theoretic paradoxes. Today, lambda calculus is at the heart of functional programming and has had a big influence in compilation representations as well as in reasoning representation. The lambda definability made possible developing functional programming (like LISP), also creating efficient compilers for functional languages. And related to the computer algebra, mathematical proofs by computers can be done more efficiently by lambda terms. Also in 1936, Church realized that lambda calculus could be used to express every function that could be computed by a machine. Perhaps Gödel had broken the gold dream of logical perfection but people like Church still worked on developing more powerful logical systems. From 1936 to 1938, Turing had studied with Church at Princeton University. Some years later, the mathematician, John von Neumann, developed his idea of a computer architecture after the ideas of Turing, leading to the so-called "von Neumann architecture", that has been implemented in the majority of actual computers today.

In 1928, David Hilbert and Wilhelm Ackermann posed the question in the form outlined above for a formal language. Before the question could be answered, the notion of "algorithm" had to be formally defined. This was done by Alonzo Church in 1935 with the concept of "effective calculability" based on his λ-calculus, and by Alan Turing the next year with his concept of Turing machines. Turing immediately recognized that these are equivalent models of computation. The works of both Church and Turing were heavily influenced by Kurt Gödel's earlier work on his incompleteness theorem, especially by the method of assigning numbers (Gödel numbering) to logical formulas in order to reduce logic to arithmetic. The basic claim of logicism is that mathematics is really a branch of logic. This is sometimes expressed by saying that mathematics (in this case, arithmetic) is reducible to logic. What must someone show in order to show that one theory is, in this sense, reducible to another?

Allen Newell and Herbert Simon trace the roots of their hypothesis back to Frege, Russell, and Whitehead but, of course, Frege and company were themselves heirs to a long atomistic, rationalist tradition. Descartes already assumed that all understanding consisted in forming and manipulating appropriate representations, that these representations could be analyzed into primitive elements (*naturas simplices*), and that all phenomena could be understood as complex combinations of these simple elements. Moreover, at the same time, Hobbes implicitly assumed that the elements were formal elements related by purely syntactic operations, so that reasoning could be reduced to calculation. "When a man reasons, he does nothing else but conceive a sum total from the addition of parcels", Hobbes wrote, "for REASON... is nothing but reckoning". Finally, Leibniz, working out the classical idea of Mathesis, *the formalization of everything*, sought support to develop a universal symbol system, so that "we can assign to every object its determined characteristic number". According to Leibniz, in understanding we analyze concepts into more simple elements. In order to avoid a regress of simpler and simpler elements, there must be ultimate simples in terms of which all complex concepts can be understood. Moreover, if concepts are to apply to the world, there must be simple features that these elements represent. Leibniz envisaged "a kind of alphabet of human thoughts".

Ludwig Wittgenstein, who also drawn on the work of Frege and Russell, stated the pure form of this syntactic, representational view of the relation of the mind to reality in his *Tractatus Logico-Philosophicus.* He defined the world as the totality of logically independent atomic facts: The world is the totality of facts, not of things. Facts, in turn, were exhaustively analyzable into primitive objects, such as "an atomic fact is a combination of objects", etc. or "If all objects are given, then thereby all atomic facts are given". These facts, their constituents, and their logical relations were represented in the mind. And according to Turing, "conjectures" are important to scientific research, which, contrary to popular opinion, doesn't "proceed inexorably from well-established fact to well-established fact". *Tractatus Logico-Philosophicus,* whether or not it proves to give the ultimate truth on the matters with which it deals, certainly deserves, by its breadth and scope and profundity, to be considered an important event in the philosophical world. Starting from the principles of symbolism and the relations that are necessary between words and things in any language, it applies the result of this inquiry to various departments of traditional philosophy, showing in each case how traditional philosophy and traditional solutions arise out of ignorance of the principles of symbolism and out of misuse of language.

In symbolism, Wittgenstein is concerned with the conditions which would have to be fulfilled by a logically perfect language. There are various problems regarding language.

1. There is the problem of what actually occurs in our minds when we use language with the intention of meaning something by it; this problem belongs to psychology.
2. There is the problem as to what the relation subsisting between thoughts, words, or sentences, and that which they refer to or mean; this problem belongs to epistemology.
3. There is the problem of using sentences so as to convey truth rather that falsehood; this belongs to the special sciences dealing with the subject matter of the sentences in question.
4. Then, there is the question: what relation must one fact (such as a sentence) have to another in order to be capable of being a symbol for that other?

Equality is the most important relation in mathematics and functional programming. In principle, problems in FOL with equality can be handled by, for example, with resolution theorem provers. Equality is theoretically difficult: First-order functional programming is Turing-complete.

- But: resolution theorem provers cannot even solve problems that are intuitively easy.
- Consequence: to handle equality efficiently, knowledge must be integrated into the theorem prover.

H implies that there is no uniform algorithm for testing polynomial diophantine equations (polynomial equation, usually involving two or more unknowns, such that the only solutions of interest are the integer ones) for solvability in positive integers, that is, that Hilbert's tenth problem is unsolvable.

> It is reasonable to hope that the relationship between computation and mathematical logic will be as fruitful in the next century as that between analysis and physics in the last. The development of this concern demands a concern for both applications and mathematical elegance.
>
> **— John McCarthy (1963)**

In 1931, Kurt Gödel created a stir in the World of Mathematics and Logic when he revealed that it was impossible to embrace mathematics within a single system of logic. He accomplished this by proving, first, that any consistent system, that includes the arithmetic of whole numbers, is incomplete. In other words, there are true statements or concepts within the system that cannot be deduced from the postulates that make up the system. Next, he proved even though such a system is consistent its consistency cannot be demonstrated within the system. Gödel's two incompleteness theorems are among the most important results in modern logic and have deep implications for various issues. They concern the limits of provability in formal axiomatic theories. The first incompleteness theorem states that any consistent FS *F* within which a certain amount of elementary arithmetic can be carried out is incomplete, that is, there are statements of the language of *F* that can neither be proved nor disproved in *F*. According to the second incompleteness theorem, such an FS cannot prove that the system itself is consistent (assuming it is indeed consistent). These results have had a great impact on the philosophy of mathematics and logic. Thus, he showed that there exists an effective procedure to prove any true statement in the FOL of *Frege and Russell*, but that FOL could not capture the principle of mathematical induction needed to characterize the natural numbers. In 1931, he showed that real limits do exist. His *incompleteness theorem* showed that *in any language expressive enough to describe the properties of the natural numbers, there are true statements that are undecidable in the sense that their truth cannot be established by any algorithm*. This fundamental result can also be interpreted as showing that there are some functions on the integers that cannot be represented by an algorithm, that is, they cannot be computed. This motivated *Alan Turing* to try to characterize exactly which functions are capable of being computed. This notion is actually slightly problematic, because the notion of a computation or effective procedure really cannot be given a formal definition. However, the *Church-Turing thesis*, which states that the Turing machine is capable of computing any computable function, is generally accepted as providing a sufficient definition. Turing also showed that there were some functions that no Turing machine can compute. For example, *no machine can tell in general whether a given program will return an answer on a given input or run forever*.

In the early 1950s, as calculating machines were coming into their own, a few pioneer thinkers began to realize that digital computers could be more than number crunchers. At that point, two opposed visions of what computers could be, each with its correlated research program, emerged and struggled for recognition. One faction saw computers as a system for manipulating mental symbols: the other, as a medium for modeling the brain. One sought to use computers to instantiate a formal representation of the world; while the other, to simulate the interactions of neurons. One took problem-solving as its paradigm of intelligence; the other, learning. One utilized logic, the other statistics. One school was the heir to the rationalist, reductionist tradition in philosophy; the other viewed itself as idealized, holistic neuroscience.

> The rallying cry of the group who saw that computers was a system for manipulating mental symbols was that both minds and digital computers were physical symbol systems. By 1955 Allen Newell and Herbert Simon, working at the RAND Corporation, had concluded that strings of bits manipulated by a digital computer could stand for anything – numbers, of course, but also features of the real world. Moreover, programs could be used as rules to represent relations between these symbols, so that the system could infer further facts about the represented objects and their relations. As Newell put it recently in his account of the history of issues in AI: The digital-computer field defined computers as machines that manipulated numbers. The great thing was, adherents said, that everything could be encoded into numbers, even instructions. In contrast, the scientists in AI saw computers as machines that manipulated symbols. The great thing was, they said, that everything could be encoded into symbols, even numbers.
>
> **– Allen Newell**

This way of looking at computers became the basis of a way of looking at minds. Newell and Simon hypothesized that the human brain and the digital computer, while totally different in structure and mechanism, had, at the appropriate level of abstraction, a common functional description. At this level, both the human brain and the appropriately programmed digital computer could be seen as two different instantiations of a single species of device in which one generated intelligent behavior by manipulating symbols by means of formal rules. Newell and Simon stated their view as a hypothesis called the *Physical Symbol System Hypothesis*. A physical symbol system has the necessary and sufficient means for general intelligent action. By "necessary", we mean that any system that exhibits general intelligence will prove upon analysis to be a physical symbol system. By "sufficient" we mean that any physical symbol system of sufficient size can be organized further to exhibit general intelligence. The hypothesis states that cognition is based upon patterns of information, that these patterns of information can be expressed as symbols, and these symbols can be manipulated (Figure 1.20).

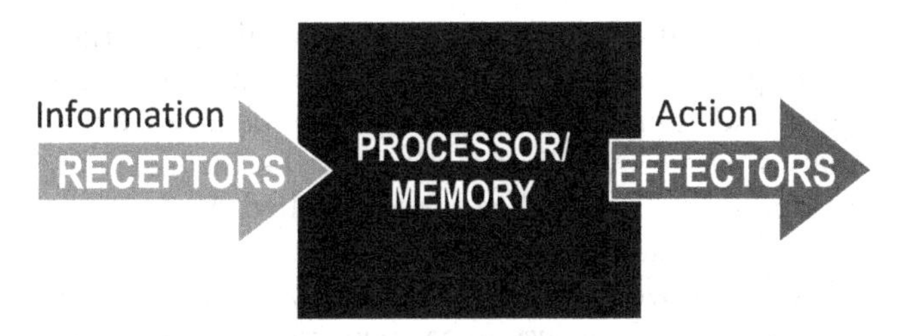

FIGURE 1.20
Newell and Simon's information-processing system diagram.

Tarski had a significant impact on analytic philosophy and philosophical logic, particularly from the 1950s onwards, in subjects such as modal logic, temporal logic, deontic logic, and relevance logic. But not only did Church establish a strong relationship between logic and computability, and in 1960, Haskell Curry first noticed what is a "bit", later was called the Curry–Howard Isomorphism. The Curry–Howard isomorphism is a striking relationship connecting two seemingly unrelated areas of mathematics – type theory and structural logic. The fundamental ideas of Curry–Howard are that proofs are programs, that formulas are types, that proof rules are type-checking rules, and that proof simplification is operational semantics. One of the seminal influences between computer science and logic was the Curry–Howard correspondence, which is the simple observation that two at-the-time-seemingly unrelated families of formalisms, the proof systems on one side and the models of computation on the other were in fact structurally the same kind of objects. As a result, the ideas and observations about logic were ideas and observations about programming languages. Nevertheless, we must go back to the year 1954. On that date, Martin Davis carried out one of the first computational logical experiments; he programmed and ran the Presburger's Decision Procedure for the first-order theory of integer addition. He proved that the sum of two even numbers was itself an even number. Next, the *halting problem* is a prominent example of an undecidable problem and its formulation and undecidability proof that is usually attributed to Turing's 1936 landmark paper. Similarly, Martin Davis proved Hilbert's tenth problem (providing a computing algorithm that will tell of a given polynomial diophantine equation with integer coefficients whether or not it has a solution in integers) was impossible. A year later, in 1955, Evert Beth and Jaakko Hintikka (independently) described a version of the basic proof procedure, being a computationally powerful technique.

Dag Prawitz in 1960 made another important (re) discovery: the resolution rule, described something obscurely 30 years before by Jacques Herbrand and coined in the actual form by John Alan Robinson in 1963. Herbrand interpretation is an interpretation in which all constants and function symbols are assigned very simple meanings. Prawitz developed a general calculus for predicate logic based on contradiction. The process of its creation consisted of combining a known inference rule with unification. In 1957, Aridus Wedberg taught a first-year logic course at the University of Stockholm. On one occasion he mentioned to the class the possibility of proving mathematical theorems in FOL on a machine. This remark raised the interest of one of the students in the class, namely, Dag Prawitz, who decided to realize this idea in practice. He first developed a general procedure for the predicate calculus. An immediately obvious drawback of the first procedures was their treatment of substituting constants in a stupid systematic manner. Prawitz was the first in proposing a unificational method instead which used metavariables and substituted constants by need rather than according to some fixed sequence. The unification was computed by way of a system of resulting equations to be solved under certain restrictions. These restrictions are derived from the well-known variable conditions in Gentzen-type systems.

Cognitive Science is *the study of the human mind and brain*, focusing on how the mind represents and manipulates knowledge and how mental representations and processes are realized in the brain. Cognitive Architecture is a specification of the structure of the brain at a level of abstraction that explains how it achieves the function of the mind. A goal of cognitive architecture is to use the research of cognitive psychology to create a complete computer model of cognition. A cognitive architecture refers to both a theory about the structure of the human mind and to a computational instantiation of such a theory used in the fields of AI and computational cognitive science.

Ironically, the new mechanical efficiency also brought a cognitive opacity. According to Robinson, resolution was more machine-oriented than human oriented. Although resolution was to be a great contribution, the paper in which it was described remained unpublished in a reviewer's desk for a year. AI and formal logic converged into a main research field, although the MIT view was still being built. This same critical process led to Larry Wos and George Robinson (at Argonne) to the development of a new rule, which they called paramodulation. This rule was implemented into several famous Argonne theorem provers, being perhaps one of the most famous being William W. McCune's OTTER. Theorem proving was one of the first fields in which symbolic (as opposed to numeric) computation was automated. Most importantly they introduced factoring, the unit preference, the set of support strategy, and implemented resolution with these additional features on a Control Data 3600 computer system. By then, Argonne was then the world champion in theorem proving! All these results made possible a functional automated deduction technology, useful for several fields of academy as well as of industry.

The area of mathematical logic is also called the mathematics of reasoning, and the first study was taken up by the Greek philosopher Aristotle (384–322 BC). Logic is an interdisciplinary field that studies truth and reasoning. Informal logic seeks to characterize valid arguments informally, for instance by listing varieties of fallacies. Formal logic represents statements and argument patterns symbolically, using FSs such as FOL. Within formal logic, *mathematical logic* studies the mathematical characteristics of formal logical systems, while philosophical logic applies them to philosophical problems such as the nature of *meaning*, *knowledge*, and *existence*. There are two types of knowledge: declarative and procedural. Declarative knowledge is knowledge about facts and it is easy to verbalize declarative knowledge. Procedural knowledge is knowledge about how to do something.

Frege's *Begriffsschrift* is a book of logic and started out to develop a formal language modeled on arithmetic for pure thought. His goal was to develop a formal approach to logic that resulted in the employment of logical calculus as a foundation for mathematics. This calculus contains the first appearance of quantified variables and is essentially classical bivalent second-order logic with identity. It is bivalent in that sentences or formulas denote either true or false, and second order because it includes relation variables in addition to object variables that allow quantification over both. The modifier "with identity" specifies that the language includes the identity relation. Mathematical notation, first introduced by Francois Viète at the end of the 16th century, consists of using symbols for representing: operations, "unspecified" numbers, relations, and any other mathematical objects. It is used to assemble these notations into expressions and formulas. Symbols are the basis for mathematical notation. They perform similarly to the role of words in natural language, analogous to the way verbs, adjectives, and nouns play different roles in a sentence. Alphabet letters are symbols, used for naming mathematical objects and for other purposes such as operations ($+$, $-$, $/$, $\oplus$, ...), for relations ($=, <, \leq, \sim, \equiv, \ldots$), for logical connectiveness ($\Rightarrow, \wedge, \vee, \ldots$), for quantifiers ($\forall, \exists, \ldots$), and for other purposes. Thus, an *expression* is a finite combination of symbols that is well-formed according to rules that depend on its context. It denotes a mathematical object and therefore plays in the language of mathematics. Frege's work was proven to be inconsistent, but Russell and Whitehead reestablished the foundations of pure mathematics in logical terms.

The Greek word logos (traditionally meaning *word, thought, principle, or speech*) has been used among both philosophers and theologians, such that "thought" and "word" are inextricable from each other. Logic is a formal language together with a relationship. A logic language together with a syntactically defined relation is called a *calculus*. Formally, a calculus is defined as a pair A, R, where A is a set of formulae and R is a set of inference rules. The application of the inference rule to the premises produces the conclusion. We also say that the conclusion is derived from the premises (by the inference rule). Because of the deduction theorem, weak properties are often enough to show a calculus has good properties. This is why *automated reasoning* is often referred to as *theorem proving*; however, those good properties are not enough to ensure that a calculus is suitable for implementation. The development of automated reasoning tools started shortly after computers were available when the first Turing-complete programmable digital computer, the ENIAC, was operational in 1946 and in 1954, a program written by Martin Davis produced a proof for the Presburger arithmetic (published in 1957). Later, George Collins implemented parts of Tarski's decision procedure for elementary algebra on an IBM 704.

Logic programming (LP) is a programming paradigm that is largely based on *formal logic*. Any program written in an LP language is a set of sentences in logical form, expressing facts and rules about some problem domain. Major LP language families include PROLOG, Answer Set Programming (ASP), and Datalog. The language called PROLOG is built on the foundation of logic and the word PROLOG is an abbreviation for programming in logic. The first PROLOG implementation was the work of Colmerauer in 1970, designed to facilitate natural language processing. Logic is used in Natural Language Processing, Theorem Proving, Game Theory, Automatic Answering Systems, Ontologies, Control Systems, and Graphical User Interfaces. A very high-level programming language always uses the same problem-solving method in the programs it creates. Since a very high-level programming language knows how it will solve any problem it is given, all it requires from a programmer is domain knowledge. Since a very high-level programming language knows the knowledge requirements of its problem-solving method, it knows the kind of access it needs to the knowledge it elicits from domain experts. Soon programmers won't have to know how to program. A representation has four parts to it:

- Lexical Part: this determines what symbols are allowed.
- Structural Part: this determines how the symbols can be arranged.
- Semantic Part: this attaches meanings to the descriptions.
- Procedural Part: this gives access functions that allow you to write and read descriptions into the structure, to modify them and to interrogate them.

The basic claim of logicism is that mathematics is really a branch of logic. This is sometimes expressed by saying that mathematics (in this case, arithmetic) is reducible to logic.

Logicism

Logicism is commonly defined as the thesis that mathematics reduces to, or is an extension of, logic. Exactly what "reduces" means here is not always made entirely clear. In mathematical logic and automated theorem proving, resolution is a *rule of inference* leading to a refutation complete theorem-proving technique for sentences in propositional logic and FOL. For propositional logic, systematically applying the resolution rule acts as a decision procedure for formula unsatisfiability, solving the (complement of the) Boolean satisfiability problem. For FOL, resolution can be used as the basis for a semi-algorithm for the unsatisfiability problem of FOL, providing a more practical method than one following from GÖdel's completeness theorem. Resolution is a theorem-proving technique that proceeds by building refutation proofs, that is, proofs by contradictions. It was invented by a mathematician, John Alan Robinson, in the year 1965. Resolution is used, if there are various statements are given, and we need to prove a conclusion of those statements. Mathematical logic is the mathematic modeling of mathematical reasoning. As such it provides methods for formally modeling structure and abstract properties of particular application domains. The theory of knowledge representation is firmly based on mathematical logic.

An important part of mathematical logic is concerned with proof theory. Algorithmic aspects of proof theory form the basis of automated theorem provers and mechanized approaches to the synthesis of executable programs from abstract specifications. Computation can be modeled as deduction in a logical system. That point of view is adopted in LP and relational databases.

David Hume's *A Treatise of Human Nature* (1739) proposed what is now known as the principle of *induction*: that general rules are acquired by exposure to repeated associations between their elements. Building on the work of *Ludwig Wittgenstein* and *Bertrand Russell*, the famous Vienna Circle, led by *Rudolf Carnap*, developed the doctrine of *logical positivism*. This doctrine holds that all knowledge can be characterized by logical theories connected, ultimately, to *observation sentences* that correspond to sensory inputs. The *confirmation theory* of Carnap and Carl Hempel attempted to understand how knowledge can be acquired from experience. Carnap's book *The Logical Structure of the World* (1928) defined an explicit computational procedure for extracting knowledge from elementary experiences. It was probably the first *theory of mind as a computational process*. The final element in the philosophical picture of the mind is *the connection between knowledge and action*. This question is vital to AI, because intelligence requires action as well as reasoning. Moreover, only by understanding how actions are justified can we understand *how to build an agent* whose actions are justifiable (or rational). Aristotle argued that actions are justified by a logical connection between goals and knowledge of the action's outcome. Aristotle's algorithm was implemented 2,300 years later by Newell and Simon in their GPS program. Goal-based analysis is useful but does not say what to do when several actions will achieve the goal, or when no action will achieve it completely.

The operational semantics of programming languages are frequently defined by inference rules such that execution of a program is formalized by deduction with the inference rules. A central subject of theoretical computer science is complexity theory, the theory about resources required for solving problems on a computer. Mathematical logic provides us with natural means of characterizing complexity classes by fundamental problems of logic reasoning.

- Predicate logic (syntax, semantics, several proof systems, and fundamental results in model theory) is a fundamental workhorse that many logic formalisms are based upon.

- Horn fragment of First-Order-Logic (FOL) underlies Logic Programming (LP) and Datalog, a query language for relational databases.

The third part is devoted to program verification using temporal logics such as linear temporal logic (LTL) and computation tree logic (CTL). In the treatment of the various subjects, we will in particular concentrate on the algorithmic aspects of how to implement specific deductive problems, usually accompanied with an analysis of the computational complexity.

- Propositional logic and with the basic notions (syntax and semantics) of FOL.
- Recursion theory and complexity theory.

In propositional logic, as the name suggests, propositions are connected by logical operators. The statement "the street is wet" is a proposition, as is "it is raining". These two propositions can be connected to form the new proposition if it is raining the street is wet. Written more formally:

$$\text{it is raining} \longrightarrow \text{the street is wet.}$$

LP is essentially a declarative programming paradigm based on formal logic. LP has its roots in automated theorem proving, where the purpose is to test whether or not a logic program can prove a logical formula, or query, ψ, that is, $\Gamma \models \psi$ or $\Gamma \not\models \psi$. For computational efficiency, the language used in logic programs is typically restricted to a subset of FOL (e.g., Horn clauses). From a reasoning perspective, LP serves as an inference engine. Hobbs et al. propose an Interpretation as Abduction framework, where natural language understanding is formulated as abductive theorem proving. In this context, a logic program is a commonsense knowledge base (e.g., {bird(Tweety), ∀ xbird(x) → fly(x)}) and a query ψ will be a question that is of interest (e.g., fly(Tweety)). It is important to remember that conventional LP cannot represent uncertainty in knowledge. "Naive Physics" is a term coined by Hayes in 1978, to describe his approach to developing a "large-scale formalism" of commonsense knowledge about the world. This concern with real-world knowledge can be related to a general awareness among AI workers that future progress in AI depends on intensive knowledge being made available to reasoning systems. The aim of naive physics as stated in is to formally describe the world in the way that most people think about it, rather than describing it in the way that physicists think about it. This description should attempt reasonable completeness – that is, it should describe a significant portion of the way we understand the world, rather than just small pieces. Implementations of proof calculi on computers are called theorem provers. One of the oldest resolution provers was developed at the Argonne National Laboratory in Chicago. Based on early developments starting in 1963, OTTER was created in 1984. Above all, OTTER was successfully applied in specialized areas of mathematics. The development of modern logic falls into roughly five periods:

1. The *embryonic period* from Leibniz to 1847, where the notion of a logical calculus was discussed and developed, but no schools were formed, and isolated periodic attempts were abandoned or went unnoticed.
2. The *algebraic period* from Boole's Analysis to Schroder's *Vorlesungen*. In this period, there were more practitioners, and a greater continuity of development.
3. The *logicist period* from the *Begriffsschrift* of Frege to the *Principia Mathematica* of Russell and Whitehead. The aim of the "logicist school" was to incorporate the logic of all mathematical and scientific discourse in a single unified system which, taking as a fundamental principle that all mathematical truths are logical, did not accept any non-logical terminology. The major logicists were Frege, Russell, and the early Wittgenstein. It culminates with the *Principia*, an important work that includes a thorough examination and attempted solution of the antinomies which had been an obstacle to earlier progress.
4. The *metamathematical period* from 1910 to the 1930s, saw the development of metalogic, in the finitist system of Hilbert, and the non-finitist system of Löwenheim and Skolem, the combination of logic and metalogic in the work of Gödel and Tarski. Gödel's incompleteness theorem of 1931 was one of the greatest achievements in the history of logic. Later in the 1930s, Gödel developed the notion of set-theoretic constructability.
5. The *period after World War II*, when mathematical logic branched into four inter-related but separate areas of research: model theory, proof theory, computability theory, and set theory, and these ideas and methods began to influence philosophy.

There are many useful logics: temporal, modal, relevance, intuitionistic, etc. They differ in what concepts are being considered and in what the basic features of these concepts are thought to be. For example, should the passage of time play a formal role (as in temporal logic), or not? Must existence assertions have constructive content (as in intuitionistic logic), or not? In the family of formal logics, one is central: classical logic. It is the most widely used logic, the logic underlying mathematics as it is generally practiced, and the logic on top of which many others have been built. Indeed, for most people who have occasion to use formal methods, logic is synonymous with classical logic. As a summary:

1. *Propositional logic* (also called Boolean Logic), by which we here mean two-valued propositional logic, arises from analyzing connections of given sentences *A*; *B*, such as *A* and *B*, *A* or *B*, not *A*, if *A* then *B*. These connection operations can be approximately described by two-valued logic.
2. Mathematics and some other disciplines such as computer science often consider domains of individuals in which certain relations and operations are singled out. When using the language of propositional logic, our ability to talk about the properties of such relations and operations is very limited. Thus, it is necessary to refine our linguistic means of expression, in order to procure new possibilities of description. To this end, one needs not only logical symbols but also variables for the individuals of the domain being considered, as well as a symbol for equality and symbols for the relations and operations in question. *FOL*, sometimes also called predicate logic, is the part of logic that subjects properties of such relations and operations to logical analysis.
3. A predicate is a generalization of a propositional variable and a symbol that represents a property or a relation. *Predicate Logic* deals with predicates, which are propositions containing variables. Predicate logic uses atomic formulas, that is, predicates with arguments, as atomic operands and the operators of propositional logic, plus the two quantifiers, "for all" and "there exists".
4. Rules can be considered a subset of predicate logic. They have become a popular representation scheme for expert systems (also called rule-based systems). They were first used in the General Problem Solver system in the early 1970s (Newell and Simon). Rules have two component parts: a left-hand side (LHS) referred to as the antecedent, premise, condition, or situation, and a right-hand side (RHS) known as the consequent, conclusion, action, or response. The LHS is also known as the "if part" and the RHS as the "then part" of the rule. Rules take the form of *if–then* statements.
5. *Model Theory* (Normal Forms or Resolution Method) to produce a practical method allowing to derive consequences and prove theorems by using computers. Model theory may be described as the union of logic and universal algebra. In general, this task is not feasible because of its enormous computational complexity. Still, for problems of a "practical size" (arising, for example, in deductive databases and other AI systems, or, trying to formalize real mathematical proofs), such methods are possible and some of them are already implemented successfully. This field of research is called *automated reasoning*, or automated theorem proving.

Logicism was advocated by Richard Dedekind, developed by Gottlob Frege, and extended by Bertrand Russell together with Alfred North Whitehead. Dedekind was able to reduce the theory of real numbers to the rational number system (by means of set theory). This convinced him that arithmetic, algebra, and analysis were all reducible to the natural numbers plus the logic of sets. In 1872, Dedekind concluded that natural numbers themselves were reducible to sets and mappings. Frege developed logicism through three works. The first, the *Begriffsschrift (Concept Script)* first published in 1879, is a technical work, introducing the reader to a formal logical system. The second work, the *Grundlagen* (*Foundations of Arithmetic*) was first published in 1884, is philosophical. The third, the *Grundgesetze der Arithmetik*, was originally published in two volumes, in 1893 and 1903. It would also be translated as *Foundations of Arithmetic*, but these are formal foundations, not philosophical ones. Whitehead and Russell continued the logicist project and published *Principia Mathematica* in three volumes. This is a technical work developing a formal theory of types, which, they argue, is pure logic. Russell also published more philosophical works: *The Principles of Mathematics* (1903) and *Introduction to Mathematical Philosophy* (1919). The major philosophical question the logicist tries to answer is: what is the essence of mathematics? As the word "logicist" suggests, the answer is that mathematics, or part of it, is essentially logic. Logicism can be either a realist philosophy of mathematics or an anti-realist philosophy of mathematics. Frege was a realist. Frege's logicist believes that mathematical truths are independent of human beings. Frege's version of logicism is epistemologically realist. That is, human beings discover, or fail to discover, mathematical truths.

More contemporary logicians, including Boole, Frege, and Tarski, have an ambition to developing the "language of thought". There can be seen a direct line through mathematics and philosophy to modern AI. Inference derives new information from stored facts and axioms can be very compact, for example, most of mathematics can be derived from the logical axioms of set theory. However, there are limitations. Not all intelligent behavior is mediated by logical deliberation (but much appears not…). Finally, a logical system consists of procedures for manipulating symbols. In propositional logic, the symbols are taken to represent propositions (i.e., sentences) and connections (e.g., and, or, if–then). Generally, there is a clear goal in such manipulation. (Logical) representation of knowledge underlying intelligence is quite nontrivial such as emphasized in the area of "knowledge representation" and also bringing in probabilistic representations, for example, Bayesian networks. So, what is the purpose of thinking and what thoughts should I have? Researchers have found that one of the special requirements for AI is the ability to easily manipulate symbols and lists of symbols rather than processing numbers or strings of characters.

Computability

Computability is the ability to solve a problem in an effective manner. It is a key topic of the field of computability theory within mathematical logic and the theory of computation within computer science. We shall proceed to define a class of (mathematical) functions that we propose to identify with the effectively calculable functions, that is, with those functions for which an algorithm that can be used to compute their values exists. Such a computing machine is deterministic in the sense that, while it is in operation, its entire future is completely specified by its status at some one instant. The computability of a problem is closely linked to the existence of an algorithm to solve the problem. A mathematical problem is *computable* if it can be solved in principle by a computing device. Some common synonyms for "computable" are "solvable", "decidable", and "recursive". Hilbert believed that all mathematical problems were solvable, but in the 1930s Gödel, Turing, and Church showed that this is not the case. There is an extensive study and classification of which mathematical problems are computable and which are not. In addition, there is an extensive classification of computable problems into computational complexity classes according to how much computation, as a function of the size of the problem instance, is needed to answer that instance.

In the 1930s, well before there were computers, various mathematicians from around the world invented precise, independent definitions of what it means to be computable. Alonzo Church defined the lambda calculus, Kurt Gödel defined recursive functions, Stephen Kleene defined FSs, Markov defined what became known as Markov algorithms, Emil Post and Alan Turing defined abstract machines now known as Post machines and Turing machines. Surprisingly, all of these models are exactly equivalent: anything computable in the lambda calculus is computable by a Turing machine and similarly for any other pairs of the above computational systems. After this was proved, Church expressed the belief that the intuitive notion of "computable in principle" is identical to the above precise notions. This belief, now called the "Church-Turing thesis", is uniformly accepted by mathematicians. Part of the impetus for the drive to codify what is computable came from the mathematician David Hilbert. Hilbert believed that all of mathematics could be precisely axiomatized. Once this was done, there would be an "effective procedure", that is, an algorithm that would take as input any precise mathematical statement, and, after a finite number of steps, decide whether the statement was true or false. Hilbert was asking for what would now be called a *decision procedure* for all of mathematics.

If the brain is literally a kind of computer, anything could be in principle used as a computer; all that matters is that the system thus used has a complex structure to be able to interpret its states as being stages in the program. The philosophical view that the mind is a computer has been dubbed "Computationalism". Despite the many different formulations of it, *computationalism* is rooted in the basic analogy between mental states and computational states, the different formulations are the result of different definitions and conceptions of these two types of states. Most often, computational states are defined in terms of the stored-program digital serial computer, or the mathematical theory of computation. For example, the atomic structure of the "wall" of this study is complex enough for there to be some configuration of events taking place within it, at the micro-level, that could be interpreted as the implementation of a word processing program; in a sense, this "wall" is therefore "running" a Word Perfect. Computationalism is the thesis that cognition is computable. But what does this mean? What is meant by "cognition" and by "computable"? "Cognition" is a vague term to cover the phenomena that others have called, equally vaguely, "mental states and processes", "thinking", "intelligence", "mentality", or simply "the mind". More specifically, this includes such things as language use, reasoning, conceiving, perceiving, planning, and so on the topics of such cognitive disciplines as linguistics, cognitive psychology, and the philosophy of mind, among others in general, of cognitive science. Roughly, a function is computable if and only if there is an "algorithm" that computes it, that is, an algorithm that takes as input the first elements of the function's ordered pairs, manipulates them (in certain constrained ways), and returns the appropriate second elements. To say that it is an algorithm that does this is to say that there is an explicitly given, "effective procedure" for converting the input into the output.

At this point, it would be useful to look at increasingly complex Turing machines, which compute increasingly complex functions and languages. Although Turing machines are very simple devices, it turns out that they can perform very sophisticated computations. In fact, any computation that can be carried out by a modern digital computer (even one with an unlimited amount of memory) can be carried out by a Turing machine. Although it is not something that can be proved, it is widely believed that anything that can reasonably be called "computation" can be done by a Turing machine. This claim is known as the Church-Turing thesis. In our premise, an FS of logic is used for knowledge representation.

The *open-world assumption* (OWA) is the assumption that the truth value of a statement may be true irrespective of whether or not it is *known* to be true. It is the opposite of the *closed-world assumption,* which holds that any statement that is true is also known to be true.

An OWA was first developed by ancient Greek philosophers as a means to explain varying degrees of validity among mathematical and philosophical concepts proposed at the time of inception. The OWA codifies the informal notion that in general, no single agent or observer has complete knowledge, and therefore cannot make the closed-world assumption. The OWA limits the kinds of inference and deductions an agent can make to those that follow from statements that are known to the agent to be true. In contrast, the closed-world assumption allows an agent to infer from the lack of knowledge of a statement being true that the statement being false.

The fundamental premise of the theory of computation is that the computer obeys certain laws, and therefore, certain unbreakable limitations. A "proof" of a theorem is an argument that convinces an intelligent person who has never seen the theorem before and cannot see why it is true with having it explained. An alphabet is any non-empty finite set, whose elements we call symbols or characters. When you want to prove that an algorithm can solve a problem, brazenly assume your algorithm has a magic ability that you can use to cheat. Then prove that adding the magic ability does not improve the fundamental power of the algorithm. Now that we know the halting problem is undecidable, we can use that as a shoehorn to prove other languages are undecidable, without having to repeat the full diagonalization. Computability focuses on which problems are computationally solvable in principle and computational complexity focuses on which problems are solvable in practice.

Although *undecidability* and *noncomputability* are important to an understanding of computation, the notion of *intractability* has had a much greater impact. A problem is called intractable if the time required to solve instances of the problem grows exponentially with the size of the instances. Besides logic and computation, the third great contribution of mathematics to AI is the theory of *probability.* The Italian *Gerolamo Cardano* first framed the idea of probability, describing it in terms of the possible outcomes of gambling events. Probability quickly became an invaluable part of all the *quantitative sciences,* helping *to deal with uncertain measurements and incomplete theories.* Pierre Fermat, Blaise Pascal, James Bernoulli, Pierre-Simon Laplace, and others advanced the theory and introduced new statistical methods. *Thomas Bayes* proposed a rule for updating probabilities in the light of new evidence. Bayes' rule and the resulting field called *Bayesian analysis* form the basis of most modern approaches to uncertain reasoning in *AI* systems.

Computability theory is the branch of theoretical computer science that studies which problems are computationally solvable using different computation models. Computability theory differs from the related discipline of computational complexity theory in asking whether a problem can be solved at all, given any finite but arbitrarily large amount of resources. A common model of computation is based on an abstract computer, the *Turing machine. Computationalism* is the theory that the human brain is essentially a computer, although presumably not a stored-program, digital computer, like the kind Intel makes. AI is a field of computer science that explores computational models of problem-solving, where the problems to be solved are of the complexity of problems solved by human beings (Figure 1.21).

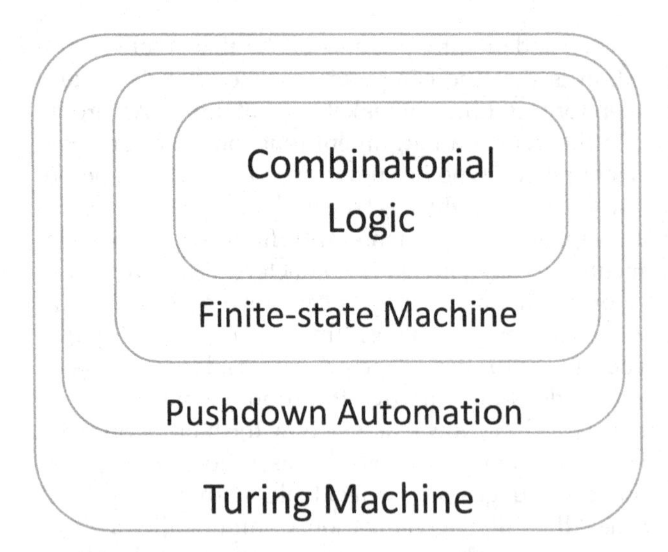

FIGURE 1.21
Logic diagram.

To summarize, logic is a parent discipline of computer science, as well as a basic constituent of computer science curricula. At the same time, logic has provided computer sciences with several tools, methods, and theories. But more than a similarity there is also an equivalence relation between programs and logical formula: a program is a sequence of symbols constructed according to formal syntactic rules and it has a meaning which is assigned by an interpretation of the elements of the language. These symbols are called statements or commands and the intended interpretation is an execution on a machine rather than an evaluation of a truth value. Its syntax is specified using FSs, but its semantics is usually informally specified. Then, a statement in a programming language can be considered a function that transforms the state of a computation. Going further, we can affirm that there is no line between software and mathematics. Translation problems between them are trivial because they are equivalent. Therefore, by the Church-Turing thesis, we can establish an identity between *programs*, *logics*, and *mathematics*. There is a common nature between these domains, with different historical backgrounds and main objectives, but at the end, they speak the same language. In the *CTM*:

1. Natural Computation is NOT AI.
2. It Processes symbols.
3. Arrangements of matter have both representational and causal properties.
4. If interpretation of input symbols is TRUE, then output symbols are also TRUE!
5. Intelligence is computation!

Logic Theorist is a computer program written in 1956 by Allen Newell, Herbert A. Simon, and Cliff Shaw. It was the first program deliberately engineered to perform automated reasoning and is called "the first artificial intelligence program". The inspiration was based on the heuristic teachings of mathematician George Pólya, who taught problem-solving, they aimed to replicate Pólya's approach to logical, discovery-oriented decision-making with more intelligent machines. However, none of the aforementioned procedures allowed functional symbols. Davis-Putnam, in 1960, introduced the Skolem functions and the Herbrand universe into the world of automated deduction. It also introduces clausal form. Finally, they proposed the unit resolution rule (which is there called "rule for the elimination of one-literal clauses"). In mathematical logic and automated theorem proving, resolution is a rule of inference leading to a refutation complete theorem-proving technique for sentences in propositional logic and FOL. The IPL (Information-Processing Language) languages in AI and their contemporary FORTRAN in numerical computing settled once and for all the essentiality of higher-level languages for sophisticated programming. The IPLs were designed to meet the needs for flexibility and generality: flexibility, because it is impossible in these kinds of computations to anticipate before run time what sorts of data structures will be needed and what memory allocations will be required for them; generality, because the goal is not to construct programs that can solve problems in particular domains, but to discover and extract general problem-solving mechanisms that can operate over a range of domains whenever they are provided with an appropriate definition for each domain. IPL is a computer language tailored for AI programming. Pure is a functional programming language based on term rewriting. Its core is actually purely algebraic and purely functional, but the name can also be taken as a recursive acronym for the "Pure Universal Rewriting Engine".

Finally, computer science concerns itself with the automatic processing of information (or at least with physical systems of symbols to which information is assigned) by means of such things as computers. From the beginning, computer programmers have been able to develop programs that permit computers to carry out tasks for which organic beings need a mind. A simple example is multiplication. It is not clear whether computers could be said to have a mind. Could they, someday, come to have what we call a mind? This question has been propelled into the forefront of much philosophical debate because of investigations in the field of AI. AI can be thought of as the attempt to find the primitive elements and logical relations in the subject (man or computer) that mirror the primitive objects and their relations that make up the world. Logic and Computation would be the foundations for the symbolic approach to AI.

2

Robotics

Robots

A robot, by definition, is a machine, especially one that is programmable by a computer and capable of carrying out a complex series of actions automatically. Badīʿ az-Zaman Abu l-ʿIzz ibn Ismāʿīl ibn ar-Razāz al-Jazarī (a.k.a. Al-Jazari) was an Arab Muslim scholar, inventor, and mechanical engineer during the Islamic Golden Ages (Middle Ages) and is known as the "father of robotics". However, the boundaries of robotics cannot be clearly defined. In the 1980s, Japan classified machine shop tools as robots. A robot can be guided by an external control device, or the control may be embedded within it. Robots may be constructed to evoke human form such as with a humanoid, but most robots are task-performing machines, designed with an emphasis on functionality, rather than expressive aesthetics. The concept of a robot can be synonymous with an unmanned vehicle as a mobile robot. As a result, the focus of mobile robots is on navigation. When we typically describe robots, practitioners focus on mobile robots, manipulators, a mobile manipulator, and humanoids. The typical justification for robotics and unmanned vehicles is to satisfy the 3 Ds or the "Dull, Dirty, and Dangerous", a metaphor for boring work, work in a dirty environment, or operations in a hazardous environment such as in space, nuclear radiation, underwater, or underground. A robot's "core" ideas, concepts, and algorithms are being applied in an ever-increasing number of "external" applications, and vice versa.

After a long history of automatons in Egypt, Japan, and Europe, the start of the 20th century signaled the arrival of unmanned vehicles that were controlled remotely, starting with Nikola Tesla (radio-control boat), Elmer Sperry (gyrostabilized seaplane), and Lawrence Sperry (aerial torpedo). The technology evolved from an unmanned aerial vehicle (UAV) that was remotely piloted to upgraded versions using an autopilot. Autopilot is short for automatic pilot. However, the control systems of World War II were required to transform rockets into accurate guided missiles. The word missile comes from the Latin verb mittere, literally meaning "to send". Missiles are basically rockets that are meant for destructive purposes only and differ from rockets by virtue of a guidance system that steers them toward a pre-selected target. The first autopilots were used on airliners in the mid-1930s and the earliest guidance system was the Mischgerät analog computer aboard Germany's V-2 rockets, controlling both horizontal direction and altitude in 1944. In October 1947, Northrop Aircraft ordered the BINAC (binary automatic computer) from the Electronic Control Company. The BINAC consisted of two identical serial computers operating in parallel, with mercury delay-line memories and magnetic tape as secondary memories and auxiliary input devices. However, in the late 1950s, electronic computers became small enough to be used onboard the aircraft, while the Cold War nuclear missile arsenals demanded even more sophisticated guidance technology. Airborne computers did not become feasible until the 1960s, when miniaturized solid-state transistorized components became available. In 1961, the American Minuteman missiles used transistorized computers to continuously calculate their position in flight. When the Minuteman I was decommissioned, universities used their computers for general-purpose computing. The Minuteman guidance computer presented unique design challenges. It had to be rugged and fast, with circuit design and packaging able to withstand a missile launch. The military's high standards for its transistors pushed a manufacturer's quality control processes.

For a manipulator (robotic arm), the teleoperator was developed during World War II to handle radioactive materials by Raymond Goertz. In 1949, while working for the Atomic Energy Commission at Argonne National Laboratory, Goertz filed a patent for an early master–slave manipulator (U.S. Patent 2,632,574) in order to handle radioactive material. Goertz recognized the value of electrically coupling manipulators and laid the foundations of modern tele-robotics and bilateral force-reflecting positional servos. An operator was separated from a radioactive task: the slave arm is located within a nuclear hot-cell and the master arm is located in a separate control room. Whenever the master arm is manipulated, the motion is reproduced precisely by the slave arm. The initial master–slave manipulator device was a seven-degree-of-freedom bilateral (symmetrical) metal tape transmission pantograph device, which

DOI: 10.1201/9781032673134-2

FIGURE 2.1
Raymond Goertz master–slave robot.

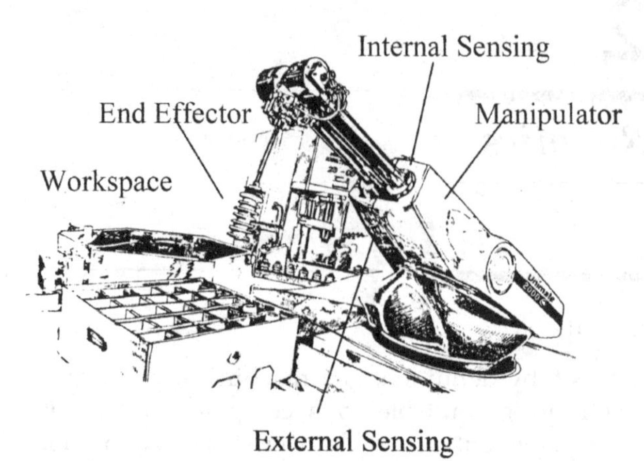

FIGURE 2.2
Unimate manipulator.

was operated through a leaded glass wall. The motion of the slave arm must be coupled to the master arms such that the position and the direction of the two arms correspond. The coupling of the two arms must be bilateral and this means that forces at the slave end must be reflected at the master end and displacements produced at the slave end must be able to produce a displacement at the master end, with the slave arm being able to align itself in response to the constraints imposed by the task being done (Figure 2.1).

The earliest industrial manipulators as we know them were created in the early 1950s by George C. Devol, an inventor from Louisville, Kentucky. He invented and patented a reprogrammable manipulator called "Unimate", (from "Universal Automation"). Unimate Manipulator is shown in Figure 2.2. For the next decade, he attempted to sell his product in the industry, but did not succeed. In the late 1960s, businessman/engineer Joseph Engleberger acquired Devol's robot patent and was able to modify it into an industrial robot and form a company called Unimation in order to produce and market the robots. Unimate was the first industrial robot, which worked in a General Motors assembly line at the Inland Fisher Guide Plant in Ewing Township, New Jersey, in 1961. The machine undertook the job of transporting die casting from an assembly line and welding these parts on auto bodies, a dangerous task for workers, who might be poisoned by toxic fumes or lose a limb if they were not careful. The original Unimate consisted of a large computer-like box, joined to another box and was connected to an arm, with systematic tasks stored in a drum memory.

An industrial robot is a robot system used for manufacturing. Industrial robots are automated, programmable and capable of movement on three or more axes. Typical applications of robots include welding, painting, assembly, disassembly, pick and place for printed circuit boards, packaging and labeling, palletizing, product inspection, and testing; all accomplished with high endurance, speed, and precision. Robot arms (or manipulators) can assist in material handling.

- Early machines used rule-based systems to control systems.
- Early rule-based systems could not deal with noise.
- By using fuzzy logic (FL), it is possible to specify "fuzzy" rules that occur in a variety of circumstances.
- A membership function then measures the degree of similarity between input variables and the fuzzy sets.
- Behaviors can be encoded as sets of fuzzy rules.

Elektro was a humanoid on exhibit at the 1939 World's Fair in New York City, USA and reappeared at that fair in 1940, with a new robot, Sparko, a dog that could bark, sit, and beg. Like some radio programs, Elektro did his talking by means of transcriptions. His speech usually lasted about 1 minute and used only 75 words. The Elektro was constructed from aluminum on a steel frame and was 210 cm tall and weighing 120 kg. It was capable of performing 26 routines (movements) and had a vocabulary of 700 words. Sentences were formulated by a series of 78 RPM record players connected to relay switches. Elektro had no remote control, instead responding to voice commands using a telephone handset connected to its chest. The chest cavity even lit up as it recognized each word. His fingers, arms, and turntable for talking were operated by nine motors, while another small motor operated the bellows so the giant could smoke. With a loud electrical whine, Elektro would walk about the stage in a slow slide that betrayed the rollers on his feet. Exciting things that Elektro could do were move his head and arms, count on his fingers, recognize colors (his photoelectric "eyes" could distinguish red and green light), smoke cigarettes, and talk. Talking to Elektro was like dialing an automatic telephone, using light impulses instead of numbers to cause the relays to act.

FL is based on the observation that people make decisions based on imprecise and non-numerical information. Fuzzy models or fuzzy sets are mathematical means of representing vagueness and imprecise information (hence the term fuzzy). These models have the capability of recognizing, representing, manipulating, interpreting, and using data and information that are vague and lack certainty. FL is an approach to variable processing that allows for multiple possible truth values to be processed through the same variable. It attempts to solve problems with an open, imprecise spectrum of data and heuristics that makes it possible to obtain an array of accurate conclusions. FL is based on fuzzy, if-then linguistic rules, representing prior knowledge that may be subject to modification through learning. The integration of fuzzy systems into neural networks offers some of the advantages of both by hybridizing subsymbolic and symbolic processing (Figure 2.3).

An important component of artificial intelligence (AI) is the ability of computers to recognize partial truths, instead of simply true or false. FL, pioneered by Lotfi Zadeh, is based on Bayesian probability. FL is an extension of classical logic by having values between true and false. Logic is required for deciding how to respond when something is neither true nor completely false. This is equivalent to the idea of "responding within a range". FL, on the other hand, uses multiple bits to represent degrees of order based on the relationships of the various objects.

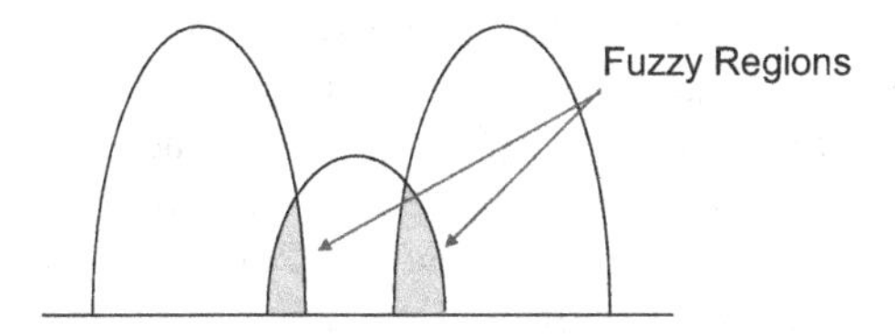

FIGURE 2.3
Lotfi Zadeh fuzzy logic diagram

It is a method of reasoning that closely resembles human reasoning. The approach of FL imitates the way of decision-making in humans that involves all intermediate possibilities between digital values YES and NO. FL is a form of many-valued logic or probabilistic logic; it deals with reasoning that is approximate rather than fixed and exact. Compared to traditional binary sets (where variables may take on true or false values) FL variables may have a truth value that ranges in degree between 0 and 1. FL has been extended to handle the concept of partial truth, where the truth value may range between completely true and completely false. This allows for more accurate reasoning based on a wider variety of situations, in turn, more closely resembling human thought. Multiple variables can be combined to produce a single heuristic, and a better decision can be made based on that comparison. Behaviors can be encoded as sets of fuzzy rules. As a result, many different behaviors can contribute to the control of the robot, according to how the input is evaluated by the membership function. So, the robot can have multiple goals and blend rules from a variety of behaviors together with strong arbitration.

Fuzzy inferencing is the core element of a fuzzy system. Fuzzy inferencing combines the facts obtained from the fuzzification with the rule base, and then conducts the fuzzy reasoning process. Fuzzy inference is also known as approximate reasoning. Fuzzy inference is a computational procedure used for evaluating linguistic descriptions. Two important inferring procedures are:

- Generalized Modus Ponens (GMP).
- Generalized Modus Tollens (GMT).

Many different behaviors can contribute to the control of the machine, according to how the input is evaluated by its membership function. A deliberate robot utilizes the sense-plan-act (SPA) control paradigm. To sense, the robot needs the ability to sense important things about its environment, like the presence of obstacles or navigation aids. What information does your robot need about its surroundings, and how will it gather that information? To plan, the robot needs to take the sensed data and figure out how to respond appropriately to it, based on a pre-existing strategy. Do you have a strategy? Does your program determine the appropriate response, based on that strategy and the sensed data? And finally, to act, the robot must actually act to carry out the actions that the plan calls for. Have you built your robot so that it can do what it needs to, physically? Does it actually do it when told?

Industrial Robots

The general characteristics of industrial work situations that tend to promote the substitution of robots for human labor are the following:

- Hazardous work environment for humans (Dangerous in the 3 Ds)
- Repetitive work cycle (Dull in the 3 Ds)
- Difficult handling for humans (potentially Dirty in the 3 Ds)
- Multi-shift operation (a robot can work around the clock if required)
- Infrequent changeover
- Part position and orientation are established in the workcell (structured environment).

Robots are being used in a wide field of applications in industry. Most of the current applications of industrial robots are in manufacturing. The applications can usually be classified into one of the following three categories: (1) Material handling applications (either mobility or manipulation), (2) Processing operations (typically manipulation), or (3) Assembly (typically manipulation). Material handling applications for industrial robots include material transfer and machine loading and/or unloading. Examples of processing operations are spot welding, continuous arc welding, spray coating, and other process applications.

There are three types of automation for an industrial factory: fixed automation (sequence of processing (or assembly) operations is fixed by the equipment configuration), flexible automation (system is capable of changing over from one job to the next with little lost time between jobs), or programmable automation (capability to change the sequence of operations through reprogramming to accommodate different product configurations). In open-loop control, the objective is to command the automation, while closed-loop control is to regulate the process. Continuous processes have states that can be described by a continuous (analog) variable (temperature, voltage, speed, etc.). Between the industrial process input and plant output, there exists a fixed relation that can be described by a continuous model or transfer function. The transfer function may be described by a differential equation, simplified to a Laplace or a z-transform when the system is linear. The principal control task in relation with a continuous process is its *regulation* (maintaining the state on a determined level or trajectory). A discrete plant is modeled by well-defined, exhaustive and non-overlapping states, and by abrupt transitions from one state to the next caused by events. Discrete plants are described by finite state machines, Petri net, state transition tables, Grafcet, or sequential function chart diagrams. The main task of a control system in relation to discrete plants is their command. Although applications differ widely, there is little difference in the overall architecture of their control systems. The biggest distinction is the domain know-how embedded in the control system (Figure 2.4).

FIGURE 2.4
Industrial manipulator.

Manipulators, including industrial manipulators or robot arms, consist of joints and links: joints provide relative motion, links are rigid members between joints, various joint types (linear and rotary), each joint provides a "degree-of-freedom", and most robot arms possess five or six degrees-of-freedom. Robot manipulator consists of two sections: body-and-arm for positioning of objects in the robot's work volume and wrist assembly for the orientation of objects. There are four robot control systems: (1) limited sequence control with pick-and-place operations using mechanical stops to set positions, (2) playback with point-to-point control (records work cycle as a sequence of points, then plays back the sequence during program execution), (3) playback with continuous path control with greater memory capacity and/or interpolation capability to execute paths (in addition to waypoints), and (4) intelligent control, which exhibits behavior that makes it seem intelligent, for example, responds to sensor inputs, makes decisions, and communicates with humans. The process controller for an industrial arm is the programmable logic controller (PLC) that uses ladder logic or similar that takes actions based on its inputs and this industrial computer replaced the more conventional relays. The PLC is a small computer, dedicated to automation tasks in an industrial environment. The function of a PLC is to measure, control (command and regulation), and protect the

TABLE 2.1
Automated Ground Vehicle (AGV) versus Autonomous Mobile Robot (AMR) Comparison

AGV vs. AMR
Automated Guided Vehicles (AGVs) – Movement is highly structured and monolithic in static environments with the help of magnetic strips or wires. Any obstacle in the path has to be removed to set the AGV in motion again. Redeployment of guided technology like magnetic strips or wires is required, with an additional cost if the facility changes through a renovation, upgrade, or shift in locations.
Autonomous Mobile Robots (AMRs) – Work in dynamic environments with autonomous navigation. AMRs create and save the locations or map of the facility to find the alternative path if there is an obstacle in the defined route. They also require minimum resets if there is any change related to the facility, as AMRs can be unboxed and put to work within one hour if they are already mapped with the facility. AMRs are extremely flexible to deploy.

various components. The PLC connects to the network for programming on workstations and makes the connection to the SCADA (Supervisory Control and Data Acquisition) function (Table 2.1).

There are real-time languages to express time such as Ada and "C", and languages for cyclic execution and also real-time used on PLCs. PLC languages that have wide-spread use in the control industry are ladder logic, function block language, instruction list, GRAFCET, and Specification and Description Language (SDL). The contact plan or "ladder logic" language allows an easy transition from the traditional relay logic diagrams to the programming of binary functions. It is well suited to express combinational logic. Also notable in industrial automation systems is the use of SCADA. A SCADA system performs four functions: (1) data acquisition, (2) networked data communication, (3) data presentation, and (4) control. Industrial robots are integrated into workcells that are very structured. There are three factors to consider when to automate: human factors, complexity, and stability. Some factories have migrated to lights-off or lights-out manufacturing, a manufacturing methodology (or philosophy) rather than a specific process. Factories that run lights out are fully automated and require no human presence on-site. Thus, these factories can be run with the lights off. In addition to manipulation in factories, mobility is important to move stock around (Figure 2.5).

FIGURE 2.5
Automated Ground Vehicle (AGV).

Computer-controlled and wheel-based, automated guided vehicles (AGVs) are load carriers that travel along the floor of a facility without an onboard operator or driver. Their movement is directed by a combination of software and sensor-based guidance systems. AGVs follow along marked long lines or wires on the floor or use radio waves, vision cameras, magnets, or lasers for navigation. Because they move on a predictable path with precisely controlled acceleration and deceleration and include automatic obstacle detection bumpers, AGVs provide safe movement of loads. They are most often used in industrial applications to transport heavy materials around a large industrial building, such as a factory or warehouse. Typical AGV applications include transportation of raw materials, work-in-process, and finished goods in support of manufacturing production lines, and storage/retrieval or other movements in support of picking in warehousing and distribution applications. Application of the automatic guided vehicle broadened during the late 20th century. In robotics, vision, physics, and any other branch of science whose subjects belong to or interact with the real world, mathematical models are developed that describe the relationship between different quantities. Some of these quantities are measured, or sensed, while others are inferred by calculation. For instance, in computer vision, equations tie the coordinates of points in space to the coordinates of corresponding points in different images. Image points are data, world points are unknowns to be computed.

Mechatronics

"The name [mechatronics] was coined by Ko Kikuchi, now president of Yaskawa Electric Co., Chiyoda-Ku, Tokyo". The word mechatronics is composed of *mecha* from mechanics and *tronics* from electronics. In other words, technologies and developed products will be incorporating electronics more and more into mechanisms, intimately and organically, and making it impossible to tell where one ends and the other begins". Mechatronics is the synergistic integration of sensors, actuators, signal conditioning, power electronics, decision and control algorithms, and computer hardware and software to manage complexity, uncertainty, and communication in engineered systems (Figure 2.6).

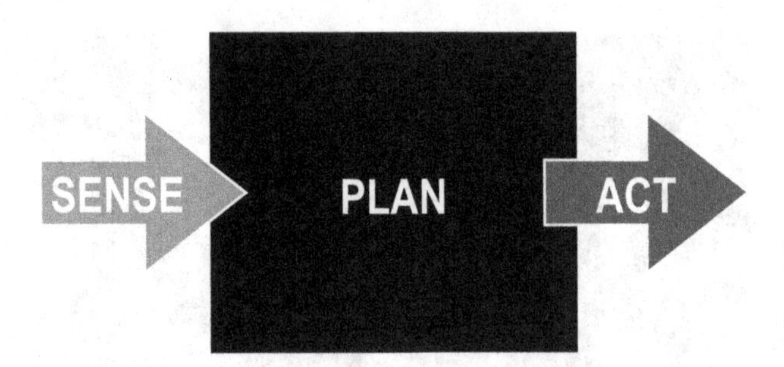

FIGURE 2.6
Mechatronics Processing Diagram is the Same as Robotics Processing Diagram.

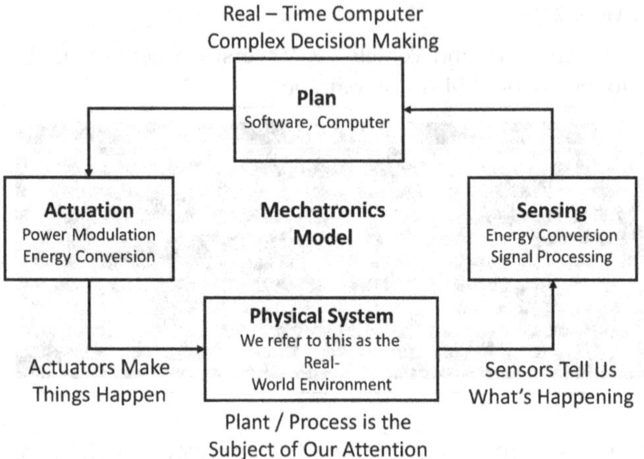

FIGURE 2.7
Mechatronics Model.

These sequential and discipline-specific concurrent design processes for product realization are at best multidisciplinary calling upon discipline specialists to "design by discipline". Mechatronics is based on:

- Designing a mechanical system or "plant".
- Selecting sensors and actuators and mounting them on the plant.
- Designing signal conditioning and power electronics.
- Designing and implementing control algorithms using electrical, electronics, microprocessor, microcontroller, or microcomputer-based hardware.

Mechatronics is interesting because there is an analogy between electrical networks and mechanical networks. Maybe the most important is the conversion between analog designs and digital designs. A necessary by-product of advancements in mechatronics has been the miniaturization of the hardware such as systems-on-chip and delivering other commercial advantages. Mechatronics utilizes the same basic SPA (or sense-plan-act) architecture introduced by Kenneth Craik. Sensors output both analog and digital signals, and actuators are typically analog commands. The ultimate mechatronic project would be a robotic humanoid and provide the entire physical body of the robot (mechanical and electrical design). Today's computer systems are capable of producing a huge amount of information, especially on the status of internal components and the world around them. The problem will not be a lack of information but finding what is needed when it is needed. A human could be less informed than before because there is a huge gap between the amount of data that could be produced and disseminated. Mechatronics delivers all the basic components needed in a robot, but without the AI piece. In mechatronics and robotics, a robot arm is modeled by equations that describe where each link of the robot is as a function of the configuration of the link's own joints and that of the links that support it. The desired position of the end effector, as well as the current configuration of all the joints, are in the data. The unknowns are the motions to be imparted to the joints so that the end effector reaches the desired target position. Today, mechatronics is no longer solely the study of mechanical and electrical interactions, but rather the study of electro-mechanical interactions with other technical systems including robotics. Mechatronics will help the robot designer to understand and design systems that feature components that are both analog and digital (Figure 2.7).

Systems Theory: Designing for its Environment

"General Systems Theory" (GST; German: *allgemeine Systemlehre*) was coined in the 1940s by Ludwig von Bertalanffy, who initially sought to find a new approach to the study of living systems. He first developed his theories via lectures beginning in 1937 and then via publications beginning in 1946. Bertalanffy promoted an embryonic form of GST as early as the 1920s and 1930s, but it was not until the early 1950s that it became more widely known in scientific circles. This is an interdisciplinary practice that describes systems with interacting components, applicable to biology, cybernetics, and other fields. For example, Bertalanffy proposed that the classical laws of thermodynamics might be applied to closed systems, but not

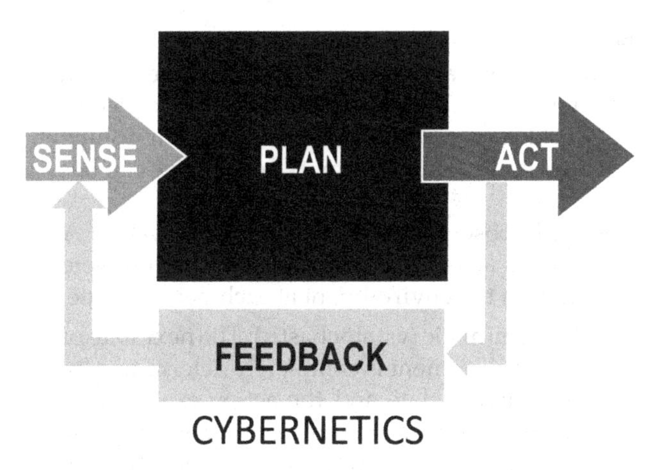

FIGURE 2.8
Cybernetics Process Diagram.

necessarily to "open systems" such as living things. GST attempted to provide alternatives to conventional models of organization and defined new foundations and developments as a generalized theory of systems with applications to numerous areas of study, emphasizing holism over reductionism and organism over mechanism. Foundational to GST are the interrelationships between elements, which all together form the whole. Systems theory identifies processes that explain how a system retains its functions while continuing to integrate new information from its environment and adjoining systems. For example, according to von Bertalanffy, the Cartesian doctrine of body-mind dualism was a product of the particular categories of thinking, such as the Aristotelian category of substance adopted also by Descartes. Substance was one of the categories adopted by the Western culture (Figure 2.8).

General system theory introduced key concepts such as open and closed systems, stressing the role and importance of context and environment, *equifinality*, or the way systems can reach the same goal through different paths, and isomorphisms or structural, behavioral, and developmental features that are shared across systems. Cybernetics introduced the concept of feedback for a closed-system. General system theory positioned itself as transdisciplinary rather than interdisciplinary. Interdisciplinary refers to interaction between disciplines, whereas transdisciplinary refers to going beyond or across disciplines. General system theory would be the common language across diverse disciplines. Central to this language was the concept of a "system", defined as a group of interacting, interdependent elements that form a complex whole. It also pointed toward a new worldview that emphasizes such key concepts as every system's embeddedness in other, larger systems, and the dynamic, ever-changing processes of self-organization, growth, and adaptation. Every system is bounded by space and time, influenced by its environment, defined by its structure and purpose, and expressed through its functioning. In our context, the robot is a system that interacts with its environment, another system (Figure 2.9).

A robot is an agent and is also a system that perceives its environment and undertakes actions that maximize its chances of being successful. The concept of agents conveys, for the first time, the idea of intelligent units working collaboratively with a common objective. This new paradigm was intended to mimic how humans work collectively in groups, organizations and/or societies. Agent theory is concerned with the question of what an agent is, and the use of mathematical formalisms for representing and reasoning about the properties of agents. Agent architectures can be thought of as software engineering models of agents; researchers in this area are primarily concerned with the problem of designing software or hardware systems that will satisfy the properties specified by agent theorists. Finally, agent languages are software systems for programming and experimenting with agents; these languages may embody principles proposed by theorists (Figure 2.10).

The environment plays an important factor in a robot in order to accomplish its job. A structured environment is essentially a space that is clearly and meticulously defined. This type of environment has no variables and is rigid, meaning a robot knows what to expect when navigating through it or picking objects up at all times. Thus, a structured environment is predictable such as with an industrial robot in a workcell. An environment that contains many obstacles and where vehicle localization is difficult is classified as unstructured or cluttered. As a result, most

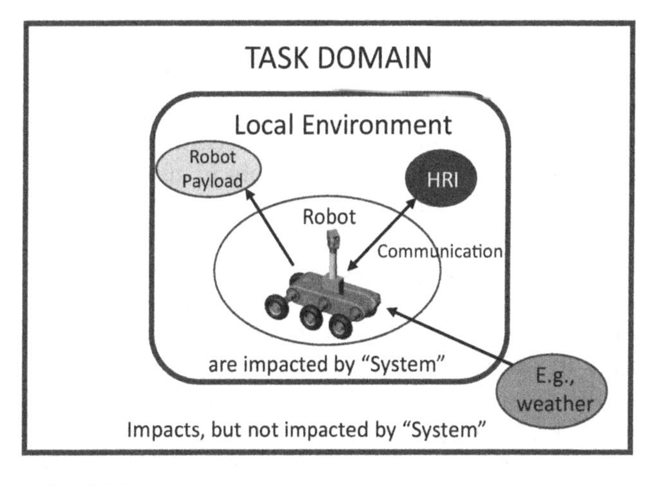

FIGURE 2.9
Robot System Model.

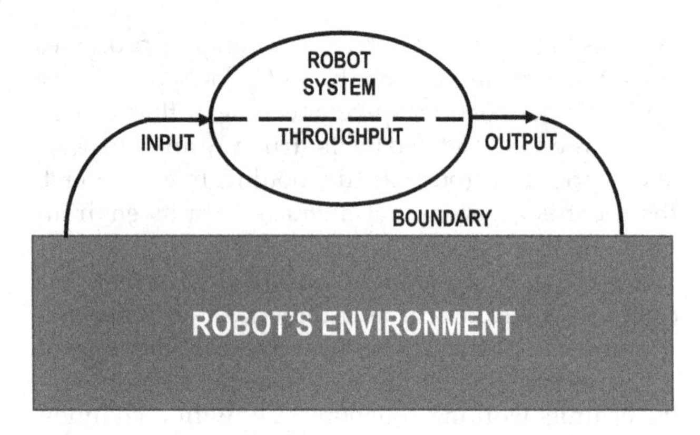

FIGURE 2.10
Robot system model.

natural environments are unstructured. A fundamental aspect of robot–environment interaction in industrial environments is given by the capability of the control system to model the structured and unstructured environment features. Industrial robots have to perform complex tasks at high speeds and have to satisfy hard cycle times while maintaining the operations extremely precise. The capability of the robot to perceive the presence of environmental objects is something still missing in the real industrial context.

Systems theory is studied in two parts: systems thinking, where the designers take a holistic look at the problem, and systems engineering, which follows a structured process to deliver a robot/AI system that meets agreed-upon requirements and specifications. A key step to system engineering is integration and verification & validation (V&V), where the design satisfies the requirements through testing. Autonomous vehicles are being used increasingly often for a range of tasks, including automated highway driving and automated flying. Such systems are typically either specialized for structured environments and depend entirely on such structure being present in their surroundings or are specialized for unstructured environments and ignore any structure that may exist. A challenging problem in robotics is to predict future observations based on previously recorded data. The deployment of autonomous robots in unstructured and dynamic environments poses a number of challenges that cannot easily be addressed by approaches developed for highly controlled environments. In unstructured environments, for example, robots cannot rely on complete knowledge about their surroundings. In fact, perceiving the environment becomes one of the key challenges. Robots have to autonomously and continuously acquire the information necessary to support decision-making. Moreover, robots cannot assume that their actions succeed reliably. Instead, they have to continuously monitor their effect on the environment and possibly react to undesired events. In contrast, many existing, well-established techniques in robotics rely on perfect knowledge of the world and perfect control of the environment. A robot's environment from its perspective can be:

- Fully observable (vs. partially observable): A robot's sensors give it access to the complete state of the environment at each point in time.
- Deterministic (vs. stochastic): The next state of the environment is completely determined by the current state and the action executed by the robot. (If the environment is deterministic except for the actions of other robots, then the environment is strategic).
- Episodic (vs. sequential): A robot's action is divided into atomic episodes. Decisions do not depend on previous decisions/actions.
- Static (vs. dynamic): The environment is unchanged while a robot is deliberating. (The environment is semi-dynamic if the environment itself does not change with the passage of time, but the robot's performance score does).
- Discrete (vs. continuous): A limited number of distinct, clearly defined percepts and actions.

Systems are not independent but exist in an environment. Thus, system's function may be to change its environment. The environment affects the functioning of the system, for example, system may require electrical supply from its environment. And the organizational, as well as the physical, environment may be important. For example, the space environment is crucial to a space robot such as a rover. There are six environmental effects to consider during a Mars Rover design, as shown in the following table (Table2.2).

TABLE 2.2
Environmental with Associated System Effect

Environmental Effect	Environment	System Effects
Vacuum	Earths' Fields, Solar-Planetary Connection	Solar UV Degradation, Contamination
Neutral	Atmospheric Physics	Aerodynamic Drag, Sputtering, Atomic Oxygen Erosion
Plasma	Ionosphere, Magnetosphere, Geomagnetic Storms	Spacecraft Charging, Arc Discharging
Radiation	Trapped Radiation Belts, Solar Proton Events, Galactic Cosmic Rays	Total Dose, Dose Rate, Single Event
Micrometeoroid/Orbital Debris	-	Hypervelocity Impact Damage
Forces	Microgravity	Small forces dominant (surface tension), degraded human performance

For example, the space environment effects can be mitigated/reduced through careful space design.

The challenges associated with unstructured environments are a consequence of the high-dimensional state space and the inherent uncertainty in mapping sensory perceptions onto specific states. It is believed that the high dimensionality of the state space represents the most fundamental challenge as robots leave the highly controlled environment of the factory floor and enter into unstructured environments. To succeed in unstructured environments, robots have to carefully select task-specific features and identify relevant real-world structures to reduce the state space without affecting the performance of the task. Robots often operate in built environments that tend to contain some underlying structure, such that newly visited locations may appear broadly similar to previously visited locations but differ in individual details. In order to autonomously explore and map unstructured and unknown indoor environments, the robot must sense and perceive its environment. In cluttered scenes, where the opponent agents are not following any driving pattern, it is difficult to anticipate their behavior and henceforth decide the robot's actions. Deep deterministic policy gradient (DDPG) enables us to propose a solution that requires only the sensor information at the current time step to predict the action to be taken. The main point is to generate a behavior-based motion model for an autonomous vehicle, which plans for every instant. While these capabilities are already commodities on ground vehicles, air vehicles seeking the same performance but face different unique challenges. For example, to describe the difficulties in achieving a fully autonomous helicopter flight, the differences between ground and helicopter robots must be highlighted that make it difficult to use algorithms developed for ground robots. Every system is bounded by space and time, influenced by its environment, defined by its structure and purpose, and expressed through its functioning. The environment has two important aspects: (1) for the body and (2) for the mind. System theory dictates the robot operates in its intended operating environment. Some physical environment of the robot includes: nuclear, underwater, space, medical, industrial, and security/civil protection. And in addition to the robot's physical environment, the mind itself must deal with its own assessment of the environment. The real world as an environment for the robot must also handle the following parameters, especially in unstructured environments:

- Inaccessible (sensors do not tell all) and need to maintain a model of the world,
- Non-deterministic (wheels slip, parts break) and need to deal with uncertainty,
- Non-episodic (effects of an action change over time) and need to handle sequential decision problems,
- Dynamic and need to know when to plan and when to reflex,
- Continuous and cannot enumerate possible actions.

As stated earlier, the real world (of the robot's environment) is a system. So, the robot can have multiple goals and must blend rules from a variety of behaviors together with strong arbitration. However, the real world is both complex and uncertain. A robot uses logic to handle all the complexities in the environment. To handle uncertainty, the robot uses probability, typically starting with a Markov model (a probabilistic tool). We can encode knowledge in the form of prior probabilities and conditional probabilities, while inference is done by calculating posterior probabilities given evidence. The "model" is the essence of knowledge and of AI (Figure 2.11).

Complexity: tends to be used to characterize something with *many parts* connected in intricate arrangements. The study of these complex linkages is the main goal of complex systems theory. The different sources of complexity are having: (1) too much information, too many components, too many constraints, too many parameters for consideration to accomplish a particular task; (2) not enough information about essential elements or components of a system or about their interfaces; and (3) not enough information about how elements or components will behave under known or unknown conditions that may lead to unintended consequences. Simple rules can lead to complex systems and is related to the concept of entropy (disorder). As a metric for intricateness is the amount of information contained in the system. Complexity

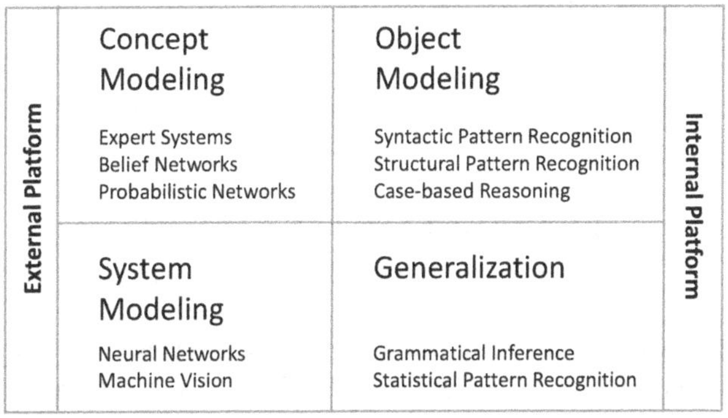

External Platform			Internal Platform
	Concept Modeling Expert Systems Belief Networks Probabilistic Networks	**Object Modeling** Syntactic Pattern Recognition Structural Pattern Recognition Case-based Reasoning	
	System Modeling Neural Networks Machine Vision	**Generalization** Grammatical Inference Statistical Pattern Recognition	

FIGURE 2.11
Concept-Object System Model.

theory attempts to reconcile the unpredictability of non-linear dynamic systems with a sense of underlying order and structure. Ideas related to complexity include:

- Size (no. of parts, lines of software code, no. of interfaces, no. of uncertainties, etc.)
- Minimum description
- Variety
- (Dis)order (midpoint between order and disorder)

Herb Simon broke down complexity in systems to include "state" and "process" descriptions. We as humans can either describe something as it is or how it works. One of Simon's points describes the idea of complex problems where he compares proving a theorem to searching through a maze and how solving the maze involves a lot of trial and error. The trial and error is not random, but rather highly selective and selective trial and error is integral to solving any problem. The difference between a more or less complex problem can be summed up by the quantity of trial and error required for a solution to be found or for progress to be made. The hierarchical organization of complex systems is defined as "a set of instructions for the construction and maintenance of the whole", or systems theory.

Complexity is related to the *NP-completeness* of some problems. In complexity theory, problems can be classified according to the computational resources required:

- running time
- storage space
- parallelism
- randomness
- rounds of interaction, communication, others.

In mathematical complexity, there are two complexity classes:

- **P**=the set of problems decidable in *polynomial time*
- **NP**=the set of problems with witnesses that can be checked in polynomial time

  ... and the notion of *NP-completeness*

In *complexity theory*, the objective is to classify problems *as easy problems* and *hard problems*, while in *computability theory*, the objective is to classify problems as *solvable problems* and non-solvable problems. Computability theory also introduces several of the concepts used in complexity theory. Easy or tractable problems can be solved in polynomial time, that is, for a problem of size n, the time or number of steps needed to find the solution is a polynomial function of n. On the other hand, hard or intractable problems require times that are exponential functions of the problem size. While a polynomial-time algorithm is considered to be efficient, an exponential-time algorithm is inefficient, because its execution time increases rapidly with the problem size. Is it easier to verify a proof than to find one? The fundamental conjecture of computational complexity is $P \neq NP$. We need a model of computation to define classes that capture important aspects of computation, and our model of computation is the Turing machine. A *Turing Machine* is a mathematical model of computation that defines an abstract machine, which manipulates symbols on a strip of tape according to a table of rules. It is believed that the Turing machine formalizes that an intuitive notion of an efficient algorithm is the "extended" Church-Turing Thesis where everything we can compute in time $t(n)$ on a physical computer can be computed on a Turing machine in time $t^{O(1)}(n)$ (a polynomial slowdown). The consequence of the extended Church-Turing Thesis is that all reasonable physically realizable models of computation can be *efficiently* simulated by a TM. An amazing fact is that there exist (natural) undecidable problems such as the famous HALT problem (refer to Turing). Back to complexity classes:

- **TIME** **($f(n)$)**=languages decidable by a multi-tape TM in at most $f(n)$ steps, where n is the input length, and f :**N**! **N**
- **SPACE** **($f(n)$)**=languages decidable by a multi-tape TM that touches at most $f(n)$ squares of its work tapes, where n is the input length, and f :**N**! **N**.

Complexity theory describes a complex system as one of multiple scales within time and space, capable of decomposition into related subcomponents. This theory provides a framework for understanding each interdependent entity of the brain in connection to the overall structure and function of the whole brain. In relation to this theory, the anatomical and functional regions of the brain that form the basis of cognitive function are divided into modules of temporal and spatial scales. Simon's research focused on bounded rationality, which emphasizes that decision-makers often face limitations in gathering and processing information, leading to satisficing (satisfactory but not optimal) decision-making rather than optimization. I think we can consider the seemingly near-decomposable properties of complex systems as analogous to linguistic determinism. The foundation of linguistic determinism is that our thoughts and understanding of the world around us are limited by the vocabulary available to us. By that logic, cultures whose languages have more words to describe different kinds of snow are more perceptive to differences in snowfall than those with fewer, for example. Complexity can manifest itself within the robot itself, its software, the robot's local environment, and the interaction between the robot and its environment.

Complexity theory targets at the heart of systems such as robots, understanding the relationship between *emergent behavior* and *intricateness of parts* (through the *non-fragmentable* property) and having a paradigm to think about systems and scales. For software complexity, especially in AI, we are not so much interested in the time and space complexity for small inputs, but for large inputs. The growth of *time* and *space* complexity with increasing input size n is a suitable measure for the comparison of algorithms. The idea behind the big-O notation is to establish an upper boundary for the growth of a function $f(x)$ for large x. Problems of higher complexity are called *intractable.* Problems that no algorithm can solve are called *unsolvable.* If the interacting system is a computer, then complexity is defined by the *execution time* and *storage required* to perform the computation. Big O could be used to specify how much space is needed for a particular algorithm; in other words, how many variables are needed. Often, there is a *time–space trade-off* where possible options would include: (1) can often take less time if willing to use more memory, or (2) can often use less memory if willing to take longer.

> Consider the problem of designing a machine to solve well-defined intellectual problems. We call a problem well-defined if there is a test which can be applied to a proposed solution. In case the proposed solution is a solution, the test must confirm this in a finite number of steps. If the proposed solution is not correct, we may either require that the test indicate this in a finite number of steps or else allow it to go on indefinitely. Since any test may be regarded as being performed by a Turing machine this means that well-defined intellectual problems may be regarded as those of inverting functions and partial functions defined by Turing machines.

The idea behind the big-O notation is to establish an *upper boundary* for the growth of a function $f(x)$ for large x. The growth of *time* and *space* complexity with increasing input size n is a suitable measure for the comparison of algorithms. A problem that can be solved with polynomial worst-case complexity is called *tractable,* problems of higher complexity are called *intractable, and* problems that no algorithm can solve are called *unsolvable.* Ultimately, Big O is a mathematical tool. Complexity can be quantified by measuring the efficiency of algorithms, analyzing the computer's time and space requirements, and performing a big-O analysis. Algorithms consume resources such as processing *time* and require memory *space.* An algorithm's efficiency is application-dependent since there is a compromise: speed up at expense of memory or go slower, but it is a more memory-efficient algorithm. We are able to measure efficiency in a computer system by timing a program using the computer's clock or by counting block or instruction iterations (Figure 2.12).

Kolmogorov complexity of an object, such as a piece of text, is the length of the shortest computer program that produces the object as output and is a measure of the computational resources needed to specify the object (also known as *algorithmic complexity).* Systems engineering is a structured design process applied to a vehicle or robot that is reliable during operations. The system process begins by defining the requirements and developing the architecture, both physically and functionally. The most important step, as it applies to robotics and AI systems, is the *integration* process. And just as important is the verification step where *testing* is used to validate the operations of the robot with its AI software. It is important to remember that complexity comes from two perspectives: a complex robot, including its processor, and an environment that can also be complex, ranging from a very structured environment up to one that is very complex. Finally, complexity is the user perspective and robot designers need to take account of emergent properties to be prepared for unexpected robot actions leading to potential safety concerns.

Human–Robot Theory and Interfaces

The origin of human–robot interaction (HRI) as a discrete problem was stated by 20th-century author Isaac Asimov in 1941, in his novel *I, Robot.* Human–computer interaction (HCI) is a field based on the design and the use of computer technology, which focuses on the interfaces between people (operators) and computers as brains. HCI researchers observe the ways humans interact with computers, and they design technologies that let humans interact with computers in novel ways. A brain–computer interface (BCI) is a direct communication pathway between an enhanced or wired brain and an external device.

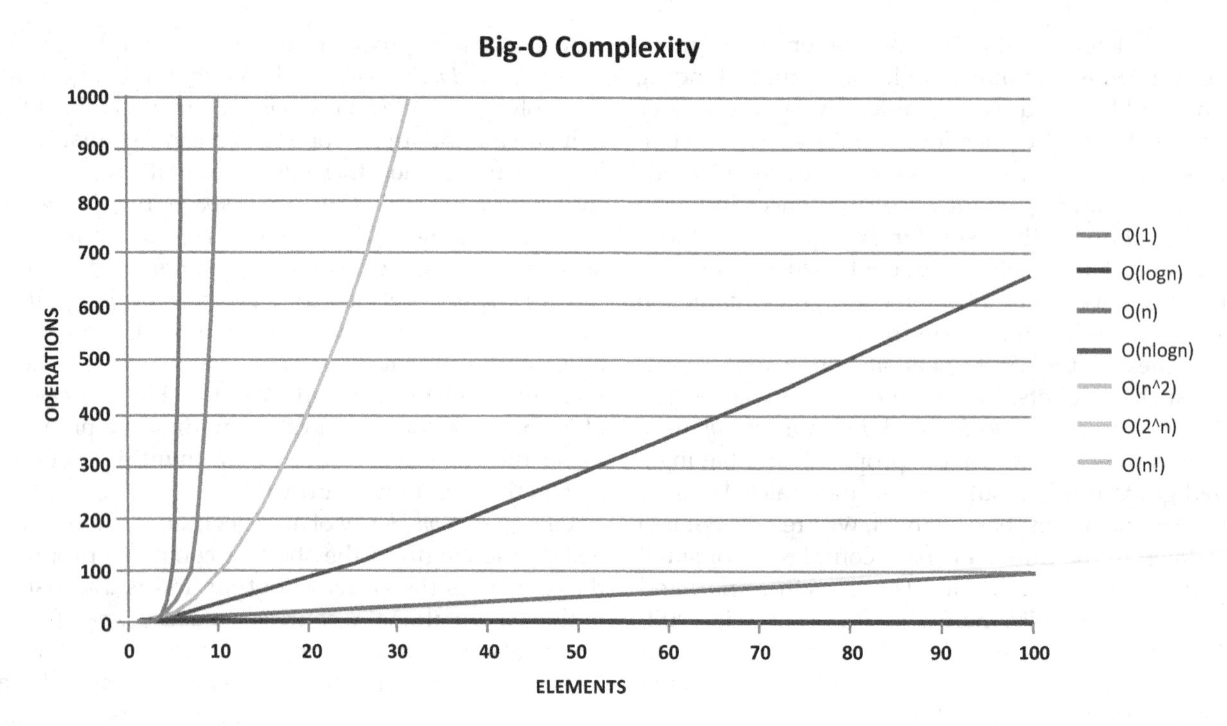

FIGURE 2.12
Big-O Complexity Notation Graph.

BCI differs from neuromodulation in that it allows for bidirectional information flow. BCIs are often directed at researching, mapping, assisting, augmenting, or repairing human cognitive or sensory-motor functions. Expanding on the previous work of Goertz, researchers Jean Vertut and Philippe Coiffet of the French Alternative Energies and Atomic Energy Commission or CEA, continued the work of substituting the operator function with a computer. W. Bill Verplank is a designer and researcher who focuses on interactions between humans and computers at MIT. The W. Verplank model examined the shared trade-off between a computer and the human operator in both a shared mode and a trading mode (Figure 2.13).

In the above figure, *L* is the load or task, *H* the human, and *C* the computer. It represents the notion of trading control or sharing control between the humans and computers. There are three different roles of humans in automation systems: as monitor, as backup, or as partner. When the human operator performs as a monitor, he or she will determine if the task is impossible, the task is dependent on information provided, would be difficult or impossible to monitor for infrequent events, if the state of information is

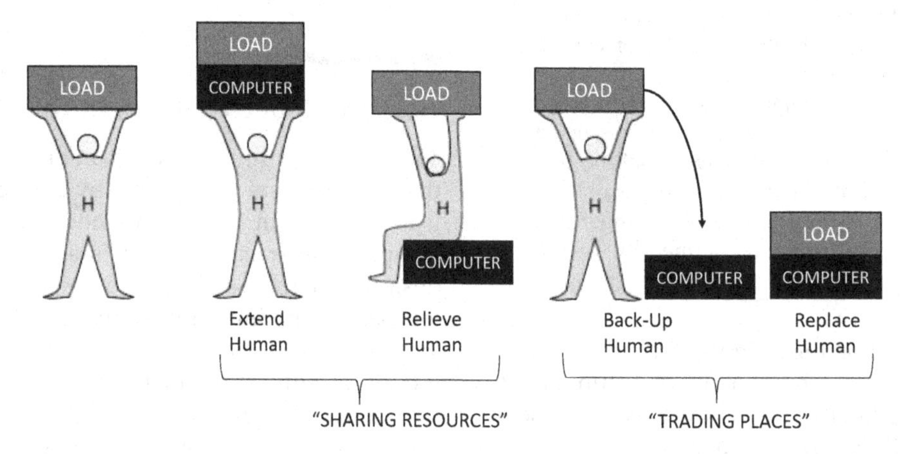

FIGURE 2.13
Bill Verplank Load Sharing Diagram between Computer and Human.

more indirect, if failures may be silent or masked, or little active behavior can lead to lower alertness and vigilance, complacency, and over-reliance. When the human operator performs as a backup, it may lead to lower proficiency and increased reluctance to intervene, limit the human's ability to practice handling "breakdown" scenarios, have fault intolerance leading to even larger errors, and make crisis handling more difficult. Or when the human operator performs as a partner, he or she may be left with miscellaneous tasks; the task may be more complex and require new tasks to be added, by taking away easy tasks, which may make difficult tasks harder and cause problems in communication between human and automation.

The "Mechanically extended man" is where the human operator supplies the initiative, the direction, the integration, and the criterion. ...mere extensions... and there was only one kind of organism, the man, and the rest was there only to help him. As a concept, J.C.R. Licklider's man-computer symbiosis is different in an important way from what John Dudley North (Boulton Paul Aircraft) has called the "mechanically extended man". In the man – machine systems of the past, the human operator supplied the initiative, the direction, the integration, and the criterion. The mechanical parts of the systems were mere extensions, first of the human arm, then of the human eye. These systems certainly did not consist of "dissimilar organisms living together..." There was only one kind of organism-man-and the rest was there only to help him. In the "mechanical replacement of man" configuration, it is an automated system of the future where human operators are responsible mainly for functions that it proved infeasible to automate.

The consequences of computers are high tech automation changing the cognitive demands of the operator. In this case, the operator is supervising rather than directly monitoring, making more complex cognitive decisions, the mode-rich system is complicated, and requiring an increased need for communication and cooperation. More consequences of computers are human-factor experts complaining about technology-centered automation. As a result, designers focus on technical issues but do not support operator tasks, leading to "clumsy" automation. For example, when mixing humans and computers, automated systems on aircraft have eliminated some types of human error and have created new ones (errors of commission versus errors of omission). It is imperative that human skill levels and his/her required knowledge may have to go up. But, the correct partnership and allocation of tasks is difficult because who has the final authority? Authority limits prevent actions that would lead to hazardous states but may prohibit maneuvers that could be needed in extreme situations (Figure 2.14).

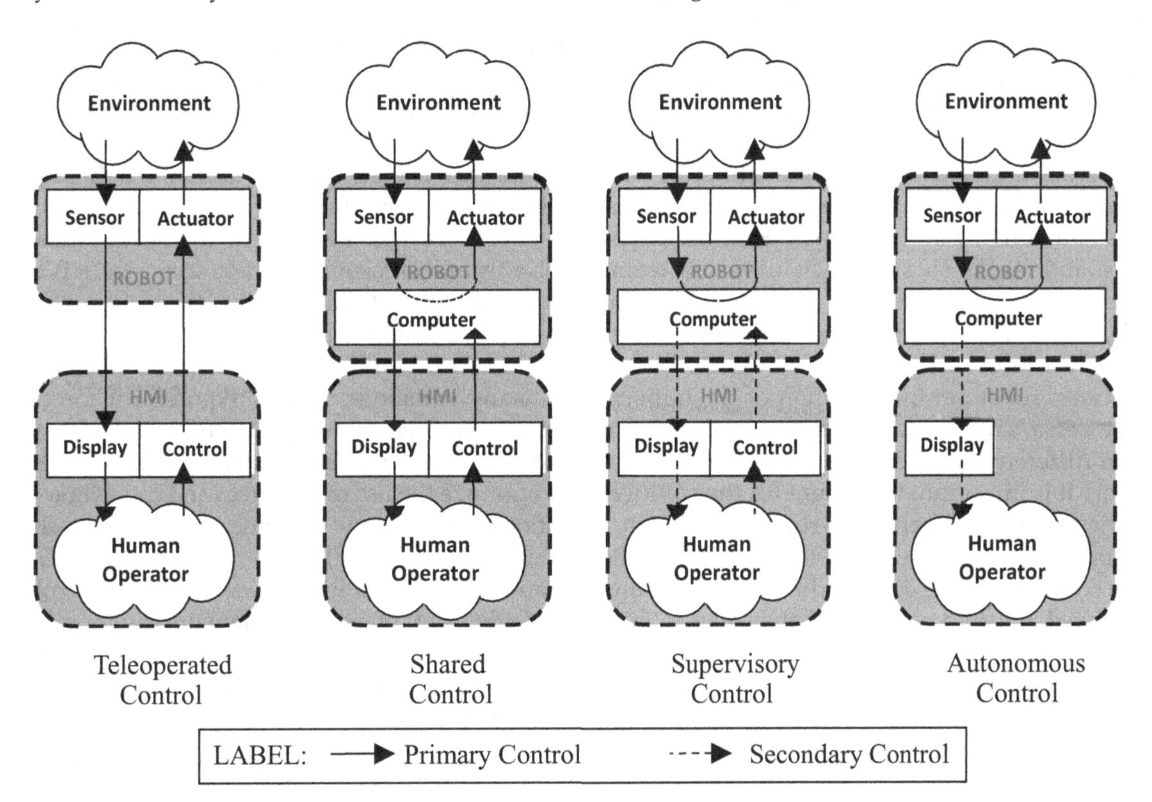

FIGURE 2.14
Tom Sheridan's Spectrum of Control Diagram.

A colleague of Bill Verplank at MIT is Professor Thomas Sheridan, a leading researcher in man – machine systems. Professor Sheridan's research interests are in experimentation, modeling, and the design of human – machine systems in air, highway and rail transportation, space and undersea robotics, process control, arms control, telemedicine, and virtual reality. Working at MIT, he developed important concepts concerning human–robot interactions, particularly regarding supervisory control and telepresence. The following figure depicts the difference between teleoperated control, shared control, supervisory control, and autonomous control of a robot as it performs in its environment. Except for the addition of the computer (or brain) to the robot, all the hardware configurations are identical. The primary differences are the primary control loops (solid arrows) and the secondary control loops (dashed arrows). Note that there are no hand controllers at the operator console in the autonomous control case.

With no difference in hardware, the difference in the various control modes is in software. In my past experience, it has been easier to implement a control mode at either end of this evolutionary scale: teleoperated since the role of the operator is clearly defined or autonomous, which has little input from the operator and the human as an observer. There is another option to implement a sliding control scheme for autonomy, as described later by Professor Anthony Stentz[1] from CMU. Autonomous robots are rarely purely autonomous; there must be some level of human interaction, although these challenges are not always considered. The issues to consider in human – robot coordination are (1) mutual understanding by modeling on both sides, (2) function allocation in human – robot teams, (3) levels of autonomy and how to allocate for different tasks, and (4) the social aspects of mixed teams with expectations of partnership and grounding. For example, can a robot understand the human (e.g., emotion or workload of human?) or for directed attention, or can a robot get the attention of the human? And vice versa. Function allocation includes how to learn the different states of all agents (both human and robot). It is important to account for the spillover boundaries from engineering to social science. In a robot-human interface, the robot's knowledge of the human's commands is needed to direct activities and any human-delineated constraints that may require command noncompliance or a modified course of action. In a robot–robot interface, the knowledge that the robots have of the commands given to them, if any, by other robots, the tactical plans of the other robots, and the robot-to-robot coordination necessary to dynamically reallocate tasks among robots if necessary. For a human's overall mission awareness, the human's understanding of the overall goals of the joint human–robot activities and the measurement of the moment-by-moment progress obtained against the goals.

Despite the advantages of automated and autonomous systems, significant problems remain in the integration of these types of systems with the human operator. With many systems, automation and autonomy have only been applied piece-meal, to tasks for which problems are identified and technological solutions available. This has been dubbed the technology-centered approach to automation. In addition to performing tasks not easily automated, the human operator usually remains in the system as a monitor to insure that the automated systems perform properly and to detect the occurrence of aberrant conditions. The out-of-the-loop performance problem is a major issue associated with automation. Human operators acting as monitors have problems in detecting system errors and performing tasks manually in the event of automation failures. These highly reported problems with automated systems can be directly attributed to lower levels of operator situation awareness (SA) that can occur with automation approaches that place people in the role of passive monitors. SA, a person's mental model of the state of a dynamic system, is central to effective decision-making and control, and is one of the most challenging portions of many operator's jobs.

Sliding autonomy can be used in teleoperation to adjusting a robot's level of local autonomy to match the user's needs or used in autonomous systems to slide down by having a human augment the autonomous system. As robots become more capable they can handle increasingly complex tasks and in highly uncertain environments, but the robotic capabilities in many domains are still insufficient to execute these tasks robustly and efficiently. In these scenarios, it is possible that robots can still accomplish the tasks with human assistance as human capabilities are often better suited for some tasks and can complement robot capabilities in many situations. A key requirement for enabling robustness and efficiency in human–robot teaming is the ability to dynamically adjust the level of autonomy to optimize the use of resources and capabilities as conditions evolve. However, much of this research is still at an early stage, and many challenges still remain in realizing this capability. These different control modes identified by Professor Sheridan[2] put different demands on the human operator and on the onboard computer with software. There are cognitive consequences of implementing a computer as the brain. They are:

- Increase memory demands.
- Require new skills and knowledge demands.
- Can complicate situational assessment.

- Can undermine people's attention management.
- Can disrupt efficient and robust scanning patterns.
- Can lead to limited visibility or changes and events, alarm and indication clutter, and extra interface management tasks.
- By increasing system reliability can provide little opportunity to practice and maintain skills for managing system anomalies.
- Force people into using tricks necessary to get tasks done that may not work in uncommon situations.

In talking to "hard-core" robotic engineers, they believe that everything is about "control" from their perspective. In a sliding autonomy model, researchers have identified six key capabilities that are essential for overcoming the human–robot issues when dialing back from full autonomy: *requesting help, maintaining coordination*, establishing *situational awareness*, enabling interactions at different levels of *granularity, prioritizing* team members, and *learning* from interactions. Autonomous systems are efficient but sometimes unreliable. In domains where reliability is paramount, efficiency is sacrificed by putting an operator in control via teleoperation by sliding down the Sheridan model. A related model is an adjustable autonomy, where a robot can *dynamically self-adjust their own level of autonomy based on the situation*. In fact, many of these applications will not be deployed, unless reliable reasoning is a central component. At the heart of adjustable autonomy is the question of whether and when a robot should make autonomous decisions and when it should transfer decision-making control to other entities (e.g., human operators).

Traditional human factors design has focused on function allocation, the division of tasks between man and machine. In the era of automation, the focus has been on determining high-workload tasks that operators need assistance with. These tasks are then allocated to automation in order to reduce the human tasks to more manageable levels. There is evidence that indicates that this under-riding principle involved in this automation strategy is flawed. Human workload does not always respond to automation as predicted. Many recent studies are beginning to confirm a lack of correspondence between workload and automation or people's use of automation. A researcher has found that a subject's choice to use automation in a task was not related to the workload level of the task, but rather to factors such as reliability, trust and risk. It appears that monitoring itself may induce high workload, and in tasks where people must provide sustained attention as monitors over a period of time, it induces considerable fatigue. Attention is the selection of important information. The human mind is bombarded with millions of stimuli and it must have a way of deciding which of this information to process. Attention is sometimes seen as a spotlight, meaning one can only shine the light on a particular set of information. For the sections to follow in this primer, the plan is to focus on autonomous robots.

GOMS: is an acronym for *Goals, Operators, Methods, and Selection rules* and is a method derived from human – computer interaction (HCI) to construct a description of human performance. GOMS is a theory of the cognitive skills involved in human–computer tasks. It is based upon an information processing framework that assumes a number of different stages or types of memory (e.g., sensory store, working memory, LTM/Long-Term Memory) with separate perceptual, motor, and cognitive processing. The level of granularity will vary based on the needs of the analysis. According to the GOMS model, cognitive structure consists of four components:

- The *g*oal is what the user wants to accomplish.
- The *o*perator is what the user does to accomplish the goal.
- The *m*ethod is a series of operators that are used to accomplish the goal.
- *S*election rules are used if there are multiple methods in order to determine how one was selected over the others.

All cognitive activities are interpreted in terms of searching a problem space. For a given task, a particular GOMS structure can be constructed and used to predict the time required to complete the task. In addition, the model can be used to identify and predict the effects of errors on task performance (Figure 2.15).

Human–AI collaboration is the study of how humans and AI work together to accomplish a shared goal. AI systems can aid humans such as in decision-making tasks. Examples of collaboration include medical decision-making aids, hate speech detection, and music generation. As AI systems are able to tackle more complex tasks, studies are exploring how different models and explanation techniques can improve human–AI collaboration. *Human–Machine Interface*, also known by the acronym HMI, refers to a dashboard or screen used to control machinery or a robot. As a rule, the information in the human–machine interface is displayed in a graphical form. The HMI systems use various icons, sounds, pictures, and colors to illustrate the machinery's current status and

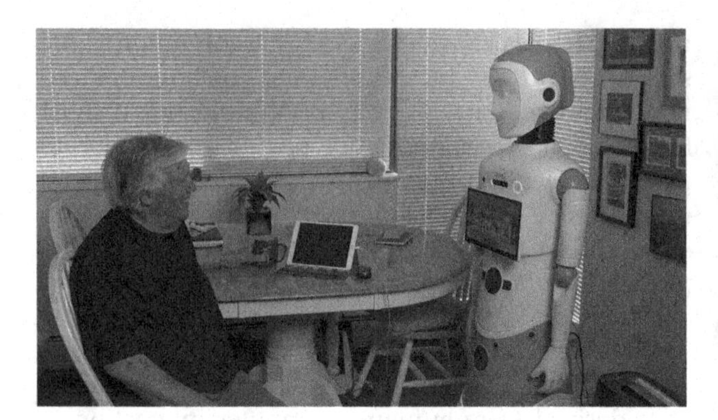

FIGURE 2.15
Interacting with Social Robot.

FIGURE 2.16
Human Operator for Controlling a Drone.

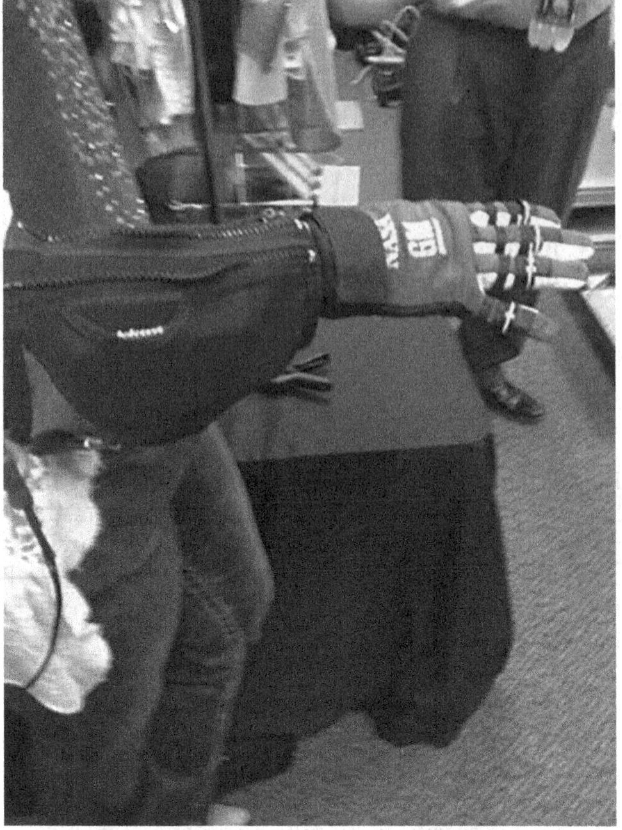

FIGURE 2.17
Human-Robot Symbiosis.

operating conditions. Of all the ways humans communicate, texting might be the most direct. Text carries less superfluous information than other ways of sending information. With text, there are no voice intonations to decipher or accents to understand, no facial gestures to interpret, and no body language to translate. Text is something computers can understand and process quickly, and it's why messaging is a great place for humans and AI to work together to serve customer needs (Figure 2.16).

Human – robot interaction is the study of interactions between humans and robots. It is often referred to as HRI by researchers. HRI is a multidisciplinary field with contributions from HCI, AI, robotics, natural-language understanding, design, and psychology. Robots are agents with capacities of perception and action in the physical world often referred to by researchers as its "workspace". Methods for perceiving humans in the environment are based on sensor information. Most methods intend to build a 3D model through vision or lidar of the environment. The proprioception sensors permit the robot to have information about its own state. This information is relative to a reference. A speech recognition system is used to interpret human desires or commands. By combining the information inferred by proprioception, sensor, and speech the human position and state (e.g., standing, seated). In this matter, natural-language processing is concerned with the interactions between computers and human (natural) languages, in particular, how to program computers to process and analyze large amounts of natural-language data. A large body of work in the field of HRI has looked at how humans and robots may better collaborate. The primary social cue for humans while collaborating is the shared perception of activity, to this end, researchers have investigated anticipatory robot control through various methods including monitoring the behaviors of human partners using eye tracking, making inferences about human task intent, and proactive action on the part of the robot. The studies revealed that the anticipatory control helped users perform tasks faster than with reactive control alone. A common approach to program social cues into robots is to first study human–human behaviors and then transfer the learning. These studies have revealed that maintaining a shared representation of the task is crucial for accomplishing tasks in groups or teams (Figure 2.17).

Notes

1 Tony is a colleague and an old friend in mobile robots at CMU since working together, starting in the mid-1980s.

2 I sat on a teleoperation latency panel with Prof. Sheridan for NASA.

3

Autonomy

Autonomous Robot Architectures

Cybernetics is an interdisciplinary science that looks at any and all systems, from molecules to galaxies, with special attention to machines, animals, and societies. The word is derived from the Greek word for steersman or helmsman, who provides the control system for a boat or ship. One important concept is the *law of requisite variety*, where the more complex the system that is being regulated, the more complex the regulator of the system must be. The *self-organizing system* is another cybernetic concept that we all see demonstrated daily. A self-organizing system is a system that becomes more organized as it goes toward equilibrium. And cybernetics also can be helpful in understanding how knowledge itself is generated. In the late 1960s cyberneticians began extending the application of cybernetics principles to understanding the *role of the observer.* This emphasis was called *"second-order cybernetics"*. First-order cybernetics deals with controlled systems, and second-order cybernetics deals with *autonomous systems.* Whereas, in the early days, cybernetics was generally applied to systems seeking goals defined for them, "second-order" cybernetics refers to systems that define their own goals (Figure 3.1).

The term robot architecture is often used to refer to two related design structures: the physical architecture, which relates hardware components and the functional architecture, which describes what the robot has to do. And, instead of functional architectures, some designers prefer to use a logical architecture (one of the views in a 4+1 architecture). Architecture is equated to "design", but at a higher level of abstraction as opposed to detail design. The architecture provides the structure for how the robot will be built, and also how it can evolve or grow in the future. Others refer to the architecture as structure, analogous to the physical architecture and style that refers to a computational concept that may be described as different communication paradigms. The architecture of a system defines its high-level structure, exposing its gross organization as a collection of interacting components. Components are needed to model a software and hardware architecture that includes components, connectors, subsystems, properties, and styles. As discussed in the previous Chapter 2, the robot is configured to implement the sense-plan-act (SPA) with sensors (some external signals to model its environment and some internal signals to the robot) inputting to the computer and outputting from the computer to actuators such as wheels, legs, or tracks for a mobile robot. The mobile robot architecture also supports a manipulator that is configured to implement the SPA with sensors (some external to model its environment and some internal to the robot) inputting to the computer and outputting from the computer to actuators on the arm. The objective of a robot system is to accomplish a specific task or goal where there are several options, including teleoperation, shared-control, supervisory control, and autonomous control. And within autonomy, there are several models beyond being a system: mechatronics, automatic, behavior-based such as subsumption, and the deliberate model with 3T (or a hybrid of deliberate and behavior-based layers). An autonomous robot can be either goal-based or utility-based.

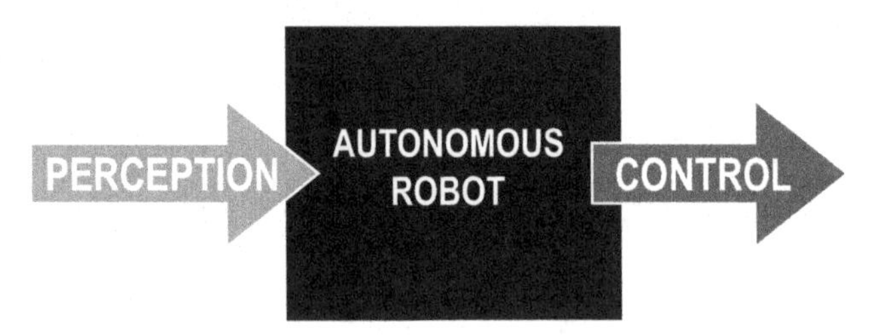

FIGURE 3.1
Autonomous robot process diagram.

And suppose we combine both the utility paradigm with the goal-based paradigm (we cannot have an autonomous vehicle that is not healthy based on its health utilities), the result is an artificially intelligent autonomous model that uses a framework that many of the weak AI technologies can fit in. The system model (A) is straightforward, where the robot takes sensor input from its environment (also referred to as perception) and is processed in the robot system, and outputs commands to the robot actuator that interacts with its environment. The mechatronics model (B) is the same as the system model with the identical physical components of sensors, computers (that process software), and actuators. The automatic model

DOI: 10.1201/9781032673134-3

(C) is ideal for operating in structured environments. Also referred to as an automaton that can only follow rules. The automatic model can be implemented as a state machine (commonly implemented in dynamically stable-legged robots) where sensors are used to estimate the state of the robot, and based on rules to follow, the robot implements an action or behavior. This is the same paradigm as a dynamically stable, legged robot using an augmented finite state machine approach. The automatic robot can be controlled by an outside force (does not consciously make its own decision) and its operations are completely observable. The behavior model (D) is a bottom-up methodology, inspired by biological studies, where a collection of behaviors acts in parallel to achieve independent goals (Figure 3.2).

An autonomous robot is an entity that acts toward achieving goals in its environment. It is important to note that there is a dependence of action on knowledge, and knowledge and action interact in two principal ways:

1. Knowledge is often required prior to taking action.
2. Actions can change what is known.

Each of these behaviors is a simple module that receives inputs from the robot's sensors and outputs actuator commands. The overall architecture consists of several behaviors reading the sensory information and sending actuator commands to a coordinator (also referred to as arbitrator) that combines them in order to send a single command to each actuator. MIT Professor Rodney Brooks explanation for a reactive paradigm:

> AI researchers... partition the problems they work on into two components. The AI component, which they solve, and the non-AI component which they don't solve. Typically, AI 'succeeds' by defining the parts of the problem that are unsolved as not AI. The principal mechanism for this partitioning is abstraction... In AI, abstraction is usually used to factor out all

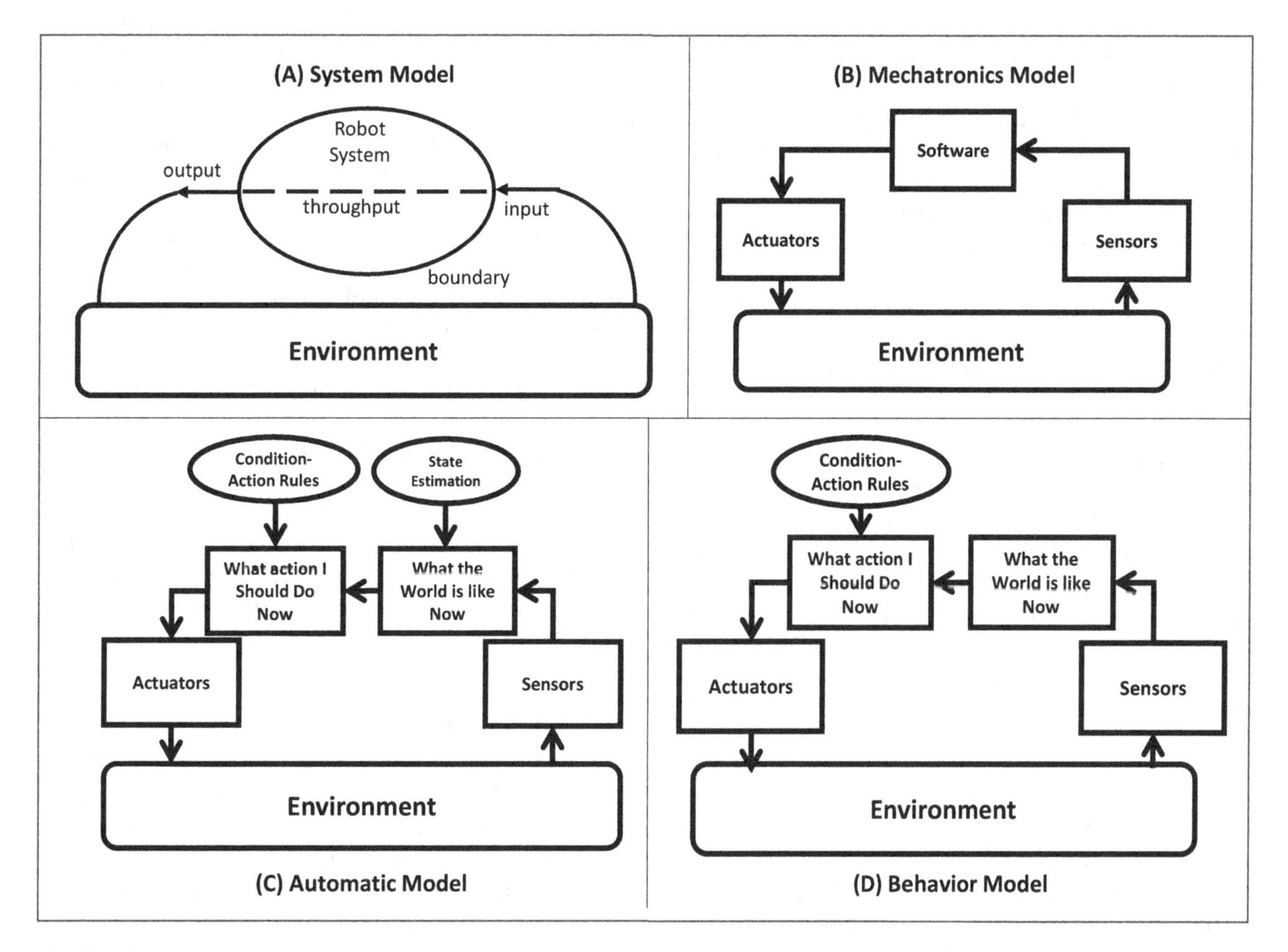

FIGURE 3.2
The Various and Different Robot Brain Models.

> aspects of perception and motor skills. I argue below that these are the hard problems solved by intelligent systems, and further that the shape of solutions to these problems constrains greatly the correct solutions of the small pieces of intelligence which remain.

The various types of behaviors include reflexive (stimulus-response), reactive (learned or muscle memory), or conscious (deliberately stringing together). It is important to remember that roboticists often use the term "reactive behavior" to mean purely reflexive and refer to reactive behaviors as "skills". MIT Professor Rodney Brooks proposed to tightly couple perception to action, and thereby, provide a reactive behavior that could deal with any unpredicted situation the robot may encounter. Moreover, Brooks[1] advocated for avoiding keeping any model of the environment in which the robot operates, arguing that "the world is its own best model". This would alleviate the bottleneck in processing that slows the robot down. This automatic behavior architecture decomposes the overall desired robot system behavior into sub-behaviors in a bottom-up fashion. In this "hierarchical" structure, the higher-level behaviors *subsume* the lower-level behaviors. In other words, the high-level behaviors can outsource smaller scale tasks to be handled by the low-level behaviors. There are two known styles of creating a reactive system: Brook's subsumption architecture and potential fields as established by Stanford University's Oussama Khatib.[2] In general, information as any stimulus, data, or content that can be "read" by an agent (robot or AI), and information is often described as a representational flow. Control is basically the robot/AI acting upon another part of the system. And as demonstrated by cybernetics, feedback is required for a closed-system where the thing controlled provides a relation to the controller.

From an implementation standpoint, this architecture can be thought of as layers of finite state machines that all connect sensors to actuators and where multiple behaviors are evaluated in parallel. An arbitration mechanism is also included to choose which of the behaviors is currently activated. While this type of architecture is much more reactive than the SPA architecture, there are also some disadvantages. The primary disadvantage of this approach is that there is no good way to do long-term planning or behavior optimization. As a result, this can make it challenging to design the system to accomplish long-term objectives. A better option is to combine both the SPA architecture with the behavior model in a hybrid model I.

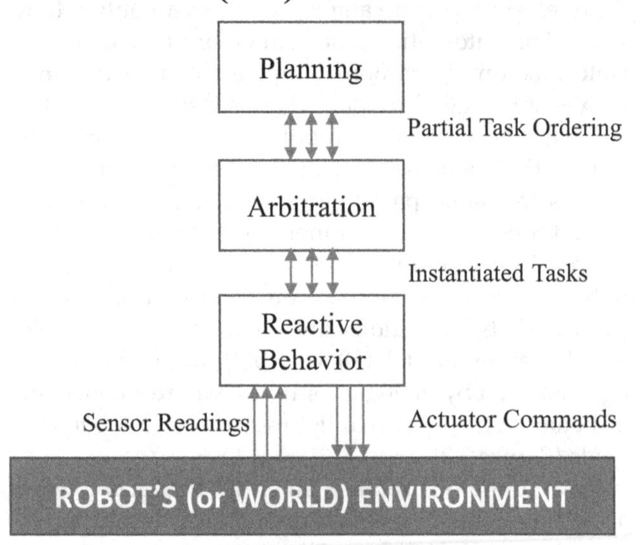

FIGURE 3.3
Three-Tier hybrid architecture model.

The reactive behaviors can be useful to implement safety protocols, while still using a world model and AI planning techniques in a deliberative architecture (Figure 3.3).

3T stands for 3-Tier, a hybrid reactive/deliberative robot architecture that consists of three distinct layers: a reactive feedback control mechanism, a reactive plan execution mechanism, and a mechanism for performing time-consuming deliberative computations. The 3T model requires a model of the robot's environment as sensed by its perception system, a model of the robot and its consequences as it mentally tests different options in order to perform planning, and condition-action rules as the basis of its reactive system. There is also a decision-making process based on (1) breath of search, (2) reducing the search space by some arbitrary criteria, and finally, (3) deciding based on certain criteria. The *three-tiered architecture* is one of the most commonly used robot architecture and contains a planning, an executive, and a behavioral control level that are hierarchically linked.

1. Planning: this layer is at the highest-level and focuses on task-planning for long-term goals.
2. Executive: the executive layer is the middle layer connecting the planner and the behavioral control layers. The executive specifies priorities for the behavioral layer to accomplish a specific task. While the task may

come directly from the planning layer, the executive can also split higher-level tasks into subtasks.

3. Behavioral Control: at the lowest level, the behavioral control layer handles the implementation of low-level behaviors and is the interface to the robot's actuators and sensors.

The primary advantage of the 3T architecture is that it combines the benefits of the behavioral-based architecture (i.e., reactive planning) with better long-term planning capabilities (i.e., resulting from the planning level). The goal-based autonomous robot and the utility-based autonomous robot use the deliberate or SPA paradigm. In the deliberate paradigm, the robot senses the world, updates its world model, generates a plan (knowledge representation+automated reasoning), and executes this planned action. However, it is this planning stage that is the bottleneck for the robot. Knowledge representation is a model of the robot's immediate (local) environment. It uses the *Closed World Assumption,* where the World model contains everything the robot needs to know. The issue with knowledge representation is the frame problem with how to model everything the robot needs to know while keeping the size of the state space manageable. The process of exploring the possible states of the world to find a plan is computationally expensive. The deliberate paradigm views sensing and acting as disconnected with no reactivity because after sensing, this model needs to update to the world model and generate the plan. Knowledge representation is based on first-order predicate logic, and planning is also logic-based. The hybrid model I includes functions such as:

- Sequencer: generates a set of behaviors to accomplish a subtask.
- Resource Manager: allocates resources to behaviors, for example, a selection of suitable sensors.
- Cartographer: creates, stores, and maintains a map or spatial. information, a global world model and knowledge representation.
- Mission Planner: interacts with the operator and transforms the commands into the robot terms.
- Performance Monitoring and Problem-Solving: it is a sort of self-awareness that allows the robot to monitor its progress.

The utility-based model (F) is similar to the hybrid architecture, but adds two new steps: what will happen if the robot postulates new actions and what will be the result of this action in keeping the robot happy by balancing resources and health using the von Neumann–Morgenstern (or VNM) utility theorem and showing that, under certain axioms of rational behavior, a decision-maker faced with risky (probabilistic) outcomes of different choices will behave as if he or she is maximizing the expected value of some function defined over the potential outcomes at some specified point in the future. This function is known as the VNM utility function. The theorem is the basis for expected utility theory. Utility is defined as a measure of the relative happiness or satisfaction (gratification) gained by consuming different bundles of goods and services. Given this measure, one may speak meaningfully of increasing or decreasing utility and thereby explain economic behavior in terms of attempts to increase one's utility. The theoretical unit of measurement for utility is the utility itself. For example, a robot cannot fully operate autonomously unless the robot is happy and has adequate resources to accomplish the task. In other words, the robot has to be healthy or it can effect or limit what the robot can do. As a utility reduces some resources needed by the robot, it must replan using a constraint-based planner such as the one developed by my colleague Dr. Nicola Muscettola[3] in LISP refer to Figure 3.4, Model F.

The goal-based autonomous model (G) is similar to (F) but with the caveat of giving the robot a mission or goal to complete in order to be successful. The goal affect its decision-making priorities in deciding what it should do next. Theoretically, a goal is the result of achievement toward which effort is directed. If the robot was a military platform, the goal would be the mission objectives. The key to meeting one's goal is to build a plan of action and manage one's time. The AI-autonomous model (H) is a hybrid model of deliberate (STP) with added behaviors (having rules to set actions of things that should be automatic), having a world model that is continuously updated by its sensors, integrating a learning capability to upgrade from previous events by understanding its past history stored in memory, do planning such as constraint-based, deciding based on utility and goal priorities, and outputting an action. The notion of having a world model and planning, learning, and decision-making play a part in reasoning. The techniques for learning, planning, and decision-making are all theories in weak AI. It is important to remember that an autonomous robot is a far-cry from having an intelligent robot that is powered by strong-AI.

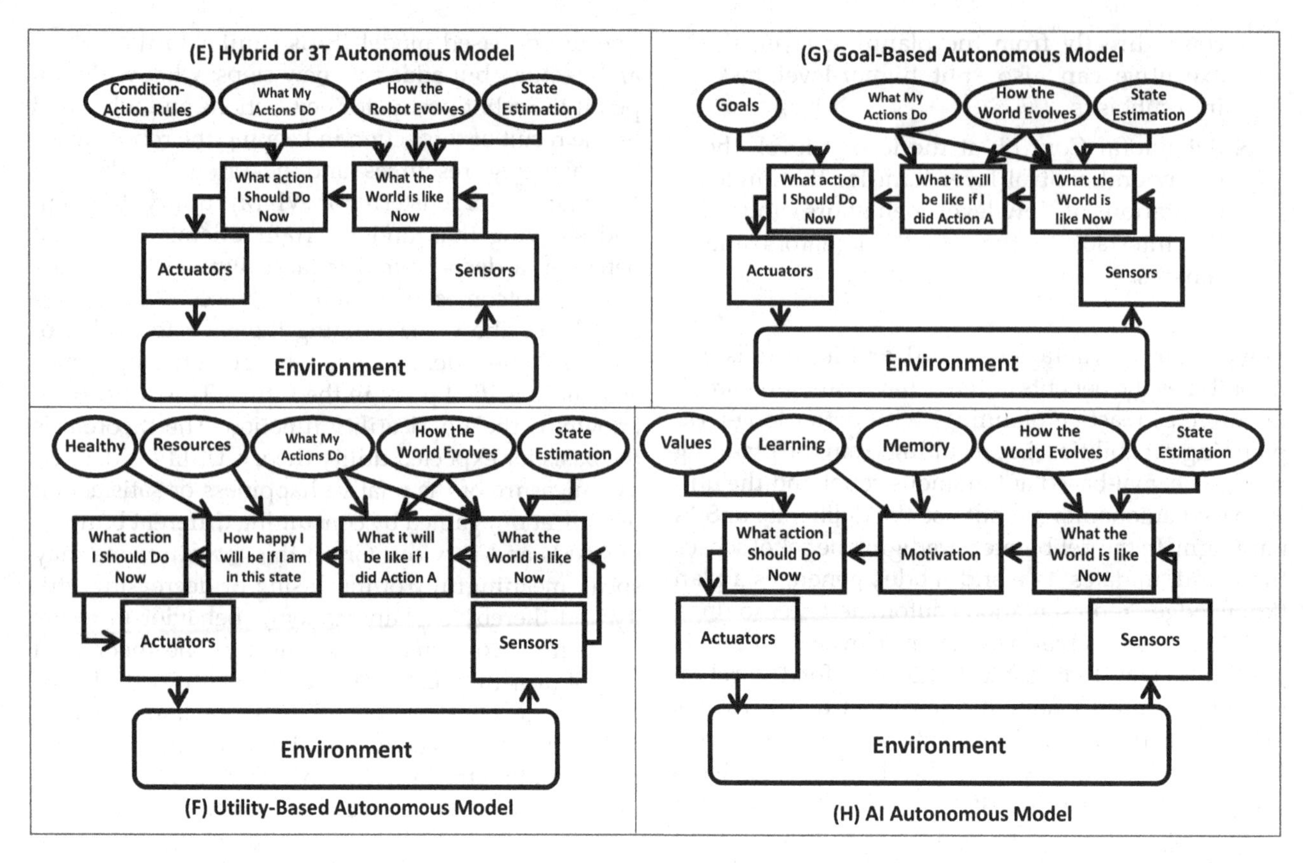

FIGURE 3.4
Additional Robot Brain Models Continued.

Decision-Making in Autonomy

The fundamental idea of decision theory is that an agent is rational if and only if it chooses the action that yields the highest expected utility, averaged over all the possible outcomes of the action. This is called the principle of maximum expected utility (MEU).

$$\text{Decision theory} = \text{probability theory} + \text{utility theory} \tag{3.1}$$

It has become commonplace that cognition involves both computation and information processing. Cognitivism is based on the manipulation of representations. On the other side, there are non-cognitivist approaches such as behaviorism and Gibsonian psychology, which reject mental representations. The thought may be that, by rebranding cognitivism as either computationalism or information-processing psychology, one is simply hinting at the divide between representational and anti-representational approaches. We use "digital computation" for whatever notion is implicitly defined by the classical mathematical theory of computation, most famously associated with Turing. By this, we do not mean to appeal to a specific formalism, such as Turing machines. But not until the invention of the digital computer has there been machines that could perform intellectual functions of even modest scope, that is, machines that could in any sense be said to be intelligent. Now "artificial intelligence" (AI) is a subdiscipline of computer science.

The components of a decision are as follows:

- Alternatives
- Criteria
- Value judgments
- Decision-maker (human or machine) preferences.

The *science of economics* began in 1776 when Scottish philosopher *Adam Smith* published *An Inquiry into the Nature and Causes of the Wealth of Nations.* While the ancient Greeks and others had made contributions to economic thought, Smith was the first to treat it as a science, using the idea that economies can be thought of as consisting of *individual agents maximizing their own economic well-being.* In economics, utility (from utility theory) is a measure of the relative happiness or satisfaction (gratification) gained by consuming different bundles of goods and services. The general public thinks of economics as being about money, but economists will say that they are really studying *how people make choices that lead to preferred outcomes or utility.* William Stanley Jevons was a British economist credited with being the first theorist to make economics a mathematical discipline. In 1860, he developed the theory of utility, which states that a commodity's degree of utility is a continuous mathematical function of the quantity available. Jevons invented the Logic Piano, which aids in calculating syllogisms: logical alphabet, logical slate, logical stamp, and logical abacus. *Decision theory,* which combines probability theory with utility theory, provides a formal and complete framework for decisions (economic or otherwise) made under uncertainty, that is, in cases where probabilistic descriptions appropriately capture the decision-makers environment. This is suitable for *"large" economies* where each agent needs to pay no attention to the actions of other agents as individuals. For *"small" economies,* the situation is much more like a game; the actions of one player can significantly affect the utility of another (either positively or negatively). VNM development of *game theory* included the surprising result that, for some games, a rational agent should act in a random fashion, or at least in a way that appears random to the adversaries.

Game theory tells us how to make good decisions in multi-agent settings, that is, such as deploying powerful game-playing agents (for chess, poker, video games, etc.). For the most part, economists did not address the third question listed above, namely, how to make rational decisions when payoffs from actions are not immediate but instead result from several actions taken in *sequence.* This topic was pursued in the field of *operations research,* which emerged in World War I1 from efforts in Britain to optimize radar installations and later found civilian applications in complex management decisions. The work of Richard Bellman (1957) [590] formalized a class of sequential decision problems called the *Markov decision processes.* Work in economics and operations research has contributed much to our notion of rational agents, and for many years, AI research developed along entirely separate paths (symbolic logic and machine learning). One reason was the apparent complexity of making rational decisions. *Herbert Simon,* the pioneering AI researcher, won the *Nobel prize* in economics in 1978 for his early work showing that models based on *satisficing-making decisions* that are "good enough", rather than laboriously calculating an *optimal decision.* The earlier gives a better description of actual human behavior.

Computers accept, store, process and present information, and the networks move information among the machines they interconnect. This assumes the robot has multiple processors, but in aerospace, we can make a single computer operate as multiple processors using ARINC 653. ARINC 653 is a software specification for space and time partitioning in safety-critical avionics for real-time operating systems such as in commercial aircraft. It allows the hosting of multiple applications of different software levels on the same hardware in the context of an integrated modular avionics architecture. Computers can manipulate information far faster than people ever will. But unlike people, machines almost never understand the messages they are manipulating. To them, information is only a deceptively uniform sequence of numbers of ones and zeros. And when comparing computers to humans:

- Humans are better than computers for data collection/entry, but comparison changing.
- For the physical size of data collected, computers surpass humans, especially as the number of data items increases.
- For speed of queries, the computer is much faster than the human. This is more evident in large data sets.

In robotics, it is often unrealistic to assume that the true state is completely observable and noise-free. The learning system will not be able to know precisely in which state it is and even vastly different states might look very similar. Thus, robotics reinforcement learning is often modeled as partially observed. The learning system must hence use filters to estimate its true state. It is often essential to maintain the information state of the environment that not only contains the raw observations but also a notion of uncertainty on its estimates (e.g., both the mean and the variance of a Kalman filter tracking the ball in the robot table tennis example).

Key to autonomy is decision-making, the act of choosing one alternative from among a set of alternatives. The decision-making process, however, is much more than this. For example, a person making the decision must have somehow recognized that a decision was necessary and identified the set of feasible alternatives before selecting one. Hence, the decision-making process includes recognizing and defining the nature of a decision situation, identifying alternatives, choosing the "best" alternative, and putting it into practice.

- A programmed decision is one that is fairly structured or recurs with some frequency.
- Nonprogrammed decisions, on the other hand, are relatively unstructured and may occur much less often.

Robots don't like ambiguity when making decisions. They need to know, very clearly, which choice to make under what circumstances. As a consequence, their decisions are always based on the answers to questions that have only two possible answers: yes or no, true or false. Statements that can be only true or false are called Boolean statements, and their true-or-false value is called a truth value. Fortunately, many kinds of questions can be phrased so that their answers are Boolean (true/false). Technically, they must be phrased as statements, not questions. So, rather than asking whether the sky is blue and getting an answer yes or no, you would state that "the sky is blue" and then find out the truth value of that statement, true (it is blue) or false (it is not blue). Note that the truth value of a statement is only applicable at the time it is checked. For example, the sky could be blue one minute and gray the next. But regardless of which it is, the statement "the sky is blue" is either true or false at any specific time. The truth value of a statement does not depend on when it is true or false, only whether it is true or false right now. Thus, "Under the hood" of all the major decision-making control structures is a simple check for the Boolean value of the (condition).

Intuition and experience also play large roles in the making of nonprogrammed decisions. In general, the circumstances that exist for the decision-maker are conditions of certainty, risk, or uncertainty. When people know with reasonable certainty what their alternatives are and what conditions are associated with each alternative, a state of certainty exists. However, few decisions are made under conditions of true certainty. The complexity and turbulence of the "real" world make such situations rare. Under a state of risk, the availability of each alternative and its potential payoffs and costs are all associated with probability estimates. Most of the significant decisions made in contemporary problems are made under a state of uncertainty, but the decision-maker does not know all the alternatives, the risks associated with each, or the consequences of each alternative it is likely to have. This uncertainty stems from the complexity and dynamism of contemporary problems and their environments. The key to effective decision-making in these circumstances is to acquire as much relevant information as possible and to approach the situation from a logical and rational perspective. Intuition, judgment, and experience always play major roles in the decision-making process under conditions of uncertainty. The notion of uncertainty leads to probabilistic robotics.

The classical decision model is a prescriptive approach that tells the robot how it should make decisions. It is grounded in the assumptions that AI machines are logical and rational and that they always make decisions that are in the best interests of the robot (goal-based). The objectives of decision-making are: (1) optimize robot productivity, (2) minimize operating costs, and (3) maximize robot state through platform health & status and resource utilization (utility-based). Finally, the decision-making process has to be safe by making decisions that never endanger inhabitants or cause damage and making the decisions should be within the range acceptable for the operations of the robot.

There are several possible decision-making approaches for robots:

- Pre-programmed decisions, such as following rules or incorporating timer-based automation.
- Reactive decision-making systems where decisions are based on condition-action rules and driven by the available facts.
- *Goal-based* decision-making systems are such that decisions are made in order to achieve a particular outcome.
- *Utility-based* decision-making systems are where decisions are made in order to maximize a given performance measure.

These decisions align with the various autonomous robot architectures. Pre-programmed decisions are used in very structured environments such as by industrial manipulators in a work cell or a drone flying in open space, which could make course corrections using global position system (GPS). Behavior-based or reactive decision-making is useful to implement safety protocols based on rules. Both goal-based and/or utility-based decisions are required in an autonomous architecture.

In decision theory, the VNM utility theorem shows that, under certain axioms of rational behavior, a decision-maker faced with risky (probabilistic) outcomes of different choices will behave as if he or she is maximizing the expected value of some function defined over the potential outcomes at some specified point in the future. This function is known as the VNM utility function. The theorem is the basis for expected utility theory. The principle of maximum expected utility (MEU) says that a rational agent should choose an action that maximizes $EU(A|E)$, requiring search or planning, because an agent needs to know the possible future states in order to assess the worth of the current state ("effect of the state on the future"). The utility function rates states and thus formalizes the desirability of a state by the robot. $U(S)$ denotes the utility of state S for the agent. A nondeterministic action A can lead to the outcome states $\text{Result}_i(A)$.

How high is the probability that the outcome state $\text{Result}_i(A)$ is reached, if A is executed in the current state with evidence E?

$$\rightarrow P\left(\text{Result}_i(A) \mid Do(A), E\right) \quad (3.2)$$

Maximum Expected Utility (MEU): the MEU principle says that a rational robot should choose an action that maximizes its expected utility in the current state I

$$EUI = \max_A \Sigma_i P\left(\text{Result}_i(A) \mid Do(A), E\right) \times U\left(\text{Result}_i(A)\right) \quad (3.3)$$

The principle of MEU says that a rational agent should choose an action that maximizes $EU(A \mid E)$.

$$P\left(\text{Result}_i(A) \mid Do(A), E\right) \quad (3.4)$$

which requires a complete causal model of the world.

- Constant updating of belief networks
- NP-complete for Bayesian networks

$$U\left(\text{Result}_i(A)\right) \quad (3.5)$$

MEU requires search or planning because an agent needs to know the possible future states in order to assess the worth of the current state ("effect of the state on the future").

Why isn't the MEU principle all we need in order to build "intelligent robots"? Because it is difficult to compute P, E or U? The MEU computational difficulties are:

- Knowing the current state of the world requires *perception, learning, knowledge representation,* and *inference.*
- Computing P(*) requires a complete *causal model of the world.*
- Computing UI often requires search or planning (calculation of utility of a state may require looking at what utilities could be achieved from that state).

All of the above can be computationally intractable; hence, one needs to distinguish between "perfect rationality" and "resource-bounded rationality" or "bounded-optimality". Also, we need to consider more than one action (one-shot decisions versus sequential decisions). Thus, rational agents can be developed on the basis of probability theory and utility theory. Robots that make decisions according to the axioms of utility theory possess a utility function. Sequential problems in uncertain environments (using a Markov decision process or MDP) can be solved by calculating a policy. Value iteration is a process for calculating optimal policies. Probabilistic graphical models (PGMs) are a rich framework for encoding probability distributions over complex domains and joint (multivariate) distributions over large numbers of random variables that interact with each other. They are also a foundational tool in formulating many machine learning problems. It describes the two basic PGM representations: Bayesian networks, which rely on a directed graph; and Markov networks, which use an undirected graph. Probability theory provides the basis for our treatment of systems that reason under uncertainty. Also, because actions are no longer certain to achieve goals, agents will need ways of weighing up the desirability of goals and the likelihood of achieving them. For this, we use utility theory. Probability theory and utility theory together constitute decision theory, which allows us to build rational agents for uncertain worlds.

Forward chaining, forward deduction, or forward reasoning is a method involving inference or logical rules (facts) for data extraction. It is a bottom-up approach performed redundantly to reach the endpoint or a goal (in order to decide). It begins with evaluating existing information, followed by manipulation based on its knowledge base. The existing information can be as facts, derivations, and conditions. Backward chaining, backward deduction or backward, is also a reasoning method that acts in the reverse direction to forward chaining. The top-down approach involves using decisions or goals to reach the facts. The backward chaining is the backtracking process of finding usage in diagnostics, debugging and prescription.

Forward-chaining production systems (invented in 1943 by Emil Post) are used as the basis for many rule-based expert systems. Production is a rule of the form: $C_1, C_2, \ldots C_n \Rightarrow A_1\ A_2 \ldots A_m$ where conditions must hold before the rule can be applied and the actions to be performed or conclusions to be performed or conclusions to be drawn when the rule is applied. The basic components of a production system are rules (unordered set of user-defined "if-then" rules), working memory (A set of "facts" consisting of positive literals defining what's known to be true about the world), and inference engine (procedure for inferring changes (additions and deletions)) to working memory and typically uses forward chaining to make inferences. The production system is a model of computation that can be applied to implement search algorithms and model human problem-solving. Facts in most production systems are basically flat tuples.

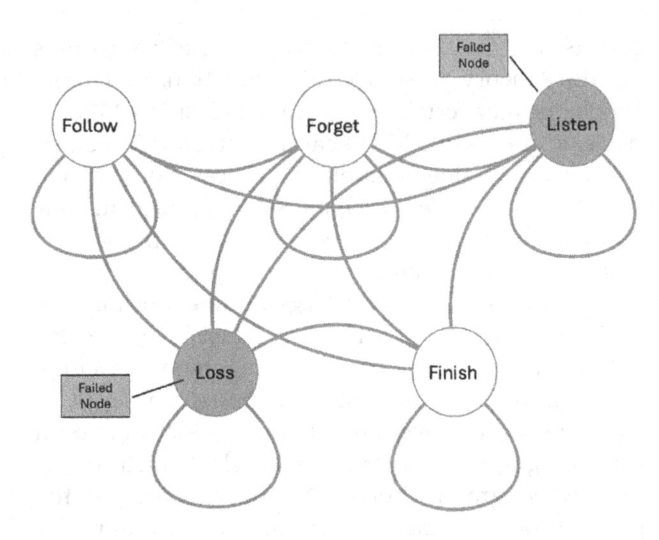

FIGURE 3.5
Forward chaining to make comparison models.

Backward chaining is an inference method described colloquially as working backward from the goal. It is used in automated theorem provers, inference engines, proof assistants, and other artificial intelligence applications. It is essentially an efficient way of implementing the Leibniz's chain rule. A simple extension supported by many is to allow simple templates using "slot-filler" pairs. While changes are made to working memory, do the following: match, conflict resolution, and act (Figure 3.5).

One potential problem with expert systems is the number of comparisons that need to be made between rules and facts in the database. In some cases, where there are hundreds or even thousands of rules, running comparisons against each rule can be impractical. The Rete algorithm is an efficient method for solving this problem and is used by a number of expert system tools, including OPS5 and Eclipse. The Rete Algorithm (Greek for "net") is the most widely used, efficient algorithm for the implementation of production systems developed by Charles Forgy at Carnegie Mellon University in 1979. The Rete is a directed, acyclic, and rooted graph. Each path from the root node to a leaf in the tree represents the left-hand side of a rule. Each node stores details of which facts have been matched by the rules at that point in the path. As facts are changed, the new facts are propagated through the Rete from the root node to the leaves, changing the information stored at nodes appropriately. This could mean adding a new fact, or changing information about an old fact, or deleting an old fact. In this way, the system only needs to test each new fact against the rules and only against those rules to which the new fact is relevant, instead of checking each fact against each rule. Rete is the only algorithm for production systems whose efficiency is asymptotically independent of the number of rules and the basis for a whole generation of fast expert system shells. There is a conflict resolution strategy of components of refraction, recency, specificity, and explicit priorities. Rational agents can be developed on the basis of a probability theory and a utility theory. Agents that make decisions according to the axioms of utility theory possess a utility function. Sequential problems in uncertain environments (MDPs) can be solved by calculating a policy. Value iteration is a process for calculating optimal policies. The Rete algorithm depends on the principle that in general, when using forward chaining in expert systems, the values of objects change relatively infrequently, meaning that relatively few changes need to be made to the Rete.

A *semantic network* is a simple representation scheme that uses a graph of labeled nodes and labeled, directed arcs to encode knowledge (usually used to represent static, taxonomic, and concept dictionaries). Semantic networks are typically used with a special set of accessing procedures that perform "reasoning". The ISA (is-a) or AKO (a-kind-of) relation is often used to link instances to classes and classes to super classes. Some links (e.g., has Part) are inherited along ISA paths. The semantics of a semantic net can be relatively informal or very formal, often defined at the implementation level. The graphical depiction associated with a semantic network is a significant reason for their popularity. Non-binary relationships can be represented by "turning the relationship into an object", which is an example of what logicians call "reification". One of the main kinds of reasoning done in a semantic net is the inheritance of values along the subclass and instance links. Thus, inference is by inheritance. Semantic networks morphed into Frame Representation Languages in the 1970s and 1980s. A frame is a lot like the notion of an object in object-oriented programming (OOP) but has more meta-data. A frame has a set of slots and a slot represents a relation to another frame (or value). A slot has one or more facets and a facet represents some aspect of the relation. Description logics provides a family of frame-like KR systems with a formal semantics. Current systems take care to keep the languages simple, so that all inference can be done in polynomial time (in the number of objects) and ensure the tractability of inference. In the context of network theory, a complex network is a graph with non-trivial topological features, features that do not occur in simple networks such as lattices or random graphs but often occur in networks representing real systems.

Description logics provides a family of frame-like KR systems with a formal semantics. An additional kind of inference done by these systems is automatic *classification* or finding the right place in a hierarchy of objects for a new description. Current systems take care to keep the languages simple, so that all inference can be done in polynomial time (in the number of objects), ensuring tractability of inference. *Abduction* is a reasoning process that tries to form plausible explanations for abnormal observations (distinctly different from deduction and induction) and is inherently uncertain. Uncertainty is an important issue in abductive reasoning. Abduction is defined as reasoning that derives an explanatory hypothesis from a given set of facts. The *inference* result is a *hypothesis* that, if true, could *explain* the occurrence of the given facts.

> *NOTE: Deduction* reasons from causes to effects, *Abduction* reasons from effects to causes, and *Induction* reasons from specific cases to general rules.

"Conclusions" are *hypotheses,* not theorems (may be false *even if* rules and facts are true). Abductive reasoning is often a hypothesize-and-test cycle. To *hypothesize,* we postulate possible hypotheses, any of which would explain the given facts (or at least most of the important facts) and *Test* (the plausibility of all or some of these hypotheses). Reasoning is *nonmonotonic.* That is, the plausibility of hypotheses can increase/decrease as new facts are collected. In contrast, the deductive inference is *monotonic;* it never changes a sentence's truth value once known. In abductive (and inductive) reasoning, some hypotheses may be discarded, and new ones formed, when new observations are made. Uncertain inputs (missing and noisy data), uncertain knowledge (multiple causes lead to multiple effects, incomplete enumeration of conditions or effects, incomplete knowledge of causality in the domain, and probabilistic/stochastic effects), and uncertain outputs (abduction and induction are inherently uncertain, default reasoning, even in deductive fashion, is uncertain, and incomplete deductive inference may be uncertain). In abductive reasoning, "conclusions" are *hypotheses* and not theorems (may be false *even if* rules and facts are true). There may be multiple plausible hypotheses: Given rules A => B and C => B and fact B, thus both A and C are plausible hypotheses. Hypotheses can be ranked by

TWEETY PENGUIN TRIANGLE PROBLEM (TPTP OR TP2)

Let's consider the set $R = \{r1, r2, r3\}$ of given rules (as known as defaults in [1]):

- *r*1: "Penguins normally don't fly" $\Leftrightarrow (p \rightarrow \neg f)$
- *r*2: "Birds normally fly" $\Leftrightarrow (b \rightarrow f)$
- *r*3: "Penguins are birds" $\Leftrightarrow (p \rightarrow b)$

The TP2 and the solutions are based first on *fallacious* Bayesian reasoning and then on Dempster–Shafer reasoning. We will then focus our analysis of this problem from the DSmT framework and the DSm reasoning. The problem with first-order logic is that universal rules are too strict ∀x bird(x)→flies(x) mean *all* birds must fly, but there are many exceptions...

If we asserted:

penguin(opus)
∀x penguin(x) → ¬flies(x)
∀x penguin(x) → bird(x)
this would make the KB *inconsistent.*

With resolution, KB ⊢ fly(tweety). This formalization of the flight attributes of penguins is insufficient. If we add the statement "Penguins cannot fly", that is, penguin(x) ⇒ ¬fly(x), ¬fly(tweety) can be derived, but fly(tweety) is still true. The knowledge base becomes therefore *inconsistent.* Refer to Figure 3.6. A logic is called monotonic if, for an arbitrary knowledge base KB and an arbitrary formula ϕ, the set of formulas derivable from KB is a subset of the formulas derivable from KB ∪ {ϕ}.

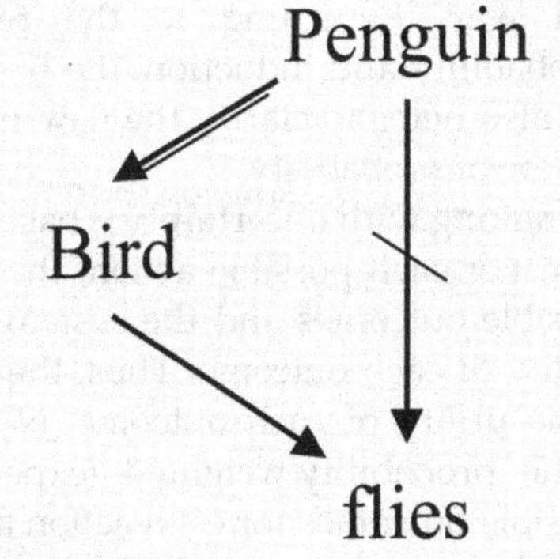

FIGURE 3.6
Tweety triangle penguin problem.

If reasoning is based on classical deductive logic, it is monotonic, while commonsense reasoning is not monotonic. The addition of new information can cause previous inferences to fail. If I tell you that Tweety is a bird, you will infer that Tweety can fly, but if I then inform you that Tweety is a penguin, the inference evaporates. Nonmonotonic (or defeasible) logic includes formalisms designed to capture the mechanisms underlying these kinds of examples.

their plausibility (if it can be determined). Reasoning is *nonmonotonic,* that is, the plausibility of hypotheses can increase/decrease as new facts are collected. In contrast, deductive inference is *monotonic, meaning* it never changes a sentence's truth value once known. In abductive (and inductive) reasoning, some hypotheses may be discarded, and new ones formed, when new observations are made.

One way of discussing notions of deductive and inductive inferences is to adopt the standpoint of a personalistic foundation of probability (as opposed, say, to the frequency interpretation of the formal probability calculus or the necessary view mentioned above). In the case of inferences expressed as propositional inferences (if (hypothesis) and {auxiliary fact) are true, then the (deduction) follows), we need to assign probabilities to each proposition of the three categories. Hume recognized the process of confirmation by positive evidence and the work described here assumes that some such process operates as the basis of sensory perception. Again, the Bayesian formula allows the process to be clearly demonstrated: the larger the deductive probability $p(D/H \cap A)$, then the larger the inductive (a posteriori) probability $p(H/D \cap A)$, if the a priori probability p(H) remains the same. The a priori (in the sense of prior to the evidential fact) probability p(H) does influence the evaluation of a hypothesis $p(H/D \cap A)$, but in an extra-evidential, extra-logical way. Returning to the relationship between probability and induction, the Bayesian formula above also helps to clarify the flaw in Carnap's necessary view of probability.

Decision-making with uncertainty is based on rational behavior. For each possible action, the AI identifies the possible outcomes and the system computes the probability of each outcome. Thus, the algorithm computes the utility of each outcome. Next, the AI computes the probability-weighted (expected) utility over possible outcomes for each action and selects the action with the highest expected utility (principle of MEU). Bayesian reasoning uses probability theory and Bayesian inference to reason diagnostically (from evidence (effects) to conclusions (causes)) or causally (from causes to effects). Bayesian networks are compact representations of probability distribution over a set of propositional random variables that

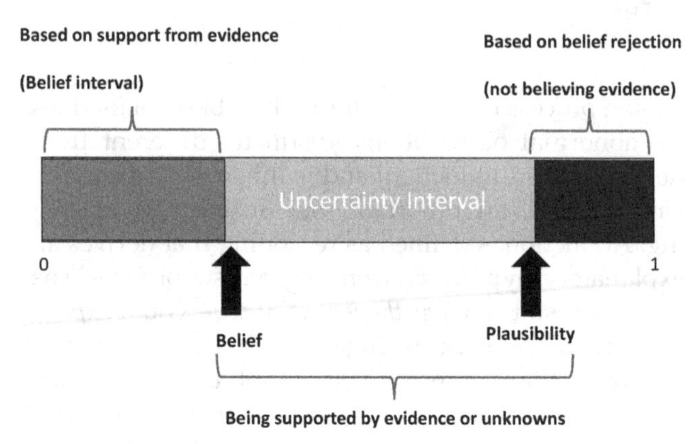

FIGURE 3.7
Decision-making with uncertainty.

take advantage of independence relationships. There are other uncertain representations such as default reasoning, rule-based methods, evidential reasoning, and fuzzy reasoning. Uncertainty is traded-off between Bayesian networks, nonmonotonic logic, certainty fact, Dempster–Shafer theory (DST) (for conflict management) and fuzzy reasoning. Belief networks are a powerful tool for representing and reasoning with uncertain knowledge. The theory of belief functions is also referred to as evidence theory or DST, and is a general framework for reasoning with uncertainty, with understood connections to other frameworks such as probability, possibility and imprecise probability theories (Figure 3.7).

The DST is designed to deal with the distinction between uncertainty and ignorance. Rather than computing the probability of a proposition, it computes the probability that the evidence supports the proposition. The DST of evidence is a powerful method for combining measures of evidence from different classifiers based on a formal calculus for combining evidence. DST is a generalization of the Bayesian *Theory of Subjective Probability.* Belief functions base degrees of belief (or confidence or trust) for one question on the subjective probabilities for a related question. The DST is based on two ideas: the idea of obtaining degrees of belief for one question from subjective probabilities for a related question and Dempster's rule for combining such degrees of belief when they are based on independent items of evidence. The DST, also known as the theory of belief functions, is a generalization of the Bayesian theory of subjective probability. Whereas the Bayesian theory requires probabilities for each question of interest, belief functions allow us to base degrees of belief for one question on probabilities for a related question. These degrees of belief may or may not have the mathematical properties of probabilities; how much they differ from probabilities will depend on how closely the two questions are related. DST offers an alternative to traditional probabilistic theory for the mathematical representation of uncertainty. The significant innovation of this framework is that it allows for the allocation of a probability mass to sets or intervals. DST does not require an assumption regarding the probability of the individual constituents of the set or interval. The DST is a generalization of the Bayesian theory of subjective probability. Whereas the Bayesian theory requires probabilities for each question of interest, belief functions allow us to base degrees of belief for one question on probabilities for a related question. This is a potentially valuable tool for the evaluation of risk and reliability in engineering applications when it is not possible to obtain a precise measurement from experiments, or when knowledge is obtained from expert elicitation.

The word "fuzzy" means "vagueness" and fuzziness occurs when the boundary of a piece of information that is not clear-cut. In the real world, there exists much fuzzy knowledge; knowledge that is vague, imprecise, uncertain, ambiguous, inexact, or probabilistic in nature. Human thinking and reasoning frequently involve fuzzy information, originating from inherently inexact human concepts. Humans can give satisfactory answers which are probably true. However, our systems are unable to answer many questions. The reason is most systems are designed based on classical set theory and two-valued logic, which is unable to cope with unreliable and incomplete information and give expert opinions (Figure 3.8).

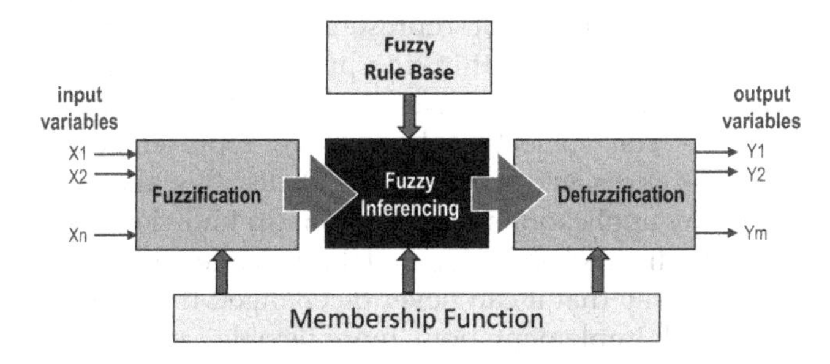

FIGURE 3.8
Block schematic of a fuzzy system model.

Fuzzy systems include fuzzy logic and fuzzy set theory. Knowledge exists in two distinct forms: (1) the objective knowledge that exists in mathematical form is used in engineering problems, and (2) the subjective knowledge that exists in linguistic form, usually impossible to quantify. Fuzzy logic can coordinate these two forms of knowledge in a logical way. Fuzzy systems can handle simultaneously the numerical data and linguistic knowledge. Fuzzy systems provide opportunities for modeling conditions that are inherently imprecisely defined. Many real-world problems have been modeled, simulated, and replicated with the help of fuzzy systems. There are many real-world problems for fuzzy systems such as information retrieval, navigation, and robot vision systems. Fuzzy inferencing is the core constituent of a fuzzy system and combines the facts obtained from the fuzzification with the fuzzy rule base and conducts the fuzzy reasoning process. Fuzzification is the process of transforming crisp values into grades of membership for linguistic terms, that is, "far", "near", "small" of fuzzy sets. Fuzzy rule base is a collection of propositions containing linguistic variables. Fuzzy Inferencing combines the facts obtained from the fuzzification process with the rule base and conducts the fuzzy reasoning process. Defuzzyfication translates results back to real-world values. A simple form of logic, called two-valued logic, is the study of "truth tables" and logic circuits. This simple two-valued logic is generalized and called fuzzy logic, which treats "truth" as a continuous quantity ranging from 0 to 1.

Goal-Oriented Action Planning: GOAP or goal-oriented action planning is a powerful STRIPS (Stanford Research Institute Problem Solver) like planning architecture designed for real-time control of autonomous character behaviors in games, but now applied to robots. GOAP will give your robot/AI choices and the ability to make intelligent decisions without having to maintain a complex finite state machine (FSM). The goal of action planning is to choose actions and ordering relations among these actions to achieve specified goals. GOAP is an artificial intelligence system for autonomous robots that allows them to dynamically plan a sequence of actions to satisfy a set goal. The sequence of actions selected by the robot is contingent on both the current state of the robot and the current state of the world, hence, despite two robots being assigned the same goal; both robots could select a completely different sequence of actions. GOAP will intelligently choose the most intuitive sequence of events based on the provided preconditions, that is, if the robot does not have access to a wrench then they must turn the bolt manually. GOAP makes the code more modular and easier to maintain, by decoupling the states from each other, each action can be worked on incognizant of the others. Furthermore, GOAP allows adding and removing actions spontaneously; an autonomous robot must simply have a defined list of actions that are automatically processed by the GOAP planner.

Plan & Goal: A plan is simply a sequence of actions that satisfy a goal wherein the actions take the robot from a starting state to whichever state satisfies the goal. A goal is any condition that a robot wants to satisfy. In GOAP, goals simply define what conditions need to be met to satisfy the goal, the steps required to reach these satisfactory conditions are determined in real time by the GOAP planner. A goal is able to determine its current relevance and when it is satisfied.

Actions: Every robot is assigned actions, which are a single, atomic step within a plan that makes a robot do something. Examples of action are playing an animation, playing a sound, altering the state, picking up flowers, etc.

Note: Every action is encapsulated and ignorant of the others.

Each defined action is aware of when it is valid to be executed and what its effects will be on the real world. Each action has both preconditions and effects attributes, which are used to chain actions into a valid plan. A precondition is the state required for an action to run and effects are the changes to the state after an action has been executed. For example, if the robot wants to close a valve using a tool, it must first acquire the tool and make sure the precondition is met. Otherwise, the robot closes the valve using its end effector. GOAP determines which action to execute by evaluating each action's cost. The GOAP planner evaluates which sequence of actions to use by adding up the cumulative cost and selecting the sequence with the lowest cost. Actions determine when to transition into and out of a state as well as what occurs in the world due to the transition.

The GOAP Planner: An agent develops a plan in real time by supplying a goal to satisfy a planner. The planner looks at an actions preconditions and effects in order to determine a queue of actions to satisfy the goal. The target goal is supplied by the AI along with the world state and a list of valid actions; this process is referred to as "formulating a plan". If the planner is successful, it returns a plan for the robot to follow. The robot executes the plan until it is completed, invalidated, or a more relevant goal is found. If at any point the goal is completed or another goal is more relevant, then the robot aborts the current plan and the planner formulates a new one. The planner finds the solution by building a tree. Every time an action is applied, it is removed from the list of available actions. Furthermore, you can see that the planner will run through all available actions to find the most optimal solution for the target goal. Remember, different actions have different costs and the planner is always looking for the solution with the cheapest cost. In most cases, GOAP is simply a more effective solution over a cumbersome FSM. Using GOAP, an AI can be very dynamic and have a large range of actions, without having to manually implement the interconnected states.

STRIPS: a simple, still reasonably expressive planning language based on propositional logic.

1. Examples of planning problems in STRIPS.
2. Planning methods.
3. Extensions of STRIPS.

Like programming, knowledge representation is still an art. Forward planning simply searches the space of world states from the initial to the goal state. In general, many actions are applicable to any given state, so the branching factor is huge. Fortunately, an accurate, consistent heuristic can be computed using planning graphs. Forward planning still suffers from an excessive branching factor. In general, there are much fewer actions that are relevant to achieving a goal than actions that are applicable to a state. A search tree uses backward planning to search a space of goals from the original goal of the problem to a goal that is satisfied in the initial state. There are often much fewer actions relevant to a goal than there are actions applicable to a state → smaller branching factor than in forward planning. However, the lengths of the solution paths are the same.

Knowledge Representation: is probably the most important ingredient for developing an AI. A representation is a layer between information accessible from outside world and high-level thinking processes. Without knowledge of representation, it is impossible to identify what thinking processes are, mainly because representation itself is a substratum for a thought. The subject of knowledge representation has been messaged for a couple of decades already. For many applications, specific domain knowledge is required. Instead of coding such knowledge into a system in a way that it can never be changed (hidden in the overall implementation), more flexible *ways of representing* knowledge and reasoning about it have been developed in the last 10 years. The need of knowledge representation was felt as early as the idea to develop intelligent systems. With the hope that readers are well conversant with the fact by now, that intelligent requires possession of knowledge and that knowledge is acquired by us by various means and stored in the memory using some representation techniques. Putting in another way, knowledge representation is one of the many critical aspects, which are required for making a computer behave intelligently. Knowledge representation refers to the data structure techniques and organizing notations that are used in AI. These include semantic networks, frames, logic, production rules and conceptual graphs. The following properties should be possessed by a knowledge representation system.

a. Representational Adequacy: It is the ability to represent the required knowledge.
b. Inferential Adequacy: It is the ability to manipulate the knowledge represented to produce new knowledge corresponding to that inferred from the original.
c. Inferential Efficiency: The ability to direct the inferential mechanisms into the most productive directions by storing appropriate guides.
d. Acquisitional Efficiency: The ability to acquire new knowledge.

Knowledge representation is a field of artificial intelligence that focuses on designing computer representations that capture information about the world that can be used for solving complex problems. A knowledge base is a collection of facts, rules, and procedures organized into schemas. The assembly of all the relevant information and knowledge about a specific field of interest. Some in the field believe that knowledge is the same as models. The knowledge representation hypothesis is:

> Any mechanically embodied intelligent process will be comprised of structural ingredients that (a) we as external observers naturally take to represent a propositional account of the knowledge that the overall process exhibits, and (b) independent of such external semantical attribution, playa formal but causal and essential role in engendering the behavior that manifests that knowledge.

The knowledge representation problem explains how to structure and retrieve information, in situations where *anything* might be relevant. The justification for knowledge representation is that conventional procedural code is not the best formalism to use to solve complex problems. Procedures (a type of routine or subroutine) simply contain a series of computational steps to be carried out. For example, SHRDLU (an early natural-language understanding computer program) represents knowledge in the form of procedures. Knowledge representation makes complex software easier to define and maintain than procedural code and can be used in expert systems. Following this hypothesis, there are two major properties that the structures in a knowledge-based system have to satisfy. First of all, it must be possible to interpret them as propositions representing the overall knowledge of the system, and second, the symbolic structures within a knowledge-based system must play a causal role in the behavior of that system, as opposed to, say, comments in a programming language. One way to represent knowledge is by using rules that express what must happen or what does happen when certain conditions are met. Rules are usually expressed in the form of IF ... THEN ... statements, such as: IF A THEN B. This can be considered to have a similar logical meaning as the following: $A \rightarrow B$. A is called the antecedent, and B is the consequent in this statement (Figure 3.9).

A symbol is something which stands for something else. With Symbolic reasoning, knowledge must be represented symbolically. A representation is a particular kind of symbol, in which the structure of the symbol itself is perceived to correspond in some way with the structure of the thing that the symbol stands for. The first and most prevalent type of knowledge to consider representing is what might be called simple facts about the world. Most current representation schemes seem to deal extensively with the concepts of "abstraction" and "generalization". Furthermore,

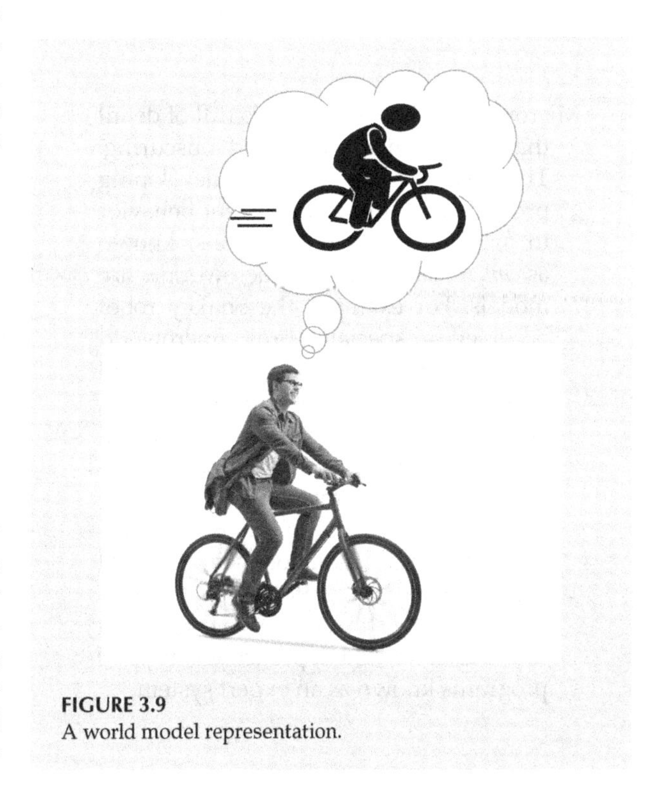

FIGURE 3.9
A world model representation.

these two terms are often used interchangeably. However, it is not clear that to do so makes sense. In fact, let us characterize them differently, as follows: an Abstraction is a relationship between individuals, one of which has been "abstracted away" from some dimension along which the other exists. Generalization is a relationship between concepts, one of which is less-well specified than the other, and which can therefore be used in describing a wider range of objects. But simply,

$$\text{Representation} = \text{Knowledge} + \text{Access} \tag{3.6}$$

The answer, in a nutshell, is that knowledge of the world cannot be captured in a finite structure. The world is too rich and agents have too great a capability for responding. Knowledge is about the world. Insofar as an agent can select actions based on some truth about the world, any candidate structure for knowledge must contain that truth. Thus, knowledge as a structure must contain at least as much variety as the set of all truths (i.e., propositions) that the agent can respond to. By distinguishing sharply between the knowledge level and the symbol level, the theory implies an equally sharp distinction between the knowledge required to solve a problem and the processing required to bring that knowledge to bear in real time and real space. Knowledge is that which makes the principle of rationality work as a law of behavior. Thus, knowledge and rationality are intimately tied together. For example, a *microworld* is construed as a set of simplified models.

Microworlds: The real world is full of detail that can be distracting and obscuring. Thus, AI research focused on developing programs capable of intelligent behavior in artificially simple situations, known as *micro-worlds*. First, microworlds are models. For example, the Shakey robot occupied a specially built microworld that consisted of walls, doorways, and a few simply-shaped wooden blocks. As another example, the micro-world approach was applied to language processing where a program could answer questions about simple stories concerning stereotypical situations where the AI program could infer information that was implicit in the story. The greatest success of the micro-world approach is a type of programs known as an expert system.

The next form we will consider, the frame description, is mainly an elaboration of the semantic network. The emphasis, in this case, is on the structure of types themselves (usually called frames), particularly in terms of their attributes (called slots). A *frame* is a data structure for representing a stereotyped situation, like being in a certain kind of living room, or going to a child's birthday party. We can think of a frame as a network of nodes and relations and the top levels of a frame are fixed and represent things that are always true about the supposed situation. The lower levels have many terminals: slots that must be filled by specific instances or data. Each terminal can specify conditions its assignments must meet. A frame can be thought of as a network of nodes and relations. However, frames cannot account for commonsense or scientific knowledge. Schank's version of frames is called "scripts". A *script* is a structure that describes appropriate sequences of events in a particular context. It is made up of slots and requirements about what can fill those slots. The structure is an interconnected whole, and what is in one slot affects what can be in another. Scripts handle stylized everyday situations. They are not subject to much change, nor do they provide the apparatus for handling totally novel situations. Thus, a script is a predetermined, stereotyped sequence of actions that defines a well-known situation (Figure 3.10).

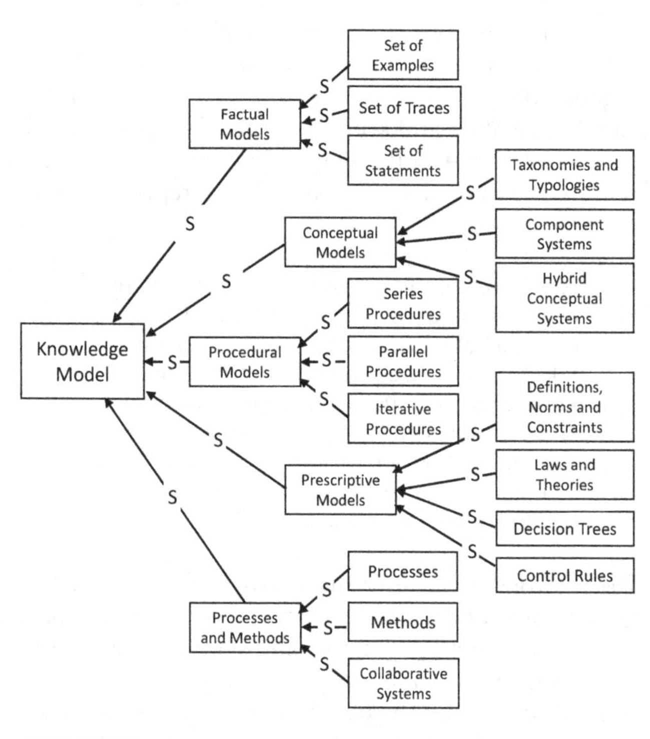

FIGURE 3.10
Knowledge model.

Psychologist George A. Miller theorized that a person's short-term memory (STM) can only hold around seven items, plus or minus two, at any given time. To deal with any information that is larger than seven items, we must first organize this information into large chunks. For example, by combining words into sentences, or combining sentences into stories, we can hold more than seven words in our short-term memory. However, our human memory still only can hold seven of these chunks at one time.

Another area of knowledge representation is the problem of commonsense reasoning. A problem for Schank's approach is the issues that "real short stories" pose. He assumes that all human practice and know-how are represented in the mind as a system of beliefs constructed from context-free primitive actions and facts. AI researchers have consistently run up against the problem of representing everyday context. For example, we cannot represent the back ground as just another object to be represented in the same sort of structure description in which every object is presented. Since intelligence must be situated, it cannot be separated from the rest of human life. One of the first realizations learned from trying to make software that can function with human natural language was that humans regularly draw on an extensive foundation of knowledge about the real world that we simply take for granted but that is not at all obvious to an artificial agent. These are the basic principles for commonsense physics, causality, intentions, etc. An example is the frame problem, that in an event-driven logic there need to be axioms that state things maintain position from one moment to the next unless they are moved by some external force. In order to make a true artificial intelligence agent that can converse with humans using natural language and can process basic statements and questions about the world, it is essential to represent this kind of knowledge.

The *frame problem*: in dynamic environments, there is the problem of knowing what changes have and have not taken place following some action. And more importantly, some changes will be the direct result of the action.

In a possible-worlds analysis of knowledge:

- Kripke introduced the idea that a world should be regarded as possible, not absolutely, but only relative to other worlds.
- The relation of one world's being a possible alternative to another is called the accessibility relation.

Much of the motivation for model-based reinforcement learning (RL) derives from the potential utility of learned models for downstream tasks, like prediction, planning, and counterfactual reasoning. Whether such models are learned from data, or created from domain knowledge, there's an implicit assumption that an agent's world model is a forward model for predicting future states. While a perfect forward model will undoubtedly deliver great utility, they are difficult to create, thus much of the research has been focused on either dealing with uncertainties of forward models or improving their prediction accuracy. Errors in the world model compound, and cause issues when used for control. If an agent has knowledge that one of its actions will lead to one of its goals, then the agent will select that action. This is known as the *Principle of Rationality*. This principle asserts a connection between knowledge and goals, on the one hand, and the selection of actions, on the other, without specification of any mechanism through which this connection is made. It connects all the components of the agent together directly.

Decision Trees: A decision tree is the denotative representation of a decision-making process. The name "decision tree" comes from the fact that the algorithm keeps dividing the dataset down into smaller and smaller portions until the data has been divided into single instances, which are then classified. Decision tree learning is one of the most widely used and practical methods for inductive inference. It is a method for approximation of discrete-valued functions, in which a tree represents the learned function. A decision tree is in a nutshell a discrete value functional mapping, and classifier. Each node in the decision tree specifies a test of some attribute of the query instance, and each branch descending from that node corresponds to one of the possible values for this attribute. To visualize the results of the algorithm, the way the categories are divided would resemble a tree and many leaves. Decision trees in artificial intelligence are used to arrive at conclusions based on the data available from decisions made in the past. Therefore, decision tree models are support tools for supervised learning and are based on the concept of search. A decision tree is the denotative representation of a decision-making process. Decision trees in artificial intelligence are used to arrive at conclusions based on the data available from decisions made in the past.

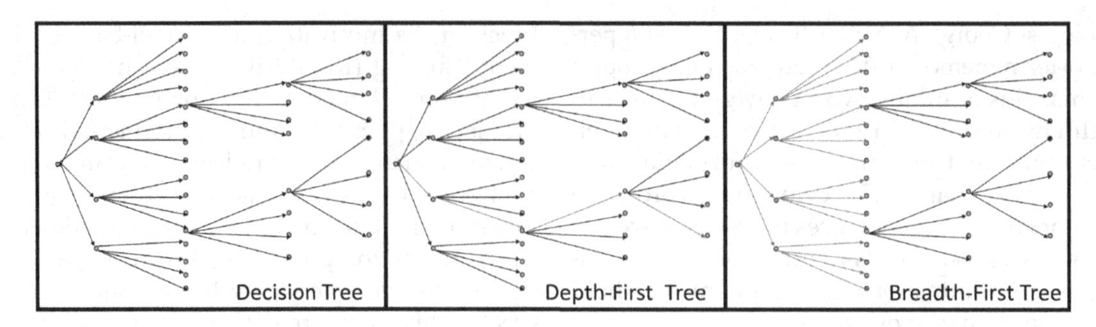

FIGURE 3.11
Decision tree models.

A decision tree isn't really an algorithm; it's a tree data structure that helps you think about a problem's solution space. Different branches through the tree represent decisions that lead to a final assignment of values of some kind, and a complete path from the root of the tree to a leaf corresponds to a complete assignment. Decision trees often grow extremely quickly. For example, if you're considering N items, then a search problem's decision tree has 2N paths leading to 2N leaf nodes representing 2N possible solutions. For example, if you can examine 10 million solutions per second (which is a pretty brisk pace), then it could take you almost 3.6 years to decide on an answer. In an exhaustive search, the algorithm examines every possible solution and picks the best one.

Backtracking is a way to envision a search through its solution space. In backtracking, the program follows branches down through the tree (or its network). After the algorithm has explored one path until it can go no farther, the program backtracks to an earlier spot in the tree and continues the search from that position by moving down other links that it hasn't yet explored. Usually, backtracking algorithms work *recursively*, and "backtracking" really means returning up the call stack to a previous call to the recursive method. The search method calls itself recursively each time it moves down a level in the tree space. To backtrack, the recursive call exits and control returns to the previous call higher up in the call stack, where it tries a different branch. The good news is that you can often *prune* the search tree so you don't need to search the entire thing. For example, *Branch and Bound* is a particularly clever way to prune decision trees. An algorithm that may produce a good result but that does not guarantee an optimal result is called a *heuristic*. An everyday example of a heuristic is the "rule of thumb". A *greedy algorithm* decides at each step that moves closer to an optimal solution. A *divide-and-conquer* approach divides the solution space into pieces, and figures out which piece contains the solution, for which it examines that piece more closely. If you can discard large pieces at each step, then you can quickly narrow the search. A *recursive algorithm* is one that calls itself. Recursive algorithms often go well with recursive data structures. For example, trees have a recursive structure. (Each node has links that lead to other nodes, which have links that lead to other nodes, which have links that. . .) As a result, that makes a recursive search a good fit for decision trees (Figure 3.11).

Decisions are typically made using a tree search via two methods: check every possible solution (exhaustive enumeration and a huge computational burden) or solve as a linear program and then round to the closest binary value (however, such an approximation but not optimally guaranteed). Tree-search methods conceptualize the problem as a huge tree of solutions, and then try some smart things to avoid searching the entire tree. The most popular solution is a tree-search method called *branch and bound*, or in combination with cutting planes. The *cutting-plane method* is any of a variety of optimization methods that iteratively refine a feasible set or objective function by means of linear inequalities, termed *cuts*.

For the *Knapsack Problem*, you have a knapsack that can hold a fixed amount of weight. You also have a collection of items, each having a weight and a value. The problem is to find the combination of items with the greatest value that will still fit in the knapsack.

Constraint satisfaction problems are typically solved by drawing a search tree whose nodes contain partial solutions. That way, the task of solving the problem becomes the task of searching a tree, with each branch representing a value that is tentatively assigned to a variable. This method is followed by a backtracking step if it is found that the chosen assignment cannot work. To make the search more intelligent, the process is to prune as many branches as possible from the tree, so that the process avoids wasting time with dead ends. There are four different strategies for pruning branches when searching for a solution. In *increasing order of power,* (but also increasing order of the amount of extra work you have to do), the strategies are:

1. Depth-first search only.
2. Depth-first search + forward checking
3. Depth-first search + forward checking + propagation through singleton domains
4. Depth-first search + forward checking + propagation through reduced domains

The basic concept underlying the branch-and-bound technique is to *divide and conquer.* Since the original "large" problem is hard to solve directly, the search is *divided* into smaller and smaller subproblems until these subproblems can be *conquered.* The *dividing (branching)* is done by partitioning the entire set of feasible solutions into smaller and smaller subsets (Figure 3.12). The *conquering (fathoming)* is done partially by:

i. Giving a *bound* for the best solution in the subset.
ii. Discarding the subset if the bound indicates that it can't contain an optimal solution.

Representation is a decision tree and there is a bias toward simple decision trees. Thus, the system initiates a "search" through the space of decision trees, from simple decision trees to more complex ones. Decision trees are statistical, algorithmic models of machine learning that interpret and learn responses from various problems and their possible consequences. As a result, decision trees know the rules of decision-making in specific contexts based on the available data. The learning process is continuous and based on feedback. This improves the outcome of learning over time. This kind of learning

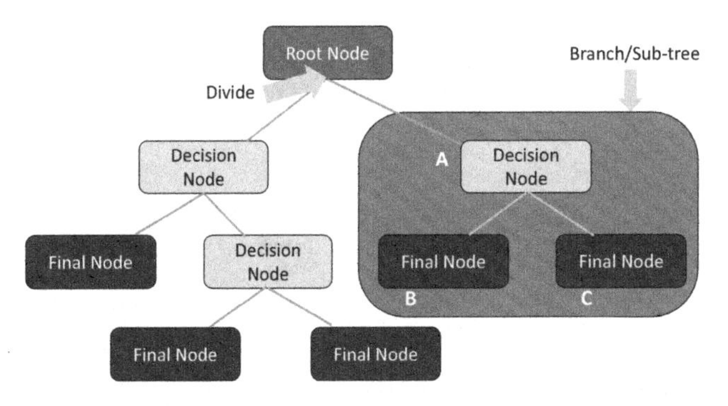

FIGURE 3.12
Decision tree with nodes.

is called supervised learning. Therefore, decision tree models are support tools for supervised learning. Decision trees are classic and natural learning models. They are based on the fundamental concept of divide and conquer. In the world of artificial intelligence, decision trees are used to develop learning machines by teaching them how to determine success and failure. These learning machines then analyze incoming data and store it. Decision trees are classic and natural learning models. They are based on the fundamental concept of divide and conquer. The decision-making approach, as listed previously is supported by specific, weak AI techniques as outlined here:

- Reactive Decision-Making → Rule-based expert system
- Goal-Based Decision-Making → Planning
- Decision theoretic Decision-Making → Belief Networks and Markov decision process
- Learning Techniques → Neural Networks and Reinforcement Learning

(Note: "→" literally means "moving" forward)

The AI-autonomous architecture takes advantage of all the above, including rules to follow in an expert system. Goal-based decision-making requires planning, and due to uncertainties in the sensors and actuators, probabilistic techniques such as Bayesian networks and Markov models are needed to reach its goals. Eugene Charniak described the best way to understand Bayesian networks is to imagine trying to model a situation in which causality plays a role but where our understanding of what is actually going on is incomplete, so we need to describe things probabilistically. Bayesian networks are directed acyclic graphs (DAGs), where the nodes are random variables, and certain independence assumptions hold. The random variables can be thought of as states of affairs, and the variables have two possible values: true and false. However, this need not be the case. The arcs in a Bayesian network specify the independence assumptions that must hold between the random variables. These independence assumptions determine what probability information is required to specify the probability distribution among the random variables in the network. To specify the probability distribution of a Bayesian network, one must give the prior probabilities of all root nodes (nodes with no predecessors) and the conditional probabilities of all nonrooted nodes given all possible combinations of their direct predecessors. Cybernetics teaches us to take advantage of feedback (where the outcomes of actions are taken as inputs for further action) and how to adapt to its environment. The benefit of incorporating learning is that it improves different aspects of various AI techniques such as perception, optimized planning, making the right decisions based on past history experiences, improving voice for command and control, and building a world model. In decision theory, mathematical analysis shows that once the sampling distributions, loss function, and sample are specified, the only remaining basis for a choice among different admissible decisions lies in the prior probabilities. Therefore, the logical foundations of decision theory cannot be put in fully satisfactory form until the old problem of arbitrariness (sometimes called "subjectiveness") in assigning prior probabilities is resolved. The principle of maximum entropy represents one step in this direction.

Decision-Making under Uncertainty: In the process model, there are sources of uncertainty in all phases of the processes. For uncertain inputs, there are missing and noisy data. Similarly, there are uncertain outputs where abduction and induction are inherently uncertain. There is also uncertainty in knowledge. With a default reasoner, even in deductive fashion, there is uncertainty. There is also uncertainty in the model with uncertain knowledge due to incomplete enumeration of conditions or effects, and incomplete knowledge of causality in the domain. Uncertainty can also arise because of incompleteness and incorrectness in the agent's understanding of the properties of the environment. Multiple causes can lead to multiple effects. The model is susceptible to probabilistic and stochastic effects (Figure 3.13).

Probabilistic Reasoning, which summarizes uncertainty from various sources, only gives probabilistic results.

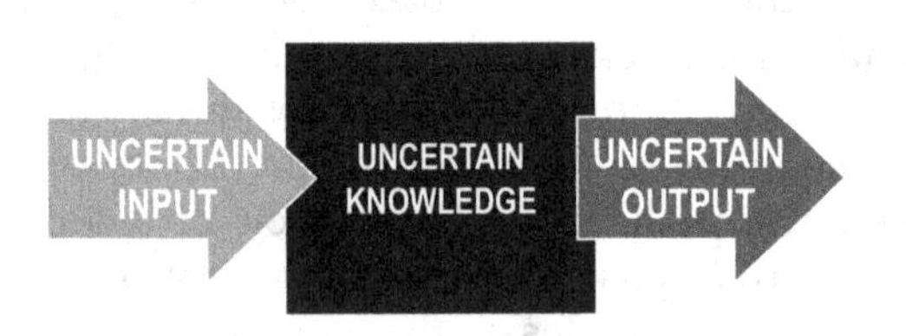

FIGURE 3.13
Sources of uncertainty processing diagram.

Monte Carlo Tree Search: Monte Carlo Tree Search (MCTS) is about approximate inference (propagation or pruning for exact inference) and is related to machine learning. MCTS is a method usually used in gaming and applicable to decision-making to predict the path (moves) that should be taken by the policy to reach the final winning solution (or make the right decision). However, before we discover the right solution, we first need to arrange the moves of the game's present state. When these moves are connected, they will look like a tree, hence the name tree search. The brute-force method considers every child's move/node requirements, but this approach requires a lot of computational power and is extremely slow. A faster approach can be achieved by using a policy, giving more importance to some nodes from others and allowing their children nodes to be searched first to reach the correct solution. MCTS uses an algorithm that calculates the best move out of a set of possible moves by using a four-step process (1. Selecting → 2. Expanding → 3. Simulating → 4. Updating), following the nodes in the tree to find the final solution. This method is repeated until it reaches the solution and learns the game's policy. *Selecting* is the process used to select a node on the tree that has the highest possibility of winning. After selecting the right node, *expanding* is used to increase the options further in the game by expanding the selected node and creating many children nodes. When *simulating*, we explore the move that will perform best and lead to the correct answer down the tree (Figure 3.14).

We use reinforcement learning (RL) to make random decisions in the game further down from every children node. Then, a reward is given to every children node by calculating how close the output of their

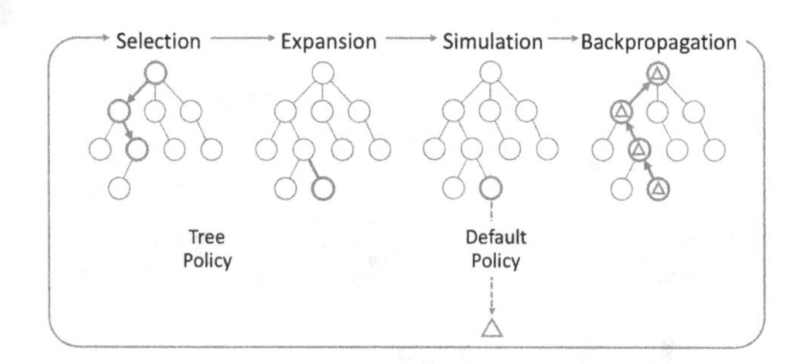

FIGURE 3.14
Monte Carlo tree search.

random decision was to the final output that we need to win the game. In RL, the agents make random decisions in their environment and learn to select the right one out of many to achieve their goal and play at a super-human level. Exploration within the MCTS algorithm uses both a policy network and a value network together. The value network assigns value/score to the state of the game by calculating an expected cumulative score for the current state. The simulation is done for every children node followed by their individual rewards due to the new nodes and their positive or negative scores in the environment. The total scores of their parent nodes must be updated by going back up the tree one by one. The new updated scores change the state of the tree and may also change the new future node of the selection process. After updating all the nodes, the loop again begins by *selecting* the best node in the tree→ expanding the selected node → using RL for *simulating* exploration → back-propagating the updated scores → then finally selecting a new node further down the tree that is actually the required final winning result.

Decision-making in light of all of these uncertainties requires *Rational Behavior.* For each possible action in rational behavior, identify the possible outcomes. Then compute the probability of each outcome. For each outcome, calculate the utility of each outcome. This is the probability-weighted (or expected) utility over possible outcomes for each action. The AI will select the action with the highest expected utility (refer to the previous principle of *Maximum Expected Utility*). Bayesian reasoning is based on probability theory and Bayesian inference. It takes advantage of independence in using probability theory and information. Bayesian inference can reason diagnostically (from evidence/effects to conclusions/causes or causally (from cause to effects)). Bayesian reasoning also requires a Bayesian network, a compact representation of probability distribution over a set of propositional random variables. It takes advantage of independence relationships.

Autonomy is a practical operational option as long as the robot understands its internal and external states, as well as having a healthy overall state. This implies that the system/robot has a health management component so that the controller has a full array of options available to the "brain". For example, if the rover loses one of its six wheels, what does the robot do? It might be possible to operate with only five wheels, but this might restrict its full repertoire of behaviors. The health management system will check the status of each of all of its components and subsystems, including having resources available to itself such as having enough power to complete its task. If there is a problem, the onboard diagnostic system can check for anomalies and make recoveries if possible. Again, if there are failures, the decision-maker can make other choices by pruning the decision tree to find a satisfactory solution. And its planning system can adjust based on the new constraint. Bottomline, a healthy robot is a "happy" robot.

Anatomy of an Autonomous Robot

Since industrial manipulators largely do not require much or any AI, so we are mostly interested in autonomous vehicles. Based on a deliberate or 3T (Three-Tier) autonomous robot architecture, robotics is the scientific area that studies the link between perception and action. Robots are devices to perform activities as humans. A *robot* is a machine, especially one that is programmable by a computer that is capable of carrying out a complex series of actions automatically or autonomously. A robot can be guided by an external control device, or its control may be embedded within its body. Robots may be constructed to evoke human form such as a humanoid, but most robots are task-performing machines, designed with an emphasis on stark functionality, rather than expressive aesthetics. The majority of robots are robot manipulators (otherwise known as an arm) and mobile robots that can make its way around its environment. Mobile robots are generally characterized by a form of mobility (wheels, tracks, or legs) and a method of navigation (teleoperated, waypoints, automatic, or autonomous). Perception is the ability to sense its environment to either react to it or to build a model (i.e., a world model). The next evolution is to combine both functionalities into a mobile manipulator, with the robot humanoid being the ultimate robot goal (Figure 3.15).

The best example of a robot that incorporates all capabilities of an autonomous robot is the warehouse robot example described by my life-long friend and colleague, Professor William J. Wolfe.[4] Consider the task as beginning with a command to go get a particular object and deliver it to some specified location. The highest levels of a functional hierarchy might break the task into three subtasks: *search* for the object; *navigate* in the environment; and *manipulate* the object. Each of these subtasks will require the robot to reason about the task and its environment in ways that are difficult to explicitly specify, and furthermore, there will be a need for the robot to communicate with the human operator and possibly with other robots (a typical scenario where the go-fetch robot might be employed). This is the typical warehouse problem. In physically searching, the robot asks where is the

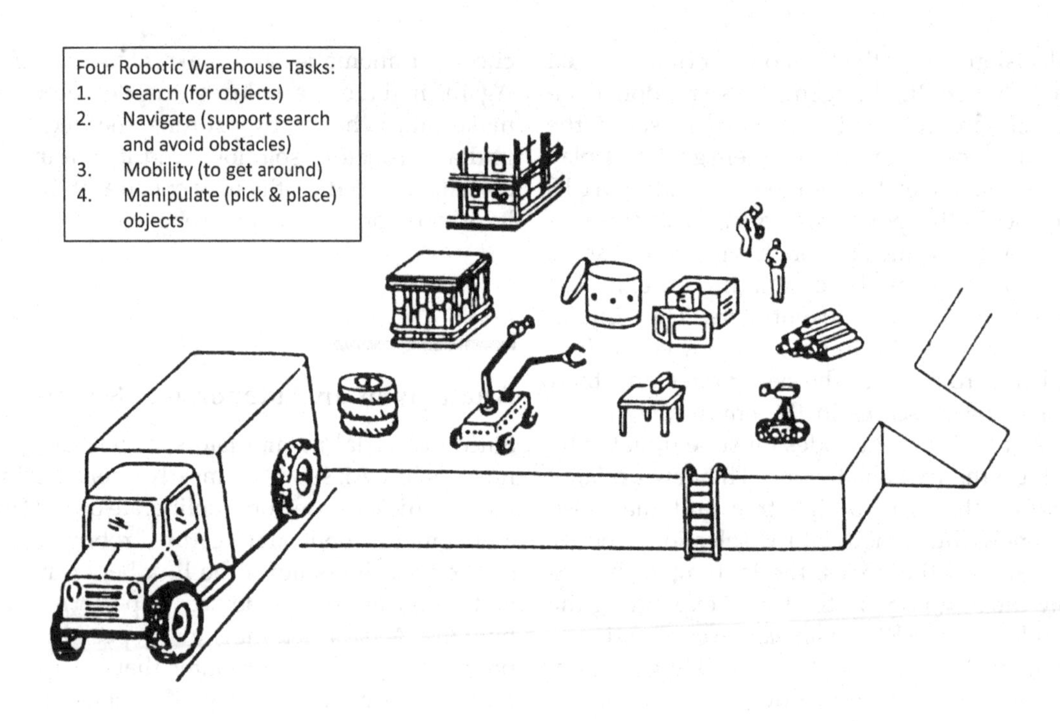

FIGURE 3.15
Wolfe's warehouse robot.

required object. To navigate, the robot needs to know how to get there. Once the robot finds the object, how does the robot manipulate the object through manipulation (Figure 3.16).

When you talk to roboticist, the focus typically goes to the topic of control. But why is robot control so hard? The reasons are many such as the complexity of high-level goals, operating a changing and dynamic environment, the robot having some dynamic constraints of its own to plan around, sensors that are noisy and uncertain, same for actuators, and handling unexpected events. Robots have to deal with multiple control problems, but at different levels. The hybrid architecture uses the deliberate paradigm of sensing the world, updating the world model and generating the plan, executing the planned action, and repeating the cycle. The world model contains everything the robot needs to do. The "global world" representation has to be up to date and has to be frequently updated to achieve a sufficient accuracy for the particular task. As stated previously, the plan is based on automated reasoning. For planning and to explore all the possible states of the world is to find a plan that is computationally expensive (an NP-complete problem). A general problem solver needs many facts about the world to search for a solution; searching for a solution in huge search space is quickly computationally intractable and this problem is related to the frame problem. Caution: planning can be very slow. This is essentially the AI search problem. As in the 3T architecture, there are different levels in the control problem: low-level or servo control level (continuous/analog problems with fast update), an intermediate level such as navigating to a destination or picking up an object (discrete or analog values with medium update loop), and high-level control (discrete problem with long time scales), for example, what is the plan for moving these boxes out of the room?

FIGURE 3.16
Mobile manipulator.

- Low-level addresses the real-time performance requirements such as sending commands to motors or processing sensory input data and avoiding states with disastrous conditions.

- Mid-level control takes the robot through a sequence of actions to achieve some simple task and must be able to deal with failures and unexpected events.
- High-level control can potentially use AI cognitive modeling architectures such as *SOAR* (Ref. Newell) or *ACT-R* (Ref. Anderson) to control robots.

ACT-R is a cognitive architecture that mainly develops and defines the basic and irreducible cognitive and perceptual operations that enable the human mind. It is basically a general system for modeling a wide range of higher-level cognitive processes. ACT-R is a cognitive architecture: a theory for simulating and understanding human cognition. Researchers working on ACT-R strive to understand how people organize knowledge and produce intelligent behavior. As the research continues, ACT-R evolves ever closer into a system that can perform the full range of human cognitive tasks, capturing in great detail the way we perceive, think about, and act on the world. SOAR is a cognitive architecture based on fixed computational building blocks that are necessary for general intelligent agents; agents that can perform a wide range of tasks and encode, use, and learn all types of knowledge to realize the full range of cognitive capabilities found in humans, such as decision-making, problem-solving, planning, and natural-language understanding. SOAR is something like a programming language having a fixed set of routines. Different kinds of tasks can be "programmed" in SOAR using these routines. Examples range from AI's favorite "toy" problems (such as the eight-puzzle) to "real-world" applications (such as configuring computer systems and robot control). SOAR solves each task given to it by creating and solving a hierarchy of subtasks. Each task (including the main one and each of the subtasks) is posed as the goal of finding a desired state in a "problem space" consisting of a set of operators that apply to a current state to produce a new state (Figure 3.17).

When we discuss a physical search within the warehouse, a robot or robots are capable of searching for a target or multiple targets by taking the place of human beings in a dangerous environment for secondary disaster avoidance. There are multiple scenarios with three modes, that is, single robot-single target, multiple robot-single target, and multiple robots – multiple targets. Accordingly, the natural scientific issues concerned with each mode, control techniques, as well as typical applications are analyzed for constructing the robot in hardware and software. Robot systems use feedback. Cybernetics is the marriage of control theory (feedback control), information science and biology in order to seek principles common to animals and machines, especially for control and communication. Situatedness is the coupling between organism and environment, or in our case, robot and environment. The feedback is the continuous monitoring of the robot sensors and reacting to their changes, resulting in self-regulation. Negative feedback acts to regulate the state/output of the system, while positive feedback acts to amplify the state/output of the system (Figure 3.18).

Model-based robot control, that is, the sensing, planning and control modules, all rely on explicit mathematical models of the robot system and its interactions with the environment. Models are used to predict the behavior of the robot; the difference between these predictions, on the one hand, and the sensor measurements on the other hand, provide information about how to adapt the models. Adaptation is necessary because, obviously, these models are always incomplete or inaccurate. In other words, they suffer from uncertainty, or incomplete information, and how Bayesian probability approach copes with this uncertainty.

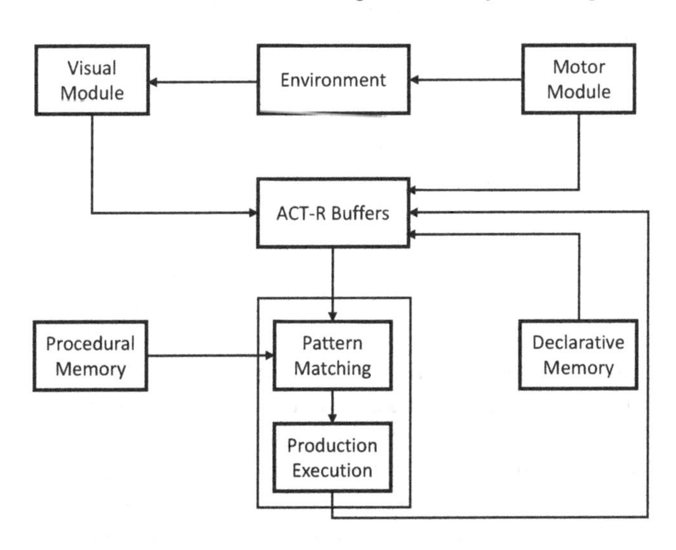

FIGURE 3.17
Anderson's ACT-R cognitive architecture.

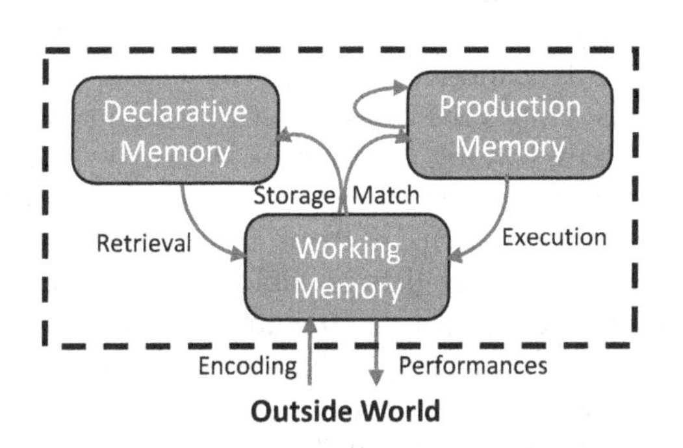

FIGURE 3.18
Memory model.

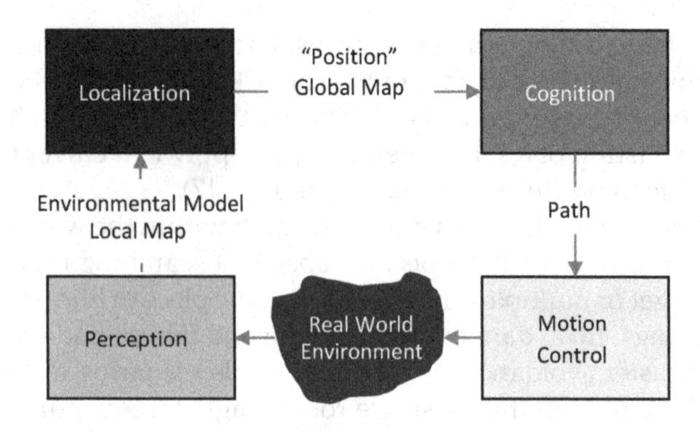

FIGURE 3.19
Robot navigation model diagram.

A robot is a system with a physical (or hardware) architecture and a functional (or software) architecture. In terms of hardware components, a robot has the following subsystems: locomotion, sensing, processing, and communication. Sensing collects information about the world. A sensor is an electrical/mechanical/chemical device that maps an environmental attribute to a quantitative measurement. Each sensor is based on a transduction principle, conversion of energy from one form to another. Without sensors, the robot can only accomplish fixed automation. Sensors are needed for unstructured environments and needed for changing worlds. Using visual and non-visual sensors, the robot can adapt to its environment (Figure 3.19).

What is data and what is unknown depends on the problem. For instance, the vision system mentioned above could be looking at the robot's arms. Then, the robot's end-effector position could be the unknown to be solved for by the vision system. Once vision has solved its problem, it could feed the robot's end-effector position as data for the robot controller to use in its own motion planning problem. Sensed data are invariably noisy, because sensors have inherent limitations of accuracy, precision, resolution, and repeatability. Consequently, the systems of equations to be solved are typically over-constrained: there are more equations than unknowns, and it is hoped that the errors that affect the coefficients of one equation are partially canceled by opposite errors in other equations. This is the basis of optimization problems: Rather than solving a minimal system exactly, an optimization problem tries to solve many equations simultaneously, each of them only approximately, but collectively as well as possible, according to some global criterion. Least squares are perhaps the most popular such criterion, and we will devote a good deal of attention to it.

Robot locomotion is needed for moving through its environment, robot sensing for measuring its environment (perception) and internally for state estimation, a processor for being the robot's brain, and communication for how the robot interfaces with its operator (if necessary). In terms of software components, an autonomous robot has the following modules: perception, sensor fusion, world modeling with data updates, logic and reasoning based on knowledge representation, mission planning, trajectory planning, control execution including fuzzy logic, decision-making with utility theory and planned goals, different search techniques, obstacle avoidance, rules and rule-based using an expert system or for reactive behaviors, and overall machine learning to improve performance. And if the human – robot interface cannot be declared early in the robot's development process, conversational AI and natural-language processing could have a more significant role. The biggest hurdle in developing the robot's "mind" will be integration and testing, especially when the system is not deterministic and the combinatorial explosion of the many test combinations makes the test seemingly impossible.

Search requires perception and object recognition. Through perception and sensor fusion, a model of the "real" world is captured in memory. A goal is given and a plan is generated, assuming the "real" world is not changing. Then, the plan is executed, one operation at a time. To aid search, the robot may need to move around or require autonomous navigation. During autonomous navigation, the robot must learn its environment through modeling (given or autonomously built), perceive and analyze the environment, find its position/situation within the environment (known as localization) and then move to the desired location in the environment with planning and execution. The three key questions in mobile robotics as summarized by Hugh Durrant-Whyte (University of Sydney) are:

- Where am I?
- Where am I going?
- How do I get there?

"Where am I" is referred to as the localization problem, "where am I going" is the navigation goal, and "how do I get there" is the trajectory planning problem. As part of the navigation problem, the robot needs to understand its physical environment, for example, it is structured or unstructured, what is known and what are the unknowns, and is the environment changing (static versus dynamic). Self-driving cars driving on roads, which are considered a structured environment, while driving off-road within woods would be unstructured.

In a planning-based controller, a model of the world is obtained, a goal is given by the operator, and a plan is generated, assuming the "real" world is not changing. The robot executes the plan, one operation at a time. Some of the problems with perception is that modeling and planning are computationally intensive. Second, the model of the "real" world must be at all times accurate, and this navigation paradigm is good for a static world only. But an unstructured world introduces uncertainties, the models are not accurate, and the real world is always changing. Bayes' theorem and Markov chain theory are used to treat the problems of uncertainty quantification and optimization (Figure 3.20).

Humans perform object manipulation in order to execute a specific task, and seldom is such an action started with no goal in mind. In contrast, traditional robotic grasping (first stage of object manipulation) seems to focus purely on getting hold of the object, neglecting the goal of the manipulation. Robot arms are used for manipulation and interacting with the real world. It is an electronically controlled mechanism, consisting of multiple segments, that performs handling tasks. They are also commonly referred to as robotic arms. Robot arms are available in multiple configurations such as a serial appendage (open chain) or a parallel mechanism (closed chain) such as a Stewart platform. The industrial arm is comprised of a robot manipulator, power supply, and controllers. The typical industrial arm has six-degrees-of-freedom. Similar in configuration to the human arm, there is a shoulder, elbow, and wrist, separated by structural arm segments. The robot wrist is used to position, the hand or tool at the manipulator's distal end in orientation (pitch, yaw, roll). The shoulder and elbow is used to position the distal end in free space via a coordinate frame such as Cartesian (x, y, z), cylindrical (r, θ, z), or spherical (r, θ, φ). The most common arm uses a series of six rotary joints for an articulated configuration. When understanding manipulators, the theory focuses on robot geometry and its accompanying kinematics (motion of objects without reference to the forces that cause the motion). To control the manipulator, the roboticist must be familiar with the physical system under consideration. This includes the inverse and forward kinematics of the arm, arm dynamics, and control (PD, PI, and PID) for either position control, rate control, or force control. An adaptive compliant controller is designed in task space and contains a nonlinear model-based term and a linear compensator action. This is obtained as a combination of its position error, a velocity error and the integral of the force error. The scheme is made adaptive with respect to the dynamic parameters of the model of the robot manipulator that ensures tracking of the unconstrained components of the desired end-effector trajectory, with regulation of the desired contact force along the constrained direction. In order to autonomously acquire a geometric 3-D object model, Near-Best-Views (NBVs) are planned based on the current models in each iteration. Thus, the geometric model generation is tightly coupled with the NBV planning (Figure 3.21).

The human operator should be able to control the manipulator process with only high-level commands. The robot manipulator determines what joint trajectories are needed to successfully achieve its goals. Task-oriented control is based on a hierarchical control structure in order to implement high-level commands over low-level commands. If we try to control a manipulator in Cartesian space, we run into difficulties since the inverse mapping from Cartesian space to joint space can sometimes

FIGURE 3.20
Stewart platform (delta) manipulator.

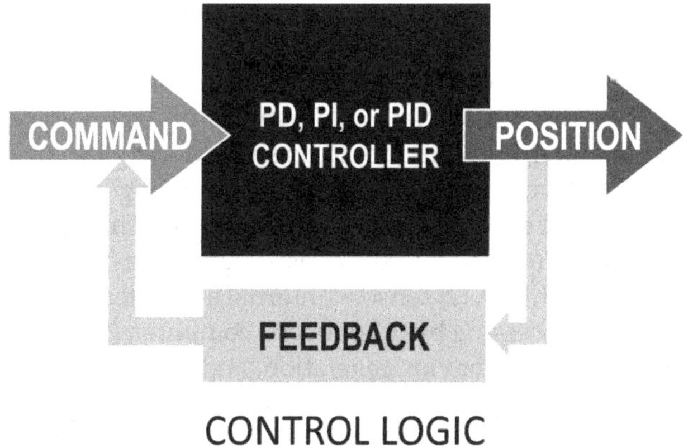

FIGURE 3.21
Control logic process diagram.

become a problem. These problem positions of the robot are referred to as singularities or degeneracies, and at a singularity, the mobility of a manipulator is reduced. Usually, the arbitrary motion of the manipulator in a Cartesian direction is lost. This is referred to as "Losing a DOF (Degree of Freedom)". There are boundary singularities during the full extension of the arm or internal (or joint-space) singularities caused by an alignment of the robots axes in space. In a joint-space singularity, an infinite number of inverse kinematic solutions may exist. For a joint-space singularity, small Cartesian motions may require infinite joint velocities, thus causing a problem. The robot controller where manipulators can orient in ways that unique solutions to matrix problems do not exist, and how to avoid them. The key to autonomous manipulation methodology lies in avoiding singular configurations with shape functions that keep the system some defined distance away from such problem states. The conservativeness of the distance can be adjusted parametrically (and simply), making this a very practical approach.

Autonomous robots are beginning to address real-world tasks in unstructured and dynamic environments. Autonomous mobile manipulation is a relatively young discipline within robotics. Today, most robots predominantly perform tasks based on mobility. However, the potential of augmenting autonomous mobility with dexterous manipulation skills is significant and has numerous important applications. In an autonomous mobile manipulator, the robot can navigate to a location prior to operating the manipulator(s). For any robot to act in uncertain environments, it is important to have a good model of the system in order to build a useful control system. The automation of a problem that is highly dimensional requires the automation of the robotic procedures to automatically generate the kinematic and dynamic equations of motion. The identification of relevant operating parameters can be performed in real time with an extended Kalman filter (EKF). An autonomous mobile manipulator robot can be organized with a hierarchical architecture such as with the NIST (National Institute of Technology and Standards) Standard Reference Model[5] to control low-level functions in real time along with a higher-level controller that is not real time. This problem is decomposed into six areas: (1) actuators (for both arm and vehicle), (2) sensors (internal and external), (3) sensory processing by the onboard computer, (4) world modeling, (5) behavior generation, and (6) value judgment in making decisions. Issues that appear repeatedly in autonomous operations fit into three target areas: vehicle navigation, vehicle positioning, and arm control systems.

Frame Problem: the frame problem is the challenge of representing the effects of action in logic without having to represent explicitly a large number of intuitively obvious non-effects. It describes an issue with using first-order logic (FOL) to express facts about a robot in the world. Representing the state of a robot with traditional FOL requires the use of many axioms that simply imply that things in the environment do not change arbitrarily. In a logical context, actions are typically specified by what they change, with the implicit assumption that everything else (the frame) remains unchanged.

The frame problem is the problem of commonsense reasoning:

- How to deal with the changing world?
- How to determine the relevant consequences of an event?

Dreyfus suggests that procedural knowledge (knowledge how) is a more fundamental knowledge underlying commonsense reasoning abilities and that the frame problem arises because classical cognitive science and AI have neglected the procedural kind of knowledge that underlies our skills.

Arm Control: Manipulation in combination with mobility thus requires algorithmic approaches that are versatile and general. The variability and complexity of unstructured environments also require algorithmic approaches that permit the robot to continuously improve its skills from its interactions with the environment. Most commercial manipulators are heavy, large, and power-hungry. Furthermore, the combination of mobility and manipulation poses new challenges for perception as well as the integration of skills and behaviors over a wide range of spatial and temporal scales. Generally, the workspace of a mechanism is a region of points that can be reached by a reference point on the manipulator extremity. The kinematic analysis of a robotic manipulator usually depends on the solution of sets of nonlinear equations. While geometric kinematics is concerned with solving the equation

FIGURE 3.22
Dual arm manipulator cooking.

of state for either given the state s or joint configuration, differential kinematics studies the relationship between the rate of change of pose and joint configuration of a robot performing a known motion. Systems of nonlinear polynomial equations generally have multiple solutions, in contrast to systems of linear equations. In a kinematic problem, the unknowns in such a system of equations represent information such as the joint angles or spatial displacements of a robot manipulator or platform. Thus, multiple solutions of the system represent different possible poses of the mechanism under the stated constraints. Bezout's Theorem states that the number of solutions of is equal to the total degree of the system, if the multiplicity of each solution is counted properly. The architecture of a robotic mechanism is concerned with the defining topology, the basic connectivity pattern of a robotic system. For both serial and parallel robots, many different architectures are possible (Figure 3.22).

The term dexterity refers to the ability of a manipulator to arbitrarily change its position and orientation or apply forces and torques in arbitrary directions. In other words, dexterity summarizes the exibility and accuracy with which the end-effector is placed in a given pose. The exibility is quantified by the dexterous workspace. The positioning accuracy of a manipulator depends on various factors, such as sensor errors, manufacturing errors or thermal errors. Manipulator singularities are particular poses of the end-effector in which either the forward or the inverse instantaneous kinematic problem becomes indeterminate, and thus, the properties effector is no longer controllable, and forces, imposed on several links, may become very large, which may cause a breakdown of the robot.

There is a tradition of mathematicians working with roboticists. Many problems in robotics, or in the disciplines that are core to what we call robotics, have attracted mathematicians to the field. As far back as the 19th century, algebraic geometers like Kemp and Tschebyshev were drawn to the beautiful mechanics of linkages (the predecessors of today's complex articulated manipulators) and geometers like Poincare were attracted by the dynamics of machines. Clearly, the development of the foundations of many classical and modern mathematical tools was spurred by technological advances in machines and mechanisms. This tradition has continued and there are several examples in the last two decades where some of the most fundamental and enduring results in robotics have come from mathematicians and their interactions with engineers. The classical general motion planning results in robotics were developed by Schwartz and Sharir, which in turn were improved on by Canny using methods from algebraic geometry. Similarly, Milgram and Trinkle have used results from modern algebraic topology to obtain an improved understanding of the configuration spaces of closed chain mechanisms, leading to improved algorithms for motion planning for such systems. Marsden, Brockett, and Sastry have used differential geometry and Lie theory to formalize the kinematics, dynamics, and control of spatial linkages. This work has also sensitized the engineers to such important mathematical ideas as frame invariance and invariance with respect to parameterization. Other areas where there are strong developing ties between mathematics and robotics include, for example, Bayesian statistics to develop algorithms for perception and learning, non-smooth optimization techniques for parameter and control of mechanical systems, the use of non-smooth analysis for developing simulation methods for mechanical systems with contact or impact, and partial differential equation formulations of image segmentation and pre-processing problems.

For perception, in order to combine different measurements in the same coordinate system, knowledge about the positions of the various sensors with respect to each other needs to be obtained. This requires coordinate transformations with matrix manipulation. In order to be able to calculate depth images, the relative poses between the cameras, as well as the intrinsic parameters of all involved cameras need to be known. Furthermore, to transform all depth measurements into a common coordinate system, the remaining static transforms have to be estimated using a calibration pattern. To carry out manipulation in industrial environments autonomously, a mobile manipulator requires knowledge about the objects to be manipulated. Here, geometric as well as appearance-based object model representations can be utilized. The different representations are needed as different recognition modules are used depending on the characteristics of the object such as size, shininess, and texture. In contrast to autonomous object modeling with an industrial robot and a laser striper, both the range images and the robot poses are significantly noisier, resulting in lower-quality object models. An appearance-based detection process takes advantage of the fact that many objects (or their parts) show similarities when slightly shifting distance, angle and/or lighting conditions. The main idea is to collect the data by taking many pictures under different lighting conditions and from varying viewpoints. In order to perform collision-free motion planning with the robot arm, a probabilistic voxel space representation is generated for each scene. Obtaining a complete 3-D map of the environment is very costly as a high-resolution model would be required for the complete site. The map-building ability of a robot is closely related to its sensing capacity. There are three steps in map building: feature extraction from raw sensor data, fusion of data from various sensor types, and automatic generation of an environment model with different degrees of abstraction.

The visual recognition of learned "object" models is a key problem, maybe "the" key problem, in computer vision. Solving it will require radical advances in object representation, image segmentation, and in our understanding of the recognition process. Probability theory and statistical inference form a promising foundation on which to build a mathematical framework for object recognition, but major conceptual and technical difficulties remain: for example, how can we define object models that are easy to learn and effectively support inference from pictures? How can we construct algorithms that will perform Bayesian inference in real time? How should we handle large numbers of objects and object classes?

To solve tasks autonomously, the robotic system needs a representation of the environment and itself to keep track of the current world state. The representation of this approach does not provide a detailed geometric description of the environment, but on a topological level, describes the relation between different kinds of world items and storing the relevant state of the world. The world model is based on a tree structure, which means that each item has to have a parent item. This structure leads to a natural representation of how manipulation is affecting the world. To solve a task such as a pick and place operation, more than one hundred functions from various modules have to be called. Most of these functions are not related to the task instance, but to the general task structure and the robot. A common problem for almost all autonomous manipulating robotic systems is to identify the objects to interact within the vast amount of sensor data from the cameras. The geometric planner is described, which can be utilized for collision-free motion planning with the robot arm based on the scene registration and by autonomously acquiring the scene model. Joint-space trajectory generation is in common usage in manipulation to provide smooth, continuous motion from one set of joint angles to another, for instance, for moving between two distinct Cartesian poses for which the inverse pose solution has yielded two distinct sets of joint angles. During a basic pick and place operation, this state is changed by robot actions or perceptions. Therefore, each operation that changes the world's state has to be tracked by the world model in order to keep the model coherent with the real world. However, the computational requirements to solve the dynamic equations of motion are complex. For example, the Newton-Euler Iterative method has the following computations, $126n - 99$ multiplication functions, and $106n - 92$ addition functions where n is the number of links. For a six degree-of-freedom manipulator with Q_1, Q_2, and Q_3 are quantized to 10 ranges each, the memory required from a look-up table is $(10\times10\times10)^6$ cells or 6×10^{18} words (or torque values) (Figure 3.23).

Vehicle Navigation: An autonomous vehicle usually has high-level goals provided to them by an operator. After receiving the goal(s), the robot must plan how to accomplish those goal(s). Autonomous vehicle needs include:

- Mission Planning: steps on how to accomplish the goal(s).
- Path Planning: how to reach a given location.

FIGURE 3.23
Intelligent AGVs.

- Sensor Interpretation: determining the environment given sensor input. Sensing is one of the most critical and underexplored aspects in autonomous operations.
- Obstacle avoidance and terrain sensing.
- Failure handling/recovery from failure (requiring a healthy vehicle).

The most common form of a ground vehicle can be broken down into four categories: (1) self-driving vehicles are automatic cars programmed to drive on roadways with marked lanes and possibly must contend with other cars; (2) all-terrain autonomous vehicles where automatic military vehicles programmed to drive off-road and must contend with different terrains with obstacles like rocks, hills, etc.; (3) all-terrain robots such as ATVs, but these can be smaller and so more maneuverable that may include robots that use tank treads instead of wheels; and (4) crawlers that use multiple legs instead of wheels/treads to maneuver. The control of autonomous ground robots involves a number of subtasks (Figure 3.24).

- Understanding and modeling of the mechanism including its kinematics, dynamics, and odometry
- Reliable control of the actuators including closed-loop control and fuzzy logic
- Generation of task-specific motions for path planning such as different trajectory planning techniques such as Voronoi diagrams, exact cell decomposition, approximate cell decomposition, potential field, freeway method, and visibility graphs
- Integration of sensors is a hard problem based on selection and interfacing of various types of sensors

FIGURE 3.24
QinetiQ talon robot.

- Coping with noise and uncertainty by filtering of sensor noise and actuator uncertainty
- Creation of flexible control policies when dealing with new situations.

Path planning for manipulator robots and, indeed, even for most mobile robots, is formally done in a representation called configuration space. Path planning addresses the task of computing a trajectory for the robot such that it reaches the desired goal without colliding with obstacles. Optimal paths are hard to compute in particular for robots that cannot move in arbitrary directions (i.e., nonholonomic robots). Shortest-distance paths can be dangerous since they always graze obstacles. For example, the paths for robot arms have to consider the entire robot (not only the end effector for an arm) but the vehicle's body.

A robot can dead-reckon until the errors build up and it needs to correct its true location. The key navigation problem is how does the vehicle/robot gets from point A to point B? Are there obstacles to avoid? Can obstacles move in the environment? Is the terrain going to present a problem? Are there other factors such as dealing with water current (autonomous sub), air current (autonomous aircraft), and blocked trails (indoor or outdoor ground robot)? This is the reason to have planning in order to reach its objective, in this case, Point B. Path planning is largely geometric and includes straight lines, following curves, and tracing walls (mostly indoors). There are additional issues to consider such as:

- How much of the path can be viewed ahead?
- Is the robot going to generate the entire path at once, or generate portions of it until it gets to the next point in the path, or just generate on the fly?
- If the robot gets stuck, can it backtrack?

Once a path is generated, the robot must follow that path, but the technique will differ based on the type of robot. For an indoor robot, path planning is often one of following the floor using a camera, find the lines that make up the intersection of floor and wall, and using these as boundaries to move down. Or for an autonomous self-driving car, path planning is similar but follows the road instead of the floor using a camera, finding the sides of the road and selecting a path down the middle of the lane. For an all-terrain vehicle, GPS must be used although this may not be 100% accurate or available (GPS-denied applications). What happens when an obstacle is detected by the sensors? It depends on the type of robot and the situation:

- For a mobile robot, it can stop, replan, and resume
- For an autonomous ground robot, it may slow down and change directions to avoid the obstacle (e.g., steer right or left) while making sure it does not drive off the road, noticing that it does not have to replan because it was in motion and the avoidance allowed it to go past the obstacle or it might stop, back up, replan, and resume
- An underwater vehicle or an air-based vehicle has an option to change depth/altitude.

For example, a fast self-driving automobile that detects an obstacle on the road needs to compute the obstacle's size (relates to sensor resolution), speed of the vehicle, which determines the vehicle look ahead distance, processing speed (size of sensor in terms of resolution+processing time+braking distance), and potential actions such as stopping before hitting the object, swerving around the object, or going over the object. While obstacle avoidance is a low-level process, it may impact higher-level processes (e.g., goals), so replanning may take place at higher levels.

As another example, the military has developed an autonomous architecture for both on-road and off-road navigation in outdoor environments. The NIST 4D/RCS (Robot Control System) architecture provides a reference model for military unmanned vehicles on how their software components should be identified, organized, and interacted with such that missions, especially those involving unknown or hostile environments, can be analyzed, decomposed, distributed, planned, and executed intelligently, effectively, efficiently and in coordination. To achieve this, the 4D/RCS reference model provides well-defined and highly coordinated sensory processing, knowledge management, cost and benefit analysis, and behavior generation functions, as well as the associated interfaces, that are based on proven scientific principles and are consistent with military hierarchical command structures. This 4D/RCS is a hierarchical deliberative architecture, that "plans up to the subsystem level to compute plans for an autonomous vehicle driving over rough or unknown terrain" (Figure 3.25).

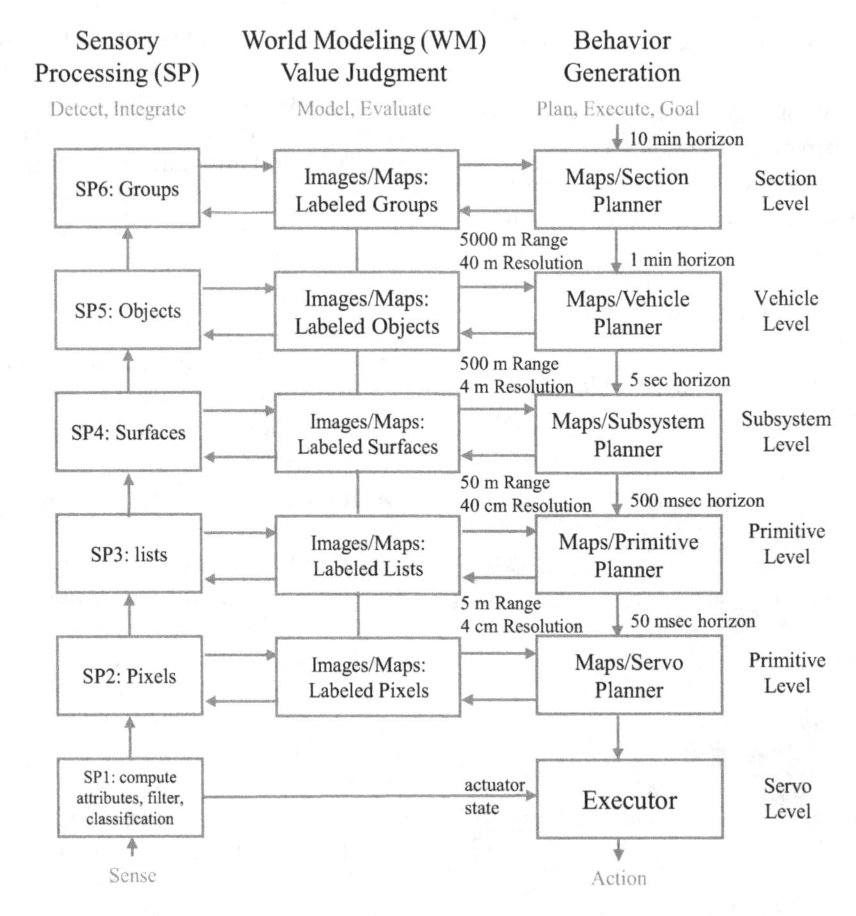

FIGURE 3.25
4D/RCS UGV architecture.

To achieve this, the 4D/RCS reference model provides well-defined and highly coordinated sensory processing, knowledge management, cost and benefit analysis, and behavior generation functions, as well as the associated interfaces, that are based on proven scientific principles and are consistent with military hierarchical command structures. This 4D/RCS is a hierarchical deliberative architecture, that "plans up to the subsystem level to compute plans for an autonomous vehicle" driving over rough or unknown terrain.

Vehicle Positioning: Robot positioning (or localization) is the process of determining where a mobile robot is located with respect to its environment. In a typical robot localization scenario, a map of the environment is available and the robot is equipped with sensors that observe the environment as well as monitor its own motion. Localization is one of the most fundamental competencies required by an autonomous robot as the knowledge of the robot's own location is an essential precursor to making decisions about future actions. The localization problem then becomes one of estimating the robot position and orientation within the map using information gathered from these sensors. Robot localization techniques need to be able to deal with noisy observations and generate not only an estimate of the robot location but also a measure of the uncertainty of the location estimate. Sensor and effector uncertainty is responsible for the difficulties of localization. Robot systems in intelligent environments must deal with sensor noise and uncertainty. For example, sensor readings are imprecise and unreliable, while actions can fail because actions have nondeterministic outcomes. The challenge for localization is noise and aliasing. If one could obtain an accurate GPS sensor to a mobile robot, then much of the localization problem would be obviated. The GPS would inform the robot of its exact position and orientation outdoors. However, this is only important if the robot has a map to use. The existing GPS network provides accuracy to within several meters, which is unacceptable for localizing human-scale mobile robots. Furthermore, GPS technologies cannot function indoors or in obstructed areas and are thus limited in their workspace.

A mobile robot equipped with sensors to monitor its own motion (e.g., wheel encoders and inertial sensors) can compute an estimate of its location relative to where it started if a mathematical model of the motion is available (Figure 3.26). This is known as odometry or dead reckoning. The errors present in the sensor measurements and the motion model makes robot location estimates obtained from dead reckoning more and more unreliable as the robot navigates in its environment. Errors in dead reckoning estimates can be corrected when the robot can observe its environment using external sensors and is able to correlate the information gathered by these sensors with the information contained in a map. The formulation of the robot localization problem depends on the type of map available as well as on the characteristics of the sensors used to observe its environment (could be part of the perception system). In one possible formulation, the map contains locations of some prominent landmarks or features present in the environment and the robot is able to measure the range and/or bearing to these features relative to the robot. The robot would triangulate on known landmarks to determine its location. Alternatively, the map could be in the form of an occupancy grid that provides the occupied and free regions of an environment and the sensors on board the robot measure the distance to the nearest occupied region in a given direction. Occupancy grid maps provide a discretized representation of an environment where each of the grid cells is classified into two categories: occupied or free. Consider the scenario where a sensor on the robot can determine the distance to the nearest occupied grid cell along a given direction. As the information from sensors is usually corrupted by noise, it is necessary to estimate not only the robot's location but also the measure of the uncertainty associated with the location estimate. Knowledge of the reliability of the location estimate plays an important role in the decision-making processes used in mobile robots as catastrophic consequences may follow if decisions

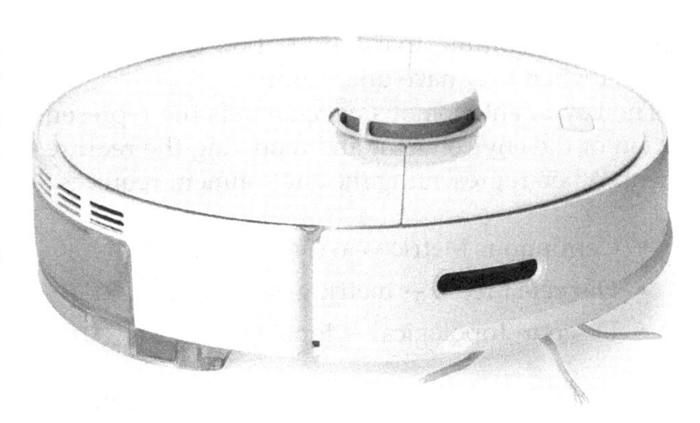

FIGURE 3.26
Mobile vacuum robot requires navigation.

are made assuming that the location estimates are perfect when they have uncertainty.

The key to autonomous navigation is the representation of the environment and modeling the requirement. When representing the environment requires:

- Continuous Metric → x, y, q
- Discrete Metric → metric grid
- Discrete Topological → topological grid

Similar to modeling the environment at multiple levels of abstraction. This starts with raw sensor data, for example, laser range data and grayscale images. This is followed by low-level features such as lines from geometric features. These features have a medium volume of data with average distinctiveness but after filtering out the useful information, there are still ambiguities in the model. Last is the high-level features including doors, a car, and the Eiffel Tower. High-level features are a low-level volume of data with high distinctiveness. At this level, useful information is filtered with few or no ambiguities, but having not enough information.

This robot may need to identify its absolute position, but its relative position with respect to key objects is equally important. Its localization task can include identifying these key objects using its sensor array and then computing the relative position between robot and object. The robot may need to acquire or build an environmental model, a map, that aids it in planning a path to the goal. Deriving a model for the whole robot's motion is a bottom-up process. Each individual wheel contributes to the robot's motion and, at the same time, imposes constraints on robot motion. Wheels are tied together based on robot chassis geometry, and therefore their constraints combine to form constraints on the overall motion of the robot chassis. But, the forces and constraints of each wheel must be expressed with respect to a clear and consistent reference frame. Once again, localization means more than simply determining an absolute pose in space; it means building a map, then identifying the robot's position relative to that map. The fundamental issue that differentiates various map-based localization systems is the issue of representation. There are two specific concepts that the robot must represent, and each has its own unique possible solutions. The robot must have a representation (a model) of the environment, or a map. What aspects of the environment are contained in this map? At what level of fidelity does the map represent the environment? These are the design questions for map representation. Clearly, the robot's sensors and effectors play an integral role in all the above forms of localization. It is because of the inaccuracy and incompleteness of these sensors and effectors that localization poses difficult challenges. Bayesian filtering is a powerful technique that could be applied to obtain an estimate of the robot's location and its associated uncertainty. Both EKF and particle filter techniques provide tractable approximations to Bayesian filtering.

Bayes' Theorem is a simple mathematical formula used for calculating conditional probabilities. Bayes' theorem describes the probability of an event, based on prior knowledge of conditions that might be related to the event. Bayes Theorem is given by:

$$P(A|B) = \frac{P(B|A) \cdot P(A)}{P(B)} \tag{3.7}$$

with A,B = events

$P(A|B)$ = probability of A given B is true
$P(B|A)$ = probability of B given A is true
$P(A)$, $P(B)$ = the independent probabilities of A and B

The process of Bayesian data analysis can be idealized by dividing it into the following three steps:

1. Setting up a *full probability model*: a joint probability distribution for all observable and unobservable quantities in a problem. The model should be consistent with knowledge about the underlying scientific problem and the data collection process.
2. Conditioning on observed data: calculating and interpreting the appropriate *posterior distribution* – the conditional probability distribution of the unobserved quantities of ultimate interest, given the observed data.
3. Evaluating the fit of the model and the implications of the resulting posterior distribution: how well does the model fit the data, are the substantive conclusions reasonable, and how sensitive are the results to the modeling assumptions in step 1? In response, one can alter or expand the model and repeat the three steps.

In that sense, a better term for the subject would be *full probability modeling* rather than *Bayesian statistics*. The basic idea is to create a full model for all the relevant quantities, then do the relevant conditional probability calculations (often using Bayes' rule) to quantify our uncertainty about the unknown quantities of interest, given the observed data. When using data, most people agree that your insights and analysis are only as good as the data you are using. Essentially, garbage data in is garbage analysis out. Data cleaning, also referred to as data cleansing and data scrubbing, is one of the most important steps for your organization if you want to create a culture around quality data decision-making. Data cleansing or data cleaning is the process of detecting and correcting corrupt or inaccurate records from a record set, table, or database and refers to identifying incomplete, incorrect, inaccurate or irrelevant parts of the data and then replacing, modifying, or deleting the dirty or coarse data. There is no one absolute way to prescribe the exact steps in the data cleaning process because the processes will vary from dataset to dataset. But, it is crucial to establish a template for your data cleaning process so you know you are doing it the right way every time. Data cleaning is the process that removes data that does not belong in your dataset. Data transformation is the process of converting data from one format or structure into another. Transformation processes can also be referred to as data wrangling, or data munging, transforming and mapping data from one "raw" data form into another format for warehousing and analyzing.

The processes of cleaning that data are:

Step 1: Remove duplicate or irrelevant observations

Step 2: Fix structural errors

Step 3: Filter unwanted outliers

Step 4: Handle missing data

Step 5: Validate and QA

Cognition/Reasoning is the ability of the robot to decide *what actions are required* to achieve a certain goal in a *given situation (a belief state).* Robot decisions can range from *what path to take* to what information on the environment to use. In comparison, today's *industrial robots* can operate *without any cognition* (reasoning) because their environment is static and very structured. In mobile robotics, *cognition and reasoning is primarily of geometric nature,* such as *picking safe path or determining where to go next.* There is a large amount of literature on cases in which *complete information about the current situation and the environment exists* (e.g., traveling salesman problem). However, in mobile robotics the *knowledge* of about the environment and situation is usually *only partially known and is uncertain.* As a result, autonomous operations are typically required, and sometimes more explicitly probabilistic robotics. This makes the task much more difficult and requires *multiple tasks running in parallel,* some for *planning* (globally) to achieve a goal and some to guarantee "survival of the robot" such as utility-based. The resultant is behaviors outputted by the cognition/reasoning system. Robot control can usually be *decomposed* into various *behaviors* or *functions,* for example, wall following, localization, path generation or obstacle avoidance. For the most part, AI-enabled robots are concerned with *path planning* and *navigation,* except the low-level motion control and localization are still required. Mobile robot navigation can generally distinguish between (*global*) *path planning* and (*local*) obstacle avoidance (Figure 3.27).

This introduces the *kidnapped robot problem,* which refers to a situation where an autonomous robot in operation is carried to an arbitrary location. The kidnapped robot problem creates significant issues with the robot's localization system, and only a subset of localization algorithms can successfully deal with the uncertainty created; it is commonly used to test a robot's ability to recover from catastrophic localization failures. By not having a map of the situation, the robot must explore while at the same time mapping its environment. Starting from an arbitrary initial point, a mobile robot should be able to autonomously explore the environment with its onboard sensors, gain knowledge about its environment, interpret the

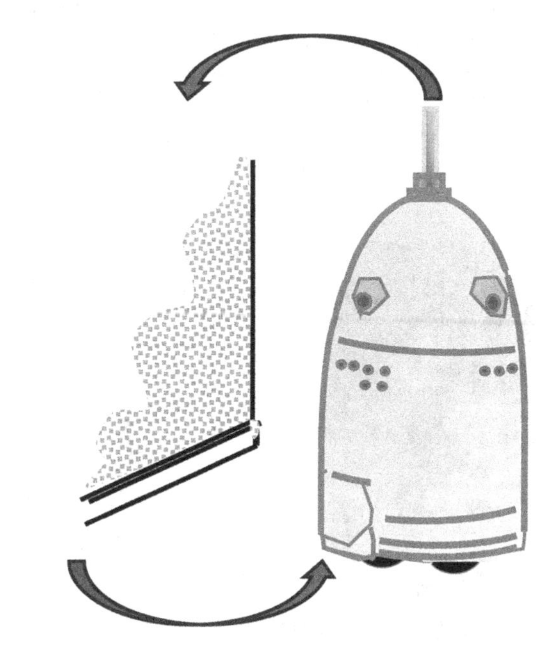

FIGURE 3.27
SLAM problem.

scene, build an appropriate map and localize itself relative to this map. When the robot arrives at a situation it recognizes, the robot can reconcile its environment or map and can now be used for future use for navigation. The solution is to incorporate simultaneous localization and mapping (*SLAM*). SLAM is a chicken-or-egg problem: A map is needed for localizing a robot or a pose estimate is needed to build a map. Thus, SLAM is (regarded as) a hard problem in robotics. The SLAM problem of environment mapping uses sensors carried by a mobile robot requires the correct localization of the sensors, and thus of the robot. On the other hand, a map is required for the correct localization of the robot. The original SLAM used an EKF approach to simultaneously estimate the location of the robot (carrying the sensors) and landmarks used to characterize the environment. The basic Kalman filter method allows multiple measurements to be incorporated optimally into a single estimate of state. The EKF is used to estimate the state (position) of the robot from odometry data and landmark observations. The EKF is usually described in terms of state estimation alone (the robot is given a perfect map). That is, it does not have the map update that is needed when using EKF for SLAM. In SLAM vs. a state estimation EKF approach, especially when the matrices are changed and can be hard to figure out in order to implement, is almost never described anywhere (Figure 3.28).

So why is robotics hard? Robotics is hard because sensors are limited and crude, while effectors are also limited and crude. Also contributing to the problem is the state (internal (robot) and external (environment)), but mostly external, which can be partially observable. This drives the need for autonomy. The environment is dynamic (constantly changing) and contains information that the robot has not taken advantage of that could be potentially useful. From a hardware and software perspective, robot hardware is hard because its physical and mechanics vary, the electrical hardware architecture changes, and each hardware component's capabilities vary from one to another. Robotic software including AI is hard because it is large and complex, has lots of diverse functions, integrates many scientific and engineering disciplines, requires real-time runtime models, has limited hardware resources and efficiencies, and has to talk to hardware (requiring integration). And finally, robotics is hard due to uncertainties in its inputs, outputs, and knowledge.

FIGURE 3.28
Mobile robot kiosk.

To understand a simple mission, a seemingly simple tasks can be difficult for a robot. For example, if I have to "Go to the Kitchen", here is a three-step process. The first step is to formulate the problem. We need a common understanding of "What is a kitchen?" and what other attributes can we use such as "If it is a room with food and appliances to cook the food..." and "What is a room?" Next, the robot must understand its environment. The things the robot needs to know are: Does the robot have a map? or How does the robot read the map? Or how does the robot build a map? Finally, the robot must execute the Task. The robot needs to know "Where is the robot now?" and "How does the robot plan a route to the kitchen?" What we take away is that there are a lot of steps for a robot to be able to perform a task. The earlier example only shows a few of the steps for a robot to perform the task. However, a robot does not have the same understanding and experience that a human does.

A cluttered and unstructured environment is known as complex. A robot is complex when it is composed of many parts that interconnect in intricate ways. A robot presents dynamic complexity when cause and effect are subtle, over time. A robot is complex when it is composed of a group of related units (subsystems), for which the degree and nature of the relationships are imperfectly known. A complex system has a set of different elements so connected or related as to perform a unique function. To be clear, the environment can be complex but so can the robot itself be complex. Complexity theory and chaos theory both attempt to reconcile the unpredictability of nonlinear dynamic systems such as a robot with a sense of underlying order and structure not performable by the elements alone. In summary, the problems encountered in robotics and vision are optimization problems. A fundamental distinction between different classes of problems reflects the complexity of the unknowns. In the simplest case, unknowns are scalars. When there is more than one scalar, the unknown is a vector of numbers, typically either real or complex.

When we discuss "complexity" in robots, we typically refer to the number of parts and the number of Software Lines of Code (SLOC). Other characteristics of complexity include the number of unknowns, number of interfaces, the issues with time, and issues with space (often described in processing speed and size of usable memory). In academia, researchers have tried to quantify complexity based on the number of interfaces and on the number of uncertainties in the system, enhanced by continuing work at the Santa Fe Institute. Herb Simon wrote the foundational article 'The Architecture of Complexity' in which the central theme is that complexity frequently takes the form of hierarchy and those hierarchic systems have some common properties that are independent of their specific content.

Finally, what do robots need? Robots need to interpret sensors for objects of interest, the terrain, and solving the symbol-ground problem. In cognitive science and semantics, the symbol-grounding problem is concerned with how it is that words get their meanings and, hence, is closely related to the problem of what meaning itself really is. Situational awareness is needed to get a grasp of the big picture. This is typically a model that we can run backward to infer a situation or progress forward to predict what will happen next. If we follow the Sheridan, Coiffert, Vertut, or Verplank models of something less than fully autonomous, the human – machine interaction requires development. Other needs include an "open world" where the mechanics of using a virtual world that the robot can explore and approach objectives freely, as opposed to a world that is more linear and structured, but not representative of the real world. Refer to the virtual twin. And when we lose our GPS, how does the robot operate in sparse areas? It has become apparent that the autonomous robot must be able to handle uncertainty, and learning would be essential to adapt to new situations and improve performance. There is a need for modeling of the robot and modeling of the robot's environment. As you can see, sensory data processing is challenging and can be computationally intensive. Also, modeling is computationally intensive and running AI programs are maybe more computationally intensive. Algorithms for search and those needed for controls will still need some processing. Why does this all matter? Because it means our robot needs a brain to do all of this processing. However, processing takes time, and potentially slows the robot down. To accommodate this need, we need better processing. Computers, in general, are improving per Moore's Law and are improving exponentially. It is postulated that AI is both a solution and a science.

Robots need AI for several reasons. First of all, the robot requires sensor interpretation where the robot can understand its environment (terrain for ground robots) in order to distinguish between a rock and a bush. This relates back to the "symbol-grounding" problem. This general understanding of the robot's environment enables situational awareness to be aware of the big picture. When the robot needs to interface with its human operator, AI is required to facilitate this interaction. In an "open world", the robot could encounter multiple faults in the system, requiring diagnostics and a plan for recovery. There is a notion that an autonomous robot requires a "healthy" robot, otherwise, contingencies are needed to accommodate any deficiencies within the robot. As an example, if GPS is lost, localization requires AI to determine its location (triangulation to natural or artificial landmarks) in sparse areas. And in general, a robot requires artificial intelligence to handle uncertainties and for learning in order to improve performance.

But what has happened in robotics to facilitate the integration of AI with a robotic system (its body)? By the 1960s, the research community had simple vision systems, simple theorem provers (using resolution), and simple path-planning methods. The next idea was to put them all together in a robot, the Shakey project. In 1970, the Shakey robot would reason about its block world at Stanford Research Institute in California. Shakey was remote-controlled by a large computer, and it hosted a clever reasoning program that fed very selective spatial data, derived from weak edge-based processing of camera and laser range measurements. Per Hans Moravec, on a very good day the robot could formulate and execute, over a period of hours, plans involving moving from place to place and pushing blocks to achieve a goal. Shakey looked promising, but it only worked in a very restricted environment. However, could this robot be extended to the real and natural worlds? So, after 50 years, the Shakey approach still did not extend to (1) visual modeling was too hard and slow, (2) nonlinear planning is intractable (NP-complete), and (3) feedback through the world model is cumbersome. As a result, researchers began to wonder if those early ideas were correct. Thus, where are we today?

Robots have evolved from early unmanned vehicles that were remote-controlled and later teleoperated based on the strength of the communication link

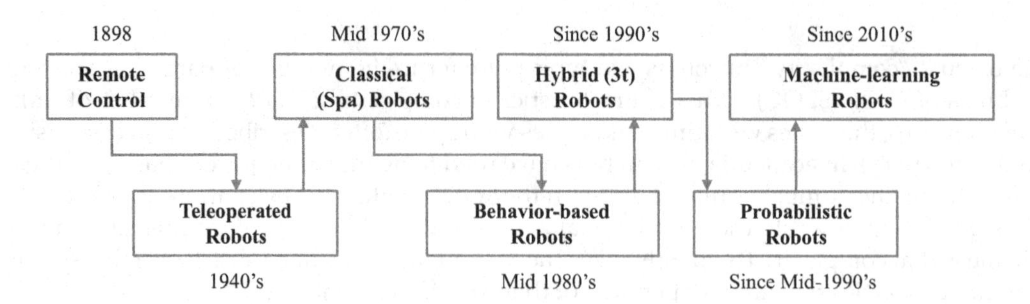

FIGURE 3.29
Mobile robot trends over time.

between the vehicle and operator. Later, with GPS, unmanned aerial vehicles could fly via waypoints. But starting with the SRI Shakey, robots were evolving into autonomous systems. Robot communication systems can be a weak point based on bad weather, time latencies, and potential jamming. Using the SPA paradigm, first discovered by Kenneth Craik, was implemented on Shakey and other robots until the mid-1980s. Rodney Brooks, Max Donath,[6] and Ron Arkin[7] looked to a change in the trend by implementing a behaviorist approach that relied on no modeling, no planning, and a series of reactive behaviors. Behaviors might be needed to be arbitrated to prioritize the various options. This strategy has led to the popularity of the Roomba vacuum. But without a model and planning, the behavior-based autonomous robot would not guarantee the robot from reaching its objective or goal. This led the research community to seek a hybrid approach, a combination of reactive behavior and planned behaviors in a 3T architecture. Robot performance progressed but to handle uncertainty in the sensor, actuator, or model, probabilistic techniques using Bayesian inference and Markov chains were implemented. The success of a probabilistic approach was emphasized with Sebastian Thrun and his Stanford University's Stanford Racing Team when winning the 2005 DARPA Grand Challenge. In parallel, processing capability and speed (measured in MIPS/seconds), as identified by Gordon Moore, would increase every couple of years and the consumer will pay less for them. As an illustration, the iPhone 4 had 512 MB of RAM as compared to the original space shuttle, which had 1 MB of RAM. Together with the introduction of the graphics processing unit (GPU), a specialized electronic circuit designed to rapidly manipulate and alter memory to accelerate the creation of images in a frame buffer, machine learning has sparked a whole explosion in AI science. A GPU is a processor but optimized for vectorized numerical code. The GPU's highly parallel structure makes them more efficient than general-purpose central processing units (CPUs) for machine learning algorithms that process large blocks of data in parallel.

Today, training a model in deep learning requires a large dataset, hence the need for a large computational operations in terms of memory. To compute the data efficiently, a GPU is an optimum choice. The larger the computations, the more the advantage of a GPU over a CPU. The evolution of robotics tends to follow the development of artificial intelligence science. Figure 3.29 depicts the evolution of robots from remote control to robots that learn. So where are we going with the technology and what is next?

Notes

1. Professor Brooks made a presentation at the First SPIE Mobile Robots conference in Boston, MA, where I chaired a session.
2. Professor Khatib was my editor on the Springer Handbook of Robotics, Second Edition.
3. Dr. Nicola Muscettola left NASA Ames Research Center, worked with me at Lockheed Martin, and eventually moved on to work at Google.
4. Professor Wolfe has been a close colleague and life-long friend since working together at Martin Marietta in the early 1980s.
5. Developed at the National Bureau of Standards for the NASA Flight Telerobotic Servicer by Dr. Jim Albus.
6. Prof. Donath at the University of Minnesota is an old colleague during the SPIE Mobile Robot conferences.
7. Prof. Arkin at Georgia Institute of Technology is another colleague during the SPIE Mobile Robot conferences.

4

Artificial Intelligence

Many older disciplines have contributed to the foundation for artificial intelligence (AI). For example:

- Philosophy: logic, philosophy of mind, philosophy of science, philosophy of mathematics
- Mathematics: logic, probability theory, theory of computability
- Psychology: behaviorism, cognitive psychology
- Computer Science & Engineering: hardware, algorithms, computational complexity theory
- Linguistics: theory of grammar, syntax, semantics

But AI is not just about robots. It represents the notion of discovery, the ability to infer, and the ability to reason. One of the primary goals of the field of AI is to solve complex tasks from unprocessed, high-dimensional, sensory input. It is also about understanding the nature of intelligent thought and action using computers as experimental devices. No one can refute a computer's ability to process logic, but to many it is unknown if a machine can *think*. And what is intelligence? Humans have natural intelligence which possesses the cognitive abilities to learn, form concepts, understand, apply logic, and reason, including the capacities to recognize patterns, comprehend ideas, plan, problem solve, make decisions, retain information, and use language to communicate by using Sensor Organs and Brain. Robots are mechanical devices that can function autonomously: robot body is as built or constructed, while functioning autonomously means the robot can sense, act, and maybe even reason; but doesn't just do the same thing over and over like automation which follows rules. Stuart Russell and Peter Norvig discussed the notion of a robot as an agent. An agent is anything that can be viewed as perceiving its environment through sensors and acting upon that environment through actuators. Human agents use eyes, ears, and other organs for sensors, hands, legs, mouth, and other body parts for actuators. Robotic agents use cameras and infrared range finders for sensors, and various motors for actuators. Formalization of a robot agent AI as an agent includes having sensors for perception of its environment and actuators to provide action on its environment. Similarly for AI to be an agent, it will use sensors and actuators to interface with its environment and maintain a model of the robot and the environment. An intelligent agent in the world will input abilities, prior knowledge, past experiences, goals/values, observations, and output actions on its environment (Figure 4.1). Cognitive Psychology suggests that Strong Artificial Intelligence should include the following capabilities (in whole or in combinations):

- having perception, both locally and globally
- learning (new functions or improving existing functions)
- understanding languages
- able to classify and identify objects

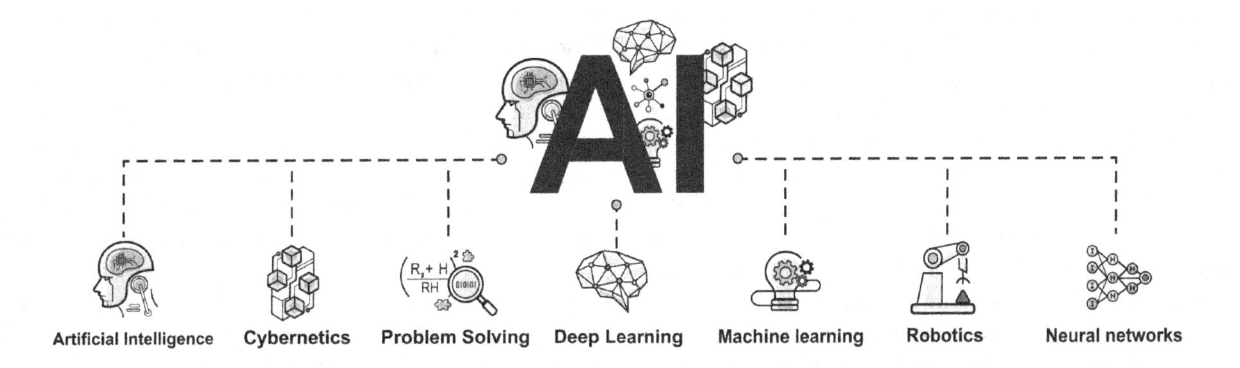

FIGURE 4.1
AI taxonomy figure.

DOI: 10.1201/9781032673134-4

- infer (what will happen next)
- reason inductively
- in forming concepts and developing new solutions
- understand situations and sense unusual events
- recognizing patterns including anomalies
- applying logic to improve understanding
- use natural language to communicate with a human
- solve problems
- make decisions
- able to plan and execute plans
- able to command actuators/action (for robotics)
- monitor health and diagnose faults.

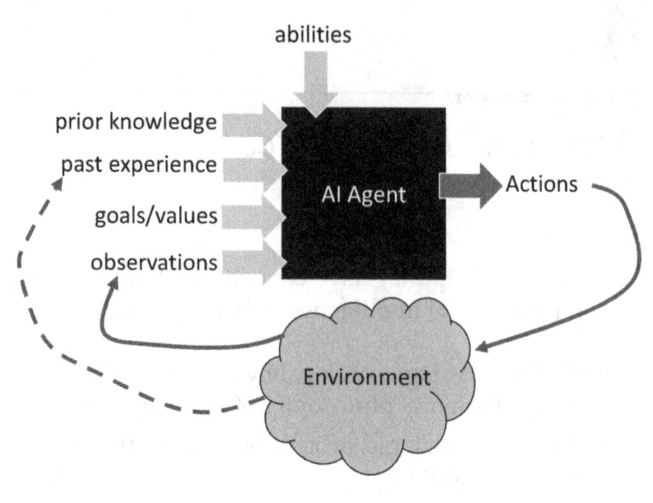

FIGURE 4.2
AI/robot agent.

Hubert Dreyfus's arguments have led him to see digital computers as limited not so much by being mindless, as by having no body. But he limits his argument to the technical question of what computers can and cannot do. He would argue that if computers could imitate man in every respect – which in fact they cannot – even then it would be appropriate, nay, urgent, to examine the computer in the light of man's perennial need to find his place in the world. Since the Greeks invented logic and geometry, the idea that all reasoning might be reduced to some kind of calculation so that all arguments could be settled once and for all has fascinated most of the Western tradition's rigorous thinkers. Socrates was the first to give voice to this vision. Like Hobbes, Boole supposed that reasoning was calculating, and he set out to "investigate the fundamental laws of those operations of the mind by which reasoning is performed, to give expression to them in the symbolic language of a Calculus . . ." Babbage designed what he called an "Analytic Engine" which, though never built, was to function exactly like a modern digital computer, using punched cards, combining logical and arithmetic operations, and making logical decisions along the way based upon the results of its previous computations. An important feature of Babbage's machine was that it was digital. There are two fundamental types of computing machines: analog and digital. Analog computers do not compute in the strict sense of the word. They operate by measuring the magnitude of physical quantities. Using physical quantities, such as voltage, duration, angle of rotation of a disk, and so forth, proportional to the quantity to be manipulated, they combine these quantities in a physical way and measure the result (Figure 4.2).

A slide rule is a typical analog computer. A digital computer as the word digit, Latin for "finger", implies represents all quantities by discrete states, for example, relays that are open or closed, a dial that can assume any one of ten positions, and so on, and then literally counts in order to get its result. Thus, whereas analog computers operate with continuous quantities, all digital computers are discrete-state machines. When, in 1944, Howard H. Aiken actually built the first practical digital computer, it was electromechanical using about 3,000 telephone relays. These were still slow, however, and it was only with the next generation of computers using vacuum tubes that the modern electronic computer was ready. Since a digital computer operates with abstract symbols which can stand for anything, and logical operations which can relate anything to anything, any digital computer (unlike an analog computer) is a universal machine. First, as Turing puts it, it can simulate any other digital computer. This special property of digital computers, that they can mimic any discrete-state machine, is described by saying that they are universal machines. In 1950, Turing wrote an influential article, "Computing Machinery and Intelligence", in which he points out that "the present interest in 'thinking machines' has been aroused by a particular kind of machine, usually called an 'electronic computer' or a 'digital computer'". A technique was still needed for finding the rules which thinkers from Plato to Turing assumed must exist as a technique for converting any practical activity such as playing chess or learning a language into the set of instructions Leibniz called a theory. Immediately, as if following Turing's hints, work got under way on chess and language.

A "Model" is a challenging concept, in part because it is both a noun and a verb, an object, and a practice, and it takes many prepositional forms – X models Y, X is a model of Y, X is a model for Y, Y is modeled on X, Y is modeled after X, and even Y models for Z. Thus, we could say that the brain was a model for the structure of the computer (or the computer was modeled on the brain) in the sense that the designers of the early computers, such as John von Neumann, treated the biological brain like an artist's model, and crafted the computer in its image. The key to understanding the development of von Neumann's ideas on the relationship between the brain and the computer is to keep clear that the computer is a representational system of a specific type for von Neumann. In the case of his automata theory, the computer is fundamentally an automatic system for representing and manipulating numbers. Numbers are not the same thing as quantities or numerals; they are abstract mathematical entities. Quantities and numerals are concrete ways of representing numbers for practical purposes. Thus, to automate mathematics it is necessary to develop a physical system that can represent numbers, and the choice between quantities and numerals is an open question before the technological possibilities are considered.

It is worthwhile to consider how the view of working models and simulations would apply to the various stated positions. It is clear that Neumann's design of the first computer memory drew heavily upon his review of research in neurophysiology, and quite explicitly upon the work of McCulloch and Pitts and their conception of logical neural networks. This much is clear from von Neumann's EDVAC design proposal. The story becomes more complex when we consider Neumann's 1948 theory of automata, his 1946 letter to Norbert Wiener about simulating the brain, and his later reflections on the relationship between the brain and the computer for his posthumously published Silliman lectures from 1958. We review some of von Neumann's thoughts on the issues involved in thinking about the computer as being modeled after the brain, and his approach to simulating the brain with a computer. The key to understanding the development of von Neumann's ideas on the relationship between the brain and the computer is to keep clear that the computer is a representational system of a specific type for von Neumann. In the case of his automata theory, the computer is fundamentally an automatic system for representing and manipulating numbers. Numbers are not the same thing as quantities or numerals; they are abstract mathematical entities. Quantities and numerals are concrete ways of representing numbers for practical purposes. Thus, to automate mathematics it is necessary to develop a physical system that can represent numbers, and the choice between quantities and numerals is an open question before the technological possibilities are considered.

On September 20, 1948, von Neumann presented a theory of automata at the Hixon Symposium on Cerebral Mechanisms in Behavior. In his presentation, "The General and Logical Theory of Automata", he spelled out the need he saw for a rigorous theory of computation and outlined a formal theory of automata (axiomatically idealized computational mechanisms). He began by distinguishing two general classes of automata by their mode of representing numbers. The class of automata built on the "analogy principle" represent numbers by analogy; that is, through certain physical quantities that they exhibit in the way that a thermometer represents temperature by the height of its mercury. If, for example, we are representing two numbers by the electrical currents in two circuits, we can add the numbers by combining the circuits appropriately and the result will be registered as the total current output from the combined circuit. It is possible to do all the basic arithmetic operations (+, –, *, ÷) in roughly this way by using the currents in a circuit and providing the appropriate configurations of the circuit's relays and switches. Automata built on the "digital principle" do not represent numbers as physical quantities, but as aggregates of numerical digits in the manner humans typically do when we write them down on paper or count on our fingers (the etymological origin of "digit" itself). Such an automata might have a dial with ten positions on it representing 0–9, or a series of such dials for the ones column, tens column, hundreds column, etc., of the decimal numbering system. The digital representation used by nearly all modern computers is a binary system in which wires carry electrical currents of two sufficiently distinct magnitudes and employ a set of canonical circuits to perform mathematical and logical operations on the binary representations.

Only in the last half-century have we had computational devices and programming languages powerful enough to build experimental tests of ideas about what intelligence is. Turing's 1950 seminal paper in the philosophy journal *Mind* is a major turning point in the history of AI. The paper crystallizes ideas about the possibility of programming an electronic computer to behave intelligently, including a description of the landmark imitation game that we know as Turing's Test. Vannevar Bush's 1945 paper in the Atlantic Monthly lays out a prescient vision of possibilities, but Turing was actually writing programs for a computer, for example, to play chess, as laid out in Claude Elwood Shannon's 1950 proposal. Early programs were necessarily limited in scope by the size and speed of memory and processors and by the relative clumsiness of the early operating systems and languages. (Memory management, for example, was the programmer's problem until the invention of garbage collection.) Symbol-manipulation languages such as LISP, IPL, and POP and time-sharing systems – on top of hardware advances in both processors and memory – gave programmers new power in the 1950s and 1960s. Nevertheless, there were numerous impressive demonstrations of programs actually solving problems that only intelligent people had previously been able to solve.

The field of machine intelligence was founded in the mid-1950s, and it was given the name AI, as opposed to the natural intelligence displayed by certain biological systems, particularly higher animals. The goal of AI was to generate machines capable of displaying human-level intelligence. Early work in AI (1950s–1960s) focused on (1) psychological modeling and (2) search techniques. Expert systems synthesize some of that work but shift the focus to representing and using knowledge of specific task areas. Early work used game playing, and reasoning about children's blocks, as simple task domains in which to test methods of reasoning. Thus, such machines are required to have the ability to reason, make plans for their future actions, and also, to carry out these actions (essence of robotics). AI research follows two distinct methods, the symbolic (or "top-down") approach, and the connectionist (or "bottom-up") approach. The top-down approach seeks to replicate intelligence by analyzing cognition independent of the biological structure of the brain, in terms of the processing of symbols, whence the symbolic label. The bottom-up approach, on the other hand, involves creating artificial neural networks (ANNs) in imitation of the brain's structure, whence the connectionist label.

Situation calculus is the name for a particular way of describing change in first-order logic. It conceives of the world as consisting of a sequence of situations, each of which is a "snapshot" of the state of the world. Situations are generated from previous situations by actions. The **situation calculus** is a logic formalism designed for representing and reasoning about dynamical domains. It was first introduced by John McCarthy in 1963, and the idea behind **situation calculus** is that (reachable) states are definable in terms of the actions required to reach them. These reachable states are called situations. What is true in a situation can be defined in terms of relations with the situation as an argument. Situation calculus can be seen as a relational version of the feature-based representation of actions.

Uncertainty = Distribution of possible errors

Situation calculus is a first-order language in which predicates that can vary in truth value over time are given an extra argument to indicate what situations they hold in, with a function "Result" that maps an agent, an action, and a situation into the situation that result from the agent's performance of the action in the first situation.

$$\textbf{Agent} = \textbf{architecture} + \textbf{program} \qquad (4.1)$$

Classical symbolic models:

1. Its representations have combinatorial syntax and semantics.
2. The principles by which representations are transformed are defined over structural properties of representations.

The ability of machines to manipulate symbols is called **symbolic AI**. Symbolic AI (or Classical AI) is the branch of AI research that concerns itself with attempting to explicitly represent human knowledge in a declarative form (i.e., facts and rules). It is an approach that trains AI in the same way human brain learns. Symbolic AI learns to understand the world by forming internal symbolic representations of its "world" where symbols play a vital role in the human thought and reasoning process. If symbolic AI is to be successful in producing human-like intelligence, then it is necessary to translate often implicit or procedural knowledge possessed by humans into an explicit form using symbols and rules for their manipulation. Symbolic AI models such things as knowledge and planning in data structures that make sense to the programmers that build them. Symbolic AI has had some impressive successes, along with some unresolved problems such as the commonsense knowledge problem. Programs with Common Sense was probably the first paper on logical AI, that is, AI in which logic is the method of representing information in computer memory and not just the subject matter of the program. **Subsymbolic AI** models intelligence at a level similar to the neuron, where we let such things as knowledge and planning emerge. Other areas that rely on procedural or implicit knowledge such as sensory/motor processes, are much more difficult to handle within the symbolic AI framework. This is where a connectionist paradigm of AI (also called non-symbolic AI) gave rise to learning and neural network-based approaches.

Distributed representations in AI are called connectionism. In a connectionist network, a distributed representation occurs when the network represents some concept or meaning. That meaning is represented by a pattern of activity across many processing units. Symbolic logic is referred to as computationalism and machine learning allows machines such as computers to learn and improve from experience automatically, but without being explicitly programmed to do so.

Connectionism is a simple network by themselves as defined by Rumelhart and McClelland. In connectionism, the mind is one big hidden-layer network and intelligence emerges when a trainer and the environment tune its connection weights. Rules and symbols are approximations for millions of streams of activation in neural connections. It is noted that humans are smarter than rats. The reasons for this are:

Reason 1: Our networks have more hidden layers between stimulus and response.

Reason 2: We live in an environment of other humans who serve as network trainers.

Another name for **connectionism** is parallel distributed processing (PDP), which emphasizes two important features (Figure 4.3). First, a large number of relatively simple processors, the neurons, operate in parallel. Second, neural networks store information in a distributed fashion, with each individual connection participating in the storage of many different items of information. Some of the earliest work in AI used networks or circuits of connected units to simulate intelligent behavior. Examples of this kind of work, called "connectionism", include Walter Pitts and Warren McCullough's first description of a neural network for logic and Marvin Minsky's work on the SNARC system. In the late 1950s, most of these approaches were abandoned when researchers began to explore *symbolic* reasoning as the essence of intelligence, following the success of programs like the Logic Theorist (LT) and the General Problem Solver.

> Church-Turing thesis: Mental processes of any sort can be simulated by a computer program whose underlying language is of power equal to that of FlooP (that is, in which all partial recursive functions can be programmed).
>
> AI thesis: As the intelligence of machines evolves, its underlying mechanisms will gradually converge to the mechanisms underlying human intelligence.

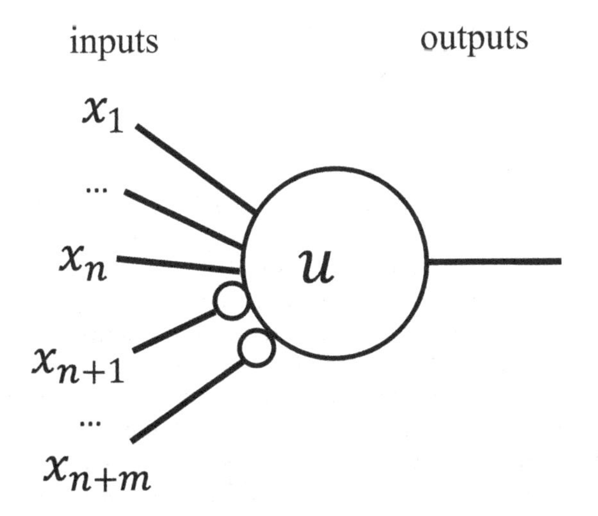

FIGURE 4.3
Neuron model.

Alan Turing was an English mathematician who is often referred to as the father of modern computer science. In 1950, English Mathematician Alan Turing published a paper entitled "Computing Machinery and Intelligence" which opened the doors to the field that would be called AI. AI as a practical science of thought mechanization could of course only begin once there were programmable computers. This was the case in the 1950s when Alan Newell and Herb Simon introduced LT, the first automatic theorem prover, and thus also showed that with computers, which actually only work with numbers, one can also process symbols. It was the first program deliberately engineered to perform automated reasoning and is called "the first artificial intelligence program". This heuristic deductive methodology worked backward, making minor substitutions to possible answers until it concluded equivalent to what had already been proven. Before this, computer programs mainly solved problems by following linear step-by-step instructions. At about the same time John McCarthy introduced, with the language LISP, a programming language specially created for the processing of symbolic structures. In the 1970s, the logic programming language PROLOG was introduced as the European counterpart to LISP. PROLOG offers the advantage of allowing direct programming using Horn clauses, a subset of predicate logic. Like LISP, PROLOG has data types for convenient processing of lists. We must distinguish here between declarative and procedural semantics of PROLOG programs. The declarative semantics is given by the logical interpretation of the horn clauses. The procedural semantics, in contrast, is defined by the execution of the PROLOG program. Different from *procedural programming languages*, such as C/C++ and Java, the focus of declarative programming is to describe a *situation* (a set of knowledge), not to describe a *solution* (a sequence of instructions). Especially, logic programming uses a language that is similar to logic, writes a program as a list of *facts* and *rules*, and treats the execution of a program as an inference process, from given facts and rules to the desired goal. Because the economic success of AI systems fell short of expectations, funding for logic-based AI research in the United States fell dramatically during the 1980s.

However, one type of connectionist work continued: the study of **perceptrons**, invented by Frank Rosenblatt, who kept the field alive with his salesmanship and the sheer force of his personality. He optimistically predicted that the perceptron "may eventually be able to learn, make decisions, and translate languages". Mainstream research into perceptrons ended abruptly in 1969, when Marvin Minsky and Seymour Papert published the book *Perceptrons*, which was perceived as outlining the limits of what perceptrons could do. Connectionist approaches were abandoned for the next decade or so. While important work, such as Paul Werbos' discovery of back-propagation (BP), continued in a limited way, major funding for connectionist projects was difficult to find in the 1970s and early 1980s. The "winter" of connectionist research came to an end in the middle of 1980s, when the work of John Hopfield, David Rumelhart, and others revived large-scale interest in neural networks. Rosenblatt did not live to see this, however, as he died in a boating accident shortly after *Perceptrons* was published. The difference between perceptron and perception is that perceptron is an element, analogous to a neuron, of an ANN consisting of one or more layers of artificial neurons while perception is organization, identification, and interpretation of sensory information. Perceptions are pattern recognition or classification devices that are crude approximations of neural networks. They make decisions about patterns by summing up evidence obtained from many small sources. They can be taught to recognize one or more classes of objects through the use of stimuli in the form of labeled training examples.

During this phase of disillusionment, computer scientists, physicists, and cognitive scientists were able to show, using computers that were not sufficiently powerful (as depicted with Moore's law), that mathematically modeled neural networks are capable of **learning** using training examples, to perform tasks which previously required costly programming. Because of the fault tolerance of such systems and their ability to recognize patterns, considerable successes became possible, especially in pattern recognition. But soon even here feasibility limits became obvious. The neural networks could acquire impressive capabilities, but it was usually not possible to capture the learned concept in simple formulas or logical rules. Attempts to combine neural nets with logical rules or the knowledge of human experts met with great difficulties. Additionally, no satisfactory solution to the structuring and modularization of the networks was found. AI as a practical, goal-driven science searched for a way out of this crisis. One wished to unite logic's ability to explicitly represent knowledge with neural networks' strength in handling uncertainty. Several alternatives were suggested. But AI is about understanding the nature of intelligent thought and action using computers as experimental devices. By 1944, for example, Herb Simon had laid the basis for the information-processing, symbol-manipulation theory of psychology:

> Any rational decision may be viewed as a conclusion reached from certain premises. The behavior of a rational person can be controlled, therefore, if the value and factual premises upon which he bases his decisions are specified for him.
>
> *(Quoted from Newell & Simon)*

AI in its formative years was influenced by ideas from many disciplines. These came from people working in engineering (such as Norbert Wiener's work on cybernetics, which includes feedback and control), biology (for example, W. Ross Ashby and Warren McCulloch and Walter Pitts's work on neural networks in simple organisms), experimental psychology (see Newell and Simon), communication theory (e.g., Claude Shannon's theoretical work), game theory (notably by John Von Neumann and Oskar Morgenstern), mathematics and statistics (e.g., Irving J. Good), logic and philosophy (e.g., Alan Turing, Alonzo Church, and Carl Hempel), and linguistics (such as Noam Chomsky's work on grammar). These lines of work made their mark and continue to be felt, and our collective debt to them is considerable. But having assimilated much, AI has grown beyond them and has, in turn, occasionally influenced them. However, both Turing and McCarthy viewed a combination of **Logic** and **Learning** as being central to the development of AI research. In the field of AI, expectations seem to always outpace the reality. Only in the last half-century have we had computational devices and programming languages powerful enough to build experimental tests of ideas about what intelligence is.

Mathematics → **Logic** → **Reasoning**
(Note: "→" literally means "moving" forward)
Mathematics → Logic → Reasoning → **Inference**
Mathematics → Logic → Reasoning → Inference → **Decision-Making**

At some point, Hilbert believed that one could prove anything using only logic, building a formal system to describe the knowledge. K. Gödel proved in his Incompleteness Theorem that within any formal system, some statements that are true could not be proven using only formal logic based on the axioms of that system. What this means is that "logic is a powerful and necessary tool in automatic reasoning, but to make useful deductions one requires domain-specific knowledge". **SAT** is a satisfiability problem. Given a logical formula involving a set of Boolean variables, is there a set of values for these variables such that the formula is true? "Several important problems can be encoded as SAT problems. For example, Henry Kautz and Bart Selman showed that generating a plan of actions can be expressed as a SAT problem". The relevance to AI is the problem of deciding if something is true in a given system (making a deduction) comes down to solving a particular SAT problem. NP-complete refers to the condition where there is no known polynomial algorithm to solve this problem, but if we find one for it, then we can solve any other NP problem. For now, a guaranteed solution is exponential. Knowledge-based systems that contain domain-specific knowledge give them more problem-solving power in the form of an expert system. And one of the most important concepts an intelligent system needs to understand is the concept of knowledge. AI systems need to understand what knowledge they and the systems or people they interact with have, what knowledge is needed to achieving particular goals, and how that knowledge can be obtained.

Data, Information, and Knowledge

Data are things given such as numbers, words, sentences, records, assumptions, etc. Information is raw data, and knowledge is interpreted data. Information is piecemeal, fragmented, and particular, whereas knowledge is structured, coherent, and often universal. Information is timely, transitory, and perhaps even ephemeral, whereas knowledge is of enduring significance. Information is a flow of messages, whereas knowledge is a stock, largely resulting from the flow, in the sense that the "input" of information may affect the stock of knowledge by adding to it, restructuring it, or changing it in any way. Today's computer systems are capable of producing a huge amount of information, especially on the status of internal components and the world around them. The problem will be not a lack of information but finding what is needed when it is needed. The information must be integrated and interpreted correctly. Knowledge is the sort of information that people use to solve problems. Knowledge includes facts, concepts, procedures, models, heuristics, and examples. Knowledge may be more specific or general, exact, or fuzzy, and procedural or declarative. Refer to Figure 4.4. A knowledge system is built around a knowledge base (KB), that is, a collection of knowledge, taken from a human, and stored in such a way that the system can *reason* with it. Information becomes knowledge once the information is verified.

Knowledge can be defined as the body of facts and principles accumulated by humankind or the act, fact, or state of knowing. Knowledge is facts, information, and skills acquired through experience or education, or the theoretical or practical understanding of a subject. Knowledge is having familiarity with language, concepts, procedures, rules, ideas, abstractions, places, customs, facts, and associations, coupled with an ability to use these notions effectively in modeling different aspects of the world. The meaning of knowledge is closely related to the meaning of intelligence. Intelligent requires the possession of and access to knowledge. A common way to represent knowledge external to a computer or a human is in the form of written language. Knowledge includes and requires the use of data and information. Knowledge combines relationships, correlations, dependencies, and the notion of gestalt with data and information. Belief is a meaningful and coherent expression. Thus, belief may be true or false. Hypothesis is defined as a belief that is backed up with some supporting evidence, but it may still be false. Knowledge is true justified belief. Epistemology is the study of the nature of knowledge. Metaknowledge is knowledge about knowledge, that is, knowledge about what we know.

$$\textbf{Knowledge} = \textbf{information} + \textbf{rules} \qquad (4.2)$$

Knowledge representation is a relationship between two domains: the real world and AI that represents information about the world in the form of a computer system. Knowledge is the collection of facts, inference rules, etc. which can be used for a particular purpose. Knowledge requires the use of data and information. It combines relationships, correlations, and dependencies with data and information. The organization of knowledge in memory is key to efficient processing. Knowledge can be organized in memory for easy access by a method known as indexing. The choice of representation can simplify the organization and access operations. For example, frames linked together in a network represent a versatile organizational structure. Each frame will contain all closely associated information about an object and pointers to related object frames making it possible to quickly gain access to this information. Subsequent processing then typically involves only a few related frames. However, knowledge is almost always incomplete and uncertain. Thus, a rule may have associated with it a confidence factor or a weight. The set of methods for using uncertain knowledge in combination with uncertain data in reasoning is called reasoning with uncertainty. A subclass of methods for reasoning with uncertainty is called "fuzzy logic", and the systems are known as "fuzzy systems".

The basic components of knowledge are:

1. A set of collected data
2. A form of belief or hypothesis
3. A kind of information.

Knowledge is different from data. Data is the collection of raw materials whereas knowledge is the collection of some well-specified inference rules and facts. Knowledge is also different from belief and hypothesis. Belief is any meaningful and coherent expression that can be represented. Belief may be true or false. A hypothesis is a justified belief that is not known to be true. A hypothesis is a belief that is backed up with some supporting evidence, but it may still be false. So, knowledge can be defined as true justified knowledge. In fact, this difference between what words convey and what perception mediates is so marked that it may suggest a distinction of two kinds of knowledge; direct knowledge of objects (acquaintance with), gained by the presentation of them in experience and immediately verifiable, and propositional knowledge or generalization (knowledge about) which concerns more than can be given at one time and thus requires some mental synthesis of what is temporally disjoined. There is no knowledge without interpretation. If interpretation, which represents an activity of the mind, is always subject to the check of further experience, how is knowledge possible at all? That the interpretation reflects the character of past experience, will not save its validity. For what experience establishes, it may destroy; its evidence is never complete. An argument from past to future at best is probable only, and even this probability must rest upon principles that are themselves more than probable. For the validity of knowledge, it is requisite that experience in general shall be in some sense orderly that the order implicit in conception may be imposed upon it. And for the validity of particular predications, it is necessary that a particular order may be ascribed to experience in advance. Thus, if there is to be any knowledge at all, some knowledge must be a priori; there must be some propositions the truth of which is necessary and is independent of the particular character of future experience (Figure 4.4).

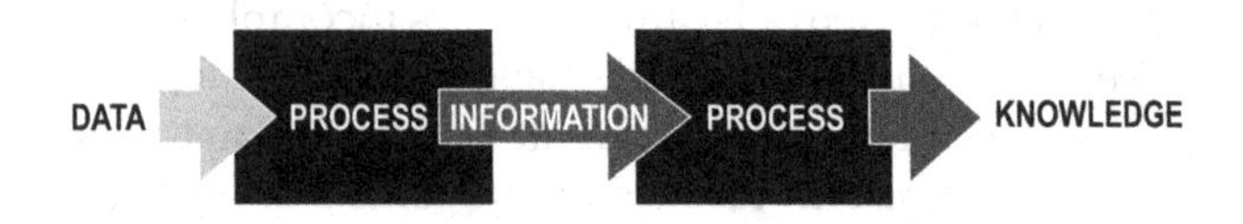

FIGURE 4.4
Information process diagram.

Knowledge-Based Systems: Knowledge-based systems get their power from the expert knowledge that has been coded into facts, rules, heuristics, and procedures. The knowledge is stored in a knowledge base separate from the control and inferencing components. Knowledge is important and essential for knowledge-based intelligent behavior.

Declarative Knowledge: It is the passive knowledge expressed as statements of facts about the world. Procedural knowledge is the compiled knowledge related to the performance of some task.

Knowledge acquisition is the gathering or collecting knowledge from various sources. It is the process of adding new knowledge to a knowledge base and refining or improving knowledge that was previously acquired. Acquisition is the process of expanding the capabilities of a system or improving its performance at some specified tasks. So, it is the goal-oriented creation and refinement of knowledge.

Knowledge Representation: Knowledge representation is probably, the most important ingredient for developing an AI. A representation is a layer between information accessible from outside world and high-level thinking processes. Without knowledge representation, it is impossible to identify what thinking processes are, mainly because representation itself is a substratum for a thought. Knowledge representation languages should have precise syntax and semantics. You must know exactly what an expression means in terms of objects in the real world.

Types of Knowledge Representation: Knowledge can be represented in different ways. The structuring of knowledge and how designers might view it, as well as the type of structures used internally are considered. Different knowledge representation techniques are:

a. Logic is a formal language, with precisely defined syntax and semantics, which supports sound inference.
b. Semantic Network is a graphical knowledge representation technique.
c. Frame is a collection of attributes and associated values that describe some entity in the world. Frames are general record-like structures that consist of a collection of slots and slot values.

d. Conceptual Graphs is a knowledge representation technique that consists of basic concepts and the relationship between them. As the name indicates, it tries to capture the concepts about the events and represents them in the form of a graph.

e. Conceptual Dependency is another knowledge representation technique in which we can represent any kind of knowledge. It is based on the use of a limited number of primitive concepts and rules of formation to represent any natural language statement. Conceptual dependency theory is based on the use of knowledge representation methodology and was primarily developed to understand and represent natural language structures.

f. Script is another knowledge representation technique. Scripts are frame-like structures used to represent commonly occurring experiences such as going to restaurant and visiting a doctor. A script is a structure that describes a stereotyped sequence of events in a particular context and consists of a set of slots. As a result, scripts add a knowledge layer.

Knowledge Manipulation: Decisions and actions in knowledge-based systems come from manipulation of the knowledge in specified ways. All forms of reasoning require a certain amount of searching and matching. In fact, these two operations by far consume the greatest amount of computation time in AI systems. For this reason, it is important to have techniques available that limit the amount of search and matching required to complete any given task. They help to limit or avoid the so called combinatorial explosion in problems that are so common in search. One of the greatest bottlenecks in building knowledge-rich systems is the acquisition and validation of the knowledge.

Depending on the type of functionality, the knowledge in AI systems is categorized as:

1. Declarative knowledge
   - The knowledge, which is based on concepts, facts, and objects, is termed as "Declarative Knowledge".
   - It provides all the necessary information about the problem in terms of simple statements, either true or false.
2. Procedural knowledge
   - Procedural knowledge derives the information on the basis of rules, strategies, agendas, and procedures.
   - It describes how a problem can be solved.
   - Procedural knowledge directs the steps on how to perform something.

     For example: Computer program.
3. Heuristic knowledge
   - Heuristic knowledge is based on thumb rule.
   - It provides the information based on a thumb rule, which is useful in guiding the reasoning process.
   - In this type, the knowledge representation is based on the strategies to solve the problems through the experience of past problems, compiled by an expert. Hence, it is also known as Shallow knowledge.
4. Metaknowledge
   - This type gives an idea about the other types of knowledge that are suitable for solving problems.
   - Metaknowledge is helpful in enhancing the efficiency of problem-solving through proper reasoning process.
5. Structural knowledge
   - Structural knowledge is associated with the information based on rules, sets, concepts, and relationships.
   - It provides the information necessary for developing the knowledge structures and overall mental model of the problem.

There are two basic components of knowledge representation, that is, reasoning and inference. It is a way of efficient computation in which thinking is accomplished. There are five types of knowledge: procedural, declarative, meta, heuristic, and structural. Procedural knowledge gives information about how to achieve something, which is also known as imperative knowledge. Declarative knowledge is about statements that describe a particular object and its attributes. Metaknowledge is knowledge about knowledge, and how to gain them. Heuristic knowledge represents knowledge of some expert in a field or subject and can be rules of thumb. Heuristic knowledge is empirical as opposed to being deterministic. Finally, structural knowledge describes what relationship exists between concepts and objects. Knowledge representation is the study of how knowledge about the world can be represented and what kinds of reasoning can be done with that knowledge. Understanding in cognition is accomplished through the interplay of multiple sources of knowledge. It is clear that we know a good deal about a large number of different standard situations. Several theorists have suggested that we store knowledge in terms of structures called variously: scripts (Schank), frames (Minsky), or schemata (Norman & Bobrow and Rumelhart). Such knowledge structures are assumed to be the basis of comprehension; however, it is important to bear in mind that most everyday situations cannot be rigidly assigned to just a single script. They generally involve an interplay between a number of different sources of information.

Scripts are another form of knowledge representation technique. Scripts are frame-like structures used to represent commonly occurring experiences such as going to restaurant, visiting a doctor. A script is a structure that describes a stereotyped sequence of events in a particular context. A script consists of a set of slots. Associated with each slot may be some information about what kinds of values it may contain as well as a default value to be used if no other information is available. Scripts are useful because in the real world, there are no patterns to the occurrence of events. These patterns arise because of clausal relationships between events. The events described in a script form a giant casual chain. The beginning of the chain is the set of entry conditions that enable the first events of the script to occur. The end of the chain is the set of results that may enable later events to occur. The headers of a script can all serve as indicators that the script should be activated. Problem-solving in AI may be characterized as a systematic search through a range of possible actions in order to reach some predefined goal or solution. The problem-solving agents decide what to do by finding sequence of action that leads to desirable states. The simplest agents which have been described below are the reflex and goal-based agents. The reflex agents use direct mapping from states to actions and are unsuitable for very large mappings. Problem-solving agents find action sequences that lead to desirable state.

There are multiple approaches and schemes to think about representation: pictures and symbols, graphs and networks, and numbers. Pictorial representations do not easily translate to useful information in computers because computers have a hard time interpreting pictures directly without complex reasoning. Graphs & networks allow relationships between objects to be incorporated where procedural knowledge can be represented by graphs. And numbers are integral part of knowledge representation and can be translated easily into computer representation. There are basically four types of knowledge representation in AI:

1. Logical representation
2. Production rule
3. Semantic networks
4. Frame representations.

Patterns are another means of representing, sharing, and reusing **knowledge**. Patterns should include information about when they are and when they are not useful. They may be represented using tabular and graphical descriptions. With too much data, researchers need techniques for reducing it to a form usable in the model. With too little data, the researchers need information that can be represented by statistical distributions. In order to give information to the robot and to get information without errors in communication, AI uses logic representation (based on truth) to represent knowledge and to perform algorithmic reasoning: propositional calculus/logic (PL) and first-order predicate logic/calculus (FOPL). Proposition (logic) is classified as a declarative sentence that is either true or false. First-order predicate calculus (FOPL) was developed by logicians to extend the expressiveness of propositional logic. It is a generalization of propositional logic that permits reasoning about world entities (objects) as well as classes and subclasses of objects. Prolog is based on FOPL. Predicate logic uses variables and quantifiers which is not present in propositional logic.

A production rule consist of<condition, action> pairs. The notion of a "production system" was invented in 1943 by Emil Post. Forward-chaining production systems are used as the basis for many rule-based expert systems and as a model of human cognition in psychology. A production is a rule of the form:

$$C1, C2, \ldots CN \Rightarrow A1, A2, \ldots AM \quad (4.3)$$

where C stands for Condition, which must hold before the rule can be applied, and A stands for Actions to be performed or conclusions to be drawn when the rule is applied. The basic components of a production system are rules, working memory, and inference engine. Mathematical inference is a set of axioms which are true assertions about the world. The Inference engine (a set of IF–THEN inference rules) allows you to: (1) Deduce new assertions from the old (forward chaining) and (2) Determine whether a given assertion is true (backward chaining). For story understanding by simulation, a simulation is a sequence of states and a state is a snapshot of the mental world of each story character and the physical world. The use of efficient procedures and rules by the Inference Engine is essential in deducting a correct, flawless solution.

Thus, rules are unordered set of user-defined "if–then" rules. **Working Memory** is a set of "facts" consisting of positive literals defining what's known to be true about the world. **The inference Engine** has the procedure for inferring changes (additions and deletions) to the working memory. It typically uses "forward chaining" to make inferences. The AI checks if a condition holds, and then gives a new situation (states). Production rules belong to and same as propositional logic. The **Rete Algorithm** is the most widely used, efficient algorithm for the implementation of production systems. The Rete Pattern-Matching Algorithm is used for matching facts against the patterns in rules to determine which rules have had their conditions satisfied. It uses temporal redundancy to save the state of the matching process from cycle to cycle, recomputing changes in this state only for the change that occurred in the fact list. Rete is the only algorithm for production systems whose efficiency is asymptotically independent of the number of rules.

Semantic networks represent knowledge in the form of a graphical network. It is a simple representation scheme that uses a graph of labeled nodes and labeled, with directed arcs to encode knowledge. Semantic networks are typically used with a special set of accessing procedures that perform "reasoning". Arcs define binary relationships that hold between objects denoted by the nodes. The ISA (is-a) or AKO (a-kind-of) relation is often used to link instances to classes, classes to super classes. Non-binary relationships can be represented by "turning the relationship into an object". This is an example of what logicians call "reification". Many semantic networks distinguish between nodes representing individuals and those representing classes, and the "subclass" relation from the "instance-of" relation. In the 1970s and 1980s, semantic networks morphed into Frame Representation Languages. A frame is a lot like the notion of an object in OOP but has more meta-data. A frame has a set of slots and a slot represents a relation to another frame (or value). A slot has one or more facets where a facet represents some aspect of the relation. A slot in a frame holds more than a value. In frame representation, frames are record-like structures that consist of a collection of slots or attributes and the corresponding slot value. Slots have names and values called facets.

In psychology and cognitive science, a **schema** (plural *schemata* or *schemas*) describes a pattern of thought or behavior that organizes categories of information and the relationships among them. It can also be described as a mental structure of preconceived ideas, a framework representing some aspect of the world, or a system of organizing and perceiving new information, such as a mental schema or conceptual model. Schemata influence attention and the absorption of new knowledge: people are more likely to notice things that fit into their schema, while re-interpreting contradictions to the schema as exceptions or distorting them to fit. Schemata tend to remain unchanged, even in the face of contradictory information. Schemata can help in understanding the world and the rapidly changing environment.

In AI and especially knowledge-based systems, the **ramification problem** is concerned with the indirect consequences of an action. The ramification problem is a hard and ever-present problem in systems exhibiting dynamic behavior. This is the frame problem in the context of actions with indirect effects. It might also be posed as how to represent what happens implicitly due to an action or how to control the secondary and tertiary effects of an action. The solution to the ramification problem offered is based on an existing predicate calculus action formalism, namely, the event calculus.

Description logics provide a family of frame-like knowledge representation systems with a formal semantics. An additional kind of inference done by these systems is automatic **classification,** where finding the right place in a hierarchy of objects for a new description. Current systems take care to keep the languages simple, so that all inference can be done in polynomial time (in the number of objects) and ensure tractability of inference. **Abduction** is a reasoning process that tries to form plausible explanations for abnormal observations. Abduction is distinctly different from deduction and induction and is inherently uncertain. Some major formalisms for representing and reasoning about uncertainty include: Mycin's certainty factors (an early representative), **Probability theory (esp. Bayesian belief networks)**, Dempster-Shafer theory, Fuzzy logic, Truth maintenance systems, and Nonmonotonic reasoning. Abduction reasoning derives an explanatory hypothesis from a given set of facts. The inference result is a **hypothesis** that, if true, could **explain** the occurrence of the given facts, for example, DENDRAL. In abductive reasoning, "conclusions" are **hypotheses,** and not theorems (may be false *even if* rules and facts are true). Reasoning is often a hypothesize-and-test cycle. Postulate possible hypotheses, any of which would explain the given facts (or at least most of the important facts) and test the plausibility of all or some of these hypotheses. For example, one way to test a hypothesis H is to ask whether something that is currently unknown but can be predicted from H, is actually true. Reasoning is **nonmonotonic**. That is, the plausibility of hypotheses can increase/decrease as new facts are collected. In contrast, deductive inference is monotonic; it never changes a sentence's truth value, once known. In abductive (and inductive) reasoning, some hypotheses may be discarded, and new ones formed, when new observations are made.

Data Mining

Data Mining is the extraction of useful patterns from data sources, for example, databases, texts, web, and images. Data Mining is typically thought of as a part of data science but shares similarities to AI and pattern recognition. The patterns in Data Mining must be: valid, novel, potentially useful, and understandable. Data mining is also called *knowledge discovery and data mining* (KDD). Basically, data mining is the process of automatically discovering useful information in large data repositories. This technology draws on ideas from machine learning, statistics, and databases. The major tasks for data mining include classification (mining patterns that can classify future data into known classes), association rules (mining any rule of form $X \rightarrow Y$, where X and Y are sets of data items), clustering (identifying a set of similarity groups in the data), sequential pattern mining (A sequential rule: $A \rightarrow B$, says that event A will be immediately followed by event B with a certain confidence), deviation detection (discovering the most significant changes in data), and data visualization using graphical methods to show patterns in data. The tasks can generally be divided into two major categories: predictive tasks and descriptive tasks. The process for data mining includes the following steps: (1) Understand the application domain, (2) Identify data sources and select target data, (3) Pre-process by cleaning and attribute selection, (4) Data mining to extract patterns or models, and (5) Post-process by identifying interesting or useful patterns, and finally incorporate patterns in real-world tasks. Predictive Data Mining (or supervised learning) starts with a training set, followed by finding or learning a "model" for the class attribute as a function of the values of the other attributes. The three different problems involved in learning are memory (memorize things), averaging (predict majority outcome), and generalization (making a plausible outcome) (Figure 4.5).

Data cleansing is the process of finding and correcting (or deleting) irrelevant, corrupt, missing, duplicate, or otherwise useless data from a data set. This is a necessary step designed to purify data, so that algorithms can work faster and make more accurate predictions. The difference between a database and a knowledge base is that a database is a collection of data representing facts in their basic form, while a knowledge base stores information as answers to questions or solutions to problems. Some data cleansing and mining are required for input to machine learning.

Pattern Recognition

Pattern Recognition a field of study developed significantly in the 1960s. It was a very much interdisciplinary subject, covering developments in the areas of statistics, engineering, AI, computer science, psychology, and physiology, among others. Some people entered the field with a real problem to solve. The large numbers of applications ranging from the classical ones, such as automatic character recognition and medical diagnosis to the more recent ones in data mining (such as credit scoring, consumer sales analysis, and credit card transaction analysis), have attracted considerable research effort, with many methods developed and advances made. Pattern recognition of the objects into a lot of categories or classes. In 1967, the nearest neighbor algorithm was conceived, which was the beginning of basic pattern recognition. This algorithm was used for mapping routes and was one of the earliest algorithms used in finding a solution to the traveling salesperson's problem of finding the most efficient route. It originally comes from cognitive neuroscience and psychology. Pattern recognition is a cognitive process that happens in our brain

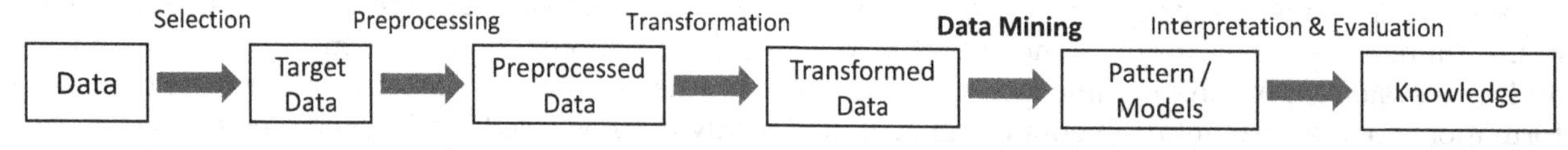

FIGURE 4.5
Data-mining process diagram.

when we match some information that we encounter with data stored in our memory. It is an integral part in most machine intelligence systems built for decision-making. The nature of the pattern recognition is engineering. But the final aim of pattern recognition is to design machines to solve the gap between application and theory. It is a process of identifying a stimulus. This process is often accomplished with incomplete or ambiguous information.

In the field of pattern recognition, the main objective is to achieve the highest possible classification accuracy. These features are extracted from data and can be of different types like continuous variables, binary values, etc. As such, a classification algorithm used with a specific set of features may not be appropriate with a different set of features. In addition, classification algorithms are different in their theories and hence achieve different degrees of success for different applications. Even though, a specific feature set used with a specific classifier might achieve better results than those obtained using another feature set and/or classification scheme, we cannot conclude that this set and this classification scheme achieve the best possible classification results. As different classifiers may offer complementary information about the patterns to be classified, combining classifiers, in an efficient way, can achieve better classification results than any single classifier (even the best one). The problem of combining multiple classifiers consists of two parts. The first part, closely dependent on specific applications, includes the problems of "How many and what type of classifiers should be used for a specific application?, and for each classifier what type of features should we use?", as well as other problems that are related to the construction of those individual and complementary classifiers. The second part, which is common to various applications, includes the problems related to the question "How to combine the results from different existing classifiers so that a better result can be obtained?".

> A pattern is the opposite of a chaos; it is an entity vaguely defined, that could be given a name.
>
> – *S. Watanabe*

Pattern recognition involves finding the similarities or patterns among small, decomposed problems that can help us solve more complex problems more efficiently.

- Finding patterns is extremely important because patterns make our tasks simpler. Problems are easier to solve when they share patterns, because we can use the same problem-solving solution wherever the pattern exists.
- The more patterns we can find, the easier and quicker our overall task of problem-solving will be.

Pattern recognition uses key features to classify input data into objects or classes. There are two methods of classification: Supervised and unsupervised classification. Using these classifications, pattern recognition applications include Computer Vision, Optical Character Recognition (Handwritten or Digital Text Documents), Radar Processing, and Speech Recognition. To find patterns in problems we look for things that are the same (or very similar) in each problem. It may turn out that no common characteristics exist among problems, but we should still look. Patterns exist among different problems and within individual problems. So, we need to look for both. The basic ingredients for problem formulation include:

- Measurement space (e.g., image intensity, pressure)
- Features (e.g., corners, spectral energy)
- Type of Classifier – soft and hard
- Decision boundary
- Training sample
- Probability of error

The result of the pattern recognition can be either class assignment, cluster assignment, or predicted variables (Figure 4.6). Therefore, there is no point in asking "what is the difference between pattern recognition and classification" – classification algorithm is a part of the supervised machine-learning problems, where the target value is a finite set of classes. Pattern recognition is a process that looks at the available data and tries to see whether there are any regularities within it. There are two main parts:

1. **Explorative part,** where the algorithms are looking for patterns in general
2. **Descriptive part,** where the algorithms start to categorize the found patterns

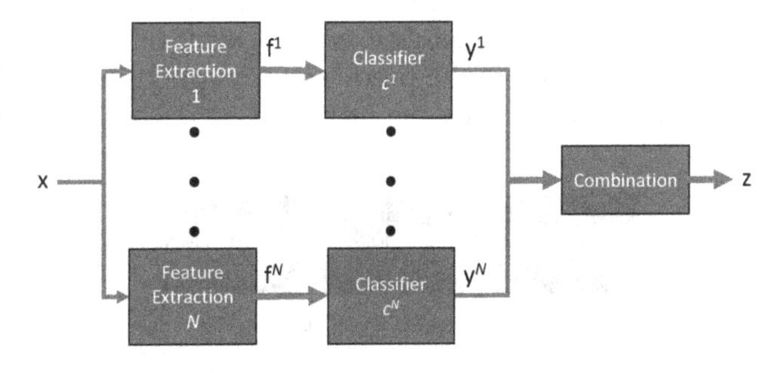

FIGURE 4.6
Multi-classifier recognition system.

Pattern recognition as it is commonly understood is only one aspect of human visual capacities. One of the primary points that we are trying to make in this thesis is that at some level of description, all aspects of perception can be at least approximately modeled by something like primitive pattern recognition events. The philosophical justification and the information-theoretic descriptions have all been more or less leading in this direction. We examine more closely what pattern recognition is, including different ways of representing the data to be classified and the logico-algebraic implications of these. This in turn leads to a discussion of Watanabe's propensity theory and the basic properties of observations or measurements. Mechanical pattern recognition does not deal with concepts or classes in terms of the logical concepts of intension and extension. Rather, like human pattern recognition, the aim is to be able to show a few paradigms (and their class affiliation) in a learning phase, and subsequently be able to classify an object (whose properties, but not class, are known), in a decision phase. This is the essential function of both human and mechanical pattern recognition: inferring a general concept from a few concrete cases. It is an inductive inference, with no necessary or logical basis. In mechanical pattern recognition, extra-logical, extra-evidential heuristic measures of similarity are often introduced to overcome the inductive ambiguity and allow the classification to proceed by logical computations. These heuristics either implicitly or explicitly express the value judgment of the computer programmer through the weight we attach to each variable, or threshold which we set, etc. While the heuristic principles have a role to play in resolving the inductive ambiguity of generalization, they are not foolproof. They can only be judged by the usefulness of the resulting classification. As well as this "paradigmatic" pattern recognition, there is pattern recognition in the sense of clustering or forming new classes (or cognition) (Figure 4.7).

Usually in pattern recognition, the number of classes and some paradigms (class samples) are given during the training period, and objects without class assignment are given during the application stage. The task of the classifier is to place these latter objects into the classes exemplified by the paradigms. Clustering is the process of grouping objects into classes where only the properties of each object are available – no information is available on possible class assignments of these objects. It is assumed that the members assigned to any particular cluster are in some sense "bonded" together more intensely than members of different subsets. The task of deciding how many clusters there should be and statistically defining the classes or clusters corresponds to Pierce's "abduction" and hence involves further inductive ambiguity above and beyond that involved in supervised pattern recognition.

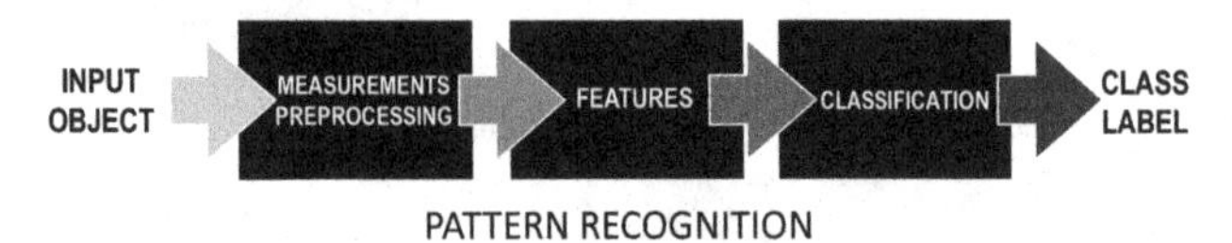

FIGURE 4.7
Pattern recognition processing diagram.

It is useful to consider pattern recognition as a process of step-by-step reduction in dimensionality. This reduction corresponds either explicitly or implicitly to making some variables more "important" than others. The more "important" variables are then referred to as "features" and can be used in the final steps of classification. In general, the only difference between "observations" and "features" is that there are fewer features, though they collectively contain most of the information required to make a classification on the basis of the original set of observations. A useful tool in the analysis of this type of pattern recognition problem is the mathematical concept of a vector space. If each pixel is considered as an independent variable, then any image can be represented as a point in a space with as many dimensions as there are numbers of pixels. Thus, a particular 150×50 image is a particular point in a 7,500-dimensional space. Most operations in this space are simply generalizations of the familiar vector and matrix operations in two and three dimensions.

Measurement theory, classification theory, pattern recognition, and discrimination theory are all based on the same fundamental idea: that of assigning labels to processes in such a way that processes that bear the same label are considered to be alike, and processes bearing different labels are considered to be different. How geometric representations can be particularly useful for understanding what is involved in pattern recognition? The really interesting notion is that in different situations, different types of geometric representations may be appropriate, and these in turn have different underlying algebraic structures. Objects in the world can be said to share features (or predicates like color, shape, and so on) with other objects. It is in the nature of most such features that they can characterize indefinitely many objects, and because of this, these features are often called universals, where particular object is an instantiation of one or more universals, usually very many. So, in this sense, a universal is a concept or a general idea.

Aristotle, also a realist (meaning that he accepted the reality of universals), recognized the reality of particular objects, which he argued, consisted of two real elements – Matter and Form. Matter endows real existence to an object and represents "potentiality" in the sense that a plank of wood has the potential to be worked

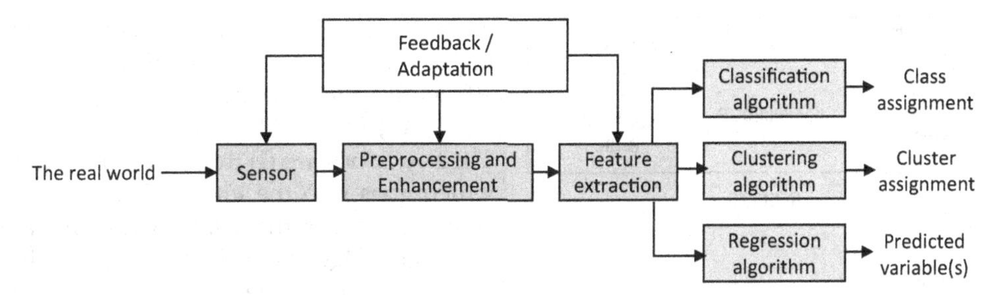

FIGURE 4.8
Feature extraction and assignment.

into a chair. Form determines the essential nature or functional meaning of an object. It represents "actuality" in the sense that regardless of the matter involved, say wood or metal, Form is what determines whether the object is a chair or a table. Unlike with computer vision, the datum for pattern recognition can be anything. Understanding means the ability to determine some new knowledge from a given knowledge. For each action of a problem, the mapping of some new actions is very necessary. Mapping the knowledge means transferring the knowledge from one representation to another representation. Knowledge representation, both the formal and informal aspects, has become a cornerstone of every AI program. John McCarthy's important 1958 paper, "Programs with Common Sense" makes the case for a declarative knowledge representation that can be manipulated easily. McCarthy has been an advocate for using formal representations, in particular extensions to predicate logic, ever since. Research by McCarthy and many others on nonmonotonic reasoning and default reasoning, as in planning under changing conditions, gives us important insights into what is required for intelligent action and defines much of the formal theory of AI. Language understanding and translation were at first thought to be straightforward, given the power of computers to store and retrieve words and phrases in massive dictionaries. Although the simple lookup methods originally proposed for translation did not scale up, recent advances in language understanding and generation have moved us considerably closer to having conversant nonhuman assistants.

Knowledge representation is the study of how knowledge about the world can be represented and what kinds of reasoning can be done with that knowledge. We discuss two different systems that are commonly used to represent knowledge in machines and perform algorithmic reasoning: propositional calculus and predicate calculus. The real knowledge representation and reasoning systems come in several major varieties and these differ in their intended use, expressivity, and features. Some major families are Logic programming languages, Theorem provers, Rule-based or production systems, Semantic networks, Frame-based representation languages, Databases (deductive, relational, object-oriented, etc.), Constraint reasoning systems, Description logics, Bayesian networks, and Evidential reasoning. The two key issues are first "how to represent the knowledge", and in our case a knowledge base, and second, what are the reasoning processes, and for us we rely on inference/reasoning. The knowledge base is a set of formal languages that represent facts about the world. In practice, the knowledge base is a database with a combination of an expert system with if–then rules. One of the main kinds of reasoning done in a semantic net is the inheritance of values along the subclass and instance links. Semantic networks differ in how they handle the case of inheriting multiple different values where all possible values are inherited, ***or*** only the "lowest" value or values are inherited (Figure 4.8).

Experts express their knowledge informally, using natural language, visual representations, and common sense, often omitting essential details that are considered obvious into a database. This form of knowledge is very different from the one in which knowledge has to be represented in the knowledge base (which is formal, precise, and complete). Remember, there are two different databases and should not be combined. One contains data and the other is a knowledge base which is a set of sentences in a formal language and these sentences are expressed using a knowledge representation language. This transfer and transformation of knowledge, from the domain expert through a knowledge engineer to the AI/robot, is long, painful, and inefficient (and is known as "the knowledge acquisition bottleneck" of the AI systems development process). The learning engine builds the knowledge base from the other database of facts or examples. In general, the learned knowledge consists of concepts, classification rules, or decision trees. The problem-solving engine is a simple one-step inference engine that classifies a new instance as being or not an example of a learned concept (Figure 4.9).

A blackboard (is a workplace) where an area of working memory is set aside for the description of a current problem and for recording intermediate results in an expert system. The knowledge base that

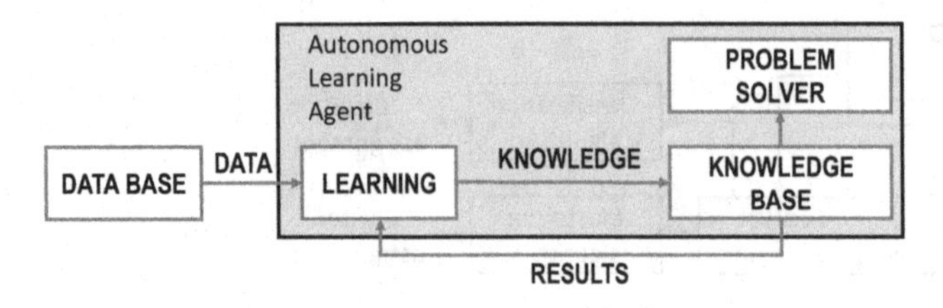

FIGURE 4.9
Knowledge representation model.

represents the world is composed of ontology and rules/cases. The knowledge base is composed of data structures that represent the objects and their relations (the object ontology) in the real world, and the general laws governing them (actions, rules, methods, cases, or elementary problem-solving) and actions that can be performed with them. The problem-solving engine implements a problem-solving method that manipulates the data structures in the knowledge base to reason about the input problem, to solve it, and to determine the actions to perform next. The problem solver can be an inference engine that represents states and actions, incorporates new percepts, updates internal representations of the world, deduces hidden properties of the world, and deduces appropriate actions. There is a clear separation between knowledge (which is contained in the knowledge base) and control (represented by the problem solver or problem-solving engine). The knowledge possessed by the AI and its reasoning processes should be understandable to humans. The agent should have the ability to give explanations of its behavior, what decisions it is making, and why. An AI/robot should be able to communicate with its users or other entities (or agents). The communication language should be as natural to the human users as possible. Ideally, it should be free natural language. The problem of natural language understanding and generation is very difficult due to the ambiguity of words and sentences, the paraphrases, ellipses, and references that are used in human communication. In order to solve "real-world" problems, an intelligent robot needs a huge amount of domain knowledge in its memory (the knowledge base). An AI usually needs to search huge spaces in order to find solutions to problems.

Intelligent AI generally attacks problems for which no algorithm is known or feasible, problems that require heuristic methods. A heuristic is a rule of thumb, strategy, trick, simplification, or any other kind of device that drastically limits the search for solutions in large problem spaces. Heuristics do not guarantee optimal solutions and are based on combinatorial rules. In fact, they do not guarantee any solution at all. A useful heuristic is one that offers solutions that are good enough most of the time. An AI is improving its competence if it learns to solve a broader class of problems and to make fewer mistakes in problem-solving. The goal is improving its performance if it learns to solve more efficiently (for instance, by using less time or space resources) the problems from its area of competence. The key issues are (1) the Representation of knowledge which goes into the knowledge base and (2) the reasoning processes that leads to inference and reasoning. The knowledge base is a set of sentences in a formal language representing facts about the world called **knowledge representation** (KR) language (Figure 4.10).

KR language candidates include a logical language (propositional/first-order) combined with a logical inference mechanism. Knowledge key aspects require a method to add sentences to the knowledge base and ability to query the knowledge base. Both tasks may involve inference, that is, how to derive new sentences from old sentences. In logical systems, inference must obey the fundamental requirement that when one asks a question to the knowledge base, the answer should *follow* from what has been told to the knowledge base previously. In other words, the inference process should not "make things" up. The knowledge base must be able to represent states, actions, etc., incorporate new percepts, update internal representations of the world, deduce hidden properties of the world, and deduce appropriate actions. The connection between sentences and facts is provided by the semantics of the language. The property of one fact following from some other facts is mirrored by the property of one sentence being entailed by some other sentences. Logical inference generates new sentences that are entailed by existing sentences. A KR language candidate should be a logical language (propositional/first-order) combined with a logical inference mechanism. The language of rational thought is logic. Thus, we need clear syntax and semantics (with well-defined meanings) and a mechanism to infer new information. The solution is to use a formal language. A representation language is preferred that is expressive, concise, unambiguous, and independent of context, having an effective procedure to derive new information when it is not easy to meet these goals. Both propositional and first-order logic meet some of the criteria for incompleteness/uncertainty and is a key contrast with programming languages. Logical representation has three components:

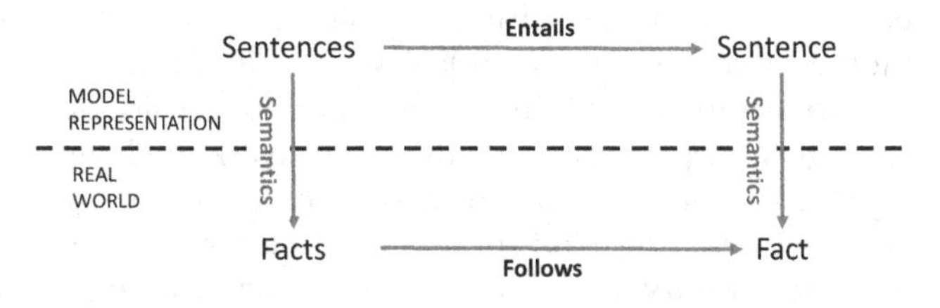

FIGURE 4.10
Modeling the real world.

1. syntax
2. semantics (link to the world)
3. proof theory ("pushing symbols")

In order to making it work, the KR language requires soundness and completeness. The key is to connect sentences to the world. However, what is somewhat misleading, formal semantics bring sentence down only to the primitive components (propositions). The tenuous link to the real world is the key issue. All the computer has been a sequence of "sentences" (hopefully reflecting the real world). Sensor data can provide some grounding in this paradigm. The hope is that the Knowledge Base Model is a unique interpretation of the real world. This leads to the "Symbol Grounding Problem". The model is a set of sentences (in the KB) that is a truth assignment in which each of the KB sentences is evaluated to be *true*. With more and more sentences, the model of KB starts looking more and more like the "real world" (or isomorphic to it).

A semantic (word description) network is a simple representation scheme that uses a graph of labeled nodes and labeled, directed arcs to encode knowledge. It is a knowledge representation technique that focuses on the relationships between objects. A directed graph or word chart is used to represent a semantic network (or net for short). Semantic networks are typically used with a special set of accessing procedures that perform "reasoning". Semantic networks were very popular in the 1960s and 1970s, but are less frequently used today, often much less expressive than other KR formalisms. The graphical depiction associated with a semantic network is a significant reason for their popularity. One of the main kinds of reasoning done in a semantic net is the inheritance of values along the subclass and instance links. Semantic networks differ in how they handle the case of inheriting multiple different values. Eventually, semantic networks morphed into Frame Representation Languages in the 1970s and 1980s. A frame is a lot like the notion of an object in OOP but has more meta-data. A frame has a set of slots and a slot represents a relation to another frame (or value). A slot has one or more facets. A facet represents some aspect of the relation. Description logics provide a family of frame-like KR systems with a formal semantics. An additional kind of inference done by these systems is automatic classification. Current systems take care to keep the languages simple, so that all inference can be done in polynomial time (in the number of objects). Abduction is a reasoning process that tries to form plausible explanations for abnormal observations. Abduction is distinctly different from deduction and induction. Abduction is inherently uncertain, and uncertainty is an important issue in abductive reasoning.

The **Symbol Grounding Problem** is related to the problem of how words (symbols) get their meanings, and hence to the problem of what meaning itself really is. The problem of meaning is in turn related to the problem of consciousness, or how it is that mental states are meaningful. According to a widely held theory of cognition, "computationalism", cognition or thinking is just a form of computation. But computation in turn is just formal symbol manipulation; symbols are manipulated according to rules that are based on the symbols' shapes, not their meanings. How are those symbols (e.g., the words in our heads) connected to the things they refer to?

Symbolic System: A symbolic system is one in which the algorithm is divided into two components: a set of knowledge (represented as symbols such as words, numbers, sentences, or pictures) and a reasoning algorithm that manipulates those symbols to create new combinations of symbols that hopefully represent problem solutions or new knowledge. There was also an influential philosophical argument made that symbolic approaches weren't biologically plausible. The proponents argued that you can't understand how a human being plans a route by using a symbolic route planning algorithm any more than you can understand how human muscles work by studying a forklift truck. The effect was a move toward natural computing: techniques inspired by biology or other natural systems. These techniques include neural networks, genetic algorithms, and simulated annealing. It is worth noting, however, that some of the techniques that became fashionable in the 1980s and 1990s were invented much earlier. Neural networks, for example, predate the symbolic era; they were first suggested in 1943 (McCulloch and Pitts, 1943). Unfortunately, the objective performance of some of these techniques never matched the evangelizing rhetoric of their most ardent proponents. Gradually, mainstream AI researchers realized that the key ingredient of this new approach was not so much the connection to the natural world, but the ability to handle uncertainty and the importance it placed on solving real-world problems. They understood that techniques such as neural networks could be explained mathematically in terms of a rigorous probabilistic and statistical framework. Free from the necessity for any natural interpretation, the probabilistic framework could be extended to find the core of modern statistical AI that includes Bayes nets, support vector machines (SVMs), and Gaussian processes.

Probabilistic Reasoning: The form of probabilistic reasoning described here is based on the Bayesian method introduced by the clergyman Thomas Bayes in the 18th century. This form of reasoning depends on the use of conditional probabilities of specific events when it is known that other events have occurred. For two events H and E with the probability PI>0, the conditional probability of event H, given that event E has occurred, is defined as:

$$P\big((E|H)\big) = P(H \& E)/P(H) \tag{4.4}$$

Frames were first introduced by Marvin Minsky as a data structure to represent a mental model of a stereotypical situation such as driving a car, attending a meeting, or eating in a restaurant. Knowledge about an object or event is stored together in memory as a unit. Then, when a new situation is encountered, an appropriate frame is selected from memory for use in reasoning about the situation. Frames are general record-like structures that consist of a collection of slots and slot values. The slots may be of any size and type. Slots typically have names and values or subfields called facets. Facets may also have names and any number of values. *Scripts* are another structured representation scheme introduced by Roger Schank. They are used to represent sequences of commonly occurring events. They were originally developed to capture the meanings of stories or to "understand" natural language text. In that respect, they are like a script for a play. A script is a predefined frame-like structure that contains expectations, inferences, and other knowledge that is relevant to a stereotypical situation. Scripts are constructed using basic primitive concepts and rules of formation, somewhat like conceptual graphs.

Commercial systems for translation, text understanding, and speech understanding now draw on considerable understanding of semantics and context as well as syntax. Another turning point came with the development of knowledge-based systems in the 1960s and early 1970s. Ira Goldstein and Seymour Papert described the demonstrations of the **DENDRAL** program in the mid-1960s as a "paradigm shift" in AI toward knowledge-based systems. Prior to that, logical inference, and resolution theorem proving in particular, had been more prominent. In the decades after the 1960s, the demonstrations have become more impressive. And our ability to understand their mechanisms has grown. Considerable progress has been achieved in understanding common modes of reasoning that are not strictly deductive, such as case-based reasoning, analogy, induction, reasoning under uncertainty, and default reasoning. Knowledge representation and inference remain the two major categories of issues that need to be addressed, as they were in the early demonstrations. Ongoing research on learning, reasoning with diagrams, and integration of diverse methods and systems will likely drive the next generation of demonstrations.

Unlike most of the early work in AI, game researchers were interested in developing high-performance, real-time solutions to challenging problems. This led to an ends-justify-the-means attitude: the result, a strong chess program, was all that mattered, not the means by which it was achieved. In contrast, much of the mainstream AI work used simplified domains, while eschewing real-time performance objectives. Checker playing requires modest intelligence to understand and considerable intelligence to master. Samuel's checkers-playing program is all the more impressive because the program learned through experience to improve its own checker-playing ability, from playing human opponents and playing against other computers. In addition, chess has long been considered a game of intellect, and many pioneers of computing felt that a chess-playing machine would be the hallmark of true AI. Chess is quite obviously an enterprise that requires thought.

Claude Shannon described two approaches to computer chess: Type-A programs, which would use pure brute force, examining thousands of moves, and using a min-max search algorithm. Or, Type-B, programs that would use specialized heuristics and "strategic" AI, examining only a few, key candidate moves. Perhaps the best-known Type-A program is IBM's Deep Blue. In 1997 Deep Blue challenged and defeated the then world chess champion Gary Kasparov. Deep Blue evaluated around 200 million positions a second (or simply 200 million moves per second) and averaged 8–12 ply search depth, while humans on the other hand are generally thought to examine nearly 50 moves to various depths. When we correlate computing power into strength, chess has a branching factor of about 40, while the game of Go typically has a branching factor of 200 (Valid moves a player can make at any point in time.). Chess was widely used as a vehicle for studying inference and representation mechanisms in the early decades of AI work. If you ask if Deep Blue uses AI, the short answer is "no". Earlier computer designs that tried to mimic human thinking weren't very good at it. No formula exists for intuition, and thus Deep Blue relies more on computational power and a simpler search and evaluation function.

Before long, AI researchers found new ways to make software programs. Newell and Simon began to talk about the possibility of teaching machines to think. In their "General Problem Solver" system, built in the late 1950s, Allen Newell, J.C. Shaw, and Herbert A. Simon showed ways to describe processes in terms of statements like "If the difference between what you have and what you want is of kind D, then try to change that difference by using method M". This and other ideas led to what we call "means-ends" and "do if needed" programming methods. Such programs automatically apply rules whenever they're needed, so the programmers don't have to anticipate when that will happen. This started an era of programs that could solve problems in ways their programmers could not anticipate, because the programs could be told what sorts of things to try, without knowing in advance which would work. Everyone knows that if you try enough different things at random, eventually you can do anything. But when that takes a million billion trillion years, like those monkeys hitting random typewriter keys, it's not intelligence, just evolution. The new systems didn't do things randomly but used "advice" about what was likely to work on each kind of problem. So, instead of wandering around at random, such programs could sort of feel around, the way you'd climb a hill in the dark by always moving up the slope. Since then, much Al research has been aimed at finding more "global" methods, to get past different ways of getting stuck, by making programs take larger views and plan ahead. Still, no one has discovered a "completely general" way to always find the best method and no one expects to. Instead, many Al researchers aim toward programs that will match patterns in memory to decide what to do next. The basic goal was to find perfect definitions for ordinary words and ideas.

Just prior to the Dartmouth workshop, Newell, Shaw, and Simon had programmed a version of Logic Theorist (LT) on a computer at the RAND Corporation called the JOHNNIAC (named in honor of John von Neumann). As a breakthrough, it was the first program in symbolic AI, which uses symbols or concepts, rather than data, to train AI to think like a person. This became the predominant approach to AI up until the 1990s. Later papers described how it proved some of the theorems in symbolic logic that were proved by Bertrand Russell and Alfred Whitehead in Volume I of their classic work, *Principia Mathematica.* LT worked by performing transformations on Russell and Whitehead's have axioms of propositional logic, represented for the computer by "symbol structures", until a structure was produced that corresponded to the theorem to be proved. Because there are so many different transformations that could be performed, ending the appropriate ones for proving the given theorem involves what computer science people call a "search process". To describe how LT and other symbolic AI programs work, I need to explain first what is meant by a "symbol structure" and what is meant by "transforming" them. In a computer, symbols can be combined in lists, such as (A; 7; Q). Symbols and lists of symbols are the simplest kinds of symbol structures. More complex structures are composed of lists of lists of symbols, such as ((B; 3); (A; 7; Q)), and lists of lists of lists of symbols, and so on. Because such lists of lists, etc. can be quite complex, they are called "structures". Computer programs can be written that transform symbol structures into other symbol structures. For example, with a suitable program the structure "(the sum of seven and five)" could be transformed into the structure "(7+5)", which could further be transformed into the symbol "12".

Newell and Simon worked on it all: complex information processing, symbolic computation, heuristic methods, human problem-solving, a programming language, and empirical exploration. Examination of the work of De Groot had led Newell and Simon to begin to use protocols, the talking evidence of problem-solving processes. The protocol of solving logic problems proved to be a key in human problem-solving: means-ends (M-E) analysis. In M-E analysis the problem solver compares the current situation with the goal situation; finds a difference between them; finds in memory an operator that experience has taught reduces differences of this kind; and applies the operator to change the situation. The idea of M-E analysis led to the General Problem Solver (Newell, Shaw, and Simon), a program that could solve problems in a number of domains after being provided with a problem space (domain representation), operators to move through the space, and information about which operators were relevant for reducing which differences. Herbert Gelernter was a professor in computer science at Stony Brook University and his research encompassed the areas of theoretical physics, AI, expert systems, and machine learning. In 1958, Gelernter teamed with Nathaniel Rochester to consider the case of a machine that can prove theorems in elementary Euclidean plane geometry.

At the same time at the 1959 Paris conference where Gelernter presented his program, Allen Newell, J. C. Shaw, and Herb Simon gave a paper describing their recent work on mechanizing problem-solving. Their program, which they called the "General Problem Solver (GPS)", was an embodiment of their ideas about how humans solve problems. Indeed, they claimed that the program itself was a theory of human problem-solving behavior. Newell and Simon were among those who were just as interested (perhaps even more interested) in explaining the intelligent behavior of humans as they were in building intelligent machines. They wrote "It is often argued that a careful line must be drawn between the attempt to accomplish with machines the same tasks that humans perform, and the attempt to simulate the processes humans actually use to accomplish these tasks". The GPS was designed to solve a variety of problems that could be formulated as a set of objects and operators, where the operators were applied to the objects to transform them into a goal object through a sequence of applications.

Information about the problem (the nature of the states, the cost of transforming from one state to another, the promise of taking a certain path, and the characteristics of the goals) can sometimes be used to help guide the search more efficiently. This information can often be expressed in the form of a heuristic evaluation function f(n,g), a function of the nodes n and/or the goals g. GPS maximally confuses the two approaches "with mutual benefit". GPS was an outgrowth of their earlier work on the LT in that it was based on manipulating symbol structures (which they believed humans did also). But GPS had an important additional mechanism among its symbol-manipulating strategies. Like Gelernter's geometry program, GPS transformed problems into subproblems, and so on. GPS's innovation was to compute a "difference" between a problem to be solved (represented as a symbol structure) and what was already known or given (also represented as a symbol structure). However, GPS failed to solve complicated problems. The program was based on formal logic and therefore could generate an infinite number of possible operators, which is inherently inefficient. The amount of computer time and memory that GPS required to solve real-world problems led to the project being abandoned. Gelernter soon began a research project to develop a geometry theorem-proving machine. Gelernter's program exploited two important ideas. One was the explicit use of subgoals (sometimes called "reasoning backward" or "divide and conquer"), and the other was the use of a diagram to close off futile search paths.

One of the hallmarks of an expert system is the use of specific knowledge of its domain of application, applied by a relatively simple inference engine. The phrase "knowledge programming" has been used to denote the relative emphasis of the effort of building an expert system. The single most important representational principle is the principle of declarative knowledge enunciated by McCarthy in the formative years of Al. Simply put, this principle states that knowledge must be encoded in an intelligent program explicitly, in a manner that allows other programs to reason about it. Arbitrary Fortran or LISP procedures, for example, cannot be explained or edited by other programs (although they can be compiled and executed), while stylized attribute value pairs, record structures, or other, more complex, stylized data structures can be. To a certain extent, a knowledge base is a database. The essential differences between knowledge bases and databases are flexibility and complexity of the relations. Current research on Al and Databases (sometimes called Expert Database Systems) is reducing these differences. It requires an organizational paradigm plus data structures for implementation. These two parts together constitute the representation of knowledge in an Al program.

Elements of knowledge needed for problem-solving may be organized around either the primary objects (or concepts) of a problem area or around the actions (including inferential relations) among those objects. For each type of representation, one may identify the primitive unit and the primitive action. The search for new logical formalisms that are more powerful than predicate calculus reflects the tension between simple, well-understood formalisms and expressive power. In brief, logic indicates an approach but does not provide a complete solution. Numerous extensions must be made to express some of the concepts that are frequently used in applications: uncertainty, strategy knowledge, and temporal relations. Experience has shown that declarative, modular representations are useful for expert systems. Some information is more difficult to encode in the action-centered paradigm, and other information is more difficult in the object-centered paradigm. McCarthy's long-term objective was to formalize common sense reasoning, the prescientific reasoning that is used in dealing with everyday problems.

Cyc is a large AI project started and run by Douglas Lenat. It consists of a knowledge base of hand-coded "common sense" facts and an inference engine to deduce further facts. Lenat started the Cyc project in 1984 after being frustrated by the difficulty of hand-coding domain-specific knowledge from his previous AI project. The objective of the Cyc project is to codify, in machine-usable form, the millions of pieces of knowledge that compose **human common sense**. This entailed, along the way, (1) developing an adequately expressive representation language, CycL, (2) developing an ontology spanning all human concepts down to some appropriate level of detail, (3) developing a knowledge base on that ontological framework, comprising all human knowledge about those concepts down to some appropriate level of detail, and (4) developing an inference engine exponentially faster than those used in then-conventional expert systems, to be able to infer the same types and depth of conclusions that humans are capable of, given their knowledge of the world.

In 2002, a subset of the knowledge base and functionality was released to the public, and in 2012, OpenCyc had around 240,000 concepts and 2,000,000 assertions. Cyc uses a declarative language called CycL based on first-order logic. CycL was written in LISP and has a similar syntactical appearance. The main lexical component is a set of concepts, also called constants, which begin with #$. These constants can be specific objects, collections of objects, or relations between objects. Cyc enables a semantic reasoner to perform human-like reasoning and be less "brittle" when confronted with novel situations. The Cyc project aims to assemble a comprehensive ontology and knowledge base that spans the basic concepts and rules about how the world works. Hoping to capture common sense knowledge, Cyc focuses on implicit knowledge that other AI platforms may take for granted. The inference engine would verify the truth of the parenthetical statements, and then the implication. Cyc can also provide a translation of these structures into a natural English sentence.

The Information-Processing Language (IPL) language is tailored to AI and is the contemporary to FORTRAN in numerical computing and settled once and for all the essentiality of higher-level languages for sophisticated programming. The IPLs were designed to meet the needs for flexibility and generality: flexibility, because it is impossible in these kinds of computations to anticipate before run time what sorts of data structures will be needed and what memory allocations will be required for them; generality, because the goal is not to construct programs that can solve problems in particular domains, but to discover and extract general problem-solving mechanisms that can operate over a range of domains whenever they are provided with an appropriate definition for each domain. To achieve this flexibility and generality the IPLs introduced many ideas that have become fundamental for computer science in general, including lists, associations, schemas (frames), dynamic memory allocation, data types, recursion, associative retrieval, functions as arguments, and generators (streams). In 1958, LISP embedded these list-processing ideas in the lambda calculus, improved their syntax, and incorporated a "garbage collector" to recover unused memory, soon became the standard programming language for AI. A closely related set of ideas that Newell and Simon developed at about the same time, out of concern with the program control problem, led to a decentralized system in which independent processes add information to a common memory ("blackboard") and obtain information they need from that memory. The blackboard idea has achieved wide use in speech recognition, vision programs, and elsewhere.

The LT introduced several concepts that would be central to AI research:

1. Reasoning as Search: LT explored a search tree: the root was the initial hypothesis; each branch was a deduction based on the rules of logic. Somewhere in the tree was the goal: the proposition the program intended to prove. The pathway along the branches that led to the goal was a proof, a series of statements, each deduced using the rules of logic, that led from the hypothesis to the proposition to be proved.
2. Heuristics: Newell and Simon realized that the search tree would grow exponentially and that they needed to "trim" some branches, using "rules of thumb" to determine which pathways were unlikely to lead to a solution. They called these *ad hoc* rules "heuristics", using a term introduced by George Pólya. Heuristics would become an important area of research in AI and remains an important method to overcome the intractable combinatorial explosion of exponentially growing searches.

3. List Processing: To implement LT on a computer, the three researchers developed a programming language, IPL, which used the same form of symbolic list processing that would later form the basis of McCarthy's LISP programming language, an important language still used by AI researchers.

Probability theory is used to discuss events, categories, and hypotheses about which there is not 100% certainty. We might write A→B, which means that if A is true, then B is true. If we are unsure whether A is true, then we cannot make use of this expression. In many real-world situations, it is very useful to be able to talk about things that lack certainty. Bayes' theorem can be used to calculate the probability that a certain event will occur or that a certain proposition is true. Bayesian statistics lie at the heart of most statistical reasoning systems.

The most promising work at that time, *probabilistic reasoning* works with conditional probabilities for propositional calculus formulas. Since then, many diagnostic and expert systems have been built for problems of everyday reasoning using *Bayesian networks*. The success of Bayesian networks stems from their intuitive comprehensibility, the clean semantics of conditional probability, and from the centuries-old, mathematically grounded probability theory. The weaknesses of logic, which can only work with two truth values, can be solved by *fuzzy logic*, which pragmatically introduces infinitely many values between zero and one. Probabilistic AI will be addressed when we discuss probabilistic robotics.

1. Mathematicians promoted mathematical logic and deductive reasoning as the language of rational thought ~ symbolic AI.
2. Subsonic AI – Perceptron (multiple numerical inputs and one output) – information processing in neurons can be simulated by a computer program. Rosenblatt stated that all its knowledge is encoded in numbers making up its weight and thresholds. It is possible to learn via conditioning (also known as supervised learning).
3. Deductive reasoning where programs extract statistics from data and use probabilities to deal with uncertainty.
4. Taking inspiration from biology and psychology to create brain-like programs from the clean semantics of conditional probability, and from the centuries-old, mathematically grounded probability theory.

Classical AI emphasized computational intelligence: knowledge bases and semantic formation processing. However, problems arose with hard-coded, top-down control because the robot did not interact with the real world. Hubert discovered that the "Computer simulation assumes incorrectly that explicit rules can govern intellectual processes". However, human intelligence breaks rules (Rules are only for elementary capabilities and later, rules are routinely broken). The coming change was the behaviorist approach. Thus, the robot should be spending less time deliberating and more time responding to the real world that is in constant flux (Agre & Chapman). Thus, no hard-core algorithms. This phase of robotics led to the Three Tier-architecture, combining the deliberate approach with the behaviorist or reactive approach.

Piano Mover Problem (1996–2004): In the early 1990s, when deterministic approaches were not able to cope with the combinatorial complexity inherent in the so-called 'piano mover' problem, new paradigms of resolution appeared, first and foremost at Stanford and Utrecht universities. This is the beginning of probabilistic methods. He contributes to the movement in collaboration with his colleague T. Siméon, both theoretically (introduction of an original mechanism for controlling random sampling methods) and practically through the development of a generic software platform, a platform that was to be enhanced in 2000 with the creation of the start-up Kineo CAM. The company he initially managed develops and markets software components dedicated to motion planning in the field of PLM (Product Lifecycle Management), mainly in the automotive and aeronautical sectors. It was acquired by Siemens in 2012.

There should be skepticism. In 1770, Wolfgang von Kempelen constructed an automaton chess-player, which was referred to in modern times as the Turk. This machine appeared to play a strong game of chess against a human opponent. However, it was discovered fraudulent where a human operator sat inside the machine behind some clockwork of gears and cogs. The Turk is shown in Figure 4.11, and served as an inspiration for speaking machines like Alexander Graham Bells' telephone and a precursor

FIGURE 4.11
Racknitz's engraving of the Turk.

to chess playing automatons like IBM's Deep Blue. In 1957, Frank Rosenblatt, a psychologist at the Cornell Aeronautical Laboratory in Buffalo, New York, began work on neural networks under a project called PARA (Perceiving and Recognizing Automaton). He was motivated by the earlier work of McCulloch and Pitts and of Hebb and was interested in these networks, which he called perceptrons, as potential models of human learning, cognition, and memory. A perceptron consists of a network of these neural elements, in which the outputs of one element are inputs to others.

Distributed AI, DAI, has been an active area of research since about 1985. One of its goals is the use of parallel computers to increase the efficiency of problem solvers. It turned out, however, that because of the high computational complexity of most problems, the use of "intelligent" systems is more beneficial than parallelization itself. More than nearly any other science, AI is interdisciplinary, for it draws upon interesting discoveries from such diverse fields as logic, operations research, statistics, control engineering, image processing, linguistics, philosophy, psychology, and neurobiology. The human brain has many interesting properties. Raj Reddy speculated that there are about one hundred billion neural cells in the human brain and the brain might be performing 200 trillion operations per second if not faster than that. In problem domains such as vision, speech, and motor processes, "it is more powerful than 1,000 supercomputers; however, for simple tasks such as multiplication, it is less powerful than a four bit microprocessor". These processing events taking place in the brain require little conscious effort and awareness on the part of humans and they are extremely difficult for machines to emulate.

DENDRAL was a project in AI of the 1960s, and the computer software expert system that it produced. Its primary aim was to study hypothesis formation and discovery in science. For that, a specific task in science was chosen: help organic chemists in identifying unknown organic molecules, by analyzing their mass spectra and using knowledge of chemistry. It was done at Stanford University by Edward Feigenbaum, Bruce G. Buchanan, Joshua Lederberg, and Carl Djerassi, along with a team of highly creative research associates and students. It began in 1965 and spans approximately half the history of AI research. The software program DENDRAL is considered the first expert system because it automated the decision-making process and problem-solving behavior of organic chemists. The project consisted of research on two main programs **Heuristic DENDRAL** and **Meta-DENDRAL**, and several sub-programs. It was written in the LISP programming language, which was considered the language of AI because of its flexibility.

Among the pursuers of the GOFAI approach were those who used logical representations and logical reasoning methods, ideas pioneered by John McCarthy. These people were sometimes called "logicists". What AI needed was more of this or that alternative to AI's reigning paradigm, the paradigm John Haugeland called "good-old-fashioned AI" or GOFAI. GOFAI, of course, had as its primary rationale Newell and Simon's belief that a "physical symbol system has the necessary and sufficient means for intelligent action". But GOFAI seemed to be running out of steam during the 1980s, making it vulnerable to challenges by AI researchers themselves, challenges that had to be taken more seriously than those of Searle, Dreyfus, Penrose, and others outside of the field. Drew McDermott, a professor at Yale University (who received his Ph.D. from MIT), was one of those who began to have doubts about the role of logic in AI. This fact was significant because McDermott himself had been a prominent logicist, but in an influential 1987 paper he concluded that the premise that ". . . a lot of reasoning can be analyzed as deductive or approximately deductive, is erroneous". Dreyfus's main point is that intelligence in humans derives from their "being in the world" and not because they are guided by rules. The use of rules in AI programs (as in humans) might allow competent behavior but not expert behavior.

In the early 1980s, expert systems were believed to represent the future of AI and of computers in general. To date, however, they have not lived up to expectations. Many expert systems help human experts in such fields as medicine and engineering, but they are very expensive to produce and are helpful only in special situations. Now, the hottest area of AI is with neural networks, which are proving successful in a number of disciplines such as voice recognition and natural language processing. No computers exhibit full AI (i.e., are able to simulate human behavior). The greatest advances have occurred in the field of games playing where the best computer chess programs are now capable of beating humans. In May 1997, an IBM super-computer called Deep Blue defeated world chess champion Gary Kasparov in a chess match. Static programs with explicit rules and knowledge bases shifted the robot away from the real world. The plan was to incorporate machine-learning techniques, that is, ANNs (supervised), genetic algorithms, reinforcement learning (RL; unsupervised) to improve AI. The goal for intelligence was to encourage emergent behavior. The issue was computer programs are goal-seeking, while humans are value-seeking. The next new paradigm was to look to neuroscience, cognitive psychology, and biology for new insights into AI. It is important to recognize that human action is derived from motor skills and implicit behavior (that lies beneath the layer of conscious awareness). The main focus of current AI is to develop methods to match or exceed human performance in certain domains, possibly by very different means.

Narrow AI, also known as weak AI, is an application of AI technologies to enable a high-functioning system that replicates, and perhaps surpasses, human intelligence for a dedicated purpose. Narrow AI is often contrasted with general AI, sometimes called strong AI; a theoretical AI system that could be applied to any task or problem or forms the underpinning of the four categories in the field of AI: thinking humanly, acting humanly, thinking rationally, and acting rationally. Expert systems were probably the "hottest" topic in AI at that time. Previously in trying to find solutions to problems, AI researchers tended to rely on search techniques or computational logic. These techniques were successfully used to solve elementary or toy problems or very well-structured problems such as games. However, real complex problems are prone to have the characteristic that their search space tends to expand exponentially with the number of parameters involved. For such problems, these older techniques have generally proved to be inadequate and a new approach was needed. This new approach emphasized knowledge rather than search and has led to the field of Knowledge Engineering and Expert Systems. Figure 4.12 provides a global map of Artificial Intelligence, beginning with the foundational sciences (philosophy, neuroscience, psychology, and logicism) to Artificial Intelligence as both a science and engineering solution, and producing a diverse set of AI capabilities. The following sections describe these "weak AI" capabilities in greater detail.

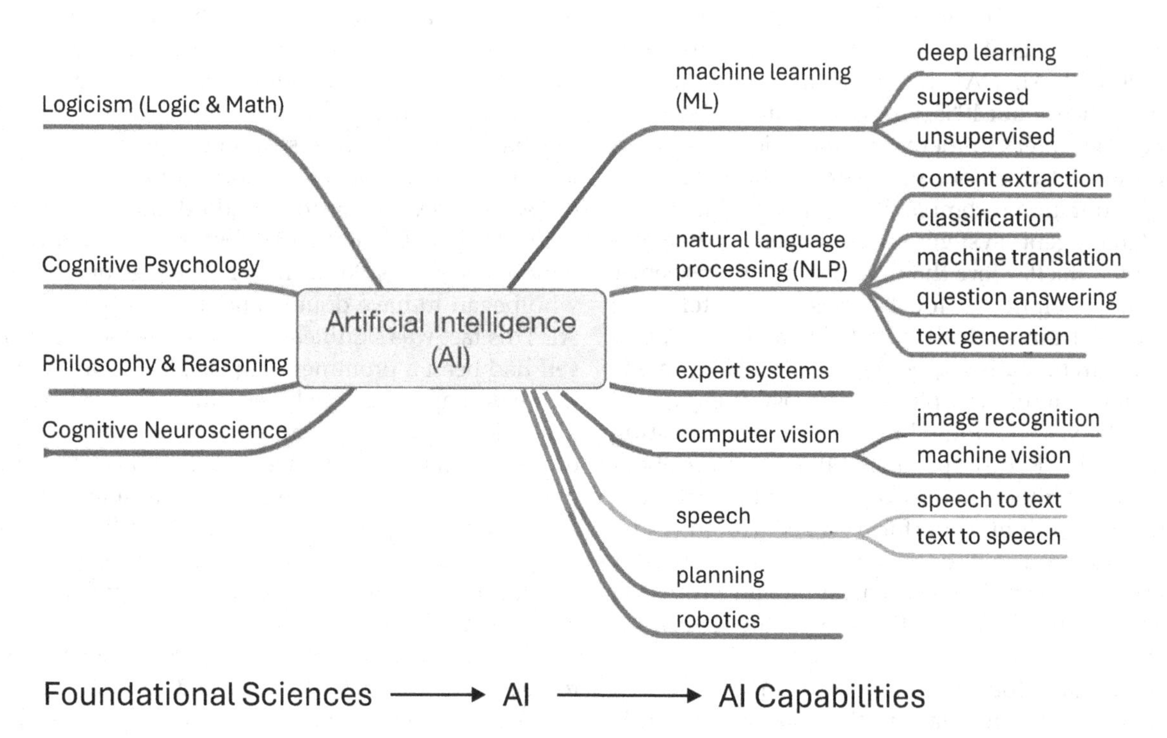

FIGURE 4.12
What makes up AI?

Expert System

An expert system is a particular kind of knowledge-based system and is a set of programs that manipulate encoded knowledge to solve problems in a specialized domain that normally requires human expertise. Essentially, expert systems capture the knowledge of an expert. Knowledge is represented by rules (if–then rules), semantic nets (Hierarchy) and frames (Shared characteristics). An expert system is a problem-solving method (inference engine) operating on a knowledge base separating the problem-solving method from the knowledge it uses making expert systems easy to develop and maintain. By attending to the characteristics of the problem-solving method, it's possible to automate the task of collecting and encoding knowledge. An expert is an experienced practitioner in his/her field. More than that, he/she is a highly effective problem solver and decision-taker in that field. Experts have three qualities: make good decisions, make those decisions quickly, and able to cope with a wide range of problems. Expert systems mimic the decision-making process of the human brain. The task that an expert system performs will generally be regarded as difficult. An expert system almost always operates in a rather narrow field of knowledge. The field of knowledge is called the knowledge domain of the system. An AI-robot system contains an internal representation of its external application domain (also referred to as the robot's environment), where relevant elements of the application domain (objects, relations, classes, laws, and actions) are represented as symbolic expressions. The industry adopted them on a relatively large scale, but many such projects failed. In 1958, Friedberg's machine evolution (now better known today as hill-climbing) using mutations was tried, but it failed to find good solutions.

Only in the last half-century have we had computational devices and programming languages powerful enough to build experimental tests of ideas about what intelligence is. Expert systems, as a subset of AI, first emerged in the early 1950s when the Rand-Carnegie team developed the GPS to deal with theorems proof, geometric problems, and chess playing. Allen Newell, J. Clifford Shaw, and Herb Simon were also writing programs in the 1950s that were ahead of their time in vision but limited by the tools. Their LT program was another early tour-de-force, startling the world with a computer that could invent proofs of logic theorems, which unquestionably requires creativity as well as intelligence. Newell and Simon acknowledge the convincingness of Oliver Selfridge's early demonstration of a symbol-manipulation program for recognition (see Feigenbaum and Feldman). Selfridge's work on learning and a multiagent approach to problem-solving (later known as blackboards), plus the work of others in the early 1950s, were also impressive demonstrations of the power of heuristics. The early demonstrations established a fundamental principle of AI to which Simon gave the name "satisficing":

> In the absence of an effective method guaranteeing the solution to a problem in a reasonable time, heuristics may guide a decision maker to a very satisfactory, if not necessarily optimal, solution.

Expert systems (also referred to as Rule-based) is a knowledge base(s) consisting of hundreds or thousands of rules of the form:

- IF (condition) THEN (action).
- Use rules to store knowledge ("rule-based").
  - The rules are usually gathered from experts in the field being represented ("expert system").
- Most widely used knowledge model in the commercial world.

Expert systems are computer programs aiming to model human expertise in one or more specific knowledge areas. They attempt to capture the knowledge of a human expert and make it available through a computer program. Discovery that detailed knowledge of a specific domain can help control search and lead to expert-level performance for restricted tasks. They usually consist of three basic components: (1) a knowledge database with facts and rules representing human knowledge and experience; (2) an inference engine processing consultation and determining how inferences are being made; and (3) an input/output interface for interactions with the user. Expert systems can be characterized by:

- using symbolic logic rather than only numerical calculations.
- the processing is data-driven.
- a knowledge database containing explicit contents of certain area of knowledge; and

- the ability to interpret its conclusions in the way that is understandable to the user.

Expert systems, as a subset of AI, first emerged in the early 1950s when the Rand-Carnegie team developed the GPS to deal with theorems proof, geometric problems, and chess playing. The Basic Components of an Expert System are (1) A problem statement with a goal, (2) A knowledge base, (3) A selector that identifies relevant, a strategy that determines which knowledge to apply in which order, (4) An inference procedure that applies the chosen knowledge, (5) A working memory for storing parts of the solution as they develop, and (6) A friendly interface that helps the user. In contrast to humans, expert systems can provide permanent storage for knowledge and expertise; offer a consistent level of consultation once they are programmed to ask for and use inputs; and serve as a depository of knowledge from potentially unlimited expert sources and thereby providing comprehensive decision support. Knowledge acquisition is usually considered as a way to discover static facts of the world and the relationships of various events that human uses in solving real-life problems. The problem-solving skills in humans oftentimes are far more complicated and complex than what knowledge collection can achieve.

Work on expert systems emphasizes problems of commercial or scientific importance, as defined by persons outside of AI. Expert systems continue to build on and contribute to Al research by testing the strengths of existing methods and helping to define their limitations. The process is a continuing one. In the 1970s, expert systems work developed the use of production systems, based on the early work in psychological modeling. In the 1980s, fundamental work on knowledge representation has evolved into useful object-oriented substrates. Hardware developments in the last decade have made a significant difference in the commercialization of expert systems. Some expert systems use a theorem prover to determine the truth or falsity of propositions and to bind variables so as to make propositions true. Others use their own interpreters in order to incorporate more than a theorem prover provides, but most importantly, capabilities for controlling the order of inferences, strategic reasoning, and reasoning under uncertainty. Most fielded rule-based expert systems have used specialized rule interpreters, not based directly on logic (Figure 4.13).

The principle distinction between expert systems and traditional problem-solving programs is the way in which the problem-related expertise is coded. In conventional applications, problem expertise is encoded in both program and data structures. In the expert system approach, all of the problem-related expertise is encoded in data structures only, none is in programs. Generally, in expert systems, the use of knowledge is

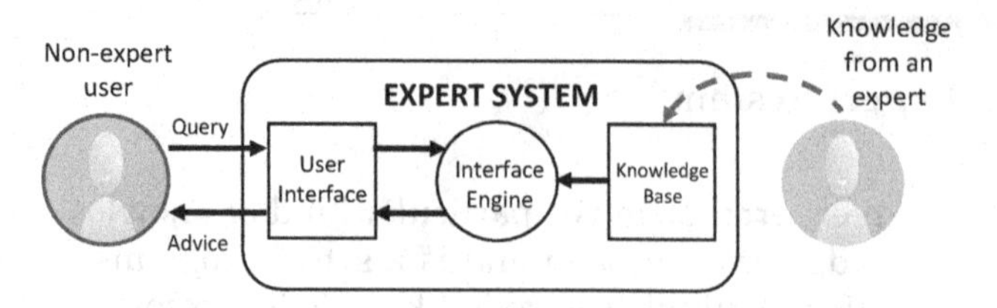

FIGURE 4.13
Expert system.

vital. But in conventional system, data is used more efficiently than knowledge. Conventional systems are not capable of explaining a particular conclusion for a problem. These systems try to solve in a straightforward manner. But expert systems are capable of explaining how a particular conclusion is reached and why requested information is needed during a process. However, the problems are solved more efficiently than a conventional system by an expert system. Generally, in an expert system, it uses the symbolic representations for knowledge, that is, the rules, different forms of networks, frames, scripts, etc. and performs their inference through symbolic computations. But conventional systems are unable to express these terms. They just simplify the problems in a straightforward manner and are incapable to express the "how, why" questions. Also, the problem-solving tools that are present in expert system are purely absent in conventional systems. The various types of problems are always solved by the experts in an expert system. So, the solution of the problem is more accurate than a conventional system. By the definition, an expert system is a computer program that simulates the thought process of a human expert to solve complex decision problems in a specific domain. The expert system's knowledge is obtained from expert sources which are coded into most suitable form. The process of building an expert system is called knowledge engineering and is done by a knowledge engineer.

Rule-based, or expert, systems use the following: IF (condition) THEN (action). We use rules to store knowledge ("rule-based"). The rules are usually gathered from experts in the field being represented ("expert system"). Most widely used knowledge model in the commercial world. Knowledge system does not learn from its experience. For example, case-based reasoning (CBR) theory that focuses on solving new problems based on similar past problem solutions seems to be able to eliminate the complex task of maintaining rules and facts through the use of adaptive acquisition of problem-solving techniques. The ability of expert system to derive correct output is often compromised by the lack of precision in rules and inputs. Inference engine is built upon algorithms that manipulate knowledge in the form of decision tree which is not designed to handle uncertainty. It was a perception that would later bring an inevitable skeptical backlash. A rule-based system is a system in which knowledge is represented completely in terms of rules (e.g., a system based on production rules). Expert systems are becoming increasingly more important in both decision support which provides options and issues to decision makers, and decision-making where people can make decisions beyond their level of knowledge and experience. They have distinct advantages over traditional computer programs. In contrast to humans, expert systems can provide permanent storage for knowledge and expertise; offer a consistent level of consultation once they are programmed to ask for and use inputs; and serve as a depository of knowledge from potentially unlimited expert sources and thereby providing comprehensive decision support. However, everyone involved has referred to the process of putting knowledge into a knowledge base as a "bottleneck" in building expert systems.

Several methods are used in expert systems to deal with uncertainty arising from (1) uncertainty of the data, (2) less than certain associations between data and conclusions, and (3) combinations of these. The major methods for addressing these issues are listed next.

1. Abstraction: assume that the uncertainty is small and can be safely ignored.
2. Bayes' Theorem: use prior and posterior probabilities to represent less than certain data and associations; then compute new probabilities with some variation of Bayes' Theorem.
3. Fuzzy Logic: represent the uncertainty of propositions such as "John is tall" with a distribution of values, then reason about combinations of distributions.
4. Criterion Tables: assign categories or weights to clauses in rules based on their relative importance in drawing conclusions (e.g., major and minor findings associated with a disease); then allow a conclusion to be drawn if sufficient numbers of clauses in each category are true.
5. Certainty Factors (CFs): assign single numbers to propositions, and to associations among propositions, representing either probabilities or a combination of probabilities and utilities; then use MYCIN'S formulas to determine CFs for inferred beliefs.

R1 (or XCON) is a Rule-based system written in OPS5 in 1978 to assist in the ordering of Digital Equipment Corporation's (DEC) VAX computer systems by automatically selecting the computer system components based on the customer's requirements. DEC was the main hardware vendor for the AI community during the 1970s, a period characterized by the birth and development of expert systems in academic research laboratories. XCON (for eXpert CONfigurer) is used for inventory management. As a background from 1975, DEC offered 50 types of central processors with 400 core operations. The estimated possible number of configurations at that time was already in the millions. So, system configuration was the key process in DEC's flexibility strategy for it converted a customer order into fully configured system that was designed, checked, and ready for delivery. This process involves three separate reviews of each order. R1 is a rule-based system that has much in common with other domain-specific systems that had been developed over the past several years at that time. It differed from these systems primarily in its use of Match rather than Generate-and-Test as its central problem-solving method, rather than exploring several hypotheses until an acceptable one is found; it exploits its knowledge of its task domain to generate a single acceptable solution.

R1's domain of expertise is configuring DEC's VAX-11/780 systems. Its input is a customer's order and its output is a set of diagrams displaying the spatial relationships among the components on the order; these diagrams are used by the technician who physically assembles the system. Since an order frequently lacks one or more components required for system functionality, a major part of R1's task is to notice what components are missing and add them to the order. R1 was used on a regular basis by DEC's manufacturing organization. Two lessons emerged from this effort to develop a computer system configuration:

1. "An expert system can perform a task simply by recognizing what to do, provided that it is possible to determine locally (i.e., at each step) whether taking some particular action is consistent with acceptable performance on the task".
2. "When an expert system is implemented as a production system, the job of refining and extending the system's knowledge is quite easy".

An expert system's knowledge is obtained from expert sources and coded in a form suitable for the system to use in its inference or reasoning processes. A system that integrates knowledge from experts is called a knowledge-based decision support system (KBDSS) or intelligent decision support system (IDSS). The expert knowledge must be obtained from specialists or other sources of expertise, such as texts, journals, articles, and databases. This type of knowledge usually requires much training and experience in some specialized field such as medicine, geology, system configuration, or engineering design. Once a sufficient body of expert knowledge has been acquired, it must be encoded in some form, loaded into a knowledge base, then tested, and refined continually throughout the life of the system. The complexity of human intelligence had been underestimated before, especially in the expert systems field. Technological limitations and managerial challenges still remain in the development of expert systems. Expert systems gave us the terminology still in use today where AI systems are divided into a knowledge base, with facts about the world and rules, and an inference engine, which applies the rules to the knowledge base in order to answer questions and solve problems. In these early systems, the knowledge base tended to be a fairly flat structure, essentially assertions about the values of variables used by the rules.

Search

Search is one of the operational tasks that characterize AI programs best. Almost every Al program depends on a search procedure to perform its prescribed functions. Problems are typically defined in terms of states, and solutions correspond to goal states. Solving a problem then amounts to searching through the different states until one or more of the goal states are found. It is customary to represent a search space as a diagram of a directed graph or a tree. Each node or vertex in the graph corresponds to a problem state, and arcs between nodes correspond to transformations or mappings between the states. The immediate successors of a node are referred to as children, siblings, or offspring, and predecessor nodes are ancestors. An immediate ancestor to a node is a parent. Search can be characterized as finding a path through a graph or tree structure. Searching the knowledge base leads to inference.

Problem-solving is basically a state-space search, that is, primarily an algorithm and not a representation. Modern AI researchers were inspired by earlier work by Alan Turing, Claude Shannon, and Norbert Wiener on tree search for playing chess. From the AI workshop at Dartmouth, tree search for game playing, for proving theorems, for reasoning, for perceptual processes such as vision and speech and for learning became the dominant mode of thought and continues to be so today for large parts of the academic enterprise called AI. Search algorithms are mostly intrinsically serial, which contrasts with the ways in which large parts of animal nervous systems work, where there are both feedforward and feedback signals, but over very low-diameter networks. The speed of the individual neurons is so slow compared with the speed of overall action of the creature that very different intrinsic computations must be at play. There have been numerous flirtations with neurally inspired networks, but most of these are used as component-learning black boxes within a more traditional framework of serial computation. A state can be expanded by generating all states that can be reached by applying a legal operator to the state. State space can also be defined by a successor function that returns all states produced by applying a single legal operator (based on propositional logic). A search tree is generated by generating search nodes by successively expanding states starting from the initial state as the root. A search node in the tree can contain corresponding state, parent node, operator applied to reach this node, length of path from root to node (depth), and path cost of path from initial state to node.

The standard search techniques include blind search, AND/OR graphs, heuristics, minimax, and alpha-beta pruning. The search algorithm is the easiest way to implement various search strategies and to maintain a queue of unexpanded search nodes. Different strategies result from different methods for inserting new nodes in the queue. Search strategies have different properties such as completeness, time complexity, space complexity, and optimality. There are two major strategies: uniformed search strategies (blind, exhaustive, brute force) that do not guide the search with any additional information about the problem and informed search strategies (heuristic, intelligent) that use information about the problem (estimated distance from a state to the goal) to guide the search. The typical search tools include **breath-first**, which expands search nodes level by level, all nodes at level d are expanded before expanding nodes at level $d+1$. The search continues by implementing and adding new nodes to the end of the queue (FIFO queue). Since this method eventually visits every node to a given depth, it is guaranteed to be complete. Also, optimal provided path cost is a nondecreasing function of the depth of the node (e.g., all operators of equal cost) since nodes are explored in depth order. **Breadth first** is another option. Assume there are an average of b successors to each node, called the branching factor. Therefore, to find a solution path of length d, we must explore nodes plus needs b^d nodes in memory to store leaves in queue. For example, assuming we can expand and check 1,000 nodes/sec, we would need 100 bytes/node storage. Note memory is a bigger problem than time. In a **Uniform Cost Search algorithm**, like breadth-first except, is always an expand node of least cost instead of least depth (i.e., sort new queue by path cost). Do not recognize goal until it is the least cost node on the queue and removed for goal testing. Therefore, guarantees optimality as long as path cost never decreases as a path increases (non-negative operator costs).

The **100 prisoners problem** is a mathematical problem in probability theory and combinatorics. In this problem, 100 numbered prisoners must find their own numbers in one of 100 drawers in order to survive. The rules state that each prisoner may open only 50 drawers and cannot communicate with other prisoners. At first glance, the situation appears hopeless, but a clever strategy offers the prisoners a realistic chance of survival.

For depth-first search, always expand node at deepest level of the tree, that is, one of the most recently generated nodes. When hit a dead-end, backtrack to last choice. Implemented by adding new nodes to front of the queue. Depth-First Properties Include Not guaranteed to be complete since might get lost following infinite path, Not guaranteed optimal since can find deeper solution before shallower ones explored, Time complexity in worst case is still $O(bd)$ since need to explore entire tree. But if many solutions exist may find one quickly before exploring all of the space, Space complexity is only $O(bm)$ where m is maximum depth of the tree since queue just contains a single path from the root to a leaf node along with remaining sibling nodes for each node along the path, and can impose a depth limit, l, to prevent exploring nodes beyond a given depth. Prevents infinite regress, but incomplete if no solution within depth limit. Iterative deepening conducts a series of depth-limited searches, increasing depth limit each time. Seems wasteful since work is repeated, but most work is at the leaves at each iteration and is not repeated. Basic search methods may repeatedly search the same state if it can be reached via multiple paths. Three methods for reducing repeated work in order of effectiveness and computational overhead are: (1) do not follow self-loops (remove successors back to the same state), (2) do not create paths with cycles (remove successors already on the path back to the root), O(d) overhead, and (3) do not generate any state that was already generated. Requires storing all generated states (O(bd) space) and searching them (usually using a hash-table for efficiency).

The search for a solution in an extremely large search tree presents a problem for nearly all inference systems. Search Trees expand exponentially. From the starting state, there are many possibilities for the first inference step. As a result, we must "trim the branches" with ad hoc rules such as heuristics or "branch and bound". For each of these possibilities, there are again many possibilities in the next step, and so on. Even in the proof of a very simple formula from with three Horn clauses, each with at most three literals, the search tree for SLD resolution has the following shape:

> The tree was cut off at a depth of 14 and has a solution in the leaf node marked by ∗. It is only possible to represent it at all because of the small branching factor of at most two and a cutoff at depth 14. For realistic problems, the branching factor and depth of the first solution may become significantly bigger.

Recall that a **heuristic** is a rule of thumb or judgmental technique that leads to a solution some of the time but provides no guarantee of success. It may in fact end in failure. Heuristics play an important role in search strategies because of the exponential nature of most problems. Heuristics is intuitive knowledge learned from experience. By using heuristics, we do not have to rethink completely what to do every time we encounter a similar problem. AI methods use heuristics to reduce the complexity of problem-solving. They help to reduce the number of alternatives from an exponential number to a polynomial number and, thereby, obtain a solution in a tolerable amount of time. When exhaustive search is impractical, it is necessary to compromise for a constrained search which eliminates many paths but offers the promise of success some of the time. Here, success may be considered to be finding an optimal solution a fair proportion of the time or just finding good solutions much of the time. Informed methods add domain-specific information to select the best path along which to continue searching. Define a heuristic function $h(n)$ that estimates the "goodness" of a node n. The heuristic function is an estimate of how close we are to a goal, based on domain-specific information that is computable from the current state description. In this regard, any policy that uses as little search effort as possible to find any qualified goal has been called a sacrificing policy. Heuristics are the quick mental strategies that people use to solve problems. These are often referred to as "rule of thumb" strategies, and they allow for a person (or software in our case) to make a fast and efficient decision without having to stop and deliberate over what the next course of action will be. Heuristics are problem-solving strategies which in many cases find a solution faster than uninformed search. However, this is not guaranteed. Even though heuristics are oftentimes very helpful, they can also lead to errors, which are referred to as biases. Heuristic search could require a lot more time and can even result in the solution not being found. Heuristic decisions are closely linked with the need to make real-time decisions with limited resources. In practice, a good solution found quickly is preferred over a solution that is optimal, but very expensive to derive.

Of the various search algorithms for uninformed search, iterative deepening is the only practical one because it is complete and can get by with very little memory. However, for difficult combinatorial search problems, even iterative deepening usually fails due to the size of the search space. Heuristic search helps here through its reduction of the effective branching factor. The IDA algorithm, like iterative deepening, is complete and requires very little memory. Heuristics naturally only give a significant advantage if the heuristic is "good". When solving difficult search problems, the developer's actual task consists of designing heuristics which greatly reduce the effective branching factor.

Game playing is perhaps the most visible success associated with search. Games from tic-tac-toe through to chess have been conquered by brute-force search. More elaborate models of how humans play games have been tried, but in each case, performance has soon been overtaken by the unstoppable march of Moore's law. The game of Go, however, has so many possible moves that it has remained impervious to brute-force search. Whereas it seems unlikely that humans play chess by brute-force search, and investigation of methods of play other than search are attractive, go enforces these investigations with a vengeance. Even when we have computers with the same level of processing power as the human brain, they will not be able to play a good game of Go using brute-force search alone. Search, however, has also been successful in other areas. It is used in proving theorems, mathematical manipulation systems, systems of speech understanding and natural language understanding.

Machine Learning

Herbert Simon defines learning "as any process by which a system improves its performance from experience". It is important to note that machine learning is possible only because there are regularities in the world. And without learning, everything is new; a system that cannot learn is not efficient because it re-derives each solution and repeatedly, making the same mistakes. Learning is a hallmark of intelligence, and many would argue that a system that cannot learn is not intelligent. Learning is a process in which the acquisition of knowledge or skills through study, experience, or being taught. Meanwhile adaptation refers to the act or process of adapting and adjusting to environmental conditions. An adaptive system is a set of interacting or interdependent entities, real or abstract, forming an integrated whole that together is able to respond to environmental changes or changes in the interacting parts. Feedback loops represent a key feature of adaptive systems, allowing the response to changes; examples of adaptive systems include: natural ecosystems, individual organisms, human communities, human organizations, and human families. Some artificial systems can be adaptive as well; for instance, robots employ control systems that utilize feedback loops to sense new conditions in their environment and adapt accordingly.

Machine Learning (ML) is based on the expert design of precise and efficient prediction algorithms. These algorithms cause ML to perform two main functions: induction (classification of data) and transduction (labeling of data). In general, learning is any process by which a system improves "performance" from "experience". Training a model simply means learning (i.e., determining) good values for all the weights and therefore the bias from labeled examples. ML creates machines that exhibits learning capability that improves its performance from experience without being explicitly programmed. ML focuses on the creation of computer programs that can access data and use it to learn from themselves. In other words, it is the study of algorithms that improve their performance (*P*) at some tasks (*T*) with some experience (*E*). ML is needed because it provides smart alternatives to analyzing huge volume of data (i.e., big data). Due to the rapid development of efficient algorithms and data-driven models for real-time processing of datasets, ML can assist to handle the classification and categorical problems. The original Linear Threshold Gates used by McCulloch and Pitts had no learning capability (Figure 4.13).

The phrase "machine learning" dates back to the middle of the last century where Arthur Samuel in 1959 defined machine learning as "the ability to learn without being explicitly programmed" and coined the term. The major vantage of a learning system is its ability to adapt to a changing environment. Of course, the existing machine-learning techniques are still far from enabling computers to learn nearly as well as people. Yet algorithms have been invented that are effective for certain types of learning tasks. Learning by memorization is the simplest form of learning. It requires the least amount of inference and is accomplished by simply copying the knowledge in the same form that it will be used directly into the knowledge base. For example, we use this type of learning when we memorize multiplication tables. A slightly more complex form of learning is by direct instruction. This type of learning requires more inference than rote learning since the knowledge must be transformed into an operational form before being integrated into the knowledge base. We use this type of learning when a teacher presents a number of facts directly to us in a well-organized manner. Machine learning has been around for decades but has made great strides in recent years, notably due to improvements in computer power and developments in deep learning, a specific technique based on neural networks, which draws on knowledge of the human brain, statistics and applied math. However, machine learning is a form of induction, and, thus, suffers from the problems of induction.

> Learning is any process by which a system improves performance from experience.
>
> **– Herbert Simon**
>
> Machine Learning is a field of study that gives computers the ability to learn without being explicitly programmed.
>
> **– Arthur Samuel**

The third type, analogical learning, is the process of learning a new concept or solution through the use of similar known concepts or solutions. We use this type of learning when solving problems on an exam where previously learned examples serve as a guide or when we learn to drive a truck using our knowledge of driving a car. We make frequent use of analogical learning. This form of learning requires still more inferring than either of the previous forms since difficult transformations must be made between the known and unknown situations. The fourth type of learning is also one that is used frequently by humans. It is a powerful form of learning which, like analogical learning, also requires more inferring than the first two methods. This form of learning requires the use of inductive inference, a form of invalid but useful inference. We use inductive learning when we formulate a general concept after seeing a number of instances or examples of the concept. For example, we learn the concepts of color or sweet taste after experiencing the sensations associated with several examples of colored objects or sweet foods. The final type of acquisition is deductive learning. It is accomplished through a sequence of deductive inference steps using known facts. From the known facts, new facts or relationships are logically derived. However, most of the machine-learning algorithms require a special training phase whenever information is extracted (knowledge generalization), which makes online adaptation (sustained learning) difficult. As described by Tom Mitchell, Machine Learning is the study of algorithms that:

- improve their performance *P*
- at some task *T*
- with experience *E*

Learning Theory: A computer program is said to learn from experience *E* with respect to some class of tasks *T* and performance measure *P* if its performance at tasks in *T*, as measured by *P*, improves with experience *E*. For example, an automated facial expression classifier which classifies facial expressions in terms of user defined interpretation labels, improves its performance as measured by its ability to accomplish user-defined interpretations at the class of tasks involving classification of facial expressions, through experience obtained by interacting with the user on the meanings that he/she associates with different facial expressions. In general, in a well-defined learning problem, these three features must be identified (i.e., the class of tasks *T*, the measure of performance to be improved *P*, and the source of experience *E*). A well-defined learning task is given by <*P*, *T*, *E*>.

Different target knowledge (hypotheses space) representations are appropriate for learning different kinds of target functions. For each of these hypothesis representations, the corresponding learning algorithm takes advantage of a different underlying structure to organize the search through the hypotheses space. Therefore, deciding about the issues involves searching a very large space of alternative approaches to determine the one that best fits the defined learning problem. In order to decide a machine-learning algorithm that will perform best for the given problem and the given target function, it is useful to analyze the relationships between the size of the hypotheses space, the completeness of it, the number of training examples available, the prior knowledge held by the learner, and the confidence we can have that a hypothesis that is consistent with the training data will correctly generalize to unseen examples (Figure 4.14).

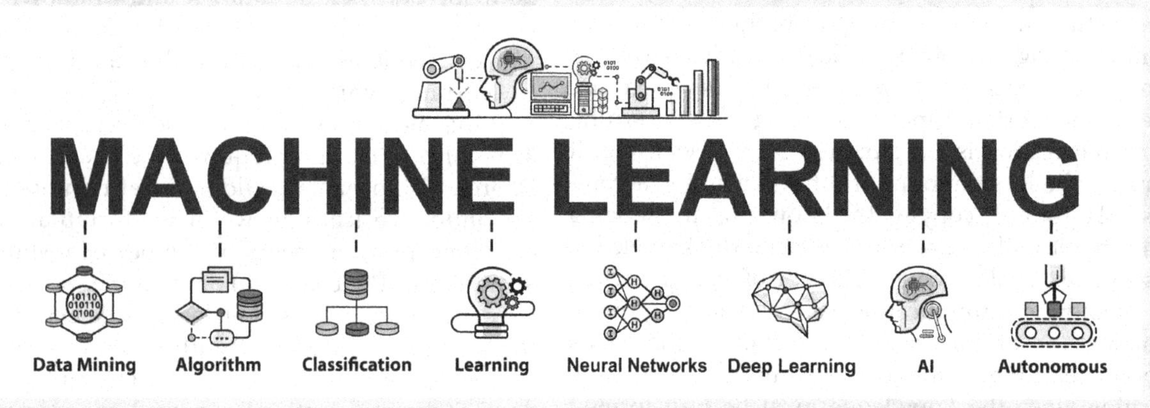

FIGURE 4.14
Machine-learning taxonomy figure.

Frank Rosenblatt developed the perceptron. It was inspired by the way neurons work together in the human brain; the perceptron is a single-layer neural network where an algorithm classifies its input into two possible categories. However, this binary classification technique is only appropriate for linear classification, and it was Geoff Hinton who introduced nonlinear classification. These recent advances have created both important opportunities and challenges for the development of autonomy in weapon systems. Recent advances in machine learning have proved to be very useful for machine perception. They allow the programmer to design sensing software that features remarkable capabilities in terms of pattern recognition (whether objects, faces, or radio signals). They create improvement opportunities in all application areas of autonomy in weapon systems, from target recognition to navigation. ANN is a computational tool inspired by the network of neurons in biological nervous system. It is a network consisting of arrays of artificial neurons linked together with different weights of connection. Threshold logic (TLU) is the simplest kind of computing unit used to build ANNs. These computing elements are a generalization of the common logic gates used in conventional computing and operate by comparing their total input with a threshold. The states of the neurons as well as the weights of connections among them evolve according to certain learning rules. Practically speaking, neural networks are nonlinear statistical modeling tools that can be used to find the relationship between input and output or to find patterns in vast database. The ANN has been applied in statistical model development, adaptive control system, pattern recognition in data mining, and decision-making under uncertainty.

Hebbian learning attempts to connect the psychological and neurological underpinnings of learning. The basis of the theory is when our brains learn something new, neurons are activated and connected with other neurons, forming a neural network. Neural networks were introduced as an implementation of a stimulus-response agent. Advantages of neural networks are:

- great processing speed by making massively use of parallel processing
- even after partial failure the network is still in service (fault tolerance)
- with increasing amount of failing neurons just slow failure of entire system (graceful degradation)
- well suited for inductive learning.

Machine learning is about learning structure from data and has been associated with probabilistic techniques. Probabilistic techniques have been associated with both the learning of functions (e.g., Naïve Bayes classification) and the modeling of theoretical properties of learning algorithms.

ANN is a computational tool inspired by the network of neurons in biological nervous system. This architecture is inspired by the way biological neurons process information in the human brain. ANNs provide a general, practical method for learning real valued, discrete-valued, and vector-valued target functions from examples. It is a network consisting of arrays of artificial neurons linked together with different weights of connection. The states of the neurons as well as the weights of connections among them evolve according to certain learning rules. Practically speaking, neural networks are nonlinear statistical modeling tools which can be used to find the relationship between input and output or to find patterns in vast database. Hebb's rule is a postulate proposed by Donald Hebb and is a learning rule that describes how neuronal activities influence the connection between neurons, that is, the synaptic plasticity. It provides an algorithm to update weight of neuronal connection within neural network. Hebb's rule provides a simplistic physiology-based model to mimic the activity-dependent features of synaptic plasticity and has been widely used in the area of ANN. Different versions of the rule have been proposed to make the updating rule more realistic. The weight of connection between neurons is a function of the neuronal activity. The classical Hebb's rule indicates "neurons that fire together, wire together". A neural network is a computer representation of knowledge that attempts to mimic the neural networks of the human brain. Neural networks, or neural nets, were inspired by the architecture of neurons in the human brain. A simple "neuron" *N* accepts input from multiple other neurons, each of which, when activated (or "fired"), cast a weighted "vote" for or against whether neuron *N* should itself activate. An ANN is based on a collection of connected units or nodes called artificial neurons, which loosely model the neurons in a biological brain. Each connection, like the synapses in a biological brain, can transmit a signal from one artificial neuron to another. An artificial neuron that receives a signal can process it and then signal additional artificial neurons connected to it. About 6 years ago, scientists discovered a new type of more powerful neural network model known as a transformer. These models can achieve unprecedented performance, such as by generating text from prompts with near-human-like accuracy. A transformer underlies AI systems such as ChatGPT and Bard, for example. While incredibly effective, transformers are also mysterious: unlike with other brain-inspired neural network models, it hasn't been clear how to build them using biological components.

An **ANN** is a computational model that approximates a mapping between inputs and outputs. It is inspired by the structure of the human brain; in that it is similarly composed of a network of interconnected neurons that propagate information upon receiving sets of stimuli from neighboring neurons. A node is an artificial neuron and the weights represent the strength of the connections between neurons (or nodes). Training a neural network involves a process that employs the BP and gradient descent algorithms in tandem, where both of these algorithms make extensive use of calculus. ANNs can be considered as function approximation algorithms. More specifically, let's say that a particular artificial neuron (or a perceptron, as Frank Rosenblatt had initially named it) receives n inputs, $[x_1, \ldots, x_n]$, where each connection is attributed a corresponding weight, $[w_1, \ldots, w_n]$.

$$y = \sigma\left(\sum_i w_i x_i\right) \quad (4.5)$$

The input values could be an individual pixel and are multiplied by its corresponding weight, producing an output y. The weighted sum calculation performed is a linear operation, and a second operation is required to transform the weighted sum by a nonlinear activation function that will map a set of inputs to an activation output. Training an ANN involves the process of searching for the set of weights that model best the patterns in the data. It is a process that employs the **BP** and **gradient descent** algorithms in tandem. Both of these algorithms make extensive use of calculus. Each time that the network is traversed in the forward (or rightward) direction, the error of the network can be calculated as the difference between the output produced by the network and the expected ground truth by means of a loss function (such as the sum of squared errors (SSE)). The BP algorithm, then, calculates the gradient (or the rate of change) of this error to changes in the weights. In order to do so, it requires the use of the chain rule and partial derivatives.

Artificial Neurons are commonly called processing elements and are modeled after real neurons of humans and other animals:

- Has many inputs and one output.
- The inputs are signals that are strengthened or weakened (weighted).
- Impulses come from other neurons.
- When sum of inputs reaches a threshold, neuron fires.
- If the sum of all the signals is strong enough, the neuron will put out a signal to the next neuron output of a 1.
- Sending impulses to next level of neurons.

Neural Networks learn by using a training set and adjusting the weights on each connection. They do not have to be "told" explicit relationship rules. Neural Networks can work with partial inputs and *cannot explain* their results. Also, Neural Networks can take a long time to train. Classical RL approaches assume that we have a Markov decision process (MDP) consisting of the set of states S, set of actions A, the rewards R, and transition probabilities T that capture the dynamics of a system. When Bellman (1957) [590] explored optimal control in discrete high-dimensional spaces, he faced an exponential explosion of states and actions for which he coined the term "Curse of Dimensionality". As the number of dimensions grows, exponentially more data and computation are needed to cover the complete state–action space. For example, if we assume that each dimension of a state-space is discretized into ten levels, we have 10 states for a one-dimensional state-space, $10^3=1{,}000$ unique states for a three-dimensional state-space, and 10^n possible states for an n-dimensional state space. Evaluating every state quickly becomes infeasible with growing dimensionality, even for discrete states. Bellman originally coined the term in the context of optimization, but it also applies to function approximation and numerical integration.

The physicist John J. Hopfield invented another type of neural network. Each neural element in a Hopfield network is connected to all of the others. The weights on these connections are symmetrical; that is, the weight connecting unit I to unit j has the same value as the weight connecting unit j to unit i. The operation of the network is a dynamical process; that is, the values of the units at each time step depend on the values at the just-preceding time step. Some researchers believe that dynamical processes, similar to those exhibited by Hopfield and Boltzmann networks (and including those described by sets of differential or difference equations), underlie much of the computation performed by the brain. Automated data-gathering techniques, together with inexpensive mass-memory storage apparatus, have allowed the acquisition and retention of prodigious amounts of data. Most machine-learning methods construct hypotheses from data. Such an inference is "inductive" rather than "deductive". Deductive inferences follow necessarily and logically from their premises, whereas inductive ones are hypotheses, which are always subject to falsification by additional data. Still, inductive inferences, based on large amounts of data, are extremely useful. Indeed, science itself is based on inductive inferences. Whereas before about 1980, machine learning (represented mainly by neural network methods) was regarded by some as on the fringes of AI, machine learning has lately become much more central in modern AI. The usual AI approach to dealing with large quantities of data is to reduce the amount of it in some way. However, our growing abilities to store large amounts of data in rapid-access computer memories and to compute with these data have enabled techniques that store and use all of the data as they are needed, without any prior condensation whatsoever. That is, these techniques do not attempt to reduce the amount of data before it is actually used for some task. All of the necessary reduction, for example to a decision, is performed at the time a decision must be made. Described next are some of these memory-based learning methods.

Hopfield model is an associative memory model using the Hebb's rule for all possible pairs ij with binary units. The state variable x_i of the neuron *I* takes on either one of the two possible values: 1 or –1, which corresponds to the firing state or not firing state, respectively. Though being simple, the classical Hebb's rule has some disadvantages. Depending on the application area of neural network, some drawbacks are tolerable but some need improvement. Generally speaking, for the purpose of data processing and statistical analysis, the speed and power of computation are valued more than the resemblance between the model and the physiological realism. On the other hand, for unraveling the memory formation mechanism and harvesting the emerging properties of biological network, the models need to be built upon a certain degree of biophysical basis.

In practical supervised learning, you could divide most of the problems into three general classes:

- Clearly defined problems with small to intermediate datasets and a manageable number of features. Ensembling methods like boosted decision trees and random forests, sometimes combined with blending and stacking, are usually the best-performing models on these problems.
- Complex problems with big datasets and a huge number of features. This is the domain of deep learning. Various types of neural networks are clearly outperforming other algorithms in this area.
- Simple problems with special interpretability requirements. This is best handled by simple models like decision trees, linear regression, logistic regression, and Naïve Bayes. SVMs are known for being hard to interpret.

A **random tree** is a tree or arborescence that is formed by a stochastic process. In probability theory, a stochastic or random process is a mathematical object usually defined as a sequence of random variables in a probability space, where the index of the sequence often has the interpretation of time. Random Trees (RRTs) use the concept of sampling to generate a few samples from the configuration space and search the trajectory in the configuration space represented by the samples only. Unlike probabilistic roadmaps, the sampling here is done for a single query only, and therefore the source and goal are known a priori. This can help in focusing the search toward areas intermediate between source and goal for searching a path, rather than diversifying into the entire configuration space. As the name suggests, Random Forest is an ensemble method usually applied to Random Tree. Random forests are a combination of tree predictors such that each tree depends on the values of a random vector sampled independently and with the same distribution for all trees in the forest A Random Forest should be compared with other ensemble methods such as AdaBoost, etc. and a Random Tree with basic, simple classifiers such as Perceptron (although it is from a different family of models). Random forest is a machine-learning classifier based on choosing random subsets of variables for each tree and using the most frequent tree output as the overall classification. Random forests or random decision forests are an ensemble learning method for classification, regression, and other tasks that operate by constructing a multitude of decision trees at training time. A random recursive tree or increasingly labeled tree is generated by using a simple stochastic growth rule.

An interesting approach, which has been pursued since about 1990, is the application of machine-learning techniques to the learning of heuristics for directing the search of inference systems, which we will briefly sketch here. A heuristic is a method that might not always find the best solution but is guaranteed to find a good solution in reasonable time by sacrificing completeness it increases efficiency. The classic example of heuristic search methods is the traveling salesman problem. A resolution prover has, during the search for a proof, hundreds or more possibilities for resolution steps at each step, but only a few lead to the goal. It would be ideal if the prover could ask an oracle which two clauses it should use in the next step to quickly find the proof. There are attempts to build such proof-directing modules, which evaluate the various alternatives for the next step and then choose the alternative with the best rating. With the help of Bayesian networks, complex applications with many variables can also be modeled. Probability theory also offers the possibility of making statements about the probability of continuous variables.

A subfield of AI, called "Case-Based Reasoning" (CBR), can be viewed as a generalized kind of memory-based learning. In CBR, a stored library of "cases" is used to help in the analysis, interpretation, and solution of new cases. Next on the list of new developments in machine learning is the automatic construction of structures called "decision trees" from large databases. Decision trees consist of sequences of tests for determining a category or a numerical value to assign to a data record. Decision trees are particularly well suited for use with non-numeric as well as numeric data. One important data-mining method uses data to construct decision trees. During the 1960s, neural net researchers employed various methods for changing a network's adjustable weights so that the entire network made appropriate output responses to a set of "training" inputs.

That problem of changing weights in multiple layers was solved in the mid-1980s by the invention of a technique called "back-propagation" (backprop for short) re-introduced by David Rumelhart, Geoffrey E. Hinton, and Ronald J. Williams. The basic idea behind backprop is simple, but the mathematics is rather complicated. In response to an error in the network's output, backprop makes small adjustments in all of the weights so as to reduce that error. It can be regarded as a hill-climbing (or rather hill-descending) method, searching for low values of error over the landscape of weights. But rather than actually trying out all possible small weight changes and deciding on that set of them that corresponds to the steepest descent downhill, backprop uses calculus to precompute the best set of weight changes.

Looking at the Hebb rule, we see that for neurons with values between zero and one, the weights can only grow with time. This problem is solved quite differently by the model presented by Hopfield. It uses binary neurons but with the two values –1 for inactive and 1 for active. Using the Hebb rule we obtain a positive contribution to the weight whenever two neurons are simultaneously active. If, however, only one of the two neurons is active, w_{ij} is negative. Hopfield networks, which are a beautiful and visualizable example of auto-associative memory, are based on this idea. Patterns can be stored in auto-associative memory. To call up a saved pattern, it is sufficient to provide a similar pattern. As an application to a pattern recognition example, we apply the described learning algorithm to a simple pattern recognition example. It should recognize digits in a 10Å~10 pixel field. The Hopfield network thus has 100 neurons with a total of

$$\frac{100 \cdot 99}{2} = 4,950 \text{ weights} \tag{4.6}$$

First, the patterns of the digits 1, 2, 3, and 4 are trained. That is, the weights are calculated. Then we put in the pattern with added noise and let the Hopfield dynamics run until convergence. The following table summarizes the different classes of machine learning (Figure 4.15).

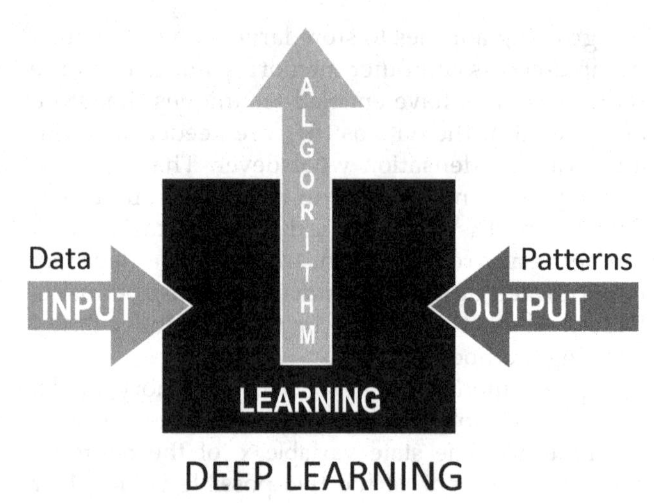

FIGURE 4.15
Deep learning process diagram.

It is noted that elements can store all their long-term knowledge in the strengths of the connections between processors. And in the last decade, there has been considerable progress in developing learning procedures for these networks that allow them to automatically construct their own internal representations. The learning procedures are typically applied in networks that map input vectors to output vectors via only a few layers of "hidden" units. Thus, the network learns to dedicate particular hidden units to particular pieces or aspects of the input vector that are relevant in determining their output. The network generally learns to use distributed representations in which each input vector is represented by activities in many different hidden units, and each hidden unit is involved in representing many different input vectors. Within the connectionist community, there has been a long and unresolved debate between those who favor localist representations in which each processing element corresponds to a meaningful concept and those who favor distributed representations. The major criticism of distributed representations has been that they cannot handle structured knowledge properly. Another criticism has been the unintelligibility of distributed representations. As soon as there are several hidden layers, it becomes very difficult to say what each hidden unit is representing. With all things being equal, it is clearly desirable to understand how a system performs a task such as medical diagnosis that arrives at a particular conclusion, and how to provide this information to the user.

Minsky and Papert showed that a two-layer feedforward network can overcome many restrictions but did not present a solution to the problem of how to adjust the weights from input to hidden units. An answer to this question was presented by Rumelhart,

Hinton, and Williams in 1986 and similar solutions appeared to have been published earlier by Werbos in 1974, Parker as learning logic in 1985, and LeCun in 1985. The central idea behind this solution is that the errors for the units of the hidden layer are determined by back propagating the errors of the units of the output layer. For this reason, the method is often called the back-propagation (BP) learning rule. BP can also be considered as a generalization of the delta rule for nonlinear activation functions and multilayer networks. A multi-layer neural network is also referred to as a deep learning system (Figure 4.15). Multilayer neural networks which are trained with back-propagation algorithms constitute the best example of today's successful gradient-based learning techniques. Given an appropriate network architecture, gradient-based learning algorithms can be used to synthesize a complex decision surface in order to classify high-dimensional patterns, such as handwritten characters with minimal preprocessing. And due to recent advances in algorithms and computer hardware, the technology has made it possible to train neural networks in an end-to-end fashion for tasks that previously required significant human expertise. For example, convolutional neural networks (CNNs) are now able to directly classify raw pixels into high-level concepts such as object categories and messages on traffic signs without using hand-designed feature extraction algorithms. Not only do such networks require less human effort than traditional approaches, but they also generally deliver superior performance. This is particularly true when very large amounts of training data are available and can produce a convincing explanation.

BP Algorithm: With the BP (reverse mode of automatic differentiation) algorithm, we now introduce the most-used neural model. The reason for its widespread use is its universal versatility for arbitrary approximation tasks. The BP learning algorithm works on multilayer feedforward networks, using gradient descent in weight space to minimize the output error. The algorithm originates directly from the incremental delta rule. In contrast to the delta rule, it applies a nonlinear sigmoid function on the weighted sum of the inputs as its activation function. Furthermore, a BP network can have more than two layers of neurons. BP is a procedure for efficiently calculating the derivatives of some output quantity of a nonlinear system, with respect to all inputs and parameters of that system, through calculations proceeding backward from outputs to inputs. It permits "local" implementation on parallel hardware. BP is defined as any technique for adapting the weights of parameters of a nonlinear system by somehow using such derivatives or the equivalent. In essence, BP allows a computer to learn from its own mistakes.

Support Vector Machines: A promising approach brings together the advantages of linear and nonlinear models, following the theory of SVMs, which we will roughly outline using a two-class problem. The SVMs support vector networks and are supervised learning models with associated learning algorithms that analyze data for classification and regression analysis. The SVM is a computer algorithm that learns by example to assign labels to objects. In essence, an SVM is a mathematical entity, an algorithm (or recipe) for maximizing a particular mathematical function with respect to a given collection of data. SVMs are among the best "off-the-shelf" supervised learning algorithm. To tell the SVM story, we'll need to first talk about margins and the idea of separating data with a large "gap". Feedforward neural networks with only one layer of weights are linear. Linearity leads to simple networks and fast learning with guaranteed convergence. Furthermore, the danger of overfitting is small for linear models. For many applications, however, the linear models are not strong enough, for example, because the relevant classes are not linearly separable. Here, multilayered networks such as BP come into use, with the consequence that local minima, convergence problems, and overfitting can occur. This algorithm learns by example to assign labels to objects. For instance, an SVM can learn to recognize fraudulent credit card activity by examining hundreds or thousands of fraudulent and nonfraudulent credit card activity reports. Alternatively, an SVM can learn to recognize handwritten digits by examining a large collection of scanned images of handwritten zeroes, ones, and so forth. Finally, SVMs have also been successfully applied to an increasingly wide variety of biological applications (Table 4.1).

All of the learning algorithms described so far, except the clustering algorithms, belong to the class of supervised learning. In supervised learning, the agent is supposed to learn a mapping from the input variables to the output variables. Here it is important that for

TABLE 4.1
Table of classes of machine learning

SUPERVISED LEARNING	LAZY LEARNING	k nearest neighbor method (classification + approximation)
		Locally weighted regression (approximation)
		Case-based learning (classification + approximation)
	EAGER LEARNING	Decision trees induction (classification)
		Learning of Bayesian networks (classification + approximation)
		Neural networks (classification + approximation)
		Gaussian processes (classification + approximation)
		Wavelets, splines, radial basis functions, . . .
UNSUPERVISED LEARNING (CLUSTERING)		Nearest neighbor algorithm
		Farthest neighbor algorithm
		k-means
		Neural networks
REINFORCEMENT LEARNING		Value iteration
		Q learning
		TD learning
		Policy gradient methods
		Neural networks

each individual training example, all values for both input variables and output variables are provided. In other words, we need a teacher or a database in which the mapping to be learned is approximately defined for a sufficient number of input values. The sole task of the machine-learning algorithm is to filter out the noise from the data and find a function that approximates the mapping well, even between the given data points. In RL the situation is different and more difficult because no training data is available. We begin with a simple illustrative example from robotics, which then is used as an application for the various algorithms.

RL is very valuable in the field of robotics, where the tasks to be performed are frequently complex enough to defy encoding as programs and no training data is available. The robot's task consists of finding out, through trial and error (or success), which actions are good in a certain situation and which are not. In many cases, we humans learn in a very similar way. For example, when a child learns to walk, this usually happens without instruction, rather simply through reinforcement. Successful attempts at walking are rewarded by forward progress, and unsuccessful attempts are penalized by often painful falls. Positive and negative reinforcement are also important factors in successful learning in school and in many sports.

Q-Learning: *Q*-learning is a machine-learning approach that enables a model to iteratively learn and improve over time by taking the correct action and is a type of RL. A policy based on evaluation of possible successor states is clearly not useable if the agent does not have a model of the world, that is, when it does not know which state a possible action leads to. In most realistic applications the agent cannot resort to such a model of the world. For example, a robot that is supposed to grasp complex objects cannot predict whether the object will be securely held in its grip after a gripping action, or whether it will remain in place. If there is no model of the world, an evaluation of an action carried out in state s_t is needed even if it is still unknown where this action leads to. Thus, we now work with an evaluation function $Q(s_t, a_t)$ for states with their associated actions (Figure 4.16).

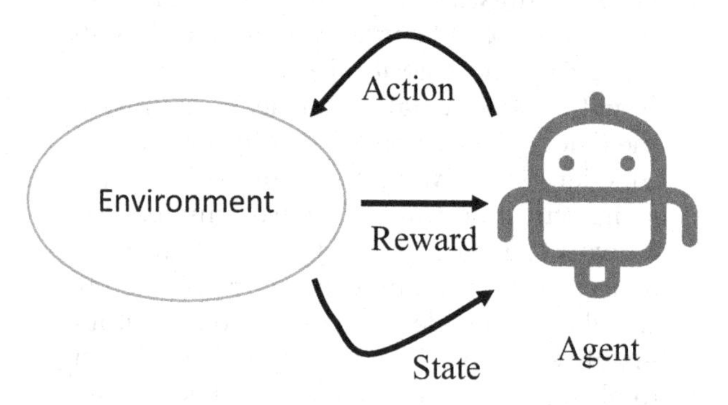

FIGURE 4.16
Learning agent model.

In active RL, the agent decides what action to take with the goal of learning an optimal policy. In Q-learning, the action-value function $Q(s,a)$ is learned using a greedy action policy, such that it directly estimates Q^*. **State–action–reward–state–action (SARSA)** is an algorithm for learning a Markov decision process policy, used in the RL area of machine learning. It was proposed by Rummery and Niranjan in a technical note with the name "Modified Connectionist Q-Learning" (MCQ-L). In SARSA, each $Q(s,a)$ is updated with a sample Q-value that can refer to a random action (i.e., "value" of action a in state s is the exponential average of the observed rewards associated to both optimal and random actions). Q-learning learns the optimal policy, however, the online (i.e., while learning) performance is not that good since the agent occasionally falls off the cliff because of the ε-greedy action selection. SARSA does not learn the optimal policy, but rather the action values that account for ε-greedy action selections, and therefore it learns the longer but safer path that minimizes the risk of $R=-100$. Acting randomly, Q-learning asymptotically converges to optimal state–action values. A PAC learning strategy places an upper bound on the probability of committing an error by placing a minimum bound on the number of examples/samples it takes to learn a target. Double Q-learning is an off-policy RL algorithm that utilizes double estimation to counteract overestimation problems with traditional Q-learning. The max operator in standard Q-learning and DQN uses the same values both to select and to evaluate an action. This makes it more likely to select overestimated values, resulting in overoptimistic value estimates. Double Q-learning was proposed for solving the problem of large overestimations of action value (Q-value) in basic Q-learning.

Reasoning under uncertainty with limited resources plays a big role in everyday situations and also in many technical applications of AI. In these areas, heuristic processes are very important. For example, we use heuristic techniques when looking for a parking space in city traffic. Heuristics alone are often not enough, especially when a quick decision is needed given incomplete knowledge, as shown in the following example. A pedestrian crosses the street and an auto quickly approaches. To prevent a serious accident, the pedestrian must react quickly. He is not capable of worrying about complete information about the state of the world, which he would need for the search algorithms. Learning of decision trees is a favorite approach to classification tasks. The reasons for this are its simple application and speed. The following taxonomy gives an overview of the most important learning algorithms and their classification (Figure 4.17).

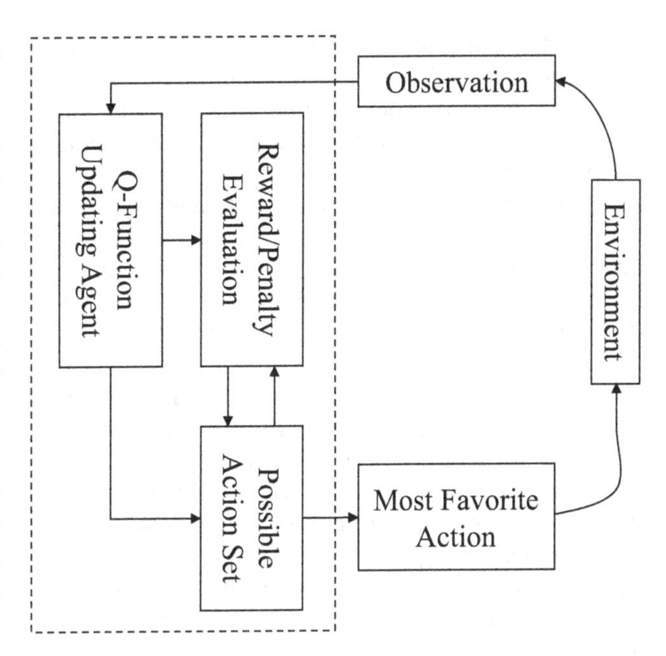

FIGURE 4.17
Q-Learning function model.

The decision tree and neural network learning methods are examples of "supervised learning", a type of learning in which one attempts to learn to classify data from a large sample of training data whose classifications are known. The "supervision" that directs learning in these systems involves informing the system about the classification of each datum in the training set. Yet, it is sometimes possible to construct useful classifications of data based just on the data alone. Techniques for doing so fall under the heading of "unsupervised learning". There is another style of learning that lies somewhat in between the supervised and unsupervised varieties. An example would be learning which of several possible actions a robot, say, should execute at every stage in an ongoing sequence of experiences given only what final result of all of its actions (Figure 4.18).

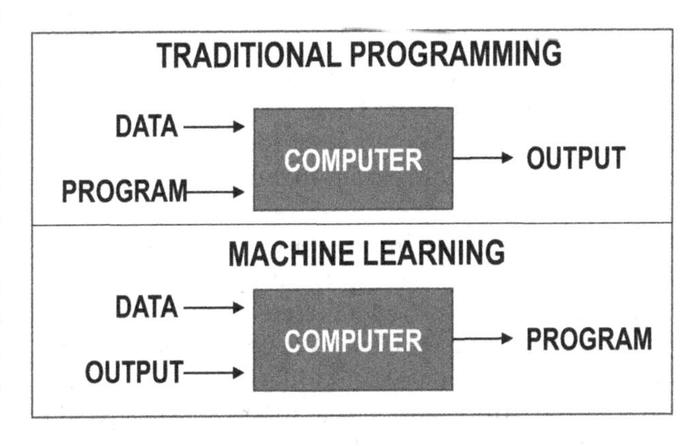

FIGURE 4.18
Traditional learning vs. machine-learning diagram.

In traditional machine learning, the learning process is supervised, and the programmer has to be extremely specific when telling the computer what types of things it should be looking for to decide if an image contains an object (*dog* or does not contain a *dog*). This is a laborious process called *feature extraction,* and the computer's success rate depends entirely upon the programmer's ability to accurately define a feature set for *dog*. The advantage of deep learning is the program builds the feature set by itself without supervision. Unsupervised learning is not only faster, but it is usually more accurate. Initially, the computer program might be provided with training data, a set of images for which a human has labeled each image *dog* or *not dog* with metatags. The program uses the information it receives from the training data to create a feature set for *dog* and build a predictive model. In this case, the model the computer first creates might predict that anything in an image that has four legs and a tail should be labeled *dog*. Of course, the program is not aware of the labels *four legs* or *tail*. It will simply look for patterns of pixels in the digital data. With each iteration, the predictive model becomes more complex and more accurate.

Graphical models use directed or undirected graphs over a set of random variables to explicitly specify variable dependencies and allow for less restrictive independence assumptions while limiting the number of parameters that must be estimated. Bayesian networks are directed acyclic graphs that indicate causal structure. On the other hand, Markov Networks are undirected graphs that capture general dependencies. And learning from data is the quantification of uncertainties. In general, the problem of Bayes Net inference is NP-hard (exponential in the size of the graph).

Deep ANNs (Artificial Neural Networks) are a set of algorithms reaching new levels of accuracy for many important problems, such as image recognition, sound recognition, recommender systems, etc. These are the graphical structures used to represent the probabilistic relationship among a set of random variables. Bayesian networks are also called Belief Networks or Bayes Nets. Belief Networks reason about uncertain domain. Belief networks are a natural way to represent conditional independence information. A computer program that uses deep-learning algorithms can be shown a training set and sort through millions of images, accurately identifying which images have dogs in them within a few minutes. A belief network is a directed acyclic graph composed of stochastic variables. We get to observe some of the variables and we would like to solve two problems: (1) the **inference** problem infers the states of the unobserved variables, and (2) the **learning** problem infers the states of the unobserved variables. If we connect binary stochastic neurons using symmetric connections, we get a Boltzmann Machine. In order to achieve an acceptable level of accuracy, deep-learning programs require access to immense amount of training data and processing power, neither of which were easily available to programmers until the era of big data and cloud computing. Because deep-learning programming can create complex statistical models directly from its own iterative output, it is able to create accurate predictive models from large quantities of unlabeled, unstructured data.

The core of RL lies in learning from experiences. The performance of the agent is hugely impacted by the training conditions, reward functions, and exploration policies. Deep Deterministic Policy Gradient (DDPG) is a well-known approach to solve continuous control problems in RL. We use DDPG with intelligent choice of reward function and exploration policy to learn various driving behaviors (Lane keeping, Overtaking, Blocking, Defensive, Opportunistic) for a simulated car in unstructured environments. Deep Learning is part of a broader family of machine-learning methods based on ANNs with representation learning. Learning can be supervised, semi-supervised, or unsupervised. Deep learning–based approaches have achieved a big leap forward over previous state-of-the-art algorithms in classification, segmentation, and recognition. Deep learning can safely be regarded as the study of models that either involve a greater amount of composition of learned functions or learned concepts than traditional machine learning does (Table 4.2).

Deep-learning methods have become increasingly popular in recent years because of their tremendous success in **image classification**, **speech recognition**, and **natural language processing** tasks. A Restricted Boltzmann Machine (RBM) is an undirected graphical model that defines a probability distribution over a vector of observed or visible, variables. The CNN or ConvNet has achieved great success in many computer vision tasks such as image classification, segmentation, and video action recognition. The specially designed architecture of the CNN is very powerful in extracting visual features from images, which can be used for various tasks. Recurrent neural networks (RNNs) are powerful concepts that allow the use of loops within the neural network architecture to model sequential data such as sentences and videos. Recurrent networks take as input a sequence of inputs and produce a sequence of outputs. Thus, such models are particularly useful for sequence-to-sequence learning. The generative adversarial network (GAN) is one of the most popular generative deep models. The core of a GAN is to play a min-max game between a discriminator D and a generator G, that is, adversarial

TABLE 4.2
Table of machine-learning approaches

Machine Learning	Description
1. Connectionist Approach	Is based on a neural network algorithm called backpropagation, which learns from user-provided samples ('correct' input/output pairs). The neural net feeds inputs (such as pictures) through layers of artificial neurons until they emit a final output, then checks whether this output is in accordance with the pre-specified ('correct') output in the training data, a process which is continued until the algorithm becomes able to correctly predict the correct output even for new inputs (in effect demonstrating inductive logic).
2. Evolutionist Approach	Believes they can build AI by emulating (and speeding up) natural selection in digital environments, through genetic algorithms.
3. Bayesians Approach	Bayesian networks (or 'belief networks') try to encode probability estimates for a large number of different competing hypotheses, with respective belief probabilities updated as new information becomes available
4. Symbolists Approach	Pursues a general-purpose learning algorithm which can freely combine rules and fills in the gaps in its knowledge.
5. Analogisers Approach	Operates on the basis of analogy, to match new cases with the most similar such situation which has been encountered in the past.

training. The discriminator D tries to differentiate if a sample is from the real world or generated by the generator, while the generator G tries to generate samples that can fool the discriminator, that is, make the discriminator believe that the generated samples are from the real world (Figure 4.19).

Deep Belief Networks or deep learning is a type of machine learning and AI that imitates the way humans gain certain types of knowledge. Deep learning is an important element of data science, which includes statistics and predictive modeling. While traditional machine-learning algorithms are linear, deep-learning algorithms are stacked in a hierarchy of increasing complexity and abstraction. Each algorithm in the hierarchy applies a nonlinear transformation to its input and uses what it learns to create a statistical model as output. Iterations continue until the output has reached an acceptable level of accuracy. The number of processing layers through which data must pass is what inspired the label *deep*. Deep learning is the science of training large ANNs. Deep neural networks (DNNs) can have hundreds of millions of parameters, allowing them to model complex functions such as nonlinear dynamics. They form compact representations of state from raw, high-dimensional, multimodal sensor data commonly found in robotic systems, and unlike many machine-learning methods, they do not require a human expert to hand-engineer feature vectors from sensor data at design time. The basic principles of linear regression were used by Gauss and Legendre, and many of those same principles still cover what researchers in deep-learning study. However, several important advances have slowly transformed regression into what we now call deep learning. First, the addition of an activation function enabled regression methods to fit to nonlinear functions. It also introduced some biological similarities with brain cells.

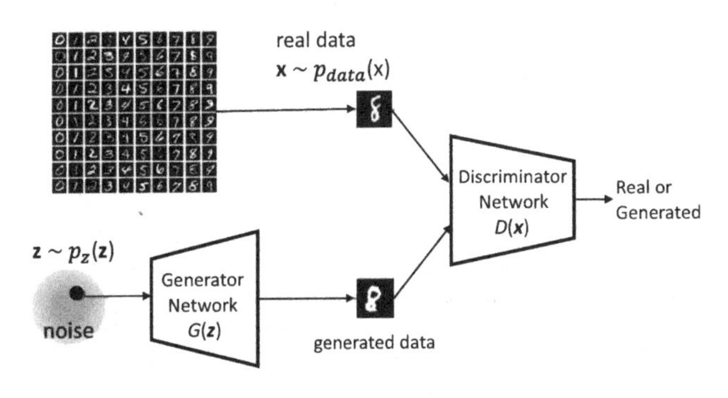

FIGURE 4.19
Deep learning generator and discriminator network.

Claude Shannon proposed that information was communicated by sending a signal through a sequence of stages or transformations. A Deep

Architecture consist of a series of layers between input and output that learn feature hierarchies/feature identification at different levels. The hidden layers act as feature detectors, which will lead to an automatic abstraction of data. The successive layers will learn high-level features. A convolution layer comprises a series of "maps" corresponding to the "S-plane" in Neocognitron, variously called feature maps or activation maps. **Neocognitron** is a hierarchical, multilayered ANN proposed by Kunihiko Fukushima. It has been used for Japanese handwritten character recognition and other pattern recognition tasks and served as the inspiration for the CNN (Convolutional Neural Networks). Each affine map has, associated with it, a learnable filter. All the maps in the previous layer contribute to each convolution. In reality, a filter is really just a perceptron, with weights and a bias (Table 4.3).

Finally, nonlinear models were stacked in "layers" to create powerful models, called multilayer perceptrons. In the 1960s a few researchers independently figured out how to differentiate multilayer perceptrons, and by the 1980s, it evolved into a popular method for training them, called BP. In many ways, BP marked the beginning of the deep-learning revolution; however, researchers still mostly limited their neural networks to a few layers because of the problem of vanishing gradients. Deeper neural networks took exponentially longer to train. Neural networks were successfully applied for robotics control as early as the 1980s. The largest bottleneck in training neural networks is a matrix-vector multiplication step, which can be parallelized using GPUs. In 2006, Hinton presented a training method that he demonstrated to be effective with a many-layered neural network. In 2011,

TABLE 4.3

Table of popular machine-learning algorithms.

Popular Machine Learning Algorithms	
Algorithm	Description
General Linear Description	GLMs can be used to construct the models for regression and classification problems by using the type of distribution which best describes the data or labels given for training the model.
K-Means Clustering	k-means clustering is a method of vector quantization, originally from signal processing, that aims to partition n observations into k clusters in which each observation belongs to the cluster with the nearest mean (cluster centers or cluster centroid), serving as a prototype of the cluster. This results in a partitioning of the data space into Voronoi cells. k-means clustering minimizes within-cluster variances (squared Euclidean distances), but not regular Euclidean distances, which would be the more difficult Weber problem: the mean optimizes squared errors, whereas only the geometric median minimizes Euclidean distances.
Principal Component Analysis (PCA)	Principal component analysis (PCA) is a popular technique for analyzing large datasets containing a high number of dimensions/features per observation, increasing the interpretability of data while preserving the maximum amount of information, and enabling the visualization of multidimensional data. Formally, PCA is a statistical technique for reducing the dimensionality of a dataset. This is accomplished by linearly transforming the data into a new coordinate system where (most of) the variation in the data can be described with fewer dimensions than the initial data.
Linear Regression	Linear Regression is a machine learning algorithm based on supervised learning. It performs a regression task. Regression models a target prediction value based on independent variables. It is mostly used for finding out the relationship between variables and forecasting.
Logistic Regression	Logistic regression is a Machine Learning classification algorithm that is used to predict the probability of certain classes based on some dependent variables. In short, the logistic regression model computes a sum of the input features (in most cases, there is a bias term), and calculates the logistic of the result.
Random Forest	Random forests or random decision forests is an ensemble learning method for classification, regression and other tasks that operates by constructing a multitude of decision trees at training time. For classification tasks, the output of the random forest is the class selected by most trees
Decision Trees	A decision tree is a non-parametric supervised learning algorithm, which is utilized for both classification and regression tasks. It has a hierarchical, tree structure, which consists of a root node, branches, internal nodes and leaf nodes.
Singular Value Decomposition	In linear algebra, the singular value decomposition (SVD) is a factorization of a real or complex matrix. It generalizes the eigen decomposition of a square normal matrix with an orthonormal eigen basis to any {\displaystyle \ m\times n\ }m x n matrix. It is related to the polar decomposition.
Classification and Regression Trees (CART)	A Classification and Regression Tree (CART) is a predictive algorithm used in machine learning. It explains how a target variable's values can be predicted based on other values. It is a decision tree where each fork is split in a predictor variable and each node at the end has a prediction for the target variable.
Gaussian Mixture Model (GMM)	Gaussian mixture models (GMMs) are a type of machine learning algorithm. They are used to classify data into different categories based on the probability distribution. Gaussian mixture models can be used in many different areas, including finance, marketing and so much more!
Learning Decision Trees	Decision tree learning is a supervised learning approach used in statistics, data mining and machine learning. In this formalism, a classification or regression decision tree is used as a predictive model to draw conclusions about a set of observations. Tree models where the target variable can take a discrete set of values are called classification trees; in these tree structures, leaves represent class labels and branches represent conjunctions of features that lead to those class labels. Decision trees where the target variable can take continuous values (typically real numbers) are called regression trees. Decision trees are among the most popular machine learning algorithms given their intelligibility and simplicity.

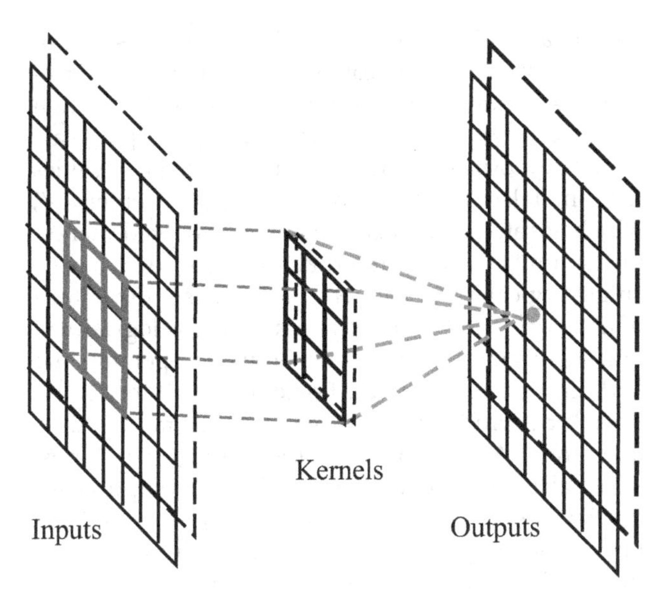

FIGURE 4.20
What is a convolution?

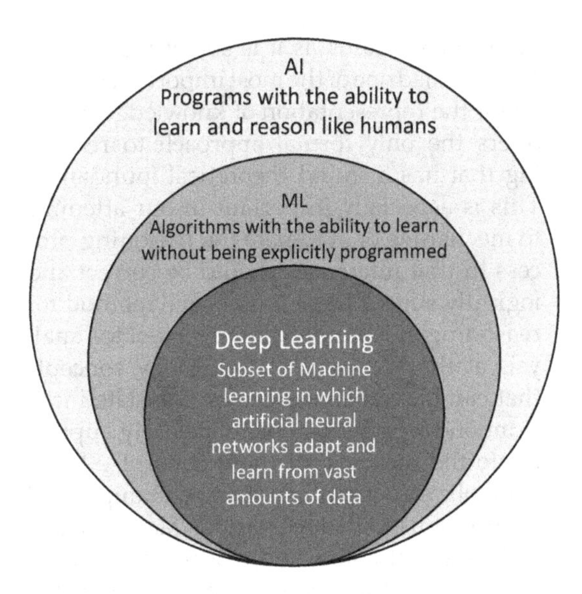

FIGURE 4.21
Hierarchy of deep learning, machine-learning, and AI diagram.

TNLDR (Temporal NonLinear Dimensionality Reduction) demonstrated that deep neural nets could effectively model both state and dynamics from strictly unsupervised training with raw images of a simulated robot. When learning a perceptron for binary classification, our goal is to determine weights by determining a line that as close as possible linearly separates the two classes. Then, we develop a gradient descent algorithm for learning these weights (Figure 4.20).

BP has been the most successful algorithm used to train ANNs. However, there are several gaps between BP and learning in biologically plausible neuronal networks of the brain. BP as the main principle underlying learning in deep ANNs has long been criticized for its biological implausibility (i.e., BP's computational procedures and principles are unrealistic to be implemented in the brain). Machine learning and data mining often employ the same methods and overlap significantly, but while machine learning focuses on prediction, based on *known* properties learned from the training data, data mining focuses on the discovery of previously *unknown* properties in the data (this is the analysis step of knowledge discovery in databases). Data mining uses many machine-learning methods, but with different goals; on the other hand, machine learning also employs data-mining methods as "unsupervised learning" or as a preprocessing step to improve learner accuracy. Machine learning also has intimate ties to optimization: many learning problems are formulated as minimization of some loss function on a training set of examples. Loss functions express the discrepancy between the predictions of the model being trained and the actual problem instances (for example, in classification, one wants to assign a label to instances, and models are trained to correctly predict the pre-assigned labels of a set of examples. Modern-day machine learning has two objectives, one is to classify data based on models that have been developed, and the other purpose is to make predictions for future outcomes based on these models. A hypothetical algorithm specific to classifying data may use computer vision of moles coupled with supervised learning in order to train it to classify the cancerous moles. For example, machine-learning algorithm for stock trading may inform the trader of future potential predictions. As another example, a multimodal learning technique can convert text to image synthesis. Figure 4.21 depicts the relationships between AI, Machine Learning, and Deep Learning.

Logic

The use of symbolic logic to represent knowledge is not new in that it predates the modern computer by a number of decades. Logic is an argument for discovering the truth. Even so, the application of logic as a practical means of representing and manipulating knowledge in a computer was not demonstrated until the early 1960s by Gilmore. Since that time, numerous systems have been implemented with varying degrees of success. Today, First-Order Predicate Logic (FOPL) or

Predicate Calculus as it is sometimes called, has assumed one of the most important roles in Al for the representation of knowledge. Logic offers the only formal approach to reasoning that has a sound theoretical foundation. This is especially important in our attempts to mechanize or automate the reasoning process in that inferences should be correct and logically sound. Logic is a formal method for reasoning. It is fundamentally a tool for analysis at the knowledge level. Many concepts that can be verbalized can be translated into symbolic representations that closely approximate the meaning of these concepts. These symbolic structures can then be manipulated in programs to deduce various facts, to carry out a form of automated reasoning. So, what are the ways for predicting properties of different information processes. Or what kinds of things do we want to predict?

The goal is to have knowledge about the world, and logically draw conclusions from it. Search algorithms generate successors and evaluate them, but do not "understand" much about the setting. If we have a knowledge base of things that we know are true, what can we conclude or infer about the situation. The correspondence between digital logic circuits and propositional logic has been known for a long time. Many problems in circuit design and verification can be reduced to automated propositional tautology or satisfiability checking ("SAT"). Logic is a great knowledge representation language for many AI problems. To use logic, it is first necessary to convert facts and rules about the real world into logical expressions using the logical operators. Propositional logic is the simple foundation and fine for some AI problems and first-order logic (FOL) is much more expressive as a knowledge representation language and more commonly used in AI. One problem with FOL, and thus with the logical agent approach, is that agents almost never have access to the whole truth about their environment. Some sentences can be ascertained directly from the agent's percepts, and others can be inferred from current and previous percepts together with knowledge about the properties of the environment. The meaning of **semantics** of a sentence determines its **interpretation.** Given the truth values of all symbols in a sentence, it can be "evaluated" to determine its **truth value** (True or False). A **model** for a KB is a *possible world*, an assignment of truth values to propositional symbols that makes each sentence in the KB True.

Logic is an "algebra" for manipulating only two values: true (T) and false (F), nevertheless, logic can be quite challenging. The simplest kind of logic is propositional logic and predicate logic (a.k.a. predicate calculus) is an extension of propositional logic. Resolution theory is a general way of doing proofs in predicate logic. Propositional logic consists of logical values true and false (T and F) and propositions: "Sentences", which are atomic (that is, they must be treated as indivisible units, with no internal structure), and have a single logical value, either true or false. Operators, both unary and binary; when applied to logical values, yield logical values and the usual operators are and, or, not, and implies. Logic, like arithmetic, has operators, which apply to one, two, or more values (operands). A truth table lists the results for each possible arrangement of operands. All logical expressions can be computed with some combination of and ($\wedge$), or ($\vee$), and not ($\neg$) operators. A world is a collection of prepositions and logical expressions relating to those prepositions. A proposition "says something" about the world, but since it is atomic (you can't look inside it to see component parts), propositions tend to be very specialized and inflexible.

Logical Reasoning: Logic is a formal language together with a relationship. And a logic language, together with a syntactically defined relation, is called a calculus. Formally, a calculus is defined as a pair <*A*,*R*>, where *A* is a set of formulae and *R* is a set of inference rules. The application of the inference rule to the premise produces the conclusion. We can also say that the conclusion is derived from the premise by the inference rule. Based on the deduction theorem, weak properties are often enough to show a calculus has good properties. This is why automated reasoning is often referred to as theorem proving. However, those good properties are not enough to ensure that a calculus is suitable for implementation. The development of automated reasoning tools started shortly after computers were made available. The first Turing-complete programmable digital computer, the ENIAC, was operational in 1946, and in 1954, a program written by Martin Davis produced a proof for the Presburger arithmetic. Soon after, George Collins implemented parts of Tarski's decision procedure for elementary algebra on an IBM 704. Most of these procedures relied on (Naïve) enumerations of proofs and/or models.

The JOHNNIAC (**JOHN** von **N**eumann **N**umerical **I**ntegrator and **A**utomatic **C**omputer) computer at

Princeton University was built by Rand Corp. It was a vacuum tube computer and was a 1024 40-bit word machine with a Mean Fight Time between Failures (MFTBF) of about 10 minutes. The JOHNNIAC's "Add Time" is 50 ms (equivalent to about 0.02 MIPS). The Logic Theory Machine (LTM) was devised and implemented on the same JOHNNIAC using a programming language devised and implemented by Newell, Shaw, and Simon. The LTM works in a backward manner in order to find subproblems from the desired axioms or known theorems. This is referred to as backward reasoning. This is analogous to working backward in which the ease in finding a needle that is backing its way out of a haystack, as opposed to the difficulty of someone finding the lone needle in the haystack. By 1956, the JOHNNIAC had its tubes replaced by transistors, its main memory had been expanded by 4,096 words and a drum, 12k of external storage had been added, and the MFTBF was improved to around 100 hours. However, none of the improvements made by contemporary researchers (Dunham, Gilmore, Gelernter, Prawitz) allowed for functional symbols. As a result, Martin Davis and Hilary Putnam introduced Skolem functions and the Herbrand Universe into automated deduction. The Davis–Putnam (D–P) algorithm was developed by Martin Davis and Hilary Putnam for checking the validity of a FOL formula using a resolution-based decision procedure for propositional logic. In 1960, John Alan Robinson and George Robinson implemented the D–P procedure in FORTRAN. The D–P algorithm was developed for checking validity of a FOL formula using a resolution-based decision procedure for propositional logic.

> Resolution Principle: is a method of theorem proving that proceeds by constructing refutation proofs, i.e., proofs by contradiction. Traditionally, a single step in deduction has been required, for pragmatic and psychological reasons, to be simple enough, to be apprehended as correct by a human being in a single intellectual act. No doubt this custom originated in the desire that each single step of a deduction should be indubitable, even though the deductions as whole may consist of a long chain of such steps. This method has been exploited in many automatic theorem provers. The resolution principle applies to first-order logic formulas in Skolemized form.

D–P is an "inside-out" approach as it assigns a truth value to atomic statements and determines the consequences of that assignment for the more complex statements composed of those atomic statements. Truth-Trees is an "outside-in" approach: it assigns truth values to complex statements and determines the consequences of that assignment for the smaller statements it is composed of. In mathematical logic and automated theorem proving, **resolution** is a rule of inference leading to a refutation complete theorem-proving technique for sentences in propositional logic and FOL. For propositional logic, systematically applying the resolution rule acts as a decision procedure for formula unsatisfiability, solving the (complement of the) Boolean satisfiability problem. For FOL, resolution can be used as the basis for a semi-algorithm for the unsatisfiability problem of FOL, providing a more practical method than one following from Gōdel's completeness theorem.

A model is an assignment of a truth value to each proposition. An expression is satisfiable. If there is a model for which the expression is true, an expression is valid if it is satisfied by *every* model. To do inference (reasoning) by computer is basically a *search* process, taking logical expressions and applying inference rules to them. Which logical expressions to use? Which inference rules to apply? Usually, you are trying to "prove" some particular statement. For example, if you have a collection of logical expressions (premises), and you are trying to prove some additional logical expression (the conclusion). If you do forward reasoning: start applying inference rules to the logical expressions you have and stop if one of your results is the conclusion you want. If you do backward reasoning: Start from the conclusion you want and try to choose inference rules that will get you back to the logical expressions you have.

Logic deals with reasoning and relatively little of the reasoning we do is mathematical, while almost all of the mathematical reasoning that nonmathematicians do is mere calculation. To have both rigor and scope, logic needs to keep its mathematical and its philosophical side united in a single discipline. The most influential figure in logical AI is John McCarthy. McCarthy has consistently advocated a research methodology that uses logical techniques to formalize the reasoning problems that AI needs to solve. The motivation for using logic is that even if the eventual implementations do not directly and simply use logical reasoning techniques like theorem proving, a logical formalization helps us to understand the reasoning problem itself. The claim is that without an understanding of what the reasoning problems are, it will not be possible to implement their solutions. McCarthy's long-term objective is to formalize common sense reasoning, the precipitous reasoning that is used in dealing with everyday problems:

- Narrative understanding
- Diagnosis
- Spatial reasoning
- Reasoning about the attitudes of other agents.

The Inference rules of propositional logic (PL) provide the means to perform logical proofs or deductions. As noted earlier, we are interested in mechanical inference by programs using symbolic FOPL expressions. One method we shall examine is called resolution. It requires that all statements be converted into a normalized clausal form. FOPL was developed by logicians to extend the expressiveness of PL. It is a generalization of PL that permits reasoning about world objects as relational entities as well as classes or subclasses of objects.

Predicate calculus includes: all of propositional logic, constants, predicates, functions, and quantifiers. A constant represents a "thing", it has no truth value, and it does not occur "bare" in a logical expression. Given zero or more arguments, a function produces a constant as its value. A predicate is like a function but produces a truth value. There are many rules in FOL, but in any case, the search space is *just too large* to be feasible. This was the case until 1970, when J. Robinson discovered resolution. Logic by use of the computer was infeasible. Why is logic so hard? If you start with a large collection of facts (predicates) and you start with a large collection of possible transformations (rules). Some of these rules apply to a single fact to yield a new fact. Some of these rules apply to a pair of facts to yield a new fact. So, at every step you must: (1) Choose some rule to apply, (2) choose one or two facts to which you might be able to apply the rule. If there are n facts, there are n potential ways to apply a single-operand rule and there are $n * (n-1)$ potential ways to apply a two-operand rule.

Default Logic is an extension of FOL that allows for exceptions. As such, FOL is monotonic such that anything that is entailed by a set of sentences will still be entailed by a superset of sentences. For example, $a \models g$ implies $a \wedge b \models g$ and bird(opus) |= flies(opus), thus bird(opus)∧penguin(opus) |= flies(opus) [necessarily]. Default Logics are nonmonotonic where old conclusions can get *retracted* due to new information such that bird(opus)∧penguin(opus) |= ¬ flies (opus). Default Logics require alternative model theory, such as possible worlds (or Kripke) semantics without going into the complexities, most default logics are based on drawing conclusions from "minimal models" that make the least assumptions necessary. The term "**nonmonotonic logic**" (in short, NML) covers a family of formal frameworks devised to capture and represent defeasible inference.

To build a logic-based representation, a user has to define a set of primitive symbols along with the required semantics. The symbols are assigned together to define legal sentences in the language for representing TRUE facts. New logical statements are formed from the existing ones. The statements which can be either TRUE or false but not both, are called propositions. A declarative sentence expresses a statement with a proposition as content. Propositional logic is a study of propositions where each proposition has either a true or a false value but not at a time. There are two types of propositions: Simple Preposition (It does not contain any other preposition) or Compound Preposition (It contains more than one prepositions).

Temporal logic is an extension of classical logic by adding operators relating to time. Modal logic has provided some of the formal foundations of temporal logic, and many of the techniques used in temporal logic are derived from their modal counterparts. Temporal logic tends to concentrate on the areas of propositional (no explicit first-order quantification), discrete (time being isomorphic to the Natural Numbers), and linear (at each moment in time have at most one successor) logic.

Next, add the new fact to your ever-expanding fact base, but be cautious, the search space is huge! This is where resolution comes in. Resolution works by transforming each of your facts into a particular form, called a clause. You apply a single rule, the resolution principle, to a pair of clauses. Clauses are closed with respect to resolution, that is, when you resolve two clauses, you get a new clause. You now add the new clause to your fact base, so that the number of facts you have grows linearly. The key is *you* still have to choose a pair of facts to resolve and *you* never have to choose a rule, because there's only one. An example is easy if it had only a very few clauses. When we have a lot of clauses, we want to *focus* our search on the thing we would like to prove. Along with models, modal logic introduces the idea of "possible worlds". And logical inference creates new sentences that logically follow from a set of sentences. The most common implications for implementing logic are used in Bayesian statistics, fuzzy logic, tense logic, and defeasible logic. Logic is a powerful and necessary tool in automatic reasoning, but to make useful deductions one requires domain-specific knowledge.

The Limitations of Logic is the "The Search Space Problem". Automated provers perform tens of thousands of inferences per second. A human in contrast performs maybe one inference per second. Although human experts are much slower on the object level (that is, in carrying out inferences), they apparently solve difficult problems much faster. A further, much more important advantage of us humans is intuition, without which we could not solve any difficult problems. The attempt to formalize intuition causes problems. Heuristics are methods that in many cases can greatly simplify or shorten the way to the goal, but in some cases (usually rarely) can greatly lengthen the way to the goal. **Heuristic Search** is important not only to logic but generally to problem-solving in AI. Heuristic algorithms aim to find good solutions to problems in a reasonable amount of time.

Heuristics make informed guesses to guide its search about the direction to a goal and typically, is not guaranteed to find the optimum, but able to find sufficiently good or near-optimal solutions.

The **Search Space Problem:** If we are solving a problem, we are usually looking for some solution that will be the best among others. The space of all feasible solutions (the set of solutions among which the desired solution resides) is called search space (also state space). Each point in the search space represents one possible solution.

Planning

The process of doing a sequence of actions to achieve a goal is called planning. It is a combination of an algorithm (search) and a set of representations, usually "situational calculus". A plan is a representation of the crude structure of the input scene by the various object labels. The process of planning is a bottom-up process to provide clues concerning which knowledge can be applied to different parts of the scene. The knowledge of the task world is represented by sets of production rules. Each rule in the bottom-up process has a fuzzy predicate that describes the properties of relations between objects. Generally, there are various agents who act to plan. The environments for an agent may be deterministic, finite, static in which change happens only when the agent acts. The discrete environment includes the time factor, objects, effects etc. These environments are called classical planning environments. On the other hand, the non-classical planning environments are partially observable and involve a different set of algorithms and agent designs. Planning refers to the process of computing several steps of a problem-solving procedure before evaluation of that problem. And AI planning systems are usually based on the assumption that, if there is an action an agent is physically able to perform, and carrying out that action would result in achievement of a goal *P*, then the agent can achieve *P*.

Planning is a combination of an algorithm (search) and a set of representations, usually "situational calculus". The planning problem in AI is about the decision-making performed by intelligent systems like robots, humans, or computer programs when trying to achieve some goal. It involves choosing a sequence of actions that will (with a high likelihood) transform the state of the world, step by step, so that it will satisfy this goal. The world is typically viewed to consist of atomic *facts* (state variables), and actions make some facts true and some facts false. Classical Planning is the planning where an agent takes advantage of the problem structure to construct complex plans of action. The planning problem finds a sequence of actions that leads to a goal state, starting from any of the initial states. The solution (obtained sequence of actions) is optimal if it minimizes the sum of action costs. Search-based problem-solving agents are a special case of planning agents. GPS was a computer program created in 1959 by Herbert Simon, J.C. Shaw, and Allen Newell intended to work as a universal problem solver machine. GPS was the first computer program that separated its knowledge of problems (rules represented as input data) from its strategy of how to solve the problems (a generic solver engine). Planning is a combination of an algorithm (search) and a set of representations, usually annotated as "situational calculus". It provides a way to "open up" the representation of initial state, goal-test, and successor functions. The planner is free to add actions to plan wherever is needed, and most parts of world (environment) are independent of other part (Figure 4.22).

Planning has the following states:

- Initial State: An arbitrary logical sentence about a situation.
- Goal State: A logical query asking for suitable situations.
- Operators: A set of descriptions of actions, using the action representation.

Theoretically, classical planning is the problem of finding a sequence of actions that maps a given initial state to a goal state, where the environment and the actions are deterministic. The computational challenge is to devise effective methods to obtain such action sequences called plans. A classical planning problem can be understood as a path-finding problem in a directed graph whose

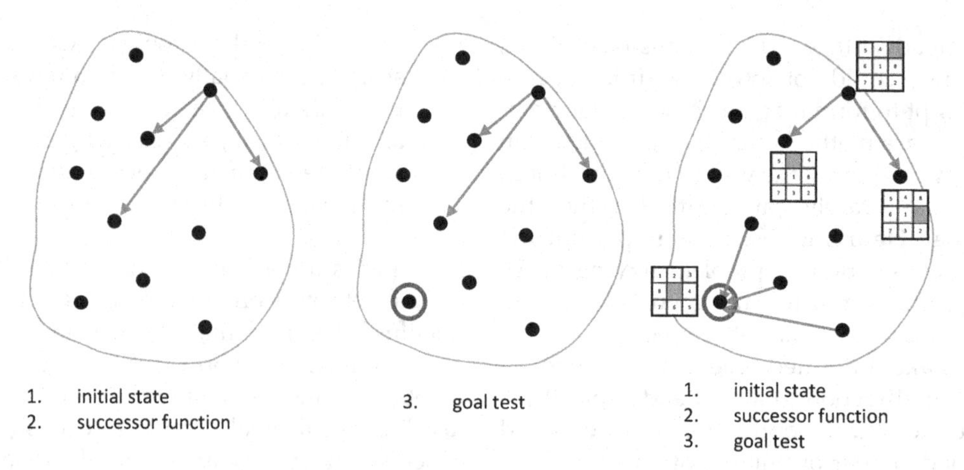

FIGURE 4.22
Planning function.

nodes represent states, and whose edges represent actions that change the state represented by the source node of the edge, to the state represented by the target node. A plan, the sequence of actions that transform the initial state into a goal state, can be understood as a path from the initial node in the graph to a node whose state is one of the goal states of the problem. The problem-solving agent can find sequences of actions that result in a goal state. But it deals with atomic representations of states and thus needs good domain-specific heuristics to perform well. The hybrid propositional logical agent can find plans without domain-specific heuristics because it uses domain-independent heuristics based on the logical structure of the problem. But it relies on ground (variable-free) propositional inference, which means that it may be swamped when there are many actions and states.

The **Missionaries and Cannibals Problem**: On one bank of a river are three missionaries and three cannibals. There is one boat available that can hold up to two people and that they would like to use to cross the river. If the cannibals ever outnumber the missionaries on either of the river's banks, the missionaries will get eaten. So, how can the boat be used to safely carry all the missionaries and cannibals across the river? This problem can be solved by searching for a solution, which is a sequence of actions that leads from the initial state to the goal state. Remember, the goal state is effectively a mirror image of the initial state and the key is a backward step. In real applications of the problem, it requires a knowledge of the world in order to solve real-world problems. Thus, this real-world problem was the basis that had led to future Expert Systems.

During planning, the agent plans after knowing what is the problem. Followed by acting which decides what action it has to take. But by learning, the actions taken by the agent make him learn new things. The focus here is on deterministic planning when the environment is fully observable and the results of actions is deterministic. By relaxing these requirements, we have to deal with uncertainty. Planning is logic-based where the system reasons about actions and change and has fewer formal representations (classical AI planning). The classical approaches to AI planning use **state-space and plan-space** to search solution plans to solve planning problems. The resulting solution plan is a sequence of actions that when applied transforms the initial world state to the goal state in one or more steps. The most basic planning problem is one instance of the general *state-transition* reachability problem for succinctly represented transition graphs, which has other important applications in Computer-Aided Verification (reachability analysis, model-checking), intelligent control, discrete event-systems diagnosis, and so on. All of the methods described below are equally applicable to all of these other problems as well, and many of these methods were initially developed and applied in the context of these other problems.

Basic Components of a Planning System: When a particular problem will be solved, at that time some specific rules regarding to that problem are to be applied. Then apply the chosen rule to compute the new problem state that arises from its application. Detect when a solution has been found and calculate the active and inactive ends of that problem. Various components of an AI planning system are described as follows:

a. States: For a planning process, the planners decompose the world into some environments. Then environments are defined by some logical conditions and states. The problems can be viewed as the task of finding a path from a given starting state to some desirable goal state. The state can be viewed as a conjunction of positive literals. For example, Rich A famous might represent the state of the best agent.

b. Goal: A goal is a specified state. To find a solution to a problem using a search procedure is to generate moves through the problem space until a goal state is reached. In the context of game-playing programs, a goal state is one in which we win. Unfortunately, for interesting games like chess, it is not usually, possible, even with a good plausible move generator, to search until a goal state is found.

c. Actions: An action is specified in terms of the pre-conditions that must be held before it can be executed and then the effects that ensue when it is executed.

Planning can also be viewed as problem-solving. The idea is to find a sequence or set of actions that can get you from the start (S) state to the goal (G) state. And more generically, the planner can generate a sequence of actions to perform tasks and achieve objectives where the algorithm searches for a solution over an abstract space of plans. It can assist humans in practical applications such as design and manufacturing, military operations, games, and space exploration. Different planning methods have strengths in different types of problems.

- Symbolic methods based on BDDs (Binary Decision Diagrams) excel in problems with a relatively small number of state variables (up to one or two hundred), with a complex but regular state space.
- Explicit state-space search is generally limited to small state spaces, but the AI planning community has been successfully applying explicit state-space search also to very large state spaces, when their structure is simple enough to allow useful heuristic distance estimates, and when there are plenty of plans to choose from.
- Methods based on logic and constraints (SAT or *satisfiability*, constraint programming) are strong on problems with relatively high numbers of state variables and not too long plans, especially when constraints about the structure of the solution plans and the reachable state-space are available.

The majority of planning uses a *state-transition system* model in which the world/system consists of a (finite or infinite) number of states, and actions/transitions change the current state to a next state. There are several possible representations for state-transition systems. A state-transition system can be represented as an arc-labeled directed multi-graph. Each *node* of the graph is a *state*, and the arcs represent actions. An "action" is usually something that can be taken in several states, with the same or different effects. All the arcs corresponding to one action are *labeled* with the name of the action. There may be more than one action that moves us from state A to state B, and this means that there are more than one arc from A to B (this is why we said that the graph is a multi-graph). In planning (similarly to other areas, including Discrete Event Systems and Computer-Aided Verification), transition systems are usually not represented explicitly as a graph, simply because those graphs would be far too big, with billions and trillions of nodes, or even far more. Instead, a *compact* (*succinct*, *factored*) representation is used. These representations are typically based on *state variables* (corresponding to (atomic) facts). A state, instead of being an atomic object, is a *valuation* of a (finite) number of state variables.

All possible actions can be represented as formulas in propositional logic. This includes all STRIPS and PDDL actions, and also further actions that cannot be represented in (deterministic variants) of PDDL. The logical representation of actions is useful for various powerful logic-based search methods. Explicit state-space search, meaning the generation of states reachable from the initial state one by one, is the earliest and most straightforward method for solving some of the most important problems about transition systems, including model-checking (verification), planning, and others, widely used since at least the 1980s. All the current main techniques for it were fully developed by the 1990s, including symmetry and partial order methods, informed search algorithms, and optimal search algorithms (A* already in the 1960s). It is relatively easy to implement efficiently, but, when the number of states is high, its applicability is limited by the necessity to do the search only one state at a time. However, when the number of states is less than some tens of millions, this approach is efficient and can give guarantees of finding solutions in a limited amount of time.

There are shortcomings to planning in general: not every action can be described with STRIPS-like operators with alternative post-conditions and not having complete knowledge about the world. The fact is the world is not stable such that re-planning must be supported and goals are not clearly defined. As an analogy, nobody plans for the solution of everyday tasks and humans have the capability to learn. Problem-solving in AI may be characterized as a systematic search through a range of possible actions in order to reach some predefined goal or solution. The problem-solving agents decide what to do by finding sequence of actions that lead to desirable states. The simplest agents which have been described below are the reflex and goal-based agents. The reflex agents use direct mapping from states to actions and are unsuitable for very large mappings. Problem-solving agents find action sequences that lead to desirable state. Search algorithms should track the paths from the start node to the goal node because these paths contain a series of operations that lead to the solution of the problem.

Symmetry reduction methods try to decrease the effective search space by recognizing symmetries in the state-space graph. In planning, symmetries are typically caused by the interchangeability of objects. If there is a plan that involves some interchangeable objects A and B, there is a symmetric plan with the roles of A and B interchanged. The standard works on symmetry reduction for state-space search are from the early 1990s, and they are directly applicable to all of the explicit state-space search algorithms. Partial-order reduction methods try to decrease the number of search steps by recognizing different forms of independence of actions/transitions. The basic idea is that if actions/transitions A and B are independent in the sense that they can be taken in either order A;B or B;A, one only needs to consider one of these orderings. Many of the existing partial-order reduction methods address deadlock detection, but there is a simple reduction from the problem of reaching a goal state to the problem of reaching a deadlock state, and therefore these methods are easily applicable to the planning problem, too (Figure 4.23).

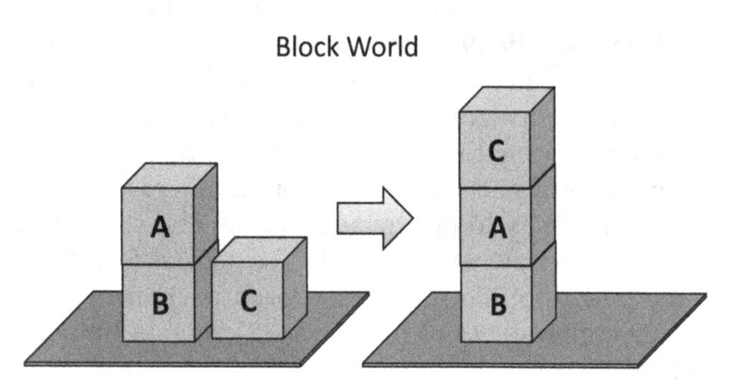

FIGURE 4.23
Block world illustration.

Block World: For a blocks world, there is a table on which some blocks are placed. Some blocks may or may not be stacked on other blocks. We have a robot arm to pick up or put down the blocks. The robot arm can move only one block at a time, and no other block should be stacked on top of the block which is to be moved by the robot arm. The aim is to change the configuration of the blocks from the Initial State to the Goal State, both of which have been specified. More complicated derivatives of the problem consist of cubes in different sizes, shapes and colors. The blocks world is chosen because **it is sufficiently simple and well behaved**, yet still provides a good sample environment to study planning (problems can be broken into nearly distinct subproblems).

Another concept that was clearly quite central was that of heuristics, the idea of problem-solving through rules of thumb, or human-like or rhetorical means. This was one of the central heuristic concepts in How to solve it: A New Aspect of Mathematical Method, by George Pólya, the mathematician who had inspired Newell in his college days at Stanford. *Heuristic Search* is the dominant approach in classical planning. In order to guide the search for a solution, heuristic search planners use a heuristic function that estimates the cost of reaching the goal of the problem from any given state. There are several ways of systematically (automatically) creating heuristics. The main approaches are different forms of *relaxation*, in which either the original problem instance is simplified, and then the simplified problem instance is solved, or the problem instance is unmodified, but the definition of a solution is simplified to be efficiently solvable. (In many cases the same thing can be achieved in either of these ways.). Given that the state space *S* of classical planning problems, it can be understood as directed graphs whose nodes represent states, and whose edges represent actions, and any graph-search algorithm can be used in order to find a plan, a path from the initial state to a goal state in the graph. Yet, blind search algorithms such as Dijkstra (do not scale up due to the size of the state space S, which can be exponential to the number of fluent of the problem). On the other hand, heuristic search algorithms have been proven to perform effectively, provided they use a heuristic function sufficiently informed to guide the search.

Best-first search (BFS) algorithms use two distinct sets of nodes for storing search nodes, the closed list and the open list, where nodes are sorted according to the evaluation function. Nodes that have not yet been expanded, that is, those whose successors have not yet been generated, are placed in the open list. Nodes that have been already expanded are placed in the closed list. Intuitively, the state space explored so far by the algorithm can be understood as a tree whose leaf nodes (search frontier) are in the open list, and the inner nodes are in the closed list. The first variant heuristic search algorithm is greedy best-first search (GBFS), using the constants $z=0$; $w=1$ rendering the evaluation function to be $f(n)=h(n)$. It always expands first the node with the lowest heuristic value in the open list, the one whose estimated cost to the goal is lower according to the heuristic estimator. Greedy best-first search is called greedy because it only pays attention to getting closer to the goal, no matter how expensive the paths to the goal are.

The **traveling salesman problem** (also called the **traveling salesperson problem** or **TSP**) asks the following question: "Given a list of cities and the distances between each pair of cities, what is the shortest possible route that visits each city exactly once and returns to the origin city?". It is an NP-hard problem (NP-hard which means that it has no efficient algorithm (unless a famous conjecture called $P=NP$ is true), but the majority of computer scientists now suspect that it is false) in combinatorial optimization, important in theoretical computer science and operations research. The traveling salesman problem is easy to state, and in theory at least, it can be easily solved by checking every round-trip route to find the shortest one. The trouble with this brute-force approach is that as the number of cities grows, the corresponding number of roundtrips to check quickly outstrips the capabilities of the fastest computers. With 10 cities, there are more than 300,000 different roundtrips, but with 15 cities, the number of possibilities balloons to more than 87 billion. So, the goal was to find a spanning tree that strikes the perfect balance between length and easy conversion into a round-trip. No efficient algorithm can uncover this perfect tree (unless $P=NP$), but a successful approximation algorithm only needs to find a pretty good one.

Planning as satisfiability is a powerful approach to domain-independent planning first proposed by Henry Kautz and Bart Selman in their SATPLAN system in the 1990s. The latest planners based on SAT rival (and often exceed) the performance of planners based on other search paradigms. Additionally, this level of performance is obtained almost completely by general-purpose SAT-solving algorithms, without the need for the kind of specialized techniques used in connection with the recent state-space search planners. We saw how the description of a planning problem defines a search problem that we can search from the initial state through the space of states, looking for a goal. One of the nice advantages of the declarative representation of action schemas is that we can also search backward from the goal, looking for the initial state. Current state-of-the-art planners solve problems, easy and hard alike, by searching and expanding hundreds or thousands of nodes. Yet, given the ability of people to solve easy problems and to explain their solutions, it seems that an essential inferential component may be missing. The inference scheme for planning below is used in the context of a forward-state search that does not appeal to a heuristic function but to the notion of causal links developed in the context of partial-order planning. SAT methods are less prone to excessive memory consumption than BDDs which represent a Boolean function. Unlike explicit state-space search, their memory consumption can be (far) less than linear in the number of visited (stored) states, and they can be implemented with a memory consumption that is linear in the length of a plan or transition sequence (as opposed to the in general exponential memory consumption of explicit state-space search). However, the currently leading algorithms for the SAT problem gain much of their efficiency by consuming more memory than theoretically optimal algorithms, and in practice consume as much memory as an explicit state-space search.

Boolean Satisfiability Problem or **SAT** problem considers whether there is an assignment of values to a set of Boolean variables, joined by logical connectives, that makes the logical formula it represents as true. SAT was the first problem to be proven NP-complete, and the first algorithms to solve it were developed in the 1960s. Many real-world problems, such as circuit design, automated theorem proving, and scheduling, can be represented and solved efficiently as SAT problems.

Now that we have shown how a planning problem maps into a search problem, we can solve planning problems with any of the heuristic search algorithms. In regression search, we start at the goal and apply the actions backward until we find a sequence of steps that reaches the initial state. It is called relevant-states search because we only consider actions that are relevant to the goal (or current state). As in belief-state search, there is a set of relevant states to consider at each step, not just a single state. Neither forward nor backward search is efficient without a good heuristic function. Recall that a heuristic function h(s) estimates the distance from a state S to the goal and that if we can derive an admissible heuristic for this distance, one that does not overestimate, then we can use AI search to find optimal solutions. An admissible heuristic can be derived by defining a relaxed problem that is easier to solve. The exact cost of a solution to this easier problem then becomes the heuristic for the original problem.

Machine Vision

Perception (from the Latin *perceptio,* meaning gathering or receiving) is the organization, identification, and interpretation of sensory information in order to represent and understand the presented information or environment. Perception is not only the passive receipt of these signals, but it's also shaped by the recipient's learning, memory, expectation, and attention. Sensory input is a process that transforms this low-level information to higher-level information (e.g., extracts shapes for object recognition). The process that follows connects a person's concepts and expectations (or knowledge), and restorative and selective mechanisms (such as attention) that influence perception. The perceptual systems of the brain enable individuals to see the world around them as stable, even though the sensory information is typically incomplete and rapidly varying. Humans can do amazing things like recognize people and objects, navigate through obstacles, understand the mood in the scene, and imagine stories. But human vision is not perfect because it suffers from illusions, ignores many details, and ambiguous description of the world. Do we care about accuracy of world? Machine vision is the ability of a computer to see; it employs one or more video cameras, analog-to-digital conversion (ADC) and digital signal processing (DSP). The resulting data goes to a computer or robot controller. Machine vision is similar in complexity to voice recognition. One or more video cameras are used with ADC and DSP. The image data is sent to a computer or robot controller. The term, computer vision, is used to designate the technology in which a computer digitizes an image, processes the data, and takes some type of action (Figure 4.24).

Vision is perhaps the most remarkable of all of our intelligent sensing capabilities. Through our visual system, we are able to acquire information about our environment without direct contact. Vision permits us to acquire information at a phenomenal rate and at resolutions that are most impressive. Once an image has been segmented into disjointed object areas, the areas can be labeled with their properties and their relationships to other objects, and then identified through model matching or description satisfaction. Once the image has been segmented into disjointed regions, their shapes, spatial interrelationships, and other characteristics can be described and labeled for subsequent interpretation. This process requires that the outlines or boundaries, vertices, and surfaces of the objects he described in some way. It should be noted,

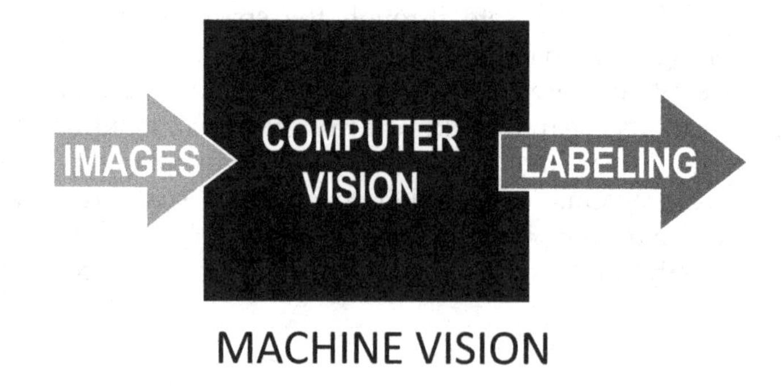

FIGURE 4.24
Machine vision process diagram.

however, that a description for a region can be based on a two- or three-dimensional image interpretation

One of the most influential papers in Computer Vision was published by two neurophysiologists, David Hubel and Torsten Wiesel, in 1959. Their publication, entitled "Receptive fields of single neurons in the cat's striate cortex", described core response properties of visual cortical neurons as well how a cat's visual experience shapes its cortical architecture. In 1959, Russell Kirsch and his colleagues developed an apparatus that allowed transforming images into grids of numbers – the binary language machines could understand. And it's because of their work that we now can process digital images in various ways. Next Lawrence Roberts' "Machine perception of three- dimensional solids" was published in 1963 and is widely considered to be one of the precursors of modern Computer Vision. The goal of the program he developed and described in the paper was to process 2D photographs into line drawings, then build up 3D representations from those lines and, finally, display 3D structures of objects with all the hidden lines removed. Lawrence Roberts processed 2D to 3D construction, followed by 3D to 2D display, which were a good starting point for future research of computer-aided 3D systems. In 1982, David Marr, a British neuroscientist, published another influential paper, "Vision: A computational investigation into the human representation and processing of visual information". Building on the ideas of Hubel and Wiesel (who discovered that vision processing doesn't start with holistic objects), Dr. Marr gave us the next important insight; he established that vision is hierarchical. The vision system's main function, he argued, is to create 3D representations of the environment so we can interact with it. He introduced a framework for vision where low-level algorithms that detect edges, curves, corners, etc., are used as stepping stones toward a high-level understanding of visual data.

Around the same time, a Japanese computer scientist, Kunihiko Fukushima built a self-organizing artificial network of simple and complex cells that could recognize patterns and was unaffected by position shifts. The network, Neocognitron, includes several convolutional layers whose (typically rectangular) receptive fields had weight vectors (known as filters). A few years later, in 1989, a young French scientist Yann LeCun applied a backprop style learning algorithm to Fukushima's CNN architecture. After working on the project for a few years, LeCun released LeNet-5, the first modern convolutional network Convnet that introduced some of the essential ingredients we still use in CNNs today. In 1997, a Berkeley professor named Jitendra Malik[1] released a paper in which he described his attempts to tackle *perceptual grouping*. Around 1999, lots of researchers stopped trying to reconstruct objects by creating 3D models of them (the path proposed by Marr) and instead directed their efforts toward *feature-based object recognition*. The paper describes a visual recognition system that uses local features that are invariant to rotation, location, and, partially, changes in illumination. These features, according to Lowe, are somewhat similar to the properties of neurons found in the inferior temporal cortex that are involved in object detection processes in primate vision. In 2001, the first face detection framework that worked in real time was introduced by Paul Viola and Michael Jones. Though not based on deep learning, the algorithm still had a deep learning flavor to it as, while processing images, it learned which features (very simple, Haar-like features) could help localize faces. In 2009, another important feature-based model was developed, the Deformable Part Model. It decomposed objects into collections of parts (based on pictorial models introduced by Fischler and Elschlager in the 1970s), enforced a set of geometric constraints between them, and modeled potential object centers that were taken as latent variables. Today, 3D reconstruction is accomplished using ray marching and volumetric rendering techniques.

In Machine Vision, the term "feature" usually refers to some measurement taken on an image, such as distances, moments, areas, lines, etc. Jain classifies features as spatial, transform, edge/boundary, shape, moments, or texture on the basis of the type of processes used to derive them. Here we will be content with the more general idea of a feature described above. The term feature is simultaneously used on the one hand for single pixels, and on the other, for particular combinations of pixel values with the same dimensionality as the original images, that is, image-like combinations of pixels with particular properties (Figure 4.25).

FIGURE 4.25
Feature boxes.

Image processing detects and enhances features using Fourier transform sampling and convolution. We use computer vision for optical flow, tracking, binocular stereo, range scanning, structured light, facial detection, and facial recognition. For robotics, the machine needs to detect objects and identify objects. General problems with recognition include variances in both external parameters (pose, illumination) and internal parameters (identify person and facial expressions). For object detection, the task starts with a given input image, and to determine if there are objects of a given class (e.g., faces, people, cars) in the image and where they are located. We start with search of the image, extract features using pixel patterns, implement a feature vector as a classifier against a training example in a database, and arriving at a classification. To help detection problems, classifiers must generalize over all exemplars of one class and the negative class consist of everything else. Unfortunately, a high accuracy is required for all applications. For identification, the task is given an image of an object of a particular class (e.g., face), identify which exemplar it is. Problems in identification include multi-class, and how the classifier must distinguish between exemplars that might look very similar, and how the classifier has to reject exemplars that were not in the training database. From an image, the computer extracts features (gray, gradient, wavelets) and uses the training data to support the classifier through a SVM to identify results.

A crucial function of vision is detecting important changes in the environment, and sensory adaptation aids in maximizing sensitivity to change. The visual system adapts to properties such as color, orientation, object and scene properties, and many others, thereby optimizing how it responds to changes in these attributes. Adaptation is a simple but powerful mechanism for leveraging past visual input to maximize change sensitivity, but there is a flip side to the coin: the physical world is largely stable and continuous over time. Objects, scenes, and physical properties tend to persist over time, making the recent past a good predictor of the present. The visual system may therefore delicately balance the need to optimize sensitivity to image changes with the desire to capitalize on the temporal continuity of the physical environment (Figure 4.26).

On a certain level computer vision is all about pattern recognition. So, one way to train a computer how to understand visual data is to feed it images, lots of images thousands, millions if possible that have been labeled, and then subject those to various software techniques, or algorithms, that allow the computer to hunt down patterns in all the elements that relate to those labels. In point of fact, pixel values are

FIGURE 4.26
Image segmentation.

almost universally stored, at the hardware level, in a *one-dimensional array*. For example, the data from the image above is stored in a manner similar to this long list of unsigned characters. This way of storing image data may run counter to your expectations, since the data certainly *appears* to be two-dimensional when it is displayed. Yet, this is the case, since computer memory consists simply of an ever-increasing linear list of address spaces. This way of storing image data may run counter to your expectations, since the data certainly *appears* to be two-dimensional when it is displayed. Yet, this is the case, since computer memory consists simply of an ever-increasing linear list of address spaces.

High-Level Processing: Before proceeding with a discussion of the final (high-level) steps in vision processing, we shall briefly review the processing stages up to this point. We began with an image of gray-level or tristimulus color intensity values and digitized this image to obtain an array of numerical pixel values. Next, we used masks or some other transform (such as Fourier) to perform smoothing and edge enhancement operations to reduce the effects of noise and other unwanted features. This was followed by edge detection to outline and segment the image into coherent regions. The product of this step is a primal sketch of the objects. Region splitting and/or merging, the dual of edge finding, can also be used separately or jointly with edge finding as part of the segmentation process. Histogram computations of intensity values and subsequent analyses were an important part of the segmentation process. There is a common saying, Semantics is Meaning (Meaning = Semantics). They help,

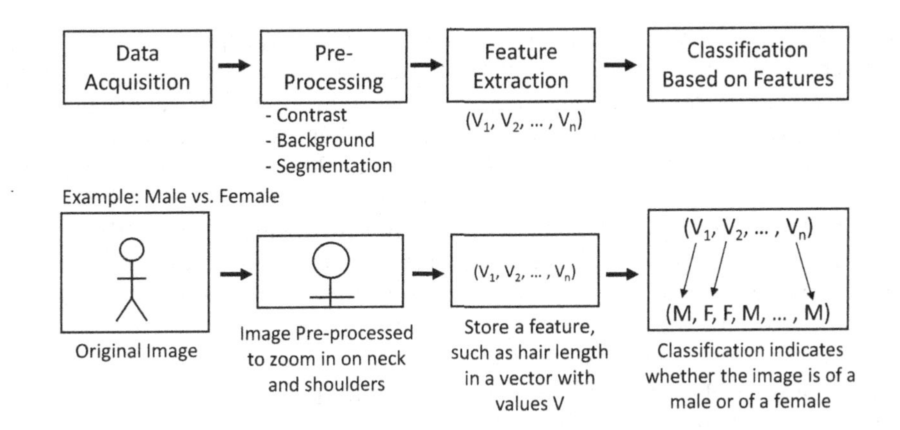

FIGURE 4.27
Low-level image processing.

to establish threshold levels which serve as cues for object separation. Other techniques such as minimum spanning tree or dynamic programming are sometimes used in these early processing stages to aid in edge finding. Following the segmentation process, regions are analyzed and labeled with their characteristic features. The result of these final steps in intermediate-level processing is a set of region descriptions (data structures). Such structures are used as the input to the final high-level image processing stage.

Low-Level Processing: Low-level processing usually involves some form of smoothing operation on the digitized arrays. Smoothing helps to reduce noise and other unwanted features. This is followed by some form of edge detection such as the application of difference operators to the arrays. Edge fragments must then be joined to form continuous contours that outline objects. Various techniques are available for these operations. The dual-of-the-edge-finding approach is region segmentation, which may be accomplished by region splitting, region growing, or a combination of both. Semantic segmentation is the task of classifying each pixel in an image to a particular label, such as person, cat, etc. Where image classification tries to assign a label to the entire image, semantic segmentation tries to isolate the distinct entities and objects in a given image, allowing for more fine-grained identification. Multi-histograms and thresholding are commonly used techniques in the segmentation process. They are applied to one or more image features such as intensity, color, texture, shading, or optical flow in the definition of coherent regions.

The end product of the segmentation process will be homogeneous regions. The properties of these regions and their interrelationships must be described in order that they may be identified in the high-level processing stage. Regions can be described by boundary segments, vertices, number of holes, compactness, location, orientation, and so on (Figure 4.27).

The levels of reasoning in vision start with an image, analyze pixels, detect edges, identify lines, and group to detect objects, and finally build a scene from multiple objects with background information. Edge detection converts a 2D image into a set of curves. Salient features of the scene are extracted, which is more compact than pixels. Edges are caused by a variety of factors: surface normal discontinuity, depth discontinuity, surface color discontinuity, and illumination discontinuity. To tell if a pixel is on an edge, the edges look like steep cliffs. If we differentiate a digital image for a discrete gradient using a Sobel operator to compare edge operators, even with the effects of noise. Smoothing is required with a Laplacian of Gaussian to produce a 2D edge detector filters. Edge detection by subtraction using the Canny Edge detector. In addition to finding edges, we also look for corners with the Harris detector. Image processing is used to extract corners and infer features of an image. A common characteristic of images is that neighboring pixels are highly correlated.

To represent the image directly in terms of the pixel values is therefore inefficient because most of the encoded information is redundant. The first task in designing an efficient, compressed code is to find a representation which, in effect, decorrelates the image pixels. This has been achieved through predictive and through transformation techniques. Pyramid, or pyramid representation, is a type of multi-scale signal representation developed in which a signal or an image is subject to repeated smoothing and subsampling. The Laplacian pyramid is a versatile data structure with many attractive features for image processing. It represents an image as a series of quasi-band passed images, each sampled at successively sparser densities. The resulting code elements, which form a self-similar structure, are localized in both space and spatial frequency. By appropriately choosing the parameters of the encoding and quantizing scheme, one can substantially reduce the entropy in the representation, and simultaneously stay within the distortion limits imposed by the sensitivity of the human visual system. In this representation, image features of various sizes are enhanced and are directly available for various image processing and pattern recognition tasks. The final stage is the knowledge application stage since the regions must be interpreted and explained. This requires task or domain-specific knowledge as well as some general world knowledge. Knowledge acquisition is the process of adding new knowledge to a knowledge base and refining or otherwise improving knowledge that was previously acquired. Acquisition is usually associated with some purpose such as expanding the capabilities of a system or improving its performance at some specified task.

More recently, machine learning provided a different approach to solving computer vision problems. With machine learning, developers no longer needed to manually code every single rule into their vision applications. Instead, they programmed "features", smaller applications that could detect specific patterns in images. They then used a statistical learning algorithm such as linear regression, logistic regression, decision trees, or SVMs to detect patterns and classify images and detect objects in them. Machine learning helped solve many problems that were historically challenging for classical software development tools and approaches. For instance, years ago, machine-learning engineers were able to create a software that could predict breast cancer survival windows better than human experts. However, building the features of the software required the efforts of dozens of engineers and breast cancer experts and took a lot of time develop. Maybe the potentially most promising developments in computer vision (and maybe even for all of AI) involves hierarchical models. There are different versions of these models and different ways to construct them, but if we stand far enough back from the details, they have similar structures and features. First, the raw pixels are aggregated spatially (and in some systems temporally) to form higher-level groupings. These groupings might constitute small edges, or corners, or other primitive components appropriate for the kinds of images being processed. At the next level of the hierarchy, the first-level groupings are aggregated again into somewhat higher-level components, and so on until, say, recognizable objects in the image are represented at the highest level.

A machine-learning algorithm can tackle a range of problems and tasks, either as part of a pure software system or when controlling a physical robot. Tasks include:

- **classification** (to which category does the data belong?)
- **regression** (how does input data relate to the output?)
- **clustering** (which data inputs are similar to each other?).

Deep learning provided a fundamentally different approach to doing machine learning. Deep learning relies on neural networks, a general-purpose function that can solve any problem represented through examples. When you provide a neural network with many labeled examples of a specific kind of data, it'll be able to extract common patterns between those examples and transform them into a mathematical equation that will help classify future pieces of information. For instance, creating a facial recognition application with deep learning only requires you to develop or choose a preconstructed algorithm and train it with examples of the faces of the people it must detect. Given enough examples (lots of examples), the neural network will be able to detect faces without further instructions on features or measurements. Deep learning is a very effective method to do computer vision. In most cases, creating a good deep learning algorithm comes down to gathering a large amount of labeled training data and tuning the parameters such as the type and number of layers of neural networks and training epochs. Compared to previous types of machine learning, deep learning is both easier and faster to develop and deploy. Next is a figure of this architecture (Figure 4.28).

RNNs are a type of neural network that are designed to process sequential data. They can analyze data with a temporal dimension, such as time series, speech, and text. RNNs can do this by using a hidden state passed from one timestep to the next. The hidden state is updated at each timestep based on the input and the previous hidden state. RNNs are able to capture short-term dependencies in sequential data, but they struggle with capturing long-term dependencies. LSTMs (Long Short-Term Memory) is a type of RNN that can detain long-term dependencies in sequential

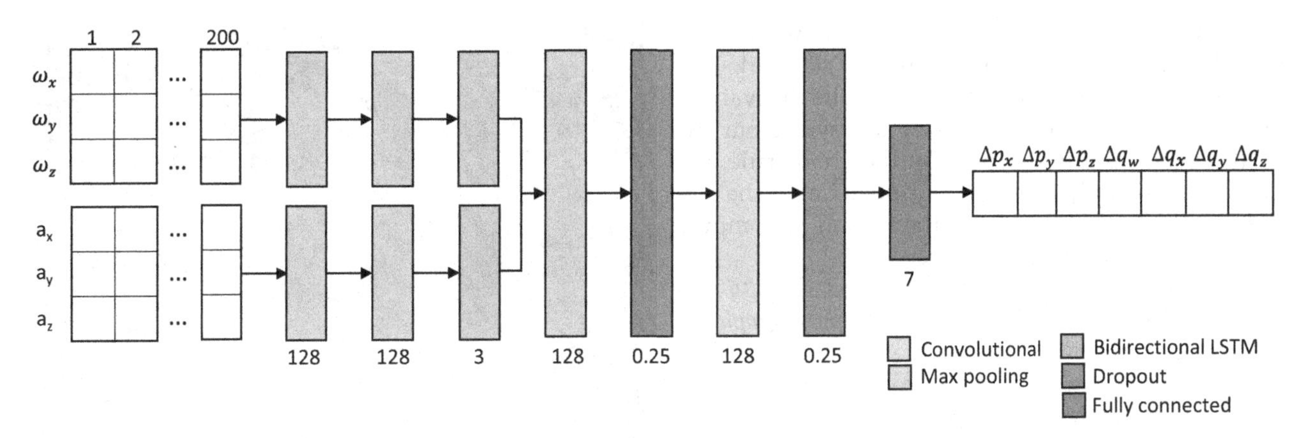

FIGURE 4.28
Convolutional layers figure.

data. LSTMs are able to process and analyze sequential data, such as time series, text, and speech and use memory cells and gates to control the flow of information, allowing them to selectively retain or discard information as needed and thus avoid the vanishing gradient problem that plagues traditional RNNs. LSTMs are long short-term memory networks that use ANNs. In contrast to normal feedforward neural networks, also known as RNNs, these networks feature feedback connections. The structure of an LSTM network consists of a series of LSTM cells, each of which has a set of gates (input, output, and forget gates) that control the flow of information into and out of the cell. The gates are used to selectively forget or retain information from the previous time steps, allowing the LSTM to maintain long-term dependencies in the input data.

Time series analysis using machine-learning techniques generally requires two stages. The first stage requires measuring the difference between the time series that you want to classify and converting that to a series of feature vectors. The second stage you use an algorithm to classify your data. LSTM stands for long short-term memory network and is used in the field of Deep Learning. It is a variation of a RNNs that are capable of learning long-term dependencies, especially in sequence prediction problems. In the network used and developed under the Convolutional and Long short-term memory, layers are used to generate the feature vectors, and the fully connected layers are used for classification. LSTM is a type of RNN that is suitable to solve problems that involve sequence processing. After two convolutional layers, a max-pooling layer of size 3 is used. The output of these layers is concatenated and fed to LSTM layers with 128 units. A bidirectional LSTM is used, so that both past 100 and future 100 IMU data readings from a mobile robot have an influence on the regressed relative pose of that robot. This is combined with a two-layer stacked LSTM model, in which a bidirectional LSTM outputs a full sequence that is the input of a second bidirectional LSTM. Finally, a fully connected layer generates the output of a relative pose.

Convolutional layers are used for feature vector engineering of the time series signal. They are powerful as they are noise-resistance, and they can extract informative, deep features automatically. The elements of the kernel get multiplied by the corresponding elements of the time series that they cover at a given point. Then the results of the multiplication are added together, and a nonlinear activation function is applied to the value. The resulting value becomes an element of a new "filtered" time series, then the kernel moves forward along the time series to produce the next value. Max pooling is applied to each of the filtered time series vectors, taking the largest value from each vector. A new vector is formed from these values, and this vector of maximums is the final feature vector that can be used as an input to the LSTM. With the regular LSTM, the input flows in one direction, either backward or forward. Bidirectionality will allow the LSTM to learn the input sequences both forward and backward, concatenating and embedding both interpretations in the hidden states. In the backward run, the bidirectional LSTM network preserves information from the future, thus adding potentially critically important context in the prediction making process. Fully Connected layers are the layers where all the inputs from the LSTM dropout layer are connected to every activation unit of the next layer; this fully connected layer generates the output. LSTM solves the vanishing gradient problem where a problem arises when a large input space is mapped into a small one, causing the derivatives to disappear. As more layers using certain activation functions are added to neural networks, the gradients of the loss function approach zero, making the network hard to train. However, gradients of neural networks are found using BP. Simply put, BP finds the derivatives of the network by moving layer by layer from the final layer to the initial one. By the chain rule, the derivatives of each layer are multiplied down the network (from the final layer to the initial) to compute the derivatives of the initial layers.

Researcher Geoffrey Hinton realized that *"If you want to do computer vision, first learn computer graphics"*. 2D vision is more widely used than 3D vision. While 3D vision can be more data and computing-intensive, it also adds depth sensing that can improve performance in some tasks. The most widely used methods of 3D vision are time-of-flight sensing, structured light, and active or passive stereoscopic vision. The most likely shift we will see to machine vision in the near future will be the offloading of much of the processing to the cloud or edge. As resolution and data gathering improve, more and more computing power is needed to quickly process it, especially as machine learning is applied to solutions.

Unfortunately, what many researchers in the computer vision research community failed to appreciate was that methods that require careful hand-engineering by a programmer who understands the domain do not scale as well as methods that replace the programmer with a powerful **general-purpose** learning procedure. And with enough computation and enough data, learning beats programming for complicated tasks that require the integration of many different, noisy cues.

Facial Recognition: Facial recognition is a way of identifying or confirming an individual's identity using their face. A **facial recognition system** is a computer vision technology capable of matching a human face from a digital image or a video frame against a database of faces. Such a system is typically employed to authenticate users through ID verification services and works by pinpointing and measuring facial features from a given image. Because computerized facial recognition involves the measurement of a human's physiological characteristics, facial recognition systems are categorized as biometrics. The earliest pioneers of facial recognition were Woody Bledsoe, Helen Chan Wolf, and Charles Bisson. In 1964 and 1965, Bledsoe, along with Wolf and Bisson, began work using computers to recognize the human face. The team used a rudimentary scanner to map the person's hairline, eyes, and nose. The task of the computer was to find matches; however, this effort wasn't successful early on. Their early facial recognition project was dubbed "man-machine" because a human first needed to establish the coordinates of facial features in a photograph before they could be used by a computer for recognition. In 1970, Takeo Kanade publicly demonstrated a face-matching system that located anatomical features such as the chin and calculated the distance ratio between facial features without human intervention. However, later tests revealed that the system could not always reliably identify facial features (Figure 4.29).

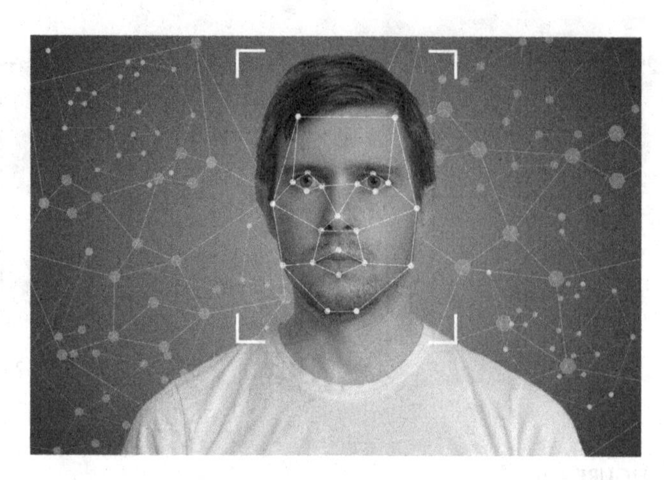

FIGURE 4.29
Facial recognition image.

Until the 1990s, facial recognition systems were developed primarily by using photographic portraits of human faces. Research on face recognition to reliably locate a face in an image that contains other objects gained traction in the early 1990s with the principle component analysis (PCA). The PCA method of face detection is also known as *Eigenface* and was developed by Matthew Turk and Alex Pentland. Purely feature-based approaches to facial recognition were overtaken in the late 1990s by the Bochum system, which used the Gabor filter to record the face features and computed a grid of the face structure to link the features. Real-time face detection in video footage became possible in 2001 with the Viola–Jones object detection framework for faces. Paul Viola and Michael Jones combined their face detection method with the Haar-like feature approach to object recognition in digital images to launch AdaBoost, the first real-time frontal-view face detector. By 2015, the Viola–Jones algorithm had been implemented using small low power detectors on handheld devices and embedded systems. Facial recognition is a challenging pattern recognition problem in computing (Figure 4.30).

FIGURE 4.30
Facial recognition – deep fake.

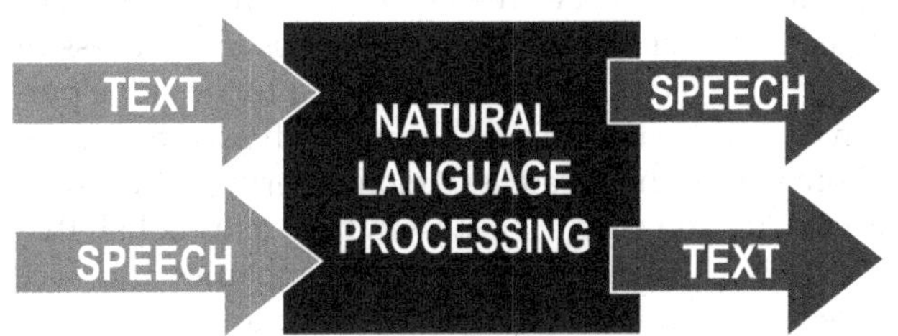

FIGURE 4.31
Natural language processing diagram.

Most systems use 2D camera technology, which creates a flat image of a face, and maps "nodal points" (size/shape of eyes, nose, cheekbones, etc.). It measures distances between eyes, nose, mouth, and jaws. These measurements are converted into a data set. The system then calculates the nodes' relative position and converts the data into a numerical code. The recognition algorithms search a stored database of faces for a match. This is an augmented replica of how the process works in our brain. 2D technology works well in stable, well-lit conditions such as passport control, but it is less effective in darker spaces and cannot deliver good results when the subjects move around. Unfortunately, it is easy to spoof with a photograph. Face detection is a binary classification problem combined with a localization problem: given a picture, decide whether it contains faces, and construct bounding boxes for the faces. And there was a bigger problem: Three-dimensional faces on living human beings, unlike two-dimensional letters on a page, are not static. Images of the same person can vary in head rotation, lighting intensity, and angle; people age and hairstyles change; someone who looks carefree in one photo might appear anxious in the next. To make the task more manageable, the Viola–Jones algorithm only detects full view (no occlusion), frontal (no head-turning), upright (no rotation), well-lit, full-sized (occupying most of the frame) faces in fixed-resolution images. A different form of taking input data for face recognition is by using thermal cameras, by this procedure the cameras will only detect the shape of the head and it will ignore the subject accessories such as glasses, hats, or makeup. For 3D facial recognition, Apple uses 3D camera technology to power the thermal infrared-based Face ID that is featured in its iPhone X. Thermal IR imagery maps the patterns of faces derived primarily from the pattern of superficial blood vessels under the skin (Figure 4.31).

This recognition problem is made difficult by the great variability in head rotation and tilt, lighting intensity and angle, facial expression, aging, etc. Some other attempts at facial recognition by machine have allowed for little or no variability in these quantities. Yet the method of correlation (or pattern matching) of unprocessed optical data, which is often used by some researchers, is certain to fail in cases where the variability is great (Figure 4.32). In particular, the correlation is very low between two pictures of the same

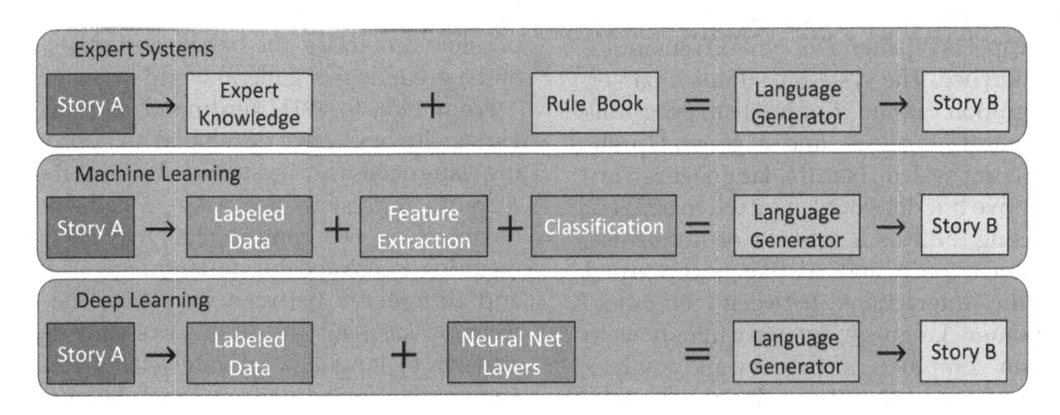

FIGURE 4.32
NLP (expert Sys vs. Deep Learning vs Mach. Learning).

person with two different head rotations. There is a crucial distinction between facial recognition and face detection. Facial recognition describes the process of scanning a face and then matching it to the same person on a database. This is the approach used to unlock phones or authenticate a person entering a building. Face detection is when a system tries to establish that a face is present. Social media companies use face detection to filter and organize images in large catalogs of photographs. The tools used to train the two systems are different and the desired levels of accuracy vary too.

Facial recognition for identification purposes needs to score more highly than any system used to organize images merely. Bottomline, the technology is improving drastically.

Natural Language Processing

Natural Language Processing (NLP) is an area of research and application that explores how computers can be used to understand and manipulate natural language text or speech to do useful things. It is a subfield of computer science and in AI that is concerned with computational processing of natural languages, emulating cognitive capabilities without being committed to a true simulation of cognitive processes. NLP is a theoretically motivated range of computational techniques for analyzing and representing naturally occurring texts at one or more levels of linguistic analysis for the purpose of achieving human-like language processing for a range of tasks or applications. It is a computerized approach to analyzing text that is based on both a set of theories and a set of technologies. It must deal in an integrated way with all of the aspects of language: syntax, semantics, and inference. The system contains a parser, a recognition grammar of English, programs for semantic analysis, and a general problem-solving system. Natural language is very descriptive but does not lend itself to efficient processing. NLP is a subfield of linguistics, computer science, and AI that is concerned with the interactions between computers and human language, in particular how to program computers to process and analyze large amounts of natural language data. NLP combines computational linguistics, the rule-based modeling of human language, with statistical, machine learning, and deep learning models. Together, these technologies enable computers to process human language in the form of text or voice data and to "understand" its full meaning, complete with the speaker or writer's intent and sentiment. NLP drives computer programs that translate text from one language to another, respond to spoken commands, and summarize large volumes of text rapidly, even in real time (Figure 4.33).

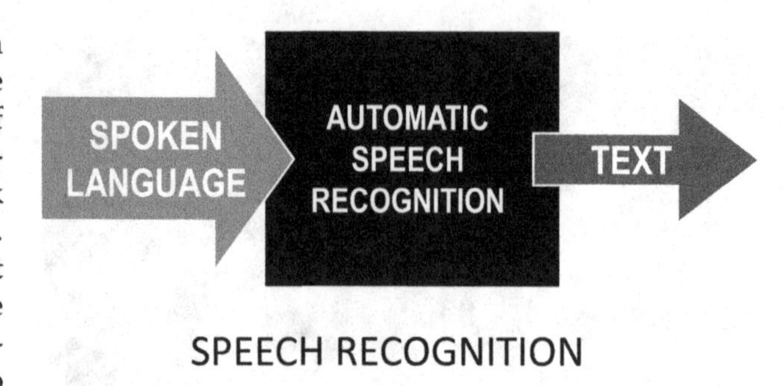

FIGURE 4.33
Automatic speech recognition process diagram.

Developing programs that understand a natural language is a difficult problem. Natural languages are large. They contain an infinity of different sentences. No matter how many sentences a person has heard or seen, new ones can always be produced. Also, there is much ambiguity in a natural language. Many words have several meanings such as can, bear, fly, and orange, and sentences can have different meanings in different contexts This makes the creation of programs that "understand" a natural language, one of the most challenging tasks in AI. It requires that a program transform sentences occurring as part of a dialog into data strictures that convey the intended meaning of the sentences to a reasoning program. In general, this means that the reasoning program must know a lot about the structure of the language, the possible semantics, the beliefs and goals of the user, and a great deal of general world knowledge.

From 1906 to 1911, Ferdinand de Saussure at the University of Geneva developed an approach describing languages as "systems". Within the language, a sound represents a concept, a concept that shifts meaning as the context changes. He argued that meaning is created inside language, in the relations and differences between its parts. The *structuralist theory of language* and linguistics says that the components of language are interrelated to one another and get their meaning from that relationship. The origins of the structuralist approach of linguistics come from Professor Saussure. Structuralism in linguistics asserts the idea that all of the parts to learning English are interwoven, and because they are all interrelated, they don't have the same meaning when in isolation. In order for the different parts of the English language to make sense, they need to be working together. The main concepts of teaching English from a structuralist approach include a focus on sentence structure, patterns of sentences, and appropriate grammar and composition. Teaching English through a structuralist approach includes a focus on four main skills:

1. Understanding the grammatical structures
2. Speaking properly, according to the rules of proper grammar and mechanics; using proper sentence structure
3. Reading properly, according to the rules of comprehension
4. Writing properly, according to the rules of proper grammar and mechanics; using proper sentence structure.

The overarching focus of these four areas is proper order. Teaching English using this methodology places emphasis on the proper order and structure of understanding, speaking, reading, and writing the English language in 1952, the Hodgkin-Huxley model showed how the brain uses neurons in forming an electrical network. These events helped inspire the idea of AI, NLP, and the evolution of computers. This process often uses higher-level NLP features, such as:

- Content Categorization: A linguistic document summary that includes content alerts, duplication detection, search, and indexing.
- Topic Discovery and Modeling: Captures the themes and meanings of text collections and applies advanced analytics to the text.
- Contextual Extraction: Automatically pulls structured data from text-based sources.
- Sentiment Analysis: Identifies the general mood, or subjective opinions, stored in large amounts of text.
- Text-to-Speech and Speech-to-Text Conversion: Transforms voice commands into text, and vice versa.
- Document Summarization: Automatically creates a synopsis, condensing large amounts of text.
- Machine Translation: Automatically translates the text or speech of one language into another.

Miller's paper on Magic Number Seven established the apparent primacy of seven digits as the number beyond which human short-term memory typically erodes in accuracy. Chomsky's paper was an early work in his project to describe innate human linguistic commonalities. In all, it presented a good cross-section of the work in cognitive psychology, linguistics, and information theory of the time. Noam Chomsky revolutionized previous linguistic concepts, concluding that for a computer to understand a language, the sentence structure would have to be changed. With this as his goal, Chomsky created a style of grammar called Phase-Structure Grammar, which methodically translated natural language sentences into a format that is usable by computers. In 1966, the NRC and ALPAC initiated the first AI and NLP stoppage, by halting the funding of research on natural language processing and machine translation. The mixing of linguistics and statistics, which had been popular in early NLP research, was replaced with a theme of pure statistics. The 1980s initiated a fundamental reorientation, with simple approximations replacing deep analysis, and the evaluation process becoming more rigorous. Until the 1980s, the majority of NLP systems used complex, "handwritten" rules. But in the late 1980s, a revolution in NLP came about. This was the result of both the steady increase of computational power and the shift to machine-learning algorithms.

In the 1990s, the popularity of statistical models for Natural Language Processes analyses rose dramatically. The pure statistics NLP methods have become remarkably valuable in keeping pace with the tremendous flow of online text. In 2001, Yoshio Bengio and his team proposed the first neural "language" model, using a feedforward neural network. The feedforward neural network describes an ANN that does not use connections to form a cycle. In this type of network, the data moves only in one direction, from input nodes, through any hidden nodes, and then on to the output nodes. The feedforward neural network has no cycles or loops and is quite different from the RNNs. In the year 2011, Apple's Siri became known as one of the world's first successful NLP/AI assistants to be used by general consumers. Within Siri, the Automated Speech Recognition module translates the owner's words into digitally interpreted concepts. The combination of a dialog manager with NLP makes it possible to develop a system capable of holding a conversation, and sounding human-like, with back-and-forth questions, prompts, and answers (Figure 4.34).

Representing the meaning of words, phrases, or even sentences using low-dimensional dense vectors is shown to be effective in a wide range of NLP tasks,

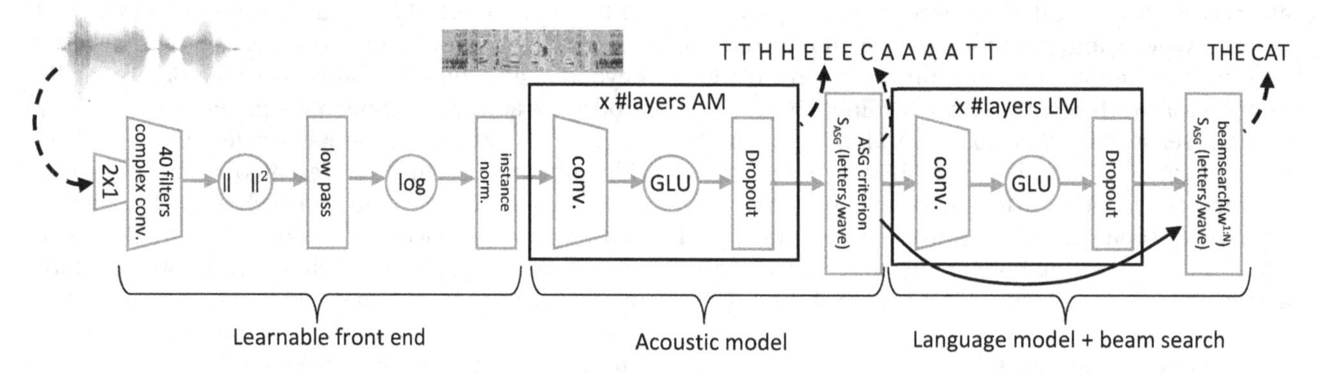

FIGURE 4.34
Model for automatic speech recognition.

such as machine translation, textual entailment, and question answering. Such representations are called distributed representations. The benefit of distributed representations is that they allow us to estimate the proximity of meaning based on vector similarity. In the field of reasoning, researchers started to leverage distributed representations in the reasoning process. Parsing is the term used to describe the process of automatically building syntactic analysis of a sentence in terms of a given grammar and lexicon. The resulting syntactic analysis may be used as input to a process of semantic interpretation. Occasionally, parsing is also used to include both syntactic and semantic analysis. The parsing process is done by the parser. The parsing performs grouping and labeling of parts of a sentence in a way that displays their relationships to each other in a proper way. The parser is a computer program that accepts the natural language sentence as input and generates an output structure suitable for analysis. A parse tree provides the relationship between the words that compose a sentence. For example, many sentences can be broken down into both a subject and a predicate. Subjects can be broken down perhaps into a noun phrase followed by a prepositional phrase and so on. Essentially, a parse tree gives the semantics that is the meaning of the sentence. The lexicon is a dictionary of words where each word contains some syntactic, some semantic, and possibly some pragmatic information. The entry in the lexicon will contain a root word and its various derivatives. The information in the lexicon is needed to help determine the function and meanings of the words in a sentence. Thus, parsing, syntax analysis, or syntactic analysis is the process of analyzing a string of symbols, either in natural language, computer languages, or data structures, conforming to the rules of a formal grammar. The term parsing comes from Latin pars, meaning part. Syntactic Analysis (or Parsing) involves analysis of words in the sentence for grammar and arranging words in a manner that shows the relationship among the words. The sentence such as "The school goes to boy" is rejected by English syntactic analyzer.

Several NLP tasks break down human text and voice data in ways that help the computer make sense of what it's ingesting. Some of these tasks include the following:

- Speech recognition, also called speech-to-text, is the task of reliably converting voice data into text data. Speech recognition is required for any application that follows voice commands or answers spoken questions. What makes speech recognition especially challenging is the way people talk – quickly, slurring words together, with varying emphasis and intonation, in different accents, and often using incorrect grammar.
- Part of speech tagging, also called grammatical tagging, is the process of determining the part of speech of a particular word or piece of text based on its use and context. Part of speech identifies "make" as a verb in "I can make a paper plane" and as a noun in "What make of car do you own?"
- Word sense disambiguation is the selection of the meaning of a word with multiple meanings through a process of semantic analysis that determines the word that makes the most sense in the given context. For example, word sense disambiguation helps distinguish the meaning of the verb "make" in "make the grade" (achieve) vs. "make a bet" (place).
- Named entity recognition, or NEM, identifies words or phrases as useful entities. NEM identifies "Kentucky" as a location or "Fred" as a man's name.

- Co-reference resolution is the task of identifying if and when two words refer to the same entity. The most common example is determining the person or object to which a certain pronoun refers (e.g., "she" = "Mary"), but it can also involve identifying a metaphor or an idiom in the text (e.g., an instance in which "bear" isn't an animal but a large hairy person).
- Sentiment analysis attempts to extract subjective qualities – attitudes, emotions, sarcasm, confusion, suspicion – from text.
- Natural language generation is sometimes described as the opposite of speech recognition or speech-to-text; it's the task of putting structured information into human language.

Natural Language Processing has been a final area of requiring a higher-level support software to be developed and in particular (parallel logic programming) by developing a new language and by developing experimental hardware to support the language. Higher-level facilities such as theorem provers, deductive databases, and natural language support assumed a secondary role, while applications assumed a tertiary role. Machine translation (MT), the subfield of computational linguistics that investigates the use of software to translate text or speech from one language to another, has seen significant improvement due to advances in machine learning. In July 2020, OpenAI unveiled GPT-3, the largest known dense language model. GPT-3 has 175 billion parameters and was trained on 570 gigabytes of text.

Computational linguistics is an interdisciplinary field concerned with the computational modeling of natural language, as well as the study of appropriate computational approaches to linguistic questions. The term "computational linguistics" is taken to be a near-synonym of NLP and language technology. These terms put a stronger emphasis on aspects of practical applications rather than theoretical inquiry. In practice, they have largely replaced the term "computational linguistics" in the NLP/ACL community, although they specifically refer to the subfield of applied computational linguistics (ACL). Computational linguistics has both theoretical and applied components. Theoretical computational linguistics focuses on issues in theoretical linguistics and cognitive science. Applied computational linguistics focuses on the practical outcome of modeling human language use. **Theoretical computational linguistics** includes the development of formal theories of grammar (parsing) and semantics, often grounded in formal logics and symbolic (knowledge-based) approaches. Areas of research that are studied by theoretical computational linguistics include:

- Computational complexity of natural language, largely modeled on automata theory, with the application of context-sensitive grammar and linearly bounded Turing machines.
- Computational semantics comprises defining suitable logics for linguistic meaning representation, automatically constructing them, and reasoning with them.

Natural Language Processing was revolutionized with the introduction of the transformer architecture. By moving to a Generative Pretrained Transformer (GPT) approach, the system deviates significantly from the original transformer by employing a decoder-only structure and this design choice was essential for its functionality in generating text. As a result, the GPT model processes text input through multiple decoder layers to predict the next word in a sequence. More on this in Volume 2.

Speech

Speech AI is the use of machine learning to help humans converse with devices, machines, and computers via speech. Its application is a complex system that integrates multiple DNNs that must work in unison to deliver a delightful user experience with accurate, fast, and natural human-to-machine voice-based interaction. The ultimate goal of *Speech AI* is to make conversing with machines indistinguishable from talking to a person. A *Speech AI* system includes two main components:

- An automatic speech recognition (ASR) system, also known as speech-to-text, speech recognition, or voice recognition system. It converts the raw speech audio signal into text for processing by subsequent components.
- A text-to-speech (TTS) system, also known as speech synthesis. It turns the text into audio.

Speech recognition is the process that enables machines to recognize spoken words and convert them to text. Speech is one of our most expedient and natural forms of communication. Speech recognition is an interdisciplinary subfield of computer science

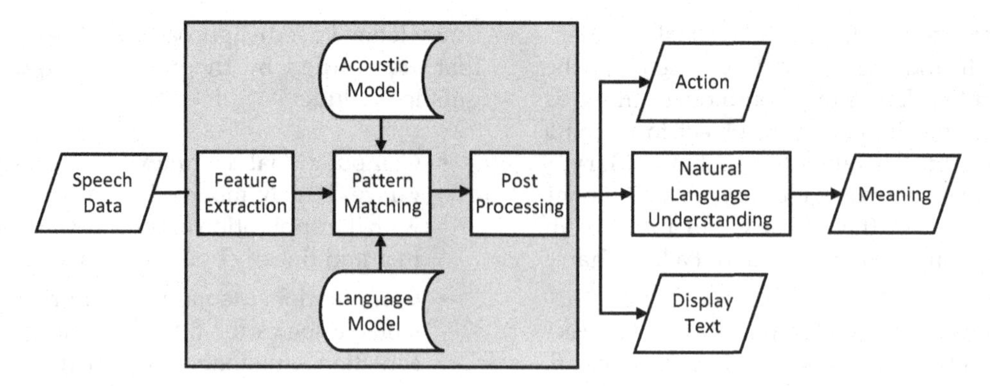

FIGURE 4.35
Speech process model.

and computational linguistics that develops methodologies and technologies that enable the recognition and translation of spoken language. It is also known as automatic speech recognition (ASR), computer speech recognition or speech to text (STT), and text to speech (TTS). TTS is also known as speech synthesis and turns the text into audio. It incorporates knowledge and research in the computer science, linguistics, and computer engineering fields (Figure 4.35).

In 1921, Homer Dudley joined Western Electric's engineering department (later to be incorporated as Bell Telephone Laboratories). He arrived at Bell at a propitious time: Harry Nyquist was about to propound his fundamental theorems about signaling speed on telegraph lines (the famous "Nyquist rate") and R.V.L. Hartley subsequently established the "transmission limits of telephone lines", an early ancestor of Shannon's information theory. With all these advances, and the "giants" behind them, Mr. Dudley invented the Vocoder (Voice Operated reCorDER). The Vocoder was a composite device consisting of an analyzer and an artificial voice. The analyzer detected energy levels of successive sound samples measured over the entire audio frequency spectrum via a series of narrow-band filters. The results of which could be viewed graphically as functions of frequency against time. The synthesizer reversed the process by scanning the data from the analyzer and supplying the results to a feedback network of analytical filters energized by a noise generator to produce audible sounds.

In speech recognition, the computer takes input in the form of sound vibrations. This is done by making use of an ADC that converts the sound waves into a digital format that the computer can understand. The first speech recognition systems were focused on numbers, not words. In 1952, Bell Laboratories designed the "Audrey" system which could recognize a single voice speaking digits aloud. Ten years later, IBM introduced "Shoebox" which understood and responded to 16 words in English. Across the globe other nations developed hardware that could recognize sound and speech. And by the end of the 1960s, the technology could support words with four vowels and nine consonants. In the 1970s, Carnegie Mellon's "Harpy" speech system came from this program and was capable of understanding over 1,000 words which is about the same as a 3-year-old's vocabulary. Also significant in the "70s was Bell Laboratories" introduction of a system that could interpret multiple voices. In 1971, DARPA funded a consortium of leading laboratories in the field of speech recognition. The project had the ambitious goal of creating a fully functional speech recognition system with a large vocabulary. The 1980s saw speech recognition vocabulary go from a few hundred words to several thousand words. One of the breakthroughs came from a statistical method known as the "Hidden Markov Model" or HMM.

The HMM is based on augmenting the Markov chain. A Markov chain is a model that tells us something about the probabilities of sequences of random variables, states, each of which can take on values from some set. These sets can be words, or tags, or symbols representing anything, like the weather. A Markov chain makes a very strong assumption that if we want to predict the future in the sequence, all that matters is the current state. The states before the current state have no impact on the future except via the current state. It's as if to predict tomorrow's weather you could examine today's weather but you weren't allowed to look at yesterday's weather. And instead of just using words and looking for sound patterns, the HMM estimated the probability of the unknown sounds actually being words.

An understanding of linguistics is not a prerequisite to the study of natural language understanding, but a familiarity with the basics of grammar is certainly important. We must understand how words and sentences are combined to produce meaningful word strings before we can expect to design successful language understanding systems. In a natural language, the sentence is the basic language element. A sentence is made up of words that express a complete thought. To express a complete thought, a sentence must have a subject and a predicate. The subject is what the sentence is about, and the predicate says something about the subject. Essentially, there have been three different approaches taken in the development of natural language understanding programs:

1. the use of keyword and pattern matching,
2. combined syntactic (structural) and semantic-directed analysis, and
3. comparing and matching the input to real-world situations (scenario representations).

Understanding and generating human language is a difficult problem. It requires a knowledge of grammar and language, syntax and semantics of what people know and believe, their goals, the contextual setting, pragmatics, and world knowledge. Recognition is the process of establishing a close match between some new stimulus and previously stored stimulus patterns. Object classification is closely related to recognition. The decision-theoretic approach is based on the use of decision functions to classify objects. Clustering is the process of grouping or classifying objects on the basis of a dose association or shared characteristics (Figure 4.36).

Speech recognition was propelled forward in the 90s in large part because of the personal computer. Faster processors made it possible for software like Dragon Dictate to become more widely used. BellSouth introduced the voice portal (VAL) which was a dial-in interactive voice recognition system. This system gave birth to the myriad of phone tree systems that are still in existence today. By the year 2001, speech recognition technology had achieved close to 80% accuracy. For most of the decade, there weren't a lot of advancements until Google arrived with the launch of Google Voice Search. Because it was an app, this put speech recognition into the hands of millions of people. It was also significant because the processing power could be offloaded to its data centers. Not only that, but Google was also collecting data from billions of searches which could help it predict what a person is actually saying. At the time Google's English Voice Search System included 230 billion words from user searches. In 2011 Apple launched Siri which was similar to Google's Voice Search. The early part of this decade saw an explosion of other voice recognition apps. And with Amazon's Alexa, Google Home we've seen consumers becoming more and more comfortable talking to machines. In 2016, IBM achieved a word error rate of 6.9%. In 2017, Microsoft usurped IBM with a 5.9% claim. Shortly after that, IBM improved their rate to 5.5%. However, it is Google that is claiming the lowest rate at 4.9%.

Matching is the process of comparing two or more structures to discover their likenesses or differences. The structures may represent a wide range of objects including physical entitles, words or phrases in some language: complete classes of things, general concepts, relations between complex entities, and the like. In its simplest form, matching is just the process of comparing two structures or patterns for equality. The match fails if the patterns differ in any aspect. Integrating new knowledge in traditional databases is accomplished by simply adding an item to its key location, deleting an item from a key-directed location, or modifying fields of an existing item with specific input information. When an item in inventory is replaced with a new one, its description is changed accordingly. When an item is added to memory, its index is computed and it is stored at the corresponding address.

ASR is the process by which a computer maps an acoustic signal containing speech to text. Automatic speech understanding is the process by which a computer maps an acoustic speech signal to some form of abstract meaning of the speech. Automatic speaker recognition is the process by which a computer recognizes the identity of the speaker based on speech samples. Automatic speaker verification is the process by which a computer checks the claimed identity of the speaker based on speech samples. All occurrences of a speech sound differ from each other even when part of the same word type and when pronounced by the same person (e.g., "*b*" in "boom" is never pronounced twice in the same way). Each speaker has his own voice characteristics. In most cases, a voice-based conversational speech pipeline consists of three stages:

- Automatic Speech Recognition
- Natural Language Processing and Dialog Manager
- Text-To-Speech

First, the raw audio is an input to the ASR system. With ASR, the audio is processed and transcribed to text. Second, the application needs to understand what the text means and act accordingly. This is the job of an NLP and dialog system which manages the conversation with the user while interacting with external fulfillment systems, also known as skills, to correctly respond to the user. Next, the output from

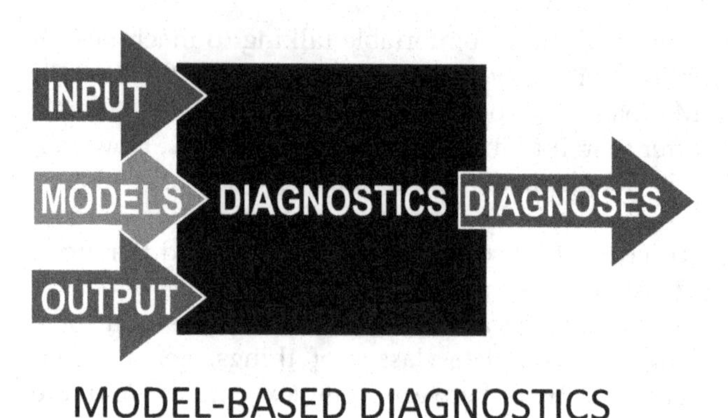

FIGURE 4.36
Model-based diagnosis process diagram.

the ASR phase is interpreted by the NLP system and acted upon by the Dialog Manager with the assistance of a relevant skill. In the final step, TTS converts the text response into speech. It is in this step that AI is used to produce human-like speech from text. Modern state-of-the-art speech systems make extensive use of DNN models trained on massive datasets. Over time, the size (i.e., number of parameters) of speech models has grown. Training such models can take weeks of intensive computing time and is usually performed using deep learning frameworks. GPUs are used both to train deep learning models and to perform inference because they can deliver 10 X higher performance than CPU-only platforms (Figure 4.36).

HMMs were widely used in early ASR experiments and were far from smooth and pleasant due to their relatively lower accuracy. Deep learning involves using DNNs consisting of millions of "neurons" and hundreds of millions of "links" between these neurons, each associated with a tunable parameter. The organization of the neurons into multiple sequential layers is the origin of the term "deep". Neural networks are known to be universal approximators, meaning that they are capable of approximating any arbitrarily complex function given sufficient parameters. Deep learning implementation in ASR significantly lowered the word error rate down to a level matching or better than human error rate (yes, humans also make mistakes when doing speech-to-text at an estimated 4%–5% error rate), making ASR a much more effective and pleasant experience. In later statistical approaches, such as HMMs, the frequency spectrum (vocal tract), fundamental frequency (voice source), and duration (prosody) of speech were modeled simultaneously by HMMs. Speech waveforms were generated from HMMs based on the maximum likelihood criterion and have better fluency than the concatenation method. Recent innovation in TTS revolves around the use of DNNs. DNNs are trained using a large amount of recorded speech to approach the naturalness of the human voice. Advances in transfer learning in recent times allow an existing speech synthesis model to be adapted to a new voice actor/actress within as little as 30 minutes of recorded speech data, making it easy to create a unique, high-quality voice personality.

A HMM includes a Markov process that is "hidden", in the sense that we cannot directly observe the state of the process. But we do have access to a series of observations that are probabilistically related to the underlying Markov model. While the formulation of HMMs might initially seem somewhat contrived, there exist a virtually unlimited number of problems where the technique can be applied. Best of all, there are efficient algorithms, making HMMs extremely practical. Another very nice property of an HMM is that structure within the data can often be deduced from the model itself. We make a brief detour to show the close relationship between dynamic programming (DP) and HMMs. The executive summary is that a DP can be viewed as an α-pass where "sum" is replaced by "max". More precisely, for π, A, and B as above, the DP algorithm, which is also known as the Viterbi algorithm.

DP is an algorithmic technique for solving an optimization problem by breaking it down into simpler subproblems and utilizing the fact that the optimal solution to the overall problem depends upon the optimal solution to its subproblems. DP is useful when:

- the goal nodes are explicit (the previous methods only assumed a function that recognizes goal nodes).
- a lowest-cost path is needed.
- the graph is finite and small enough to be able to store the *cost_to_goal* value for each node.
- the goal does not change very often; and
- the policy is used a number of times for each goal, so that the cost of generating the *cost_to_goal* values can be amortized over many instances of the problem.

But not possible yet to characterize the different sounds by (hand-crafted) rules. Instead:

- A large set of recordings of each sound is made
- Using statistical methods, a model for each sound is derived (acoustic model)

- Incoming sound is compared, using statistics, with acoustic model of a sound.

Feature extraction turns speech signals into something more manageable. The sampling of a signal transforms the voice into a digital form. The signal is chopped into small pieces (frames), and spectral analysis of a speech frame produces a vector representing the signal properties. We split utterances into basic units, for example, phonemes. The acoustic model describes the typical spectral shape (or typical vectors) for each unit and for each incoming speech segment, the acoustic model will tell us how well (or how badly) it matches each phoneme. The acoustic model returns a score for each incoming feature vector indicating how well the feature corresponds to the model, resulting with a local score. The computer calculates a score of a word, indicating how well the word matches the string of incoming features. A search algorithm looks for the best scoring word or word sequence. The language model describes how words are connected to form a sentence and limit possible word sequences. This reduces the number of recognition errors by eliminating unlikely sequences. As the speed of the recognizer increases, it has implications to real-time implementations. Key element in ASR is based on learning from observations such as huge amounts of spoken data are needed for making acoustic models and huge amounts of text data are needed for making language models. There are lots of statistics, but few rules. The flexibility and predictive power of DNNs, in particular, have allowed speech recognition to become more accessible (Figure 4.37).

Speech AI components typically form part of a larger voice-based conversational AI system, which combines various technologies such as ASR, NLP, Text-To-Speech, and Dialog Manager to understand and respond to different interactions. The technology behind speech AI is complex. It involves a multi-step process requiring a massive amount of computing power and several deep learning models that must run in tens of milliseconds to deliver human-like responses. In most cases, a voice-based conversational AI pipeline consists of three stages:

- Automatic Speech Recognition
- Natural Language Processing and Dialog Manager
- Text-To-Speech

Microsoft has established human parity in speech recognition where speech recognition systems has achieved the same error rates or less than a human.

Model-based Diagnosis

Model-based Diagnosis from the Greek *diagnOsis*, from *diagignOskein* to distinguish, from *dia-+gignOskein* to know. It is defined as the art or act of identifying a problem from its signs and symptoms. Model-based Diagnosis (MBD) is an approach to diagnosis that was proposed in the early 80s to overcome limitations of the traditional expert systems approach. The goal is for a healthy robot to "tolerate" faults so that they have no or inconsequential impact to its performance. Diagnosis should be based on an objective model of the device (system) to be diagnosed. The theory behind model-based diagnostics is as follows: Diagnosis should be based on an objective model of the device (system) to be diagnosed. More specifically, different types of models can be considered: structural (concerning the physical or logical structure of a device), functional (describing the functions of a device), behavioral (describing how a device works, i.e., how its functions are achieved), teleological (describing the purposes of the use of a device), or a combination of them. Models should be reusable, in two ways. On the one hand, the same model of a device should be used for different problem-solving tasks (such as diagnosis, simulation, reconfiguration,...). On the other hand, models should be compositional: the model of a device should be usable in all the cases where the device is used as a component of a larger system (Figure 4.38).

The process relies on models which are in most cases component-oriented. A device is described in terms of its minimal replaceable or repairable components. For each type of component, the model includes a list of its variables (interface, internal or state variables, parameters), as well as a definition of its modes of behavior (including correct and fault modes). This is based on using isolation or independence on selective components so that the misbehavior of any one component does not negatively affect the behavior of another. There is a diagnostic problem whenever the observed behavior of a device is not in accordance with the expected behavior, that is, the behavior that can be predicted from the model, assuming that all the components are in a normal mode (fault detection). When failures do occur, failure warnings and indicators are used to provide early detection of failures so that preventive actions can be taken. Faults can be either a hardware defect or a software/programming mistake. Faults and errors can spread throughout the system. If a chip shorts out power to ground, it may cause nearby chips to fail as well. Diagnosis is the process of determining the nature and cause of something. In a health management context, this "something" is a fault. Failure is a condition of the system in which it does not meet functional specifications. Faults can grow and lead to failure, or a fault may be significant enough in magnitude, or a severe enough change in configuration, such that the system will have failed (due to the occurrence of the fault). Diagnostics informs decision-making, for example which components to repair/replace, inform fault mitigation, inform fault recovery, and inform functional reallocation.

The initial detection of a fault is dependent on anomaly detection techniques, which relies on AI. Anomaly detection is **the process of identifying unexpected items or events in data sets, which differ from the norm**. And anomaly detection is often applied on unlabeled data which is known as unsupervised anomaly detection. Anomalous events occur relatively infrequently. However, when they do occur, their consequences can be quite dramatic and quite often in a negative sense. An anomaly is a pattern in the data that does not conform to its expected behavior, also referred to as outliers, exceptions, peculiarities, surprises, etc. However, we are drowning in the deluge of data that are being collected worldwide, while starving for knowledge at the same time. This is the big data problem. Defining a representative normal region is challenging and the boundary between normal and outlying behavior is often not precise.

There are three types of anomalies: point, contextual, and structural. For data labels of such, techniques such as supervised anomaly detection, semi-supervised anomaly detection, and unsupervised anomaly detection are used based on traditional classification approaches (rule-based, neural-based, Bayesian-based, or support vector machines). The main idea is to build a classification model for normal (and anomalous or rare) events based on labeled training data and can use it to classify each new unseen events. Classification models must be able to handle skewed (imbalanced) class distributions. Outputs of the anomaly detection algorithms are based on whether the data is labeled or not; machine-learning algorithms can be used for anomaly detection. If the historical data is labeled (anomaly/not), supervised techniques can be used or if the historical data isn't labeled, unsupervised algorithms can be used to figure out if the data is normal/anomalous. The result of anomaly detection leads to anomaly resolution (Figure 4.39).

As others have said, they both train feature weights w_j for the linear decision function $\sum j w_j x_j$ (decide true if above 0, false if below). The difference is how you fit the weights from training data. In NB, you set each feature's weight independently, based on how much it correlates with the label. (Weights come out to be the features' log-likelihood ratios for the different classes.). In logistic regression, by contrast, you set all the weights together such that the linear decision function tends to be high for positive classes and low for negative classes. (Linear SVMs work the same, except for a technical tweak of what "tends to be high/low" means.) The difference between Naïve Bayes (NB) and Logistic Regression (LogReg or LR) happens when features are correlated. Say you have two features that are useful predictors – they correlate with the labels – but they themselves are repetitive, having extra correlation with each other as well. NB will give both of them strong weights, so their influence is double-counted. But logistic regression will compensate by weighting them lower. This is a way to view the probabilistic assumptions of the models; namely, Naïve Bayes makes a conditional independence assumption, which is violated when you have correlated/repetitive features. One nice thing about NB is that training has no optimization step. You just calculate a count table for each feature and you're done with it – it's single pass and trivially parallelizable every which way. One nice thing about LR is that you can be sloppy with feature engineering. You can throw in multiple variations of a feature without hurting the overall model (provided you're regularizing appropriately); but in NB this can be problematic.

In *model-based* diagnostics, the reference comes from a model that explicitly describes the nominal behavior where the models can be static or dynamic and can be used for simulation. Machine learning is used for detection (classifying between nominal and non-nominal behavior) and used for isolation

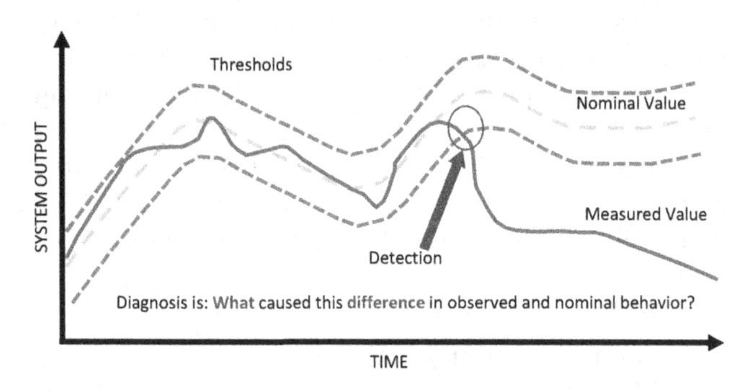

FIGURE 4.37
Anomaly detection time graph.

(classifying between different classes of faulty behavior). A fault is *detected* once we reach a diagnoser state where the label contains only faults. The fault is *isolated* once we reach a diagnoser state where the label contains only a specific fault. Diagnostic analysis can feed an extended prognostic capability. Prognostics is defined as an "Estimation of remaining life of a component or subsystem". It evaluates the current health of a component and, conditional on future load and environmental exposure, estimates at what time the component (or subsystem) will no longer operate within its stated specifications.

Problem-Solving

A problem is really a collection of information that the agent will use to decide what to do. Early researchers in AI believed that the best approach to solutions was through the development of general-purpose problem solvers, that is, systems powerful enough to prove a theorem in geometry to perform a complex robotics task, or to develop a plan to complete a sequence of intricate operations. In general, solving problems are through using "search". Problem-solving is basically a form of means-end analysis that aims at discovering a process description of the path that leads to a desired goal. To demonstrate their theories, several systems were developed including several logic theorem powers and a GPS system. Robots and AI need to perform sequences of actions in order to achieve goals. Intelligent behavior can be generated by having a lookup table or reactive policy that tells the robot what to do in every circumstance, but such a table or policy is difficult to build and all contingencies must be anticipated. A more general approach is for the AI to have knowledge of the world and how its actions affect it and be able to simulate execution of actions in an internal model of the world in order to determine a sequence of actions that will accomplish its goals. This is the general task of problem-solving and is typically performed by searching through an internally modeled space of world states. Problem-solving is basically a state-space search, and primarily an **algorithm only** and **not a representation**. It is going from an initial state, a goal-test, and a successor function. This is contrasted to planning, a combination of an algorithm (search) **and** a set of representations, usually using "situational calculus" – a logic formalism. It provides a way to "open up" the representation of the initial state, goal-test, and successor functions. The planner is free to add actions to plans wherever it is needed and most parts of world are independent from other parts.

Building a system to solve a problem requires the following steps:

- Define the problem precisely including detailed specifications and what constitutes an acceptable solution.
- Analyze the problem thoroughly for some features may have a dominant effect on the chosen method of solution.
- Isolate and represent the background knowledge needed in the solution of the problem.
- Choose the best problem-solving techniques in the solution.

A state space represents a problem in terms of states and operators that change states. For example, given an initial state of the world, a set of possible actions or operators that can be performed, and a goal test that can be applied to a single state of the world to determine if it is a goal state. The problem is to find a solution stated as a path of states and operators that shows how to transform the initial state into one that satisfies the goal test. The initial state and set of operators implicitly define a state space of states of the world and operator transitions between them, the solution may be infinite. When measuring the performance of the problem-solving activity, there is a perceived cost. There is a path cost, a function that assigns a cost to a path, typically by summing the cost of the individual operators in the path. We may prioritize finding a minimum-cost solution. There is also a search cost, the computational time, and space (memory) required to find the solution. Generally, there is a trade-off between path cost and search cost and one must satisfice and find the best solution in the time that is available.

Problem-Solving Example: Let us now place problem-solving by machine in the more precise mental context of evaluating two particular kinds of finite functions, namely:

s: Situation → Actions, and

t: Situations × Actions → Situations.

These expressions say that s maps from a set of situations (state-descriptions) to a set of actions, and that t maps from a set of situation-action pairs to a set of situations. The function symbol s can be thought of as standing for "strategy" and t as standing for "transform". To evaluate s is to answer the question: "What to do in this situation?". To evaluate t corresponds to: "If in this situation such-and-such were done, what situation would be the immediate result?". For example, if the problem-domain were bicycling, we could probably construct a serviceable lookup table of s from a frame-by-frame examination of filmed records of bicyclists in action. But t would certainly be too large for such an approach. The only way to predict the next frame of a filmed sequence would be by numerically computing t using a Newtonian physics model of the bicycle, its rider, and the terrain.

Machine representations corresponding to s and t are often called **heuristic** and **causal**, respectively. Note that they model different things. The first models a problem-solving skill but says nothing about the problem-domain. The second models the domain including its causality, but in itself says nothing about how to solve problems in it. The causal model partakes of the essence of the traditional sciences, such as physics. The physics text from school has much to say about the tension in a string suspending bananas from the ceiling, then about the string's breaking point under stress; the force added if a monkey of stated weight were to hang from a boat hook of given mass and dimensions having inserted its tip into the bunch, and so forth. How the monkey can get the bananas is left as an exercise for the reader, or the monkey. When it has been possible to couple causal models with various kinds and combinations of search, mathematical programming, and analytic methods, then evaluation of t has been taken as the basis for "high road" procedures for evaluating s. In "low road" representations, s may be represented directly in machine memory as a set of (pattern-advice) rules overseen by some more or less simple control structure. A recent pattern-directed heuristic model used for industrial monitoring and control provides for default fallback into a (computationally costly) causal-analytic model. The system thus "understands" the domain in which its skill is exercised. The pattern-based skill itself is, however, sufficiently highly tuned to short-circuit, except in rare situations, the need to refer back to that understanding.

Many AI (and non-AI) tasks can be formulated as search problems. The goal is to find a sequence of actions to solve the problem. Optimization problems are used to find a solution that achieves one or more pre-defined goals or successor functions, for example, maximization/minimization problems. When formulating an optimization problem, do the following:

- Design variables represent a solution.
- Design variables define the search space of candidate solutions.
- [Optional] Solutions must satisfy certain constraints.
- Objective function defines our goal.

This statement, that machines can have minds just as people do, would be later named "strong AI" by philosopher John Searle. It remains a serious subject of debate up to the present day. Machine language program was feasible but oh so tedious. Assembly language programming was a great advance in its day. Mnemonic commands could be *assembled* and *compiled* into the machine language coding required by the computers. But assembly language programming was still almost unbearably tedious so when John Bachus and his group created Fortran (short for Formula Translation) and John McCarthy created LISP (short for List Processing), it was a whole new day in computer technology, a new generation. Fortran was the first high-level computer programming language for wide use and Fortran stands out as a high-performance computing language utilized for the world's fastest supercomputers. Fortran became the language for routine number crunching and still is to some extent despite the development of other more sophisticated computer languages. McCarthy's LISP had an entirely different career. It was founded upon a bit of esoteric mathematics called the lambda calculus. LISP had a value at the fringes of computer technology, in particular for what became known as AI. AI researchers in the United States made LISP their standard as a high-level programming language (Figure 4.40).

Optimization can be used to evaluate the quality of solutions and the function to be optimized (maximized or minimized). Some of the interdependencies of Narrow AI and their dependencies are shown in the previous figure. The programming language LISP was designed by J. McCarthy between 1956 and 1958 at the Massachusetts Institute of Technology (MIT) as a language for the construction of "intelligent" computer systems. The name of the language, which is an acronym for "LISt Processing", readily indicates the most important and strongest feature of the language, namely, list manipulation. This language is based around the handling of lists of data. A list in LISP is contained within brackets, such as: [A B C]. LISP uses lists to represent data, but also to represent programs. Hence, a program in LISP can be treated as data. This introduces the possibility of writing self-modifying programs in LISP and allows us to use evolutionary techniques to "evolve" better LISP programs. LISP has been the origin of a new programming style known as functional programming. The language furthermore has relations with lambda calculus and term rewriting systems, important fields of research in theoretical computer science. It is possible in LISP to dynamically modify a program during its execution by interpreting data as program parts, a feature that is particularly important for the development of interpreters for formal languages. Right from its inception LISP was applied in the burgeoning field of AI, McCarthy used the language for the development of a GPS named "Advice Taker". At present, LISP is still mainly used for applications in which symbol manipulation predominates, such as in AI in general and for the development of expert systems in particular.

There are many different ways to define and optimize problem-solving. We can use greedy algorithms, constraint programming, mixed integer programming, genetic algorithms, or local search. Constraint programming is a paradigm for solving combinatorial problems where users declaratively state the constraints on the feasible solutions for a set of decision variables. The system can make a single decision or several sequential decisions. For each decision made in a sequence, the system does a restart, it predicts the best heuristics to solve the problem, or otherwise the description is refined and another iteration is made. Monte Carlo methods are a broad class of computational algorithms that rely on repeated random sampling to obtain numerical results. The underlying concept is to use randomness to solve problems that might be deterministic in principle. They are often used in problems that are difficult or impossible to use by other approaches. Monte Carlo methods can be used to solve any problem having a probabilistic interpretation.

Figure 4.38 follows the interdependencies between many of the artificial intelligent technologies described in this chapter. Pamela McCorduck also sees in the LT the debut of a new theory of the mind, the information-processing model (sometimes called computationalism). Newell and Simon would later formalize this proposal as the physical symbol systems hypothesis. AI research has both theoretical and experimental sides. The experimental side has both basic and applied aspects. There are two main lines of research:

- One is biological, based on the idea that since humans are intelligent, AI should study humans and imitate their psychology or physiology.
- The other is phenomenal, based on studying and formalizing common sense facts about the world and the problems that the world presents to the achievement of goals.

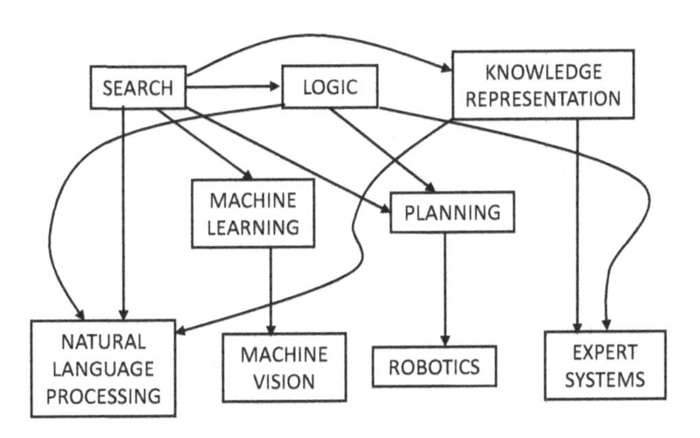

FIGURE 4.38
Interdependencies of AI technologies.

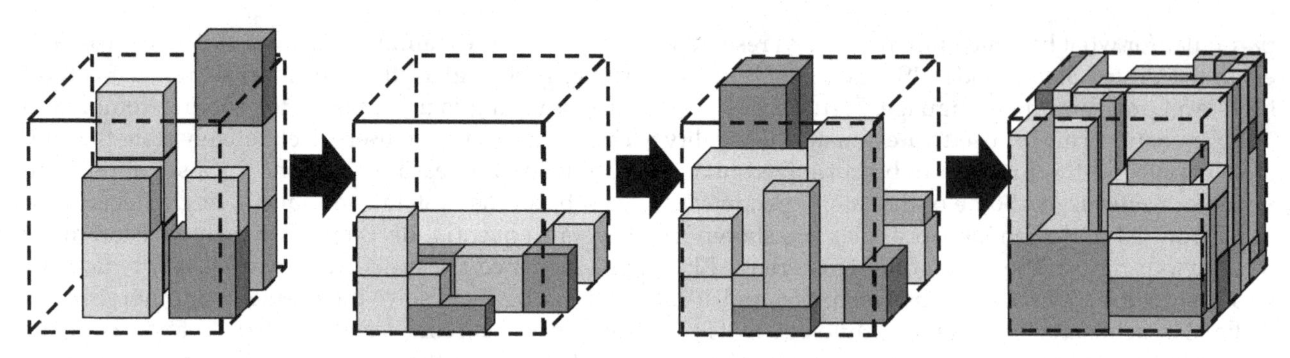

FIGURE 4.39
Bin packing problem.

Bin Packing Problem: find an assignment of items to bins that minimizes the number of bins used is an optimization problem, in which items of different sizes must be packed into a finite number of bins or containers. Each bin is of a fixed given capacity in a way that minimizes the number of bins used. Given are bins with maximum volume *V*, which cannot be exceeded. We have *n* items to pack, each with a volume and must pack all items (Figure 4.39).

The two approaches interact to some extent, and both should eventually succeed. According to John McCarthy, this is a race, but both racers seem to be walking (going slow). Since about 1990, data mining has developed as a subdiscipline of AI in the area of statistical data analysis for extraction of knowledge from large databases. Data mining brings no new techniques to AI, rather it introduces the requirement of using large databases to gain explicit knowledge. Another approach to modularization of learning is distributed learning or multiagent learning. However, there can be a "Curse of Dimensionality". When learning a humanoid robot's motor skills, up to 50 different motors must be simultaneously controlled, which results in 50-dimensional state space and also a 50-dimensional action space. To reduce this gigantic complexity, central control is replaced by distributed control. For example, each individual motor could get an individual control that steers it directly, if possible independently of the other motors. In nature, we find this kind of control in insects. For example, the many legs of a millipede are not steered by a central brain, rather each pair of legs has its own tiny "brain". Here is a dilemma. Some AI applications are so deeply complex, deployed in tasks that lie so far beyond the capacities of human intelligence in terms of speed and complexity, that only formal proofs can assure us that the answers or advice we receive are actually correct (Figure 4.41).

The programming language PROLOG can be considered to be a first step toward the practical realization of logic programming; the separation between logic and control has not been completely realized in this language. A PROLOG system consists of two components: a PROLOG database and a PROLOG interpreter. A PROLOG program, essentially a logic program consisting of Horn clauses (which however may contain some directives for controlling the inference method), is entered into the PROLOG database by the programmer. The PROLOG interpreter offers a deduction method, which is based on a technique called SLD resolution. Solving a problem in PROLOG starts with discerning the objects that are relevant to the particular problem, and the relationships that exist between them.

One of the important lessons learned by AI researchers during the 1970s and early 1980s is that knowledge is not easily acquired and maintained. It is a difficult and time-consuming process. Yet expert and other knowledge-based systems require an abundant amount of well-correlated knowledge to achieve a satisfactory level of intelligent performance.

Note

1 I met with Prof. Malik from UC-Berkeley when working on the NAHSC program with Prof. Shladover.

5

When Mind and Body (Artificial Intelligence and Robotics) Are Integrated

The desire for intelligent machines was just an elusive dream until the first computer was developed. The early computers could manipulate large databases effectively by following prescribed algorithms but could not reason about the information provided. This gave rise to the question of whether computers could ever think. Alan Turing defined the intelligent behavior of a computer as the ability to achieve human-level performance in a cognitive task. The main advances in artificial intelligence (AI) over the past 60 years have been advances in search algorithms, machine-learning algorithms, and integrating statistical analysis into understanding the world at large. In the field of AI, expectations seem to always outpace the reality. After decades of research, no computer has come close to passing the Turing Test (a model for measuring "intelligence"); expert systems have grown but have not become as common as human experts; and while we've built software that can beat humans at some games, open-ended games are still far from the mastery of computers. Intelligent systems interpret raw data according to probabilistic models and use contextual information that gives meaning to the data. We have described autonomous robots with the warehouse example as the integration of several weak AI technologies with the hardware. In Asimov and Capek's vision, the ultimate goal for robotics is a humanoid with Strong AI (Artificial General Intelligence). There are several reasons for building a humanoid: Minsky postulated that people enjoyed looking at themselves, for example, seeing oneself on a mirror but for better reasons, the world is designed to accommodate humans in terms of using tools and fitting into a world designed around people. Others have argued that a robot would be socially acceptable if the robot looks more like a person (Figure 5.1).

So, what is AI? We have making computers think to the automation activities we associate with human thinking, like decision-making, learning, etc. Or the art of creating machines that perform functions that require intelligence when performed by people or maybe the study of mental faculties using computational models. They are all true to some extent. Other definitions for AI include the study of computations that make it possible to perceive, reason, and act (SPA) and a field of study that seeks to explain and emulate intelligent behavior in terms of computational processes. More importantly, AI is a branch of computer science that is concerned with the automation of intelligent behavior and anything in Computing Science that we don't yet know how to do properly!

One of the biggest issues in robotics is the eventual integration of hardware and software (Figure 5.1). If we approach a deliberate robot, using Craik's Sense–Plan–Act paradigm, the robot requires exteroceptive (external) sensors (cameras, lidars, radars, ultrasonics, lasers, Vorad, sonar, range sensors) to sense the robot's environment, localization sensors (GPS, Vicon and Optitrack for indoors, etc.), and proprioceptive (internal) sensors (encoders, heading compass, heading gyroscope, accelerometers, Inertial Navigation Units). Sensors vary according to physical principle, resolution, bandwidth, price, and energy required. Sensors are the key components for perceiving the environment. Sensor readings are sent to the CPU (central processing unit) for processing. Outputs from the CPU are commands to the actuators (motors in an arm, drives in a mobile robot, steering commands, and gimbal commands for moving sensors around). Automatic systems follow rules and are ideal for structured environments. If the environment is complex, then something more than an automatic system is required, thus leading us to autonomy. Teleoperation

FIGURE 5.1
NASA Valkyrie humanoid.

DOI: 10.1201/9781032673134-5

and telepresence are both potential solutions for communicating with the robot, but are limited by the radio technology (long distance introduces latencies, potential for jamming, and environmental conditions could reduce the effectiveness of the communication system underwater and in radiation environments). An early limitation of AI was that it was hard to scale solutions from toy problems to more realistic ones due to difficulty of formalizing knowledge and combinatorial explosion of the search space of potential solutions.

Robotics and AI coincidentally follow the SPA paradigm. It is the intelligent connection of perception to intelligent software. Perception differs from sensing or classification in that it implies the construction of representations (models) that are the basis for recognition, reasoning, and eventually action. Intelligent software addresses issues such as spatial reasoning, dealing with uncertainty, geometric reasoning, compliance, and learning. Intelligence includes the ability to reason and learn about objects and the processes involved. AI is the challenge for robotics, where robotics severely challenges AI by forcing it to deal with real objects in the real world. The issues for AI in robotics are: (1) what knowledge is needed, (2) how to represent that knowledge, and (3) how does the AI use that knowledge on the robot. As a result, robotics need to deal with the real world, including needing detailed geometric models. As well as geometry, robotics need to represent forces, causation, and uncertainty (Figure 5.2).

The knowledge is needed for reasoning in relatively formalized and circumscribed domains such as symbolic mathematics. A key contribution of AI is the observation that knowledge should be represented explicitly and not heavily encoded; for example, numerically in ways that suppress structure and constraint. A given body of knowledge is used in many ways in thinking. Conventional data structures are tuned to a single set of processes for access and modification, and this renders them too inflexible for use in thinking. AI has developed a set of techniques such as semantic networks, frames, and production rules, that are symbolic, highly flexible encodings of knowledge, yet which can be efficiently processed. Differential equations, for example, are a representation of knowledge that, while extremely useful, are still highly limited. Forbus points out that conventional mathematical representations do not encourage qualitative reasoning, instead, they invite numerical simulation. Though useful, this falls far short of the qualitative reasoning that people are good at. AI has also uncovered techniques for using knowledge effectively. One problem is that the knowledge needed in any particular case cannot be predicted in advance. Thus, intelligent AI programs have to respond flexibly to a nondeterministic world. Among the techniques offered by Modern AI are search, structure matching, constraint propagation, and dependency-directed reasoning.

Figure 5.2
Warehouse assistant robot.

In an unknown environment (e.g., in a maze), a computer agent (e.g., a robot) takes an action (e.g., to walk) based on its own control policy. Then its state is updated (e.g., by moving forward) and evaluation of that action is given as a "reward" (e.g., praise, neutral, or scolding). Through such interaction with the environment, the agent is trained to achieve a certain task (e.g., getting out of the maze) without explicit guidance. A crucial advantage of reinforcement learning is its non-greedy nature. That is, the agent is trained not to improve performance in a short term (e.g., greedily approaching an exit of the maze) but to optimize the long-term achievement (e.g., successfully getting out of the maze). Robotics needs to represent uncertainty so that reasoning can successfully overcome it. To be intelligent, robotic programs need to be able to plan actions and reason about those plans. Surely, AI has developed the required planning technologies? Unfortunately, it seems that most, if not all, current research for planning and reasoning developed in AI require significant extensions before they can begin to tackle the problems that typically arise in robotics. Another reason is that to be useful for robotics, a representation must be able to deal with the vagaries of the real world, its geometry, inexactness, and noise. All too often, AI planning and reasoning systems have only been exercised on a handful of toy examples.

The engineering approach, which applies control theory to complex systems such as a robot, is bounded by:

$$y_{t+k} = f\left(\in(t)\right) = f\left(x_t - \tau_t\right) \tag{5.1}$$

y_{t+k}=action
f=(unknown) control function
x_t=status
τ_t=target status

Robot manipulators need to address singularities, end effectors, and collaborative manipulators. Many tasks, particularly those related to assembly, require a variety of capabilities, such as parts handling, insertion, screwing, as well as fixtures that vary from task to task. One approach is to use one arm but multiple single-DOF grippers (end effectors), or multiple arms each with a single DOF, or some combination. One problem with using multiple single-DOF grippers is that a large percentage of the work cycle is spent changing grippers. Multiple arms raise the problem of coordinating their motion while avoiding collision and without cluttering the workspace. The coordination of two arms was illustrated when two arms were combined to install a simple hinge. One of the arms performed the installation, and the other acted as a programmable fixture, presenting the hinges and other parts to the work arm. Arms are used for pick and place tasks, as well as assembly tasks. For moving tasks, a stiff manipulator is required but for fitting one part into another, compliance is required if the two parts are not perfectly aligned. Robot manipulators have evolved for positional accuracy and are designed to be mechanically stiff. High-tolerance assembly tasks typically involve clearances of the order of a thousandth of an inch. In view of inaccurate modeling of the world and limitations on joint accuracy, low stiffnesses are required to effect assemblies (Figure 5.3).

Professor Lozano-Perez introduced a representation called C-space that consists of the safe configurations of a moving object. Programmers find it relatively easy to specify motions in position space, but find it hard to specify the force-based trajectories needed for compliance. In Latombe's approach, the geometry of the task is defined by a semantic network, the initial and goal configurations of parts are defined by symbolic expressions, and the knowledge of the program is expressed as production rules. This method requires that the relationships between the surfaces of parts in contact be known fairly precisely. In general, this is difficult to achieve because of errors in sensors. Robotics have to deal with uncertainty. An expression-bounding program can be applied with equal facility to the symbolic expression for the error and the desired size of a screw hole (the specifications of the insertion task). In this example, the result is a bound on the only free variable of the problem, in this case the length of the screwdriver. As another example, current tactile sensors leave much to be desired. They are prone to wear and tear, have poor hysteresis, and have a low dynamic range. Industrially available tactile sensors typically have a spatial resolution of only about 8 points per inch. Tactile sensors are as poor as TV cameras were in the 1960s, the analogy being that they are seriously hampering the development of tactile interpretation algorithms. Geometric analysis of shape allows grasp points to be determined by the robot.

For mobile robots, the basic forms of mobility are wheels, tracks, and legs. There are two generic problems with legged locomotion: moving over uneven terrain and achieving dynamic balance without a static pyramid of support. Despite these features, robot

FIGURE 5.3
Manipulator arm.

programming is tedious, mostly because in currently available programming languages the position and orientation of objects, and sub-objects of objects, have to be specified exactly in painful detail. "Procedures" in current robot programming languages can rarely even be parameterized, due to physical assumptions made in the procedure design. Dr. Lozano-Perez calls such programming languages as robot-level. Since robotics is the connection of perception to action, AI must have a central role in robotics if the connection is to be intelligent. Robotics challenges AI by forcing it to deal with real objects in the real world, an important part of the challenge is dealing with rich geometric models (Figure 5.4).

The government has a desire in today's military to introduce autonomous robots to reduce manpower requirements and also take the soldier out of harm's way. This takes us back to the unmanned mantra of the Dull, Dirty, and Dangerous (3 Ds). An autonomous robot dictates that decisions are made based on planned goals and balancing utility theory. It is also assumed that a robot will be capable of learning from past or previous experiences. Based on the previous section, we will correlate various autonomy functions to the different narrow AI sciences. It is important to remember that a map of the robot's environment is a model that needs to be represented in a knowledge base saved in memory. Based on the discussion in the previous section, AI search is the foundation of many of the technologies required in AI techniques and subsequently autonomy. It is a foundation for logic, which is required for many of the Narrow AI capabilities that we need for autonomy:

FIGURE 5.4
Military UGV robot.

Search → Knowledge Representation
Search → **Logic**
Search → Knowledge Representation → Expert Systems → Decision-Making
Search → **Planning** → **Robotics** → Action
Search → **Planning** → Decision-Making
Search → **Logic** → **Planning** → **Robotics** → Action
Machine Learning → Computer Vision → Knowledge Representation
Machine Learning → Natural Language Processing
Search → **Logic** → Expert Systems

(Note: "→" literally means "moving" forward)

For a common sense approach, it is useful to highlight AI search and AI logic that sets the foundation for what we need in an autonomous robot with knowledge representation for models, search for decision-making, logic and inference for reasoning leading to decisions, expert system for rules, planning system for robotics to reach goal, machine learning for improved computer vision, and decision-making with the weighing of mission goals with utility functions. It was generally echoed that all "AI is Search" during the symbolic period of AI, including game theory and problem spaces. Every problem is a *feature space* of all possible (successful or unsuccessful) solutions. The key is to finding an efficient search strategy.

Brute-force search is the same as exhaustive search using a generate and test paradigm. The software systematically enumerates all possible candidates for the solution and checks which one is the best. But there is a factorial time complexity:

- 2! = 2
- 3! = 6
- …
- 10! = 3,628,800
- 20! = 2,432,902,008,176,640,000 ≈ 2.43 × 1018

If you assume that 109 permutations take 1 second, we get:

- 2!/109 = 0.000000002s
- 3!/109 = 0.000000006s
- …
- 10!/109 = 0.0036288s
- 20!/109 ≈ 2,432,902,008s ≈ **77 years**

As a result, **Brute-force works only for very small problems!** So, what do we do?

TABLE 5.1
Table of function and associated AI science

FUNCTION	AI SCIENCE	INPUT	OUTPUT
What the world is like	World Model	Exteroceptive Sensors	World Model
State Estimation	Robot Model	Proprioceptive Sensors	Internal Model
How the world evolves	Reasoning	Refreshed sensor data	Feedback
What Action Should I do	Planning	Goals, Values	Input to Decision Making
Improved Performance	Learning	Training Data	Positive Feedback
Reflexive Behaviors	Rules/Expert Systems	State Input	State Output
Past Experience	Memory (body)/ Knowledge Base (Mind)	Knowledge	Knowledge Representation
Errors in Control	Fuzzy Logic	Command	PI, PD, PID Control
Missing Input Data	Deep Learning/Inference	Incomplete Data	Inference
Errors in Localization	Probabilistic AI	Bayes Rule, Markovian Network	Absolute Position Correction
Path Planning	Cell Decomposition, Voronoi	MAP or SLAM	Trajectory Replanning
Obstacle Avoidance-Mob	Occupancy Grid,	Local Sensors	Reflexive Behavior
Obstacle Avoidance-Arm	Potential Field	Tactile Sensors Exteroceptive sensors	Reflexive Behavior (safety)
Utility		Resource Mgt.	Re-planning

This curse of dimensionality can be viewed either as the limitation on data analysis due to the large amount of data or parameters needed to analyze the data.

Another example is demonstrated by playing the game of chess.

- First Move=20 possible states
- Second Move=400 possible states

By the 7th Move=1,280,000,000 possible states or moves. This is the combinatorial explosion that limits the symbolic approach to AI (Table 5.1).

The key for autonomous and an intelligent robot is the integration of the hardware to the software with sensory-motor capabilities with deliberative, goal-oriented capabilities. As a result, we need to integrate planning to acting. For planning, it takes the integration of heterogeneous representations of space, time, kinematics and dynamics, physics of sensors, uncertainty, logical properties, and various constraints including computational limits. It is important to integrate various forms of planning, as well as to integrate planning and learning to extend sensory-motor controllers.

Autonomous Robots: For an autonomous robot, we need to specify the action space, percept space, and the robot's environment as a string of mappings from the action space to the percept space. From the robot's perspective, it views the real world as an environment that is:

- Inaccessible (because the sensors do not tell all) and the robot needs to maintain a model of the world
- Nondeterministic (wheels slip, parts break) and as a result, needs to deal with uncertainty
- Non-episodic (effects of an action change over time) and the AI needs to handle sequential decision problems
- Dynamic (meaning constantly changing) and the robot needs to know when to plan and when to use a reflexive behavior
- Continuous (cannot enumerate possible actions) (Figure 5.5).

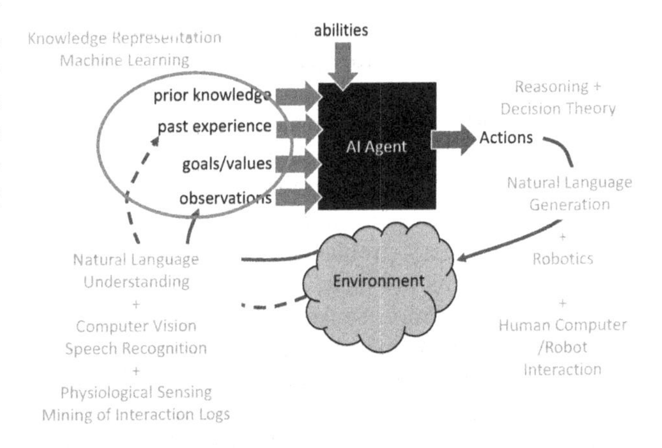

FIGURE 5.5
AI/Robot agent expanded (from previous chapter).

A problem-solving (planning) robot that has several options can first examine different possible sequences of actions to choose its best sequence. In this problem-solving environment, learning is needed if the environment is static, logic is needed to understand what is observable, the world is not deterministic due to uncertainty, and is it discrete, enabling the robot to handle uncertainty and utilize logic. Then how do these robots make decisions? Robot control is based on a set of behaviors where each behavior has a set of preconditions that either: (1) must be satisfied, or (2) are desired. A behavior is selected when all of the "musts" become true (checking the rules) and a behavior is selected from several behaviors based on how many desired conditions are true. The control tradeoff is:

- Thinking is *slow.*
- Reactions must be *fast.*
- Thinking enables *looking ahead* (planning) to avoid bad solutions.
- Thinking too long can be *dangerous* (e.g., falling off a cliff, being run over).
- To think, the robot needs (a lot of) accurate *information* → requiring ***world models.***

Robots need sensors to provide "awareness" of the robot's surroundings, for example, what's ahead, around, and out there. It allows interaction with its environment, for example, it allows the robot lawn mower to "see" the cutting of grass. Sensors are used for protection and self-preservation, including for safety, damage prevention, and potentially a stairwell sensor. It gives the robot the capability to goal-seek by finding colorful objects or to seek goals. Without sensors, there is only fixed automation. Sensors are needed for unstructured environments where the world is changing. Robot adapts to their surroundings with both visual and non-visual sensors (Figure 5.6).

Sensors are the key component for perceiving an environment. Sensors are primarily used to ensure the robot is following an appropriate path (e.g., corridor, road) and to seek out obstacles to avoid. But perception is hard. Intelligent robots interpret raw data according to probabilistic models and use contextual information that gives meaning to the data. In robotics, some of the easy problems are hard and some of the hard problems are easy. For example, creating a system with "common sense" is very hard. A common technique is to maintain a map of the environment and use this to plan safe paths to a desired goal location. As the vehicle traverses the environment, it updates its map and path based on its observations. Such an approach works well when dealing with reasonably small areas but storing and planning over maps of the entire environment is impractical when traversing long distances. When driving in unstructured environments, the system employs a global map and a planner to generate an efficient trajectory to a desired goal. More power has also enabled implementation of real-time perceptual systems, often based on neural models, that can simulate in serial computers the massive parallelism found in the brains of animals. Marr and Hildreth required 10 minutes of computer time to find the edges in a single image in the late 1970s; we now have computer vision systems that track multiple moving objects in a scene at 30 frames per second. Others can visually track the boundaries of roads and cars a few times per second. Using such a system, the "No hands across America" project, at Carnegie Mellon University, made an automated truck drive from the east to the west coast of the United States with no human control for 98% of the journey. After perception finds features, it can interact with objects in the scene (e.g., doors, humans, and cars). At the highest level is to reason about places or situations such as a kitchen. Lidar produces a point cloud, but also referred to as depth map or range map (Figure 5.6).

Proprioceptive sensors are used to record the movement of the robot for dead reckoning. However, dead reckoning with its many non-linearities (wheel slip, noise, calibration, and drift) will require the robot to periodically correct its local position (localization) with global techniques such as with GPS outdoors or well-placed targets internally that can be used to triangulate the robots global position. Lidar is popular due to producing the point cloud without requiring further processing, as opposed to stereo cameras that require stereoscopic calculations. There are multiple techniques for producing a depth map: structured

FIGURE 5.6
Manipulator perception end effector.

light, and time-of-flight, 3D range cameras. Sentences represent models using semantics to capture aspects of the real world. When aspects of the real-world change, it is followed by a change in representation where one sentence (future model) was entailed from another sentence (past model). The model of the robot and the model of the environment is represented by its state, a sufficient description of the system. State is defined as:

If it is "observable", the robot knows its state at all times and if it is Hidden/Inaccessible/Unobservable, the robot does not know its state. If the system is partially observable, the robot knows some part of its state: either discrete (e.g., up, down, blue, red) or continuous (e.g., 3.765 mph). State Space is all the possible states a system can be in. The external state is the state of the world. For example, night/day, raining/sunny, and at home as sensed using the robot's sensors. The internal state is the state of the robot. For example, the state can be happy/sad, stalled/moving, battery level, velocity, etc. as can be sensed (e.g., velocity) and can be stored/remembered (e.g., happy/sad). The robot's state is a combination of its external and internal state. The key question is how intelligent the robot appears will strongly depend on how much and quickly it can sense its environment and itself. The Internal state can be used to remember *information about the world* (e.g., remember paths to the goal, remember maps, and remember friends versus enemies) (Figure 5.7). This is called a *representation* or an *internal model, and* representations/models have a lot to do with how complex a controller is!

Cybernetics had from its beginning been interested in the similarities between autonomous, living systems and machines. As a result, second-order cybernetics emphasized less mechanistic approaches to machines such as **autonomy**, self-organization, **cognition**, and the role of the observer in modeling a system. Many of the core ideas of cybernetics have been assimilated by other scientific disciplines where they continue to influence further developments. Other important cybernetic principles seem to have been forgotten, though, only to be periodically rediscovered or reinvented in different domains. Some examples are the rebirth of neural networks, first invented by cyberneticists in the 1940s, in the late 1960s, and again in the late 1980s; the rediscovery of the importance of autonomous interaction by robotics and AI in the 1990s; and the significance of positive feedback effects in complex systems, rediscovered by economists in the 1990s. If the vehicle is not 100% autonomous, it may wait for new instructions in the case of a failure. If the vehicle is on its own it must first determine if the obstacle is going to cause the goal-level planning to fail. If so, replanning must take place at that level considering the new knowledge of an obstacle. If not, simple rules might be used to get it around the obstacle so that it can resume. If a failure is more severe than an obstacle (e.g., power outage, sensor failure, uninterpretable situation), then the ultimate failsafe is to stop the robot and have it would send out a signal for help. If the robot is a terrain vehicle, it may "pull over" or a submarine may surface and broadcast a message "help me" or what about an autonomous aircraft? The external state is the state of the robot's world. Using the perception function, it models the world dynamics and its state transition function. For the utility function, how does the robot know what constitutes "good" or "bad" behavior?

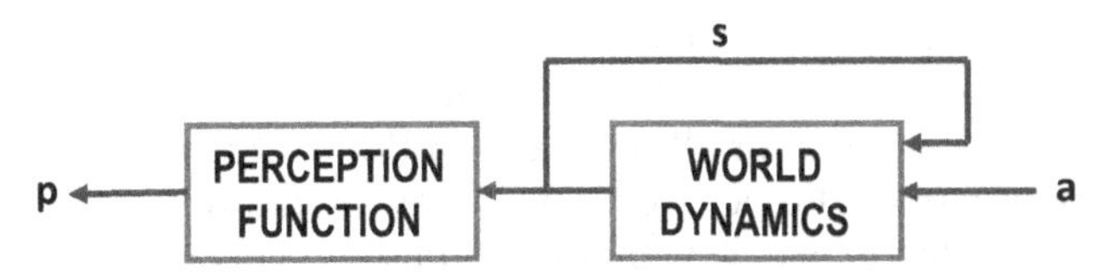

FIGURE 5.7
Perception function and world dynamics model.

The robot's environment can be: observable vs. not fully observable (does the agent see the complete state of the environment?), accessible vs. inaccessible, deterministic vs. nondeterministic (Is there a unique mapping from one state to another state for a given action?), static vs. dynamic (Can the world change while the agent is thinking?), episodic vs. sequential (Does the next "episode" depend on the actions taken in previous episodes?), or discrete vs. continuous (Are the distinct percepts and actions limited or unlimited?). For example, what is needed for driving a car or playing a game of Chinese checkers (Table 5.2).

The robot needs to perform sensor interpretation. There are many forms of interpretation:

- Simple neural network recognition (more common if we have a single source of input, e.g., camera, so that the NN can respond with "safe" or "obstacle")
- Fuzzy logic controller (can incorporate input from several sensors)
- Bayesian network and hidden Markov models (for single or multiple sensors)
- Blackboard/KB approach (post sensor input to a blackboard, let various agents work on the input to draw conclusions about the environment).

We can build maps using the robot's perceptions. However, making these maps is no use unless the robot knows where it is to add the current location

TABLE 5.2
Table of sliding autonomy

	No Autonomy, Remote Control	No Autonomy, No Sense and Respond	No Autonomy but Sense and Respond	Autonomy	Advanced Autonomy
Description	Robot is operated manually and remotely	Robot is not aware of its internal state or external surroundings	Robot can sense force and motion and respond according to pre-determined parameters	Robot can sense its external environment and calculate a response	Robot can identify characteristics of its external environment
Examples	Remote Operated surgical robot	Welding robot in automotive factory	Collaborative assembly robot in manufacturing	Pick-and-place robot feeding a machine with parts from an unsorted bin	Assistant robot that responds differently to a child than an adult
Safety	User ensures safety	Sensors on cages	Setting for force and speed	Hard code overrides probabilistic algorithms	Hard code overrides probabilistic algorithms
	No AI	(Deterministic Algorithms)		(Probabilistic Algorithms)	AI

into the map. This is a chicken-and-egg problem. The robot needs a map to know where it is, but it also needs to know where it is to make a map, and sensor fusion is all about two equations, the **predict** and **update** equations.

Since sensor interpretation needs to be real-time, we need to make sure that the approach is not overly elaborate (Figure 5.8). Sensors are primarily used to ensure the vehicle/robot is following an appropriate path (e.g., corridor, road) and to seek out obstacles to avoid. It used to be very common to equip robots with sonar or radar but not cameras because cameras were costly and vision algorithms required too much computational power and were too slow to react in real time. The technology has improved. Today, outdoor vehicles/robots commonly use cameras and lasers (if they can be afforded). Additionally, a robot might use GPS, so the robot needs to interpret input from multiple sensors.

Autonomous robotic systems perform a variety of visual tasks, including:

- image classification (describing an image as a whole)
- object recognition (identifying specific objects within an image)
- scene understanding (describing what is happening in an image)
- facial recognition (identifying individual faces, or types of features)
- gait recognition (identifying a person by the way they walk)
- pose estimation (determining the position of a human body)
- tracking a moving object (in a video)
- behavior recognition (determining emotional states and behaviors using "affective computing").

Mission Planning, literally, is largely a planning process (developing a sequence of steps). Planning a policy is the same as considering the future consequences of actions to choose the best one. Given goals, how does the robot accomplish them? This may be accomplished through rule-based planning, plan decomposition, or plans that may be provided by human controllers. In many cases, the mission goal is simple: go from point A to point B so that no navigation planning is required. For a mobile robot that is not autonomous, the goals may be more diverse with reconnaissance and monitoring, search (e.g., find enemy locations, find buried land mines, find trapped or injured people), and go from point A to point B but stealthily while monitoring its internal states to ensure that the mission is carried out.

An issue with planning is due to the rationality definition. For example, what if agent cannot compute the best action? There may be computational constraints! When we have too many states, we want a convenient way of dealing with sets of states. The sentence "It's sunny" stands for all the states of the world in which it is sunny. Logic provides a way of manipulating big collections of sets by manipulating short descriptions instead. Instead of thinking about all the ways a world could be, we're going to work in a language of

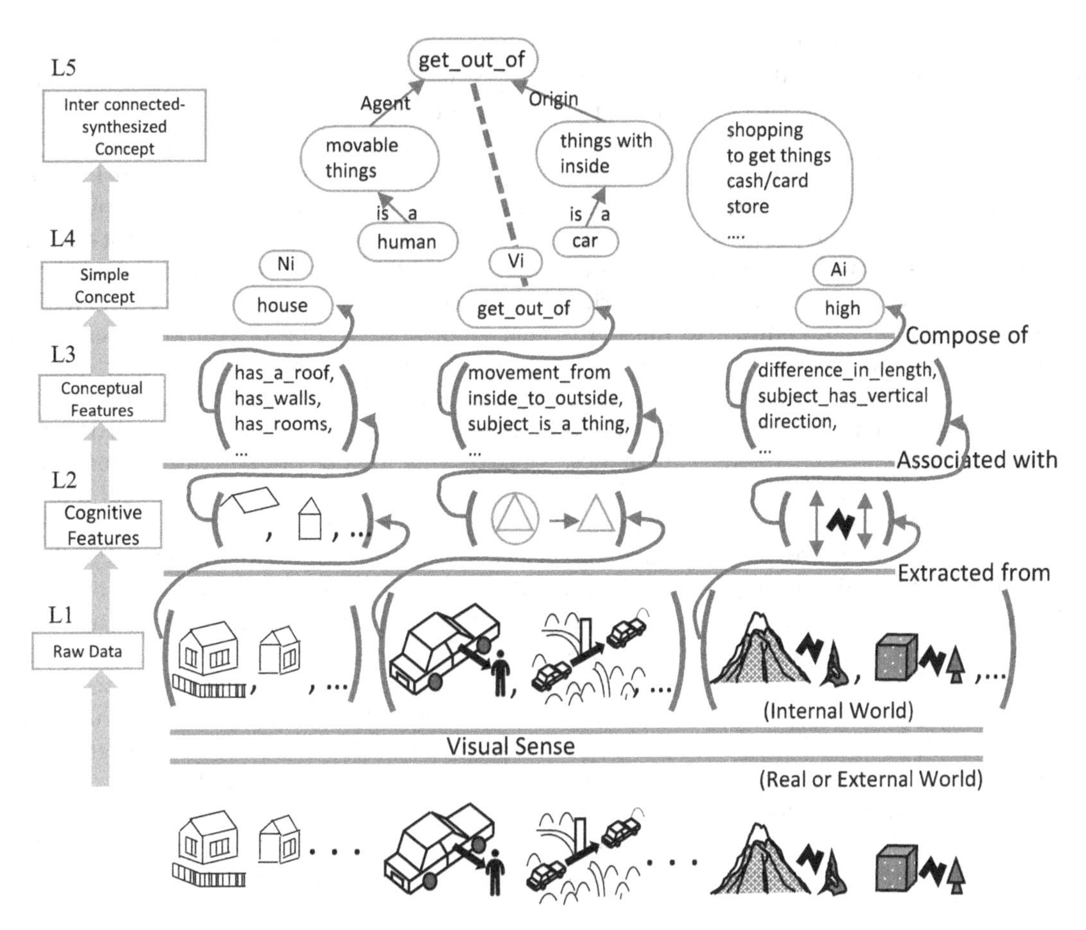

FIGURE 5.8
Visual processing hierarchy.

expressions that describe those sets. Logic is a formal language with syntax (what expressions are legal), semantics (what legal expressions mean), and proof system which is a way of manipulating syntactic expressions to get other syntactic expressions (which will tell us something new). But why proofs? There are two kinds of inferences an AI Robot might want to make: (1) Multiple percepts (conclusions about the world) and (2) Current state and operator (properties of next state).

Search is foundational to symbolic AI because "Much of symbolic AI is search", such as in game theory and problem spaces. Every problem is a feature space of all possible (successful or unsuccessful) solutions, and the trick is to find an efficient search strategy. Search is the fundamental technique of AI. All AI programs represent and use knowledge. The conceptual paradigm of problem-solving that underlies all of AI is also one of search (i.e., a program, or person, can solve a problem by searching among alternative solutions). Although immediately clear and simple, this formulation of the paradigm does little to tell us how to search a solution space efficiently and accurately. The number of possible solutions may be astronomical. Possible answers, decisions, or courses of action are structured into an abstract space, which we then search. Search can be either "blind" or "informed". For a blind (uninformed) search, we move through the space without worrying about what is coming next but recognizing the answer if we see it. In an informed search, we guess what is ahead, and use that information to decide where to look next. We may want to search for the first answer that satisfies our goal, or we may want to keep searching until we find the best answer. A search problem starts with a state space (initial state, goal test, and successor function), a goal test, and a cost for the path chosen.

The key idea behind the probabilistic robot is the explicit representation of uncertainty using the calculus of probability theory:

- Perception = state estimation
- Action = utility optimization

This is based on axioms of probability theory and using Bayes theorem. We define knowledge by connecting facts, beliefs, and knowledge through underlying and connective series called causal chains. Often the world is dynamic since actions carried out by the robot, actions carried out by other agents, or just the time passing by change the world. How can we incorporate such actions? What are typical robot actions? The robot turns its wheels to move and the robot uses its manipulator to grasp an object. Similarly, plants grow over time. However, actions are never carried out with absolute certainty. In contrast to measurements, actions generally increase the uncertainty.

To control robots to perform tasks autonomously, a number of tasks have to be addressed such as modeling of robot arm and vehicle mechanisms (Kinematics, Dynamics), robot sensor selection (active and passive proximity sensors), low-level control of actuators (Closed-loop control) and deciding on the robot's control architectures (Traditional planning architectures, Behavior-based control architectures, Hybrid architectures). Modeling the robot mechanism requires understanding the: forward kinematics describes how the robot joint angle configurations translate to locations in the world, inverse kinematics computes the joint angle configuration necessary to reach a particular point in space, and Jacobians calculates how the speed and configuration of the actuators translate into velocity of the robot. To accurately achieve a task in an intelligent environment, the robot has to be able to react dynamically to changes in its surrounding:

- Robots need sensors to perceive the environment
- Most robots use a set of different sensors (different sensors serve different purposes)
- Information from sensors has to be integrated into the control of the robot.

For the robot to move, it needs to compute its forces and inertia. The dynamic state is a function of its position and velocities. The relationship between kinematics and velocity is via a differential equation. The robot controller keeps the robot on track. A reference controller centers the robot on its reference trajectory path. The selected trajectory is optimized with a planning cost function, such as a Markov Decision Process policy. Robot control refers to the way in which the sensing and action of a robot are coordinated. The many different ways in which robots can be controlled all fall along a well-defined spectrum of control. The control approaches are reactive (Don't think and (re)act, behavior-based control (think the way you act), deliberate control (think hard and act later as in SPA), and hybrid (think and act independently and in parallel). The last step is the controller that sends commands to the actuators. As a result, you might want a custom-designed procedural language for robot control applications.

The ultimate goal of cognitive robots is to actively explore the real environment, acquire knowledge, and learn skills. To make the robot continuously develop through active exploration, the robot's learning process should be based on sensorimotor information obtained through physical interactions with its physical environment. Cognitive robots benefit from predictive coding mechanisms to infer others' actions using hierarchically organized Recurrent Neural Networks, with a variety of methods proposed to exploit this idea of prediction-error-minimization or propagation. "Higher levels" (internal representation) generate predictions about the dynamics of the "lower levels" up to the sensorimotor level. Prediction errors at the sensorimotor level are based on observations and then propagated "upwards" in a 3-T hierarchy, correcting the internal state and thus minimizing the errors. The current state-of-the-art research focuses on scaling active inference in planning tasks with high-dimensional inputs and improving representation learning through multimodal common latent space or introducing structural inductive biases, such as objects. The current approaches in robot planning use a compressed encoded representation of the world dynamics, which aids in predicting future states and in action generation. Observations obtained from the environment are high-dimensional, and compressing this information into a low-dimensional space is critical for efficient data handling, abstraction, and planning. Low-dimensional representations should have scene understanding and task meaning, such as objects, locations, and temporal events.

The term *"symbols"* refers to manipulative discrete representations that are used in symbolic AI and cognitive science. The neuro-symbolic approach attempts to integrate conventional symbolic and modern neural network-based AIs. Artificial agents benefit from **predictive coding** mechanisms for reasoning, decision-making, and planning. Predictive forward models are used to generate plans that involve a sequence of actions. In robotics, predictive coding over object symbols takes actions and effects into account in addition to the features of objects and the environment, which facilitates the formation of symbols that are likely to capture object affordances. An affordance is an action, or possibility offered to the agent by its environment. This same environment would offer different action possibilities to different agents, depending on their sensorimotor capabilities. While exploring the action possibilities, the agent can learn the effects of those actions. Most computational models of affordances in robotics rely on representations that include not only the action but also the action's effects (or in other terms, the goal). The computational techniques used for learning affordance models often overlap with those used for learning world models.

In cognitive science, cognitive functionalities such as memory, perception, and decision-making are implemented as modules in the cognitive architectures, and the specific task can be solved by activating these modules in a coordinated fashion (i.e., in ACT-R or Soar architecture). ROS has been widely used in the robotics community for bridging hardware and AI software layers. In neuroscience, **predictive coding** (also known as **predictive processing**) is a theory of brain function that postulates that the brain is constantly generating and updating a "mental model" of its environment. According to this theory, such a mental model is used to predict input signals from the sensors that are then compared with the actual input signals from those sensed. Predictive coding posits that the brain/computer actively predicts upcoming sensory input rather than passively registering it, thus predictive coding is efficient in the sense that the brain does not need to maintain multiple versions of the same information at different levels of the processing hierarchy.

Verifying and validating (V&V) autonomous robotic systems that can respond or adapt to their environment in order to ensure sufficient predictability and reliability, brings a wide array of engineering challenges. Testing normally includes computer simulations and real-world physical tests to assess the response of the system in all the different circumstances a robot may encounter. However, **it is not possible to test all the potential sensory input and actuator output combinations of the entire system for all circumstances**, or even to know what percentage of the possible outputs that one has tested. This means that it is mathematically difficult to formally verify and validate the predictability of the entire system with all its reliability issues (including its probability of failure). And when considering autonomous weapon systems, it is therefore very difficult to ensure that a weapon system is capable of being used in compliance with international humanitarian laws and all other ethical concerns, especially if the system incorporates AI and in particular machine-learning algorithms used in its control system.

There is an understanding when developing an autonomous robot that a "happy" robot is "healthy" based on the Neumann–Morgenstern utility theorem. This provides more options to choose from when making a decision. Maximum utility is the concept that the robot seeks to attain the highest level of satisfaction from its economic decision. For example, if a component fails such as losing a wheel, it limits what the robot can do. Some vehicles are self-healing depending on what the failure is in the same way satellites will detect a fault, isolate the fault, and potentially recover from the fault depending on if there are redundant components that it can switch to. In a way, this is a self-healing capability. In addition to faults, the robot needs to understand the resources available to the robot such as the amount of fuel or power available to continue operating. The NASA Remote Agent Software has three major components: a planner/scheduler, a smart executive, and the Mode Identification and Recovery (MIR) subsystem that monitors the health of the vehicle and attempts to correct any problems that occur. The MIR constantly monitors the state of the robot, identifies failures, and suggests recovery actions. This may result in replanning of its mission using its constraint-based planning feature as it communicates with the smart executive and the planner/scheduler (maximum flexibility). This is a model-based system that compares status with stored models of how components should work. System engineering is required to address all the miscellaneous functions of a fieldable robot such as availability, maintenance, safety, reliability, etc.

The increasing **adaptability in an autonomous system is generally equated with increasingly "intelligent" behavior or incorporating AI**. Definitions of AI may vary, but they are in essence computer programs that carry out tasks, often associated with human intelligence, that require cognition, planning, reasoning, or learning. What is considered AI has changed over time where autonomous systems once considered "intelligent", such as an aircraft's autopilot system, are now seen as merely automated. There is growing interest in military application of AI for purposes that include weapon systems and decision support systems and more broadly, whether for targeting or for other military applications.

Computing Hardware – Processing and Memory

At the beginning of 1950, John Von Neumann and Alan Turing made the transition from computers to 19th-century decimal logic (which thus dealt with values from 0 to 9) and machines to binary logic (which rely on Boolean algebra, dealing with more or less important chains of 0 or 1). The two researchers formalized the architecture of our contemporary computers and demonstrated that it was a universal machine, capable of executing what is programmed in high-level mental processes such as perceptual learning, memory organization, and critical reasoning.

Processing: The von Neumann architecture (also known as the Princeton architecture) has a processing unit with both an arithmetic

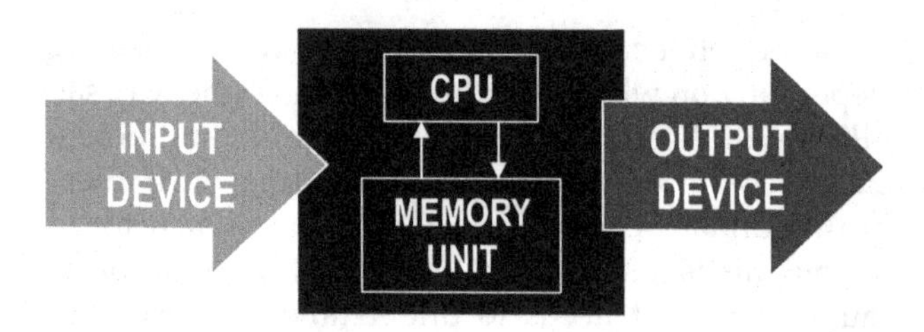

FIGURE 5.9
Von Neumann architecture process diagram.

logic unit and processor registers, a control unit that includes an instruction register and a program counter, memory, external storage, and input/output organisms. It is a stored-program computer (CPU) with instruction fetch and data operation sharing the same common bus. Instruction fetch and data operation cannot occur at the same time (also known as the von Neumann bottleneck). A stored-program digital computer keeps both program instructions and data in read-write, random-access memory (RAM). Machine Instructions Per Second and Million Instructions Per Second (MIPS) are an approximate measure of a computer's raw processing power and are tracked with Moore's Law (Figure 5.9).

A GPU (graphics processing unit) is a chip designed to accelerate the processing of multidimensional data such as an image. A GPU consists of thousands of smaller cores, intended to work independently on a subspace of the input data, that needs heavy computation. Repetitive functions that can be applied to different parts of the input, such as texture mapping, image rotation, translation, and filtering, are performed at a much faster rate and more efficiently, through the use of the GPU. A GPU has dedicated memory and the data must be moved in and out in order to be processed. A heterogeneous computing system refers to a system that contains different types of computational units, such as multicore Central Processing Units (CPUs), GPUs, Digital Signal Processors (DSPs), Field Programmable Gate Arrays (FPGAs), and Application-Specific Integrated Circuits (ASICs). The computational units in a heterogeneous system typically include a general-purpose processor that runs an operating system. Processors other than the general-purpose processor are called accelerators because they accelerate a specific type of computation by assisting the general-purpose processor. Each accelerator in the system is best suited to a different type of task (Figure 5.10).

CPU works together with a GPU to increase the throughput of data and the number of concurrent calculations within an application. GPUs were originally designed to create images for computer graphics and video game consoles, but since the early 2010s, GPUs can also be used to accelerate calculations involving massive amounts of data. A CPU can never be fully replaced by a GPU; a GPU complements CPU architecture by allowing repetitive calculations within an application to be run in parallel while the main program continues to run on the CPU. The CPU can be thought of as the taskmaster of the entire system, coordinating a wide range of general-purpose computing tasks, with the GPU performing a narrower range of more specialized tasks (usually mathematical). Using the power of parallelism, a GPU can complete more work in the same amount of time as compared to a CPU.

CPUs are general-purpose processors, while GPUs and tensor processing units (TPUs) are optimized accelerators that accelerate machine learning. A CPU is a general-purpose process that is good at solving a wide range of problems. CPU stands for CPU and is considered as the brain of the computer. It is the primary hardware of the computer that executes the instructions for computer programs. The CPU handles the basic arithmetic, logic, and control. The CPU also handles the input/output functions of the software program. It runs the operating system, continually receiving inputs and providing output to the users. A CPU contains at least one processor. The processor is an actual chip inside the CPU to perform all the calculations. While the CPU is known as the brain of the computer, and the logical thinking section of the computer, the GPU helps in displaying what is going on in the brain by rendering the graphical user interface visually.

A GPU is a graphics processor that is designed specifically for the types of calculations needed to render graphic images. It's good at this because it was designed that way and it would not perform

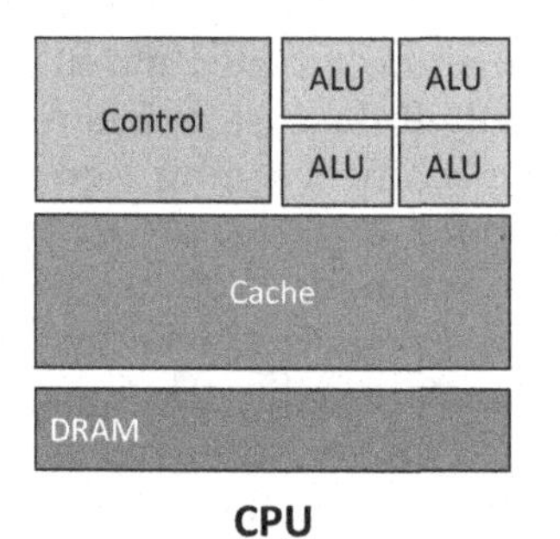

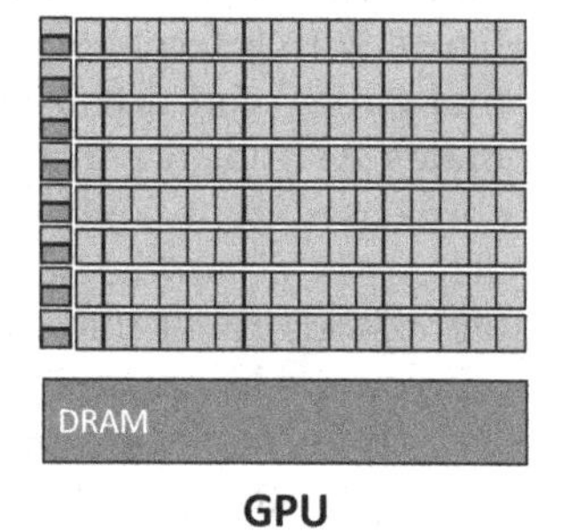

FIGURE 5.10
CPU vs. GPU comparison.

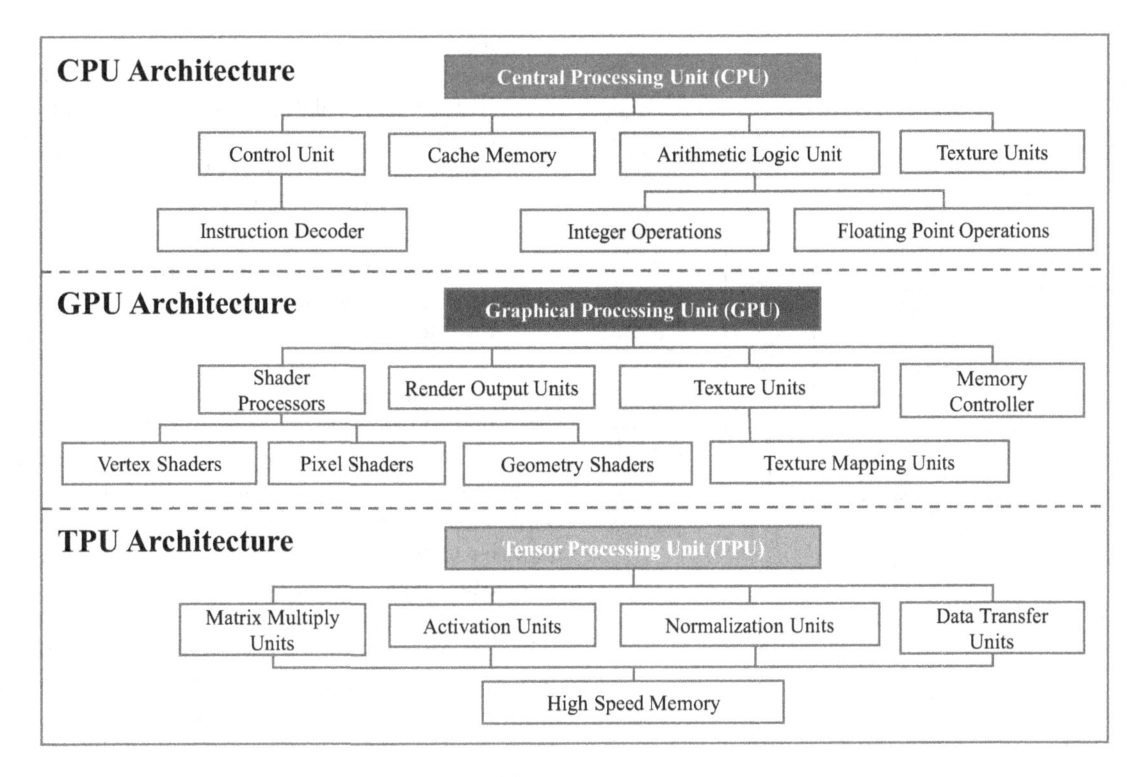

FIGURE 5.11
Models of CPUs, GPUs, and TPUs.

that well at a different sort of problem. A TPU is another specialized processor, targeted at the types of calculations needed for neural networks and it would fail to perform well at graphics processing as well as general-purpose computing. A TPU is an application-specific integrated circuit, to accelerate the AI calculations and algorithm. Google developed the TPU specifically for neural network machine learning for the TensorFlow software. TPUs are custom-built processing units to work for a specific application framework, that is TensorFlow. An open-source machine-learning platform, with state-of-the-art tools, libraries, and community, so the user can quickly build and deploy ML apps (Figure 5.11).

The difference between CPU, GPU, and TPU is that the CPU handles all the logic, calculations, and input/output of the computer; it is a general-purpose processor. In comparison, GPU is an additional processor to enhance the graphical interface and run high-end tasks. TPUs are powerful custom-built processors to run the project made on a specific framework, that is, TensorFlow. For example, the models that used to take weeks to train on GPUs or any other hardware can be put out in hours with a TPU. GPUs and TPUs both have their advantages and disadvantages. A single GPU can process thousands of tasks at once, but GPUs are typically less efficient in the way they work with neural networks than a TPU. TPUs are more specialized for machine-learning calculations and require more traffic to learn at first, but after that, they are more impactful with less power consumption. Bottom line, TPUs are 3× faster than CPUs and 3× slower than GPUs for performing a small number of predictions, and the TPU is 15–30 times faster than current GPUs.

To summarize, general-purpose computing on GPUs refers to the use of a GPU to perform general-purpose computations in addition to its traditional role of rendering graphics. GPUs are designed to handle large amounts of data in parallel, making them ideal for performing certain computations much faster than traditional CPUs. The parallel processing architecture of a GPU enables faster computation than a CPU, thus enabling a wide number of AI applications.

Memory: Human memory is the faculty of the mind by which data and information are encoded, stored, and retrieved when needed. It is often understood as an informational processing system with explicit and implicit functioning that is made up of a sensory processor, short-term (or working) memory, and long-term memory. Human memory is analogous to computer memory and is often likened to that of a filing cabinet. The computer offers an intuitively appealing model for thinking

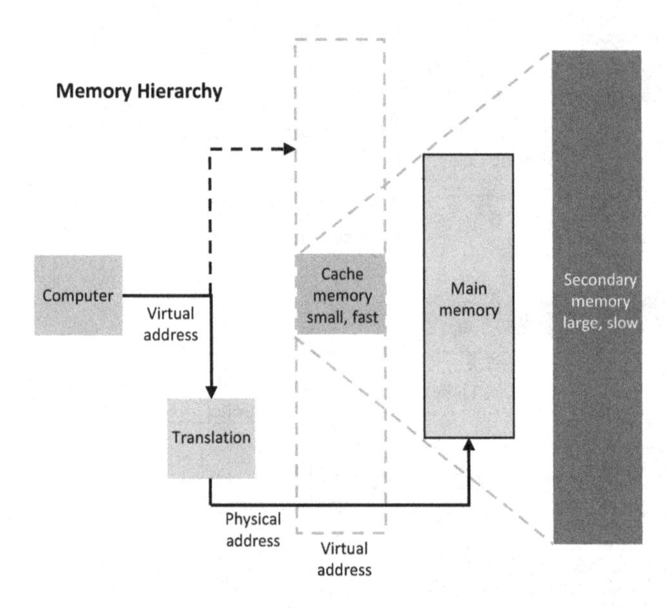

FIGURE 5.12
Memory hierarchy.

about the nature and structure of working memory. To simplify the workings of a computer, there are two means by which information is stored, the hard disk and RAM. The hard disk is the means by which information is stored permanently in a stable and reliable form; all software programs, data files, and the operating system of the computer are stored on the hard disk. To use this stored information, you must retrieve it from the hard disk and load it into RAM. Now for the analogy: the information stored in the hard disk is like long-term memory, RAM corresponds to working memory. Computer memory is the storage space in the computer, where data is to be processed and instructions required for processing are stored. The memory is divided into large number of small parts called cells. Each location or cell has a unique address, which varies from zero to memory size minus one. Computer memory is of two basic types: Primary memory (RAM and ROM) and Secondary memory. RAM is primary-volatile memory and Read-Only Memory (ROM) is primary-non-volatile memory. Bits and bytes are the basic building blocks of memory. "Bit" stands for binary digit. A bit is a one or a zero, on or off, which is how all computer information is stored. A byte is made up of eight bits. Eight bits, or a byte, was the original amount of information needed to encode a single character of text (Figure 5.12).

In terms of memory limits, suppose: you have a 2GHz CPU (or 2000000000 cycle/sec) with a 1 GB main memory. With the following conversion factors: 100 instructions/expansion, 5 bytes/node, and 1 instruction=4 bytes, and at 200,000 expansions/sec, the 1 GB memory will be filled in 100 seconds… or in **less than 2 minutes**. That is not very long for a robot.

As a comparison for different problem spaces:

This is an example of the combinatorial explosion of data in AI search problems. The computer memory contains millions of transistors which at any time can be in one of two physical states, usually labeled as 0 and 1. Each transistor carries 1 bit of information (Table 5.3).

8 bits=1 byte
210 bytes=1 Kilo byte ≡ 1024 bytes
220 bytes=1 Mega byte
230 bytes=1 Giga byte
240 bytes=1 Tera byte

How many bits are needed? For textual data, a unit is a "character"; a file consists of a number of characters.

A. Characters
English: 26 lower case: a, b, c, ….
26 upper case: A, B, C, ….
10 numerals: 1, 2, 3, ….
15 "extras" + {.,; : * etc.
Total ≈ 77

TABLE 5.3
Table of problem and Brute-Force Search

PROBLEM	Nodes	BRUTE-FORCE SEARCH (10 million Nodes/Sec)
8 Puzzle	10^5	0.01 seconds
2^3 Rubik's Cube	10^6	0.2 seconds
15 Puzzle	10^{13}	6 days
33 Rubik's Cube	10^{19}	68,000 years
24 Puzzle	10^{25}	12 billion years

Given N bits, we can create 2^N different combinations

Example: N=2, {0,0}, {0,1}, {1,0}, {1,1}

N=6 gives 2^6=64 combinations, yet too low for CHARACTERS.

In fact, use 1 byte=8 bits for each character.

B. Books

Say a book has 40 lines × 80 characters per page,

1 page=3200 bytes ≈ 3 Kilo bytes.

300 pages ≈ 1 Mega byte

For picture data, a unit is a "dot" or "pixel"; a file consists of a number of dots or pixels.

C. Pictures (or images)

Each dot on a computer image is called a Pixel. A good screen might have 1080×780 ≈ 1M Pixels. Each pixel has color and brightness specified by (about) 3 bytes. Therefore, a single image requires 3 Megabyte (Mb). (A chemical photograph contains approximately 30–40 Mb of information.)

FIGURE 5.13
JPL Spirit Martian Rover

Integers are usually stored using an integer number of bytes; hence one usually refers to 8-bit, 16-bit, 32-bit or 64-bit integers. The number of bits controls the range of integers that can be stored, for example, 8-bits allows 28=256 combinations, and so allows only 256 integers to be stored. One method for storing (8-bit) integers on the computer, that leads to a convenient binary "arithmetic", is known as the two's complement method. A computer is an entirely deterministic device, that is, it does not have access to any genuinely random process. "Random" numbers must therefore be generated from a deterministic sequence, ideally one which "appears" to be random to the casual observer (although of course is not really). "Random" numbers generated in this fashion are adequate for most practical purposes. Total memory requirements to train a deep neural network (DNN) will depend on implementation and optimizer based on memory for parameters, memory for parameter gradients, memory for momentum, memory for layer outputs, and memory for layer errors. There is also memory for implementation overhead (memory for convolutions) (Figure 5.13).

As an example, a shortage of memory on board the Spirit Mars rover is what caused it to become unresponsive on the Martian surface on January 22, 2004 (right after landing), raising fears that the Martian mission might end almost before it began in earnest. The NASA Spirit rover dedicates 32MB of its 128MB of RAM to the onboard Wind River VxWorks operating system and a host of science applications, and as the mission progresses, technicians are scheduled to periodically delete old files and directories to clear out the memory for reuse. The VxWorks operating system was embedded in a specially prepared, radiation-hardened 20-MHz PowerPC CPU (RAD750) installed on each of the rovers, along with 128MB of RAM. The hardware was cutting-edge back when it was chosen in the mid-1990s, but then it had to be hardened to ensure its reliability in the radiation environment of deep space, a process that takes 5–10 years. Technicians were eventually able to correct the problem when the rover went into a diagnostic mode. Diagnostic commands were beamed up to the machine, and a series of files and folders were deleted from a flash-memory-based file system onboard, allowing the rover to resume normal operations in 2004.

Kalman Filters: A Kalman filter is an optimal estimator, that is, infers parameters of interest from indirect, inaccurate, and uncertain observations. It is recursive so that new measurements can be processed as they arrive (cf. batch processing where all data must be present). For statistics and control theory, Kalman filtering, also known as linear quadratic estimation (LQE), is an algorithm that uses a series of measurements observed over time, including statistical noise and other inaccuracies, and produces estimates of unknown variables that tend to be more accurate than those based on a single measurement alone, by estimating a joint probability distribution over the variables for each timeframe. The algorithm works by a two-phase process. For the prediction phase, the Kalman filter produces estimates of the current state variables of a robot, along with their uncertainties. Once the outcome of the next measurement (necessarily corrupted with some error, including random noise) for navigation is observed, these estimates are updated using a weighted average, with more weight being given to estimates with greater certainty. This algorithm is recursive. It can operate in real time, using only the present input measurements and the state calculated previously and its uncertainty matrix; no additional past information is required.

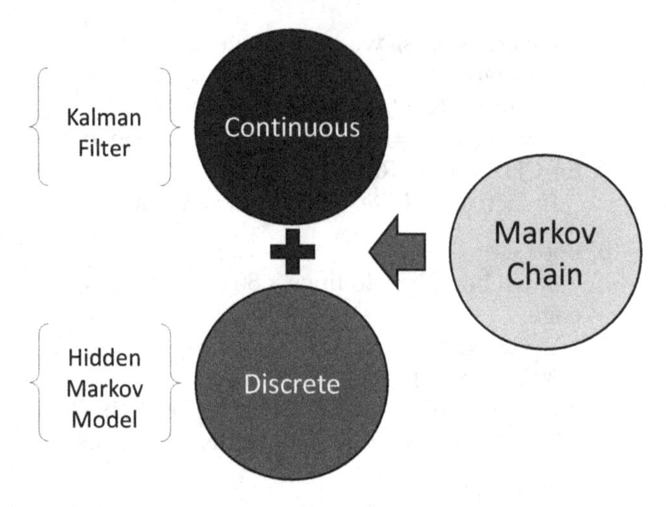

FIGURE 5.14
Markov chain.

The Kalman filter is built around one key concept: make all the densities Gaussian. Kalman filter, in its most basic form, consists of three steps. (1) Predict – Based on previous knowledge of a vehicle position and kinematic equations, we predict what should be the position of vehicle after time $t+1$. (2) Measurement – Get readings from sensor regarding position of vehicle and compare it with Prediction. (3) Update – Update our knowledge about position (or state) of vehicle based on our prediction and sensor readings. And that's it. All variants of Kalman filter are just different variants of above three steps, depending upon different Kinematic equations you want to use and different kinds of sensor reading which you want to incorporate in algorithm. The basic idea of the Kalman filter is to use a model of the system being measured and to update the model as new measurements become available. The filter works by making a prediction of the current state of the system based on the previous state estimate and the system model and then combining this prediction with a new measurement to obtain an updated state estimate (Figure 5.14).

Most robot systems have numerous sensors that estimate hidden (unknown) states based on a series of sensor measurements. One of the biggest challenges of tracking and controlling a robot is providing an accurate and precise estimation of its hidden states in the presence of uncertainty. The **Kalman Filter produces estimates of hidden variables** based on inaccurate and uncertain measurements. Also, the **Kalman Filter predicts the future system state** based on past estimations. The target parameters $[x,y,z,v_x,v_y,v_z,a_x,a_y,a_z]$, $[x,y,z,v_x,v_y,v_z,a_x,a_y,a_z]$ are called a System State. The current state is the input to the prediction algorithm and the next state (the target parameters at the next time interval) is the algorithm's output. The above set of equations is called a Dynamic Model (or a State Space Model). The Dynamic Model describes the relationship between input and output. The target motion is not strictly aligned with motion equations due to external factors. The dynamic model error (or uncertainty) is called Process Noise. The most widely used prediction algorithm is the Kalman Filter. The Kalman Filter is based on five equations: (1) The state update equations, (2) The dynamic model equations, (3) Kalman Gain, (4) Covariance Update, and (5) Covariance Extrapolation. The Kalman Gain will decrease if the measurements match the predicted system state. If the measured values say otherwise, the elements of matrix K become larger. The Kalman filter is an efficient optimal estimator (a set of mathematical equations) that provides a recursive computational methodology for estimating the state of a discrete data-controlled process from measurements that are typically noisy, while providing an estimate of the uncertainty of the estimates.

Mobile robots require a navigation system that is automatic, highly precise, and quickly responsive. In an INS/GNSS integrated navigational systems, the Kalman filter is generally used **to estimate the navigation parameters (position, velocity, and attitude angle) of a ground robot**. The Kalman filter gives a linear, unbiased, and minimum error variance recursive algorithm to optimally estimate the unknown state of a linear dynamic system from Gaussian-distributed noisy observations. The Kalman filtering process can be considered as a prediction-update formulation step. The Kalman filter is the optimal linear estimator for *linear* system models with additive independent white noise in both the transition and the measurement systems. Unfortunately, in engineering, most systems are *nonlinear*, so attempts were made to apply this filtering method to nonlinear systems. In estimation theory, the **Extended Kalman filter** (**EKF**) is the nonlinear version of the Kalman filter which linearizes about an estimate of the current mean and covariance. In the case of well-defined transition models, the EKF has been considered the *de facto* standard in the theory of nonlinear state estimation, navigation systems, and GPS. The EKF adapted techniques from calculus, namely, multivariate Taylor series expansions, to linearize a model about a working point. If the system model is not well known or is inaccurate, then Monte Carlo methods, especially particle filters, are employed for estimation. Monte Carlo techniques predate the existence of the EKF but are more computationally expensive for any moderately dimensioned state space (Figure 5.15).

Multi-sensor data fusion has found widespread application in industry and commerce. The purpose of data fusion is to produce an improved model or estimate of a system from a set of independent data sources. There are various multi-sensor data fusion approaches, of which Kalman filtering is one of the most significant. Methods for Kalman filter-based data fusion include measurement fusion and state fusion. For example, if we fuse a GPS and an IMU reading, a Kalman filter is applied to both sensors, and both sensors are converted to give similar measurements (e.g., x, y, z) by applying a Kalman filter to both sensors and returning an average of the estimates. The sensor fusion aspect is more or less a weighted average based on the process noise and measurement noise.

A summary of the competencies for autonomous navigation requires cognition and reasoning. It is the ability to decide what actions are required to achieve a certain goal in a given situation (belief state). The decisions required range from what path to take to what information on the environment to use. Today's industrial robots can operate without any cognition (reasoning) because their environment is static and very structured. Every other environment is dynamic and unstructured. In mobile robotics, cognition and reasoning are primarily of geometric nature, such as picking safe path or determining where to go next. These navigation problems have already been largely explored in literature for cases in which complete information about the current situation and the environment exists (e.g., solving the traveling salesman problem). However, in mobile robotics the knowledge of about the environment and situation is usually only partially known and is uncertain, making the task much more difficult which requires multiple tasks running in parallel, some for planning (global), some to guarantee "survival of the robot". Functions that support global planning are used to meet an autonomous goal and the idea of survival goes to decisions made on robot utility. Robot control can usually be decomposed into various behaviors or functions such as wall following, localization, path generation, or obstacle avoidance. We are concerned with path planning and navigation but cannot forget the low-level motion controller and localization. We can generally distinguish the navigation functions between (*global*) path planning and (*local*) obstacle avoidance. A simplistic definition of an "autonomous" robot is if they succeed in moving in a safe interaction with an unstructured environment, while autonomously achieving their specified tasks.

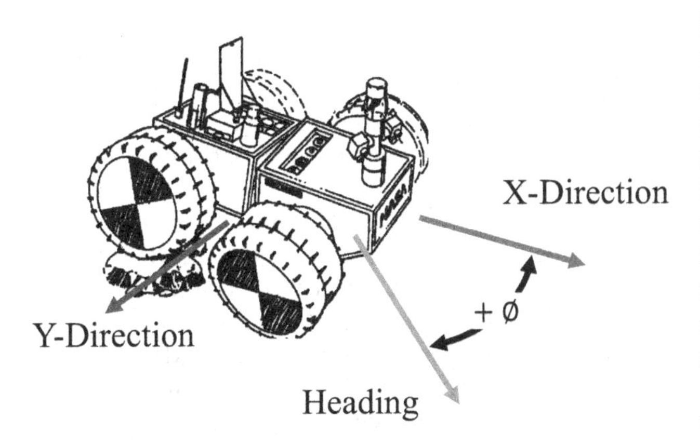

FIGURE 5.15
Ground robot coordinate frame.

Trust in AI and autonomous systems is a major area of enquiry, especially as regards their use for safety-critical applications or where they have other implications for human life and personal freedom. **"Explainability" is a fundamental problem for machine-learning algorithms** that are not transparent in the way they function and provide no explanation for why they produce a given output. Even when explanations are available, the question remains of whether one can extrapolate the trust built by analyzing specific training data to trust in analysis of a general data set after deployment. Explainable artificial intelligence (XAI) is a set of processes and methods that allows human users to comprehend and trust the results and output created by machine-learning algorithms. Explainable AI is used to describe an AI model, its expected impact and potential biases. It helps characterize model accuracy, fairness, transparency, and outcomes in AI-powered decision-making. Building trust in the model is more difficult, because the number of potential inputs in the environment may be infinite. Explainable AI is crucial for an organization in building trust and confidence in the system.

In general, **trusted autonomy** requires confidence in the software artifact itself, the processes and organizations that determine how the software is maintained and improved, and the affordances the software offers to both developers and other mission personnel that allow its behavior to be understood, tested, critiqued, and communicated. Perhaps the single most effective source of trust in complex software is a long history of use, preferably in diverse applications by numerous, distinct users. Long deployments provide time to shake out bugs, improve performance, and refine the system. Confidence of the software is built over time. They provide opportunity to analyze performance and failure modes, study and try alternative designs, and compile documentation of users' experiences, all providing concrete evidence of software reliability.

Intelligent Robots: An agent is anything that can be viewed as a device that can perceive its environment through sensors and act upon that environment through actuators. The most critical task for intelligent behavior is the ability to communicate effectively. AI programs must be able to communicate with their human counterparts in a natural way, and natural language is one of the most important mediums for that purpose. A program understands a natural language if it behaves by taking a correct or acceptable action in response to the input. For an agent to be "intelligent", it must be able to understand the meaning of information. Intelligent Robotics use AI to boost collaboration between people and devices. AI helps robots to adapt to dynamic situations and communicate naturally with people. It would use the basic narrow AI technologies described in an autonomous robot, but with the addition of National Language Processing and Speech. An implied human–robot collaboration will require a human–robot interface. So, why do Robots need AI? For sensor interpretation, that is, is the detected object a bush or big rock, symbol-ground problem, and terrain interpretation? The robot needs situation awareness to see the Big Picture. What is the human–robot interaction? Is this an "Open world" and are there multiple fault diagnosis and recovery options? For a ground robot, what happens to localization in sparse areas when GPS goes out. How do we handle uncertainty? How about including learning? Robotics is AI-complete. Robotics integrate many AI tasks such as perception (vision, sound, haptics), reasoning (search, route planning, action planning), and learning (recognition of objects/locations and exploration). Beyond autonomy, an intelligent robot will have a situational awareness capability.

An intelligent robot should employ all the capabilities of Artificial General Intelligence (AGI), a hypothetical type of intelligent agent that could learn to accomplish any intellectual task that a human being can perform. AGI has also been defined as an autonomous robot that surpasses human capabilities that surpass human capabilities in the majority of economically valuable tasks.

Situational Awareness (SA) can be defined simply as "knowing what is going on around us", or more technically, as "the perception of the elements in the environment within a volume of time and space, the comprehension of their meaning and the projection of their status in the near future". SA is being aware of what is happening around the robot in terms of where you are, where you are supposed to be, and whether anyone or anything around you is a threat to the robot's health and safety. For a person, knowledge, experience, and education enable a person to understand what is going on around him and help him to determine if it is safe. This means that everyone's SA is individual and potentially different. We can use our SA to make decisions in many environments and instruct others. It is the perception of elements in the environment, within a volume of time and space. SA includes the comprehension of their meaning, and the projection of their status into the near future. Basically, we want to answer what happened, where am I, what is happening, and what could happen.

Situational knowledge is the knowledge gained from individual experiences, such as dealing with and understanding specific situations. AI will see things and will learn all things, new and old. The system should start to identify the big issues and eventually start to recognize patterns (e.g., incorporating pattern recognition in SA). As stated previously, good SA is the basis for decision-making and follow-on actions. There are three levels to SA. Level 1 has to see and sense (also known as perception), Level 2 is used to understand what was actually seen (comprehend), and Level 3 uses what was understood in Level 2 to think ahead (project). Level 2 compares the model with the sensed data and in Level 3, thinking ahead updates the model (Figure 5.16).

Levels 1–3 provide timely and accurate information for decision-making. The output informs decision-making about protective actions such as evacuation and sheltering, and the anticipation and deployment of response resources. By understanding the situation, it triggers decision-making, followed by action, and later review. The goal is to determine what is going on around in its environment; where our robot is, its orientation, the mode the robot is

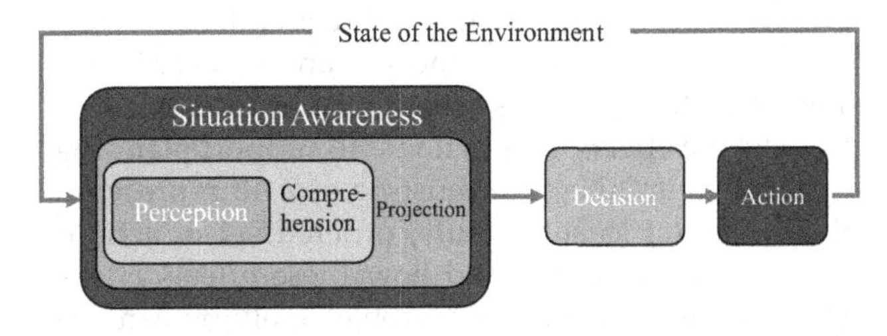

FIGURE 5.16
Situational awareness diagram.

in, and what objects/people around are doing. The model that is built is based on perceptions of the important aspects of the surroundings and comparing them with experiences and a physical model in its memory. Thus, the robot will gather information about the real world, and combine it with knowledge and experiences learned that it has stored. It is understood that an accurate understanding of the situation is essential for planning ahead and planning ahead is an initial step to decision-making. However, the SA is only as accurate as our own perception or reading of the situation, so what we think is happening may not accurately reflect the true situation. Many kinds of changes occur in physical situations. For example, objects move, collide, flow, bend, heat lip, cool down, stretch, and boil. These and other things that cause changes in objects over time are intuitively characterized as *processes*. Much of formal physics consists of characterizations of processes by differential equations that describe how the parameters of objects change over time. But the notion of process is richer and more structured than this. We often reach conclusions about physical processes based on very little information.

Intelligent robots are required to operate in complex environments. By integrating AI with robots, researchers should direct their efforts toward building intelligent systems that incorporate a broad range of capabilities, use knowledge from a variety of sources (preprogrammed, learned from experience, and acquired from other agents), and operate in complex dynamic environments that cannot be fully predicted in advance. The original aim of AI was to understand the nature of intelligence by constructing artifacts that exhibited the same breadth and depth of cognition as humans. The focus on standalone algorithms (expert systems, search, learning, logic, planning, computer vision, language processing, speech, diagnostics, problem-solving, reasoning, and inference, also known individually as human intelligence factors) has helped us understand the elements of cognition, but it has told us little about how they might work together. The goal of mind and brain is to work toward the construction of integrated cognitive systems that exhibit broad, human-level intelligence. Intelligent systems are based on first-order logic on the one hand, and on the other hand, on artificial neural networks (ANNs; also called connectionist systems), and they differ substantially. It would be very desirable to combine robust neural networking machinery with symbolic knowledge representation and reasoning paradigms like logic programming in such a way that the strengths of either paradigm will be retained.

While both Connectionist and Classical architectures postulate representational mental states, the latter but not the former are committed to a symbol-level of representation, or to a "language of thought", that is, to representational states that have combinatorial syntactic and semantic structure. Connectionist systems are networks consisting of very large numbers of simple but highly interconnected "units". Classicists and Connectionists all assign semantic content to something. Roughly, Connectionists assign semantic content to "nodes" (that is to units or aggregates of units, that is, to the sorts of things that are typically labeled in Connectionist diagrams; whereas Classicists assign semantic content to expressions, that is, to the sorts of things that get written on the tapes of Turing machines and stored at addresses in Von Neumann machines. In summary, Classical and Connectionist theories disagree about the nature of mental representation. For the former, but not for the latter, mental representations characteristically exhibit a combinatorial constituent structure and combinatorial semantics. Classical and Connectionist theories also disagree about the nature of mental processes; for the former, but not for the latter, mental processes are characteristically sensitive to the combinatorial structure of the representations on which they operate.

A cognitive architecture is a hypothesis about the fixed structures that provide a mind, whether in human or artificial systems, and how they work together, in conjunction with knowledge and skills embodied within the architecture, to yield intelligent behavior in a diversity of complex environments. A grand unified architecture integrates across (nominally symbolic) higher-level thought processes plus any other (nominally subsymbolic) aspects critical for successful behavior in human-like environments, such as perception, motor control, and emotions. A generically cognitive architecture spans both the creation of AI and the modeling of natural intelligence, at a suitable level of abstraction. A functionally elegant architecture yields a broad range of capabilities from the interactions among a small general set of mechanisms.

Cognitive Architectures: A cognitive architecture specifies the underlying infrastructure for an intelligent system. Cognitive architectures constitute the antithesis of expert systems, which provide skilled behavior in narrowly defined contexts. In contrast, architectural research aims for breadth of coverage across a diverse set of tasks and domains. A central issue that confronts the designer of a cognitive architecture is how to let agents access different sources of knowledge. Another key question is whether the cognitive architecture supports a capability directly, using embedded processes, or whether it instead provides ways to implement that capability in terms of knowledge. Recognition is closely related to categorization, which involves the assignment of objects, situations, and events to known concepts or categories. Recognition and categorization are closely linked to perception, in that they often operate on output from the perceptual system, and some frameworks view them as indistinguishable. To operate in an environment, an intelligent system also requires the ability to make decisions and select among alternatives. Prediction requires some model of the environment and the effect actions have on it, and the architecture must represent this model in memory. Once an architecture has a mechanism for making predictions, it can also utilize them to monitor the environment (Figure 5.17).

Intelligence recognizes situations or events as instances of known or familiar patterns. Planning and problem-solving can also benefit from learning. Prediction requires some model of the environment and the effect actions on it, and the architecture must represent this model in memory. Planning is only possible when the agent has an environmental model that predicts the effects of its actions. To support planning, a cognitive architecture must be able to represent a plan as an (at least partially) ordered set of actions, their expected effects, and how these effects enable later actions. Planning and problem-solving can also benefit from learning. Naturally, improved predictive models for actions can lead to more effective plans, but learning can also occur at the level of problem space search, whether this activity takes place in the agent's head or in the physical world. Such learning can rely on a variety of information sources. Problem-solving is closely related to reasoning, another central cognitive activity that lets an agent augment its knowledge state. Whereas planning is concerned primarily with achieving objectives in the world by taking action, reasoning draws mental conclusions from other beliefs or assumptions that the agent already holds. Naturally, a cognitive architecture also requires mechanisms that draw inferences using these knowledge structures. According to the mental model theory of deductive reasoning, reasoners use the meanings of assertions together with general knowledge to construct mental models of the possibilities compatible with the premises. Each model represents what is true in a possibility. A conclusion is held to be valid if it holds in all the models of the premises. Recent evidence described here shows that the fewer models an inference calls for, the easier the inference is. Deductive reasoning is an important and widely studied form of inference that lets one combine general and specific beliefs to conclude others that they entail logically. However, an agent can also engage in inductive reasoning, which moves from specific beliefs to more general ones and can be viewed as a form of learning. An architecture

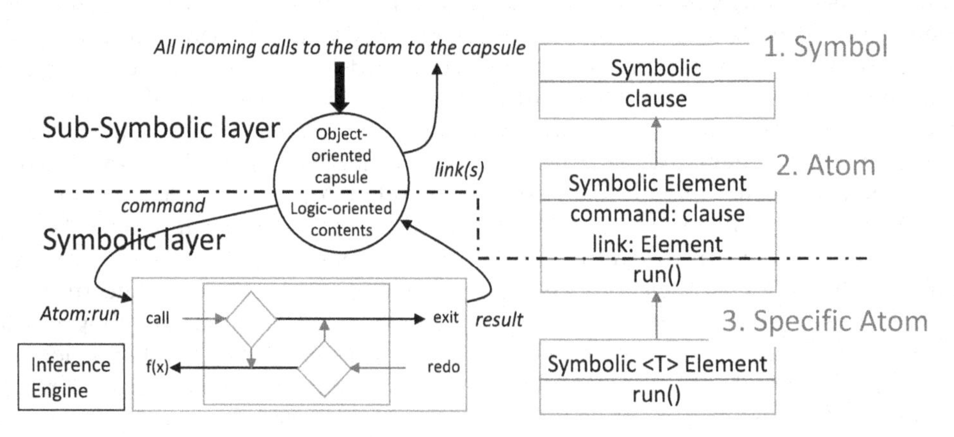

FIGURE 5.17
Symbolic intelligence diagram.

may also support abductive inference, which combines general knowledge and specific beliefs to hypothesize other specific beliefs, as occurs in medical diagnosis. **Inductive reasoning** is a method of reasoning in which a body of observations is considered to derive a general principle. It consists of making broad generalizations based on specific observations. Inductive reasoning is distinct from deductive reasoning. If the premises are correct, the conclusion of a deductive argument is *certain*; in contrast, the truth of the conclusion of an inductive argument is probable, based on the evidence given.

Polyscheme is a cognitive architecture needed to build systems that combine several different representation and inference schemes when AI "thinks". It is based on three principles. First, different aspects of situations that intelligent systems must deal with are best modeled with different schemes for representing knowledge and making inferences. Cassimatis' Polyscheme includes several "specialists" such that each models a particular aspect of the world with its own (possibly unique) representation and inference techniques. The specialists must communicate with other specialists frequently so that each specialist uses the most complete, accurate, and relevant information when it deals with a situation. Thus, specialists in Polyscheme communicate and combine information by simultaneously concentrating on the same focus of attention. And finally, because information about some aspects of a situation is more important than information about another (and because the order that specialists focus on those aspects is important), a system of focused specialists must have mechanisms that decide where to focus. Polyscheme's specialists can guide the focus of attention and thereby implement **inference** schemes using an "attraction" mechanism to specify the preferred foci. This cognitive architecture enables multiple inference techniques to be integrated when dealing with a particular situation because each inference technique can be implemented with one or more focus schemes. Cassimatis described how to implement several important inference techniques (e.g., script matching, backtracking search, reason maintenance, stochastic simulation, and counterfactual reasoning) as focus schemes.

Traditionally higher cognitive functions, such as thinking, reasoning, planning, problem-solving, or linguistic competencies have been the focus of AI, relying on symbolic problem-solving to build complex knowledge structures. These functions involve sequential search processes, while lower cognitive functions, such as perception, motor control, sensorimotor actions, associative memory recall, or categorization, are accomplished on a faster time scale in a parallel way, without stepwise deliberation. Perhaps the biggest mismatch between AI reality and popular expectations is in the language-related domains, for example in general-purpose conversational systems, developed mostly in the form of various chatterbots by commercial companies and enthusiastic individuals. Another area that poses remarkable challenge to AI is word games, and in particular the 20-question game. Word games require extensive knowledge about objects and their properties, but not about complex relations between objects. Recall that cognitive architectures are intended to support general intelligent behavior. Thus, generality is a key dimension along which to evaluate a candidate framework. We can measure an architecture's generality by using it to construct intelligent systems that are designed for a diverse set of tasks and environments, then testing its behavior in those domains. The more environments in which the architecture supports intelligent behavior, and the broader the range of those environments, the greater its generality (Figure 5.18).

Humans can predict what happens next, or what happened before, by combining observed facts with their knowledge of the world. As stated previously, one of the most fundamental properties of thought is its power of predicting events. The ability to draw a conclusion with reasoning, henceforth, inference, is one of the crucial components of future intelligent robots. Thinking is the process of using one's mind to consider or reason about something. Thought (or thinking) encompasses a flow of ideas and associations that can lead to logical conclusions. Although thinking is an activity of an existential value for humans, there is still no consensus as to how it is adequately defined or understood. Because thought underlies many human actions and interactions, understanding its physical and metaphysical origins and its effects has been a longstanding goal of many academic disciplines including philosophy, linguistics, psychology, neuroscience, AI, biology, sociology, and cognitive science.

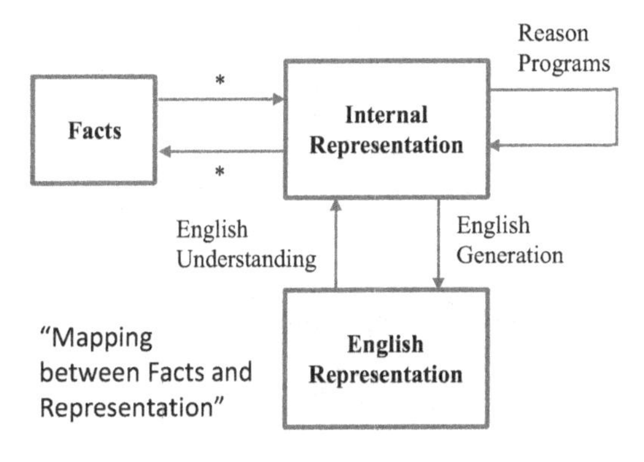

FIGURE 5.18
Facts and representation mapping.

Thinking allows humans to make sense of, interpret, represent, or model the world they experience, and to make predictions about that world. It is therefore helpful to an organism or robot with needs, objectives, and desires as it makes plans or otherwise attempts to accomplish those goals. A reasoning system is a software system that generates conclusions from available knowledge using logical techniques such as deduction and induction. Deductive reasoning is a type of propositional logic in AI, and it requires various rules and facts. It is sometimes referred to as top-down reasoning, and contradictory to inductive reasoning. In deductive reasoning, the truth of the premises guarantees the truth of the conclusion. Reasoning systems have a wide field of application that includes scheduling, rule processing, problem-solving, complex event processing, intrusion detection, predictive analytics, robotics, computer vision, and natural language processing.

The first reasoning systems were theorem provers, systems that represent axioms and statements in First-Order Logic and then use rules of logic such as modus ponens to infer new statements. Another early type of reasoning system was general problem-solvers. These were systems such as the General Problem Solver designed by Newell and Simon. General problem-solvers attempted to provide a generic planning engine that could represent and solve structured problems. They worked by decomposing problems into smaller more manageable sub-problems, solving each sub-problem, and assembling the partial answers into one final answer. In practice, these theorem provers and general problem-solvers were seldom useful for practical applications and required specialized users with knowledge of logic to utilize. The first practical application of automated reasoning (AR) was expert systems. Expert systems focused on much more well-defined domains than general problem-solving such as medical diagnosis or analyzing faults in an aircraft. Expert systems also focused on more limited implementations of logic. Rather than attempting to implement the full range of logical expressions, they typically focused on modus ponens implemented via IF-THEN rules. Focusing on a specific domain and allowing only a restricted subset of logic improved the performance of such systems so that they were practical for use in the real world and not merely as research demonstrations as most previous AR systems had been (Figure 5.19).

The term reasoning system can be used to apply to just about any kind of sophisticated decision support system needed in robotics. However, the most common use of the term reasoning system implies the computer representation of logic. Various implementations demonstrate significant variation in terms of systems of logic and formality. Most reasoning systems implement variations of propositional and symbolic (predicate) logic. These variations may be mathematically precise representations of formal logic systems (e.g., FOL), or extended and hybrid versions of those systems (e.g., Courteous logic). Reasoning systems may explicitly implement additional logic types (e.g., modal, deontic, and temporal logics). However, many reasoning systems implement imprecise and semi-formal approximations to recognized logic systems. Many reasoning systems

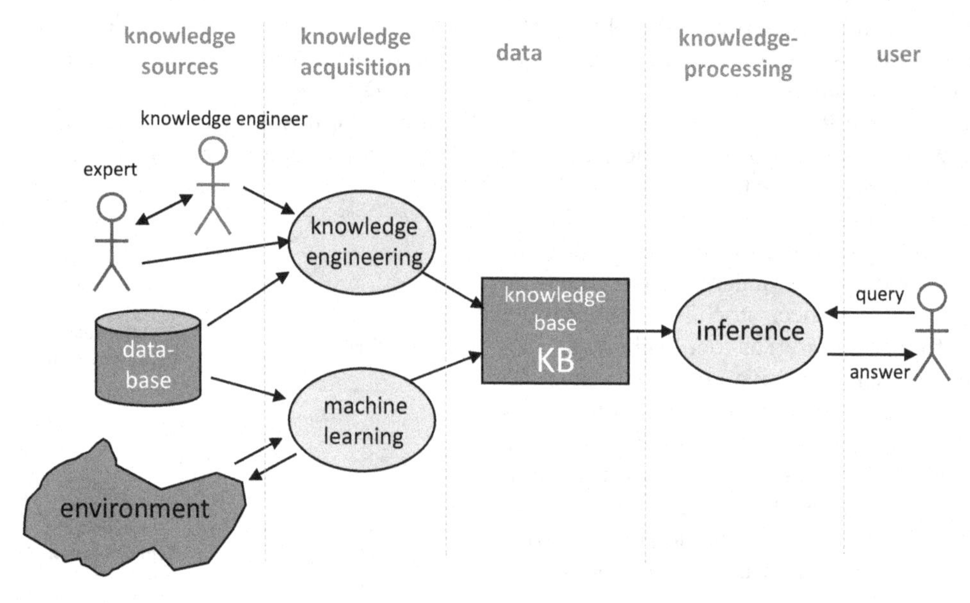

FIGURE 5.19
Reasoning system.

employ deductive reasoning to draw inferences from available knowledge. These inference engines support forward reasoning or backward reasoning to infer conclusions via modus ponens. The recursive reasoning methods they employ are termed "forward chaining" and "backward chaining", respectively. Although reasoning systems widely support deductive inference, some systems employ abductive, inductive, defeasible, and other types of reasoning. Heuristics may also be employed to determine acceptable solutions to intractable problems.

Search → **Logic** → Reasoning → Inference → Decision-Making
Search → **Logic** → Reasoning → Situational Awareness
Search → **Logic** → Reasoning

(Note: "→" literally means "moving" forward)

Reasoning: The theory of approximate reasoning is concerned with the deduction of possibly imprecise conclusions from a set of imprecise premises based on fuzzy logic in which the truth-values are linguistic, and the rules of inference are approximate rather than exact. Knowledge-based systems, which contain domain-specific knowledge giving them more problem-solving power, are called *Expert Systems*. The industry adopted them on a relatively large scale, but many such projects failed. **Heuristic reasoning is problem-solving**. The first step in devising a formalism for reasoning about knowledge is to decide what general properties of knowledge we want that formalism to capture. Production systems are similar to rule-based Expert Systems because they both use rules. Production system is a special case of rule-based systems where production (*Condition → Action*) can be considered as *if Condition then Action*. The difference is Production systems implement graph search with either goal-driven or data-driven strategy, while rule-based Expert Systems implement logic reasoning. Reasoning is based on "Search", but combinatorial explosions will be an issue based on the lack of computing power until Moore's Law reaches a minimum threshold delivering adequate computability. Planning finds a sequence of actions to accomplish some specific task. Planning systems are knowledge intensive and organize pieces of knowledge and partial plans into a solution procedure. Planning may be seen as a state space search. The techniques of graph search may be applied to find a path from the starting state to the goal state. The operators on this path constitute a "plan". Triangle Tables are data structure for organizing the sequence of actions, including potentially incompatible subgoals. Triangle tables relate the preconditions to post-conditions (combined add and delete lists) and are used to determine when a macro operator could be used.

Rule-based systems (production systems) are computer systems that use rules to provide recommendations or diagnoses, or to determine a course of action in a particular situation or to solve a particular problem. A rule-based system consists of a number of components:

- a database of rules (also called a knowledge base)
- a database of facts
- an interpreter, or inference engine

In a rule-based system, the knowledge base consists of a set of rules that represent the knowledge that the system has. The database of facts represents inputs to the system that are used to derive conclusions, or to cause actions. The interpreter, or inference engine, is the part of the system that controls the process of deriving conclusions. It uses the rules and facts and combines them together to draw conclusions. Using deduction to reach a conclusion from a set of antecedents is called forward chaining. An alternative method, backward chaining, starts from a conclusion and tries to show it by following a logical path backward from the conclusion to a set of antecedents that are in the database of facts.

Opportunistic reasoning combines selected elements of both, data-directed (forward) and goal-directed (backward) reasoning. It is useful when the number of possible inferences is very large, no single line of reasoning is likely to succeed, and the reasoning system must be responsive to new data becoming known. As new data are observed, or become known, new inferences can be drawn; and as new conclusions are drawn, new questions about specific data become relevant. An opportunistic reasoning system can thus set up expectations, which help focus the discrimination of a few data elements from among an otherwise confusing mass of data (refer to big data). Reasoning under uncertainty is essential in problem areas outside of logic and mathematics, in which information is incomplete or erroneous. In practice, reasoning is often a hypothesize-and-test cycle based on **search**.

Automated Reasoning: Model theory may be described as the union of logic and universal algebra where we produce a practical method that allows the derivation of consequences and prove theorems by using computers (also known as Automated Theorem Proving [ATP] or AR). Reasoning is a process of deriving new statements (conclusions) from other statements (premises) by argument. For reasoning to be correct, this process should generally preserve truth. That is, the arguments should be valid. AR neatly separates inference and control, from sequential to parallel organization of inferences. For centuries, mathematical proof has been the hallmark of logical validity. But there is still a social aspect as peers have to be convinced by argument. A proof is a repeatable experiment in persuasion.

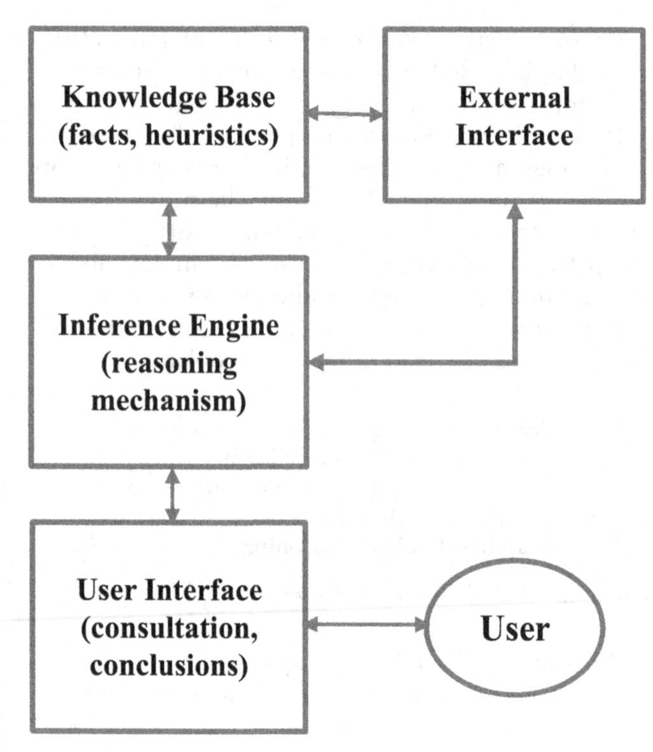

FIGURE 5.20
Expert system reasoning with inference engine.

AR refers to reasoning in a computer using logic and has been an active area of research since the 1950s. AR is used to model human reasoning. First-order logic (also known as predicate logic) can formalize fundamental mathematical concepts. To give semantics to a logical system means to define a notion of truth for the formulas. The concept of truth that we will now define for first-order logic goes back to Tarski. Inference systems Γ (proof calculi) are sets of tuples $(F_1, \ldots, F_n, F_{n+1})$, $n \geq 0$, called inferences or inference rules. Logic requires some representation of the complex knowledge about the world/its environment and uses inference to derive new information from that knowledge combined with new inputs (e.g., via perception) (Figure 5.20).

AR is an application-oriented subfield of logic in computer science and AI. It is about algorithms and their implementation on computers for reasoning with mathematical logic formulas. The reasoning is so simple that any 5-year-old child can understand it, yet most computers can't. Part of the computer's problem has to do with its lack of knowledge about day-to-day social conventions that the 5 years old has learned from her parents, such as don't scratch the furniture and don't injure little brothers. AR considers a variety of logics and reasoning tasks with applications in computer science, such as program verification, dynamic properties of reactive systems, and databases. Conventional logic uses a form of reasoning known as deduction. Programmers know how to get a computer to perform deduction, because the mathematics involved is well understood. But if you want a computer to perform the conjectural, but usually correct, common sense reasoning upon which human survival depends, you must invent a whole new kind of mathematical logic. Applications in logic-based AI are used for mathematical theorem proving, planning, diagnosis, knowledge representation (description logics), logic programming, and constraint solving. The modern notion of symbolic proof was developed in the late 19th and 20th centuries by logicians and mathematicians such as Bertrand Russell, Gottlob Frege, David Hilbert, Kurt Gödel, Alfred Tarski, and Julia Robinson. The benefit of formal logic is that it is based on a pure syntax: a precisely defined symbolic language with procedures for transforming symbolic statements into other statements, based solely on their form. No intuition or interpretation is needed, merely applications of agreed-upon rules to a set of agreed-upon formulae.

ATP concerns the development and use of systems that automate sound reasoning, or the derivation of conclusions that follow inevitably from facts. ATP (also referred to as Automated Deduction) is a subfield of AR, concerned with the development and use of systems that automate sound reasoning: the derivation of conclusions that follow inevitably from facts. ATP systems are at the heart of many computational tasks and are used commercially, for example, for integrated circuit design and computer program verification. ATP problems are typically solved by showing that a conjecture is or is not a logical consequence of a set of axioms. ATP problems are encoded in a chosen logic, and an ATP system for that logic is used to (attempt to) solve the problem. A key concern of ATP research is the development of more powerful systems, capable of solving more difficult problems within the same resource limits. ATP systems are at the heart of many computational tasks, including software verification.

Rule-based reasoning (RBR) is the most important type of legal reasoning. In RBR, you take a rule (a statute or a case holding) and apply it to a set of facts. (This is a type of deductive reasoning.) In conventional RBR, both common sense knowledge and domain-specific domain expertise are represented in the forms of plausible rules (e.g., IF<precondition (s)>THEN<conclusion (s)>). Rules are knowledge representations about which "patterns of information experts use to make decisions and what are the decisions that follow". RBR offers a set of rules that chain to a given conclusion. The reasoning architecture of rule-based systems has two major components: the knowledge base (usually consisting of a set of "IF... THEN..." rules representing domain knowledge) and the inference engine (usually containing some domain-independent inference mechanisms, such as forward ~ backward chaining). The literature suggests that rule-based systems are best suited for problem-solving when the system being analyzed is a single-purpose, specialized system and the rules for solving the problems are clear and do not change with high frequency.

The learning rules that contain variables will learn first-order Horn theories as opposed to learning sets of propositional (i.e., variable free) rules. Inductive learning of first-order rules is also referred to as inductive logic programming (ILP), because this process can be viewed as automatically inferring PROLOG1 programs from examples. Inductive logic refers to a system of inference that describes the relationship between propositions on data and propositions that extend beyond data. A variety of algorithms has been proposed for learning first-order rules. A typical example is FOIL, which is an extension of the sequential covering algorithms to first-order representations. Another approach to ILP is inverse deduction, which is based upon the simple observation that induction is just the inverse of deduction.

Rudolf Carnap was a German philosopher and an advocate of logical positivism, this theory asserts that only statements verifiable through direct observation or logical proof are meaningful in terms of conveying truth value, information, or factual content. He defined Probability-1 as a logical concept, a certain logical relation between two sentences (or, alternatively, between two propositions). It is the same as the concept of the degree of confirmation. Probability-2 is an empirical concept; it is the relative frequency in the long run of one property with respect to another. These two theories deal with two different probability concepts which are both of great importance for science. Therefore, the theories are not incompatible but rather supplement each other. In a certain sense, we might regard deductive logic as the theory of L-implication (logical implication, entailment). And inductive logic may be construed as the theory of Degree of Confirmation, which is, so to speak, partial L-implication. "e L-implies h" says that h is implicitly given with e, in other words, that the whole logical content of h is contained in e. On the other hand, "$c(h, e) = 3/4$" says that h is not entirely given with e but that the assumption of h is supported to the degree 3/4 by the observational evidence expressed in e.

Case-based reasoning (CBR) means using old experiences to understand and solve new problems. A reasoner remembers a previous situation similar to the current one and uses that to solve the new problem. CBR can mean adapting old solutions to meet new demands; using old cases to explain new situations; using old cases to critique new solutions; or reasoning from precedents to interpret a new situation or create an equitable solution to a new problem). Model-based diagnosis is an approach to diagnosis that was proposed in the early 80s to overcome limitations of the traditional expert systems approach. Diagnosis should be based on an objective model of the device (system) to be diagnosed. More specifically, different types of models can be considered: structural (concerning the physical or logical structure of a device), functional (describing the functions of a device), behavioral (describing how a device works, i.e., how its functions are achieved), teleological (describing the purposes of the use of a device), or a combination of them. To understand a new problem in terms of old experiences, CBR has two parts: recalling old experiences and interpreting the new situation in terms of the recalled experiences (Figure 5.21).

In diagnosis, a problem solver is given a set of symptoms and asked to explain them. When there are a small number of possible explanations, one can view diagnosis as a classification problem. When the set of explanations cannot be enumerated easily, we can view diagnosis as the problem of creating an explanation. A case-based diagnostician can use cases to suggest explanations for symptoms and to warn of explanations that have been found to be inappropriate in the past. To understand how CBR works, one must understand what a case is to a computer. In the simplest form, a case is a list of features that lead to a particular outcome. In its most complex form, a case is a connected set of sub-cases that form

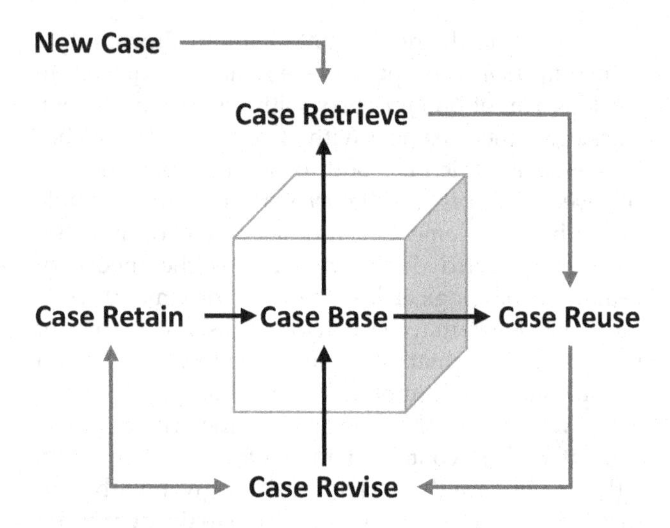

FIGURE 5.21
Case-based diagnostics.

the problem-solving task's structure, for example, the computer chip on a computer motherboard. The designs of the computer chip and the computer are made up of sub-designs, each of which could be considered a case unto itself. Taking advantage of existing techniques for extracting useful information from examples lets case-based systems avoid some of the main problems of rule-based approaches in gathering problem-solving or classification knowledge and putting it to good use. The inductive-indexing capabilities in CBR systems provide several major advantages over pattern recognition techniques. In problem-solving CBR, old solutions are used as inspiration for solving new problems. Since new situations rarely match old ones exactly, however, old solutions must be fixed to fit new situations. In this step, called adaptation, the ballpark solution is adapted to fit the new situation. There are two major steps involved in adaptation: figuring out what needs to be adapted and doing the adaptation. Inductive systems can represent and learn from a wider range of feature types. The ability to use richer feature sets for describing examples makes them at least as accurate and many times more precise.

Roger Schank is widely held to be the originator of CBR. He proposed a different view on model-based reasoning inspired by human reasoning and memory organization. He suggested that our knowledge about the world is mainly organized as memory packets holding together particular episodes from our lives that were significant enough to remember. These memory organization packets (MOPs) and their elements are not isolated but interconnected by our expectations as to the normal progress of events (called scripts). Using these observations about human reasoning process, Schank proposed memory-based reasoning model and memory-based expert systems, which are characterized as follows:

- The utilized knowledge base is derived primarily from enumeration of specific cases or experiences. This is founded upon the observation that human experts are much more capable of recalling experiences than of articulating internal rules.
- As problems are presented to a memory-based expert system to which no specific case or rule can match exactly, the system can reason from more general similarities to come up with an answer. This is founded upon the generalization power of human reasoning.

However, there are limits to CBR technology. The most important limitations relate to how cases are efficiently represented, how indexes are created, and how individual cases are generalized. This type of reasoning involves solving new problems by identifying and adapting similar problems stored in a library of past experiences/problems. The reasoning architecture of CBR consists of a case library (stored representations of previous experiences/problems solved) and an inference cycle. The important steps in the inference cycle of CBR are to find and retrieve cases from the case library that are most relevant to the problem at hand (input) and adapt the retrieved cases to the current input.

One of the key differences between rule-based and case-based knowledge engineering is that automatic case-indexing techniques drastically reduce the need to extract and structure specific rule-like knowledge from an expert, the most time-consuming part of rule-based knowledge engineering. RBR and CBR have emerged as two important and complementary reasoning methodologies in AI. For problem-solving in complex, real-world situations, it is useful to integrate RBR with CBR. It is shown that the integration of CBR and RBR is possible without altering the inference engine of the RBR.

In contrast to learning methods that construct a general, explicit description of the target function when training examples are provided, instance-based learning methods simply store the training examples. Generalizing beyond these examples is postponed until a new instance must be classified: given a new instance, its relations to the already stored examples are examined in order to assign a target function value (the classification) for the new instance. Due to this property, instance-based learning methods are also called lazy learning methods. A key advantage of instance-based learning as a

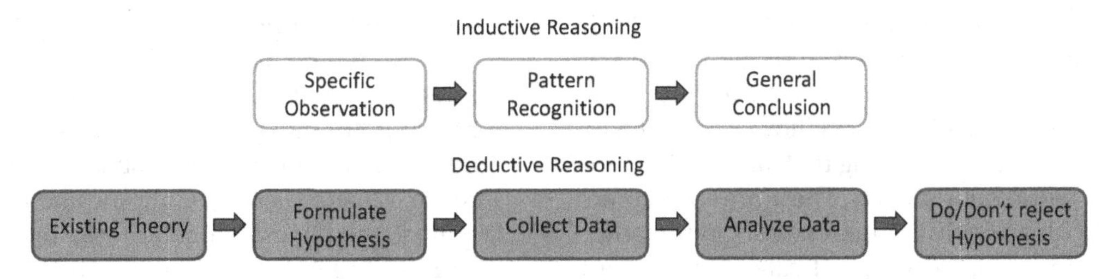

FIGURE 5.22
Inductive vs deductive reasoning.

delayed, or lazy, learning method is that instead of estimating the target function once for the entire instance space, these methods can estimate it locally and differently for each new instance to be classified. Yet, these methods are at a disadvantage because of their computation and memory/storage requirements (Figure 5.22).

Inductive Reasoning: So far, machine learning has been viewed as a viable way of avoiding the knowledge bottleneck problem in developing knowledge-based systems. For deduction such as with First-Order Logic, it can be proven that it is sound if the premises are true, and so will be the conclusion. Deductive reasoning involves using general premises to form a specific conclusion, while inductive reasoning involves starting from specific premises and forming a general conclusion. The Problem of Induction is can we justify induction; to show that the truth of the premise supported, if it did not entail, the truth of the conclusion. Induction is defined by the process of inferring a general law or principle from the observation of particular instances, in contrast with deduction where we (may) apply general laws to specific instances. Thus, Inductive Learning, also known as Concept Learning, is how AI systems attempt to use a generalized rule to carry out observations. Inductive Learning Algorithms are used to generate a set of classification rules. These generated rules are in the "If this, then that" format. These rules determine the state of an entity at each iteration step in learning and how the learning can be effectively changed by adding more rules to the existing ruleset. When the output and examples of the function are fed into the AI system, Inductive Learning attempts to learn the function for new data. When applied to robotics, robotic induction automates a repetitive manual process to reduce dependency on manual labor and increase efficiency.

Inference

is the process of deriving a specific sentence from a KB (where the sentence must be entailed by the KB). It is the process of chaining multiple rules together based on available data to cover numerous conditions, since expert knowledge cannot be represented in single rule. To make an inference is to reason about probability. It is also a method to draw conclusions about "something" from data. Forward Chaining can be used for planning. For each new piece of data, generate all new facts, until the desired fact is generated. Data-directed is reasoning. Forward Chaining is data driven with automatic, unconscious processing, for example, object recognition with routine decisions. Unfortunately, this process may do lots of work that is irrelevant to the goal. Backward Chaining is used for diagnostics. To prove the goal, this technique finds a clause that contains the goal as its head and proves the body recursively (or backtracking when you chose the wrong clause). Backward Chaining is goal-driven and goal-directed reasoning. It is appropriate for problem-solving, for example, "Where are my keys?" or "How do I start the car?" Resolution is an inference rule (with many variants) that takes two or more parent clauses and soundly infers new clauses. A special case of resolution is when the parent causes are contradictory, and an empty clause is inferred. Resolution is a general form of modus ponens (Figure 5.23).

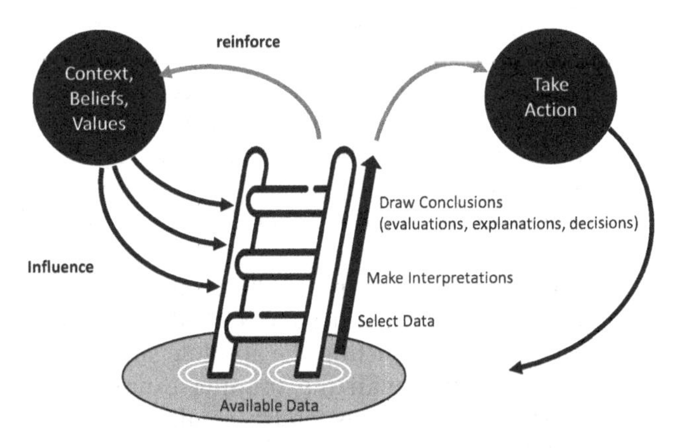

FIGURE 5.23
Inference ladder.

Reasoning in Knowledge-Based Systems uses both shallow reasoning and deep reasoning with forward and backward chaining. An alternative inference method uses metaknowledge. Shallow reasoning is also called experiential reasoning that aims to describe aspects of the world heuristically. These are short inference chains with possibly complex rules. On the other hand, Deep reasoning (also called causal reasoning) aims to building a model of the world that behaves like the "real thing". These are long inference chains with often simple rules that describe cause-and-effect relationships. There is inference by model checking. We enumerate all the KB models and check if α_1 and α_2 are true in all the models (which implies that we can only use them when we have a finite number of models). Statistical Inference learns from data.

Decision Tree Learning: A **decision tree** is one of the simplest models. In its most basic form, all the data is broken down in such a way that it can be placed in the tree. One starts at the top and at each level selects a branch based on a particular feature's value. One continues right to the base of the tree, where the final outcome – the decision – is found. A decision tree is a useful machine-learning algorithm used for both regression and classification tasks. The name "decision tree" comes from the fact that the algorithm keeps dividing the dataset down into smaller and smaller portions until the data has been divided into single instances, which are then classified. If you were to visualize the results of the algorithm, the way the categories are divided would resemble a tree and many leaves. Decision trees divide the feature space into axis parallel (hyper-) rectangles. Each rectangular region is labeled with one label or a probability distribution over labels. Decision trees can represent any Boolean function of the input attributes. In the worst case, the tree will require exponentially many nodes. Decision trees have a variable-sized hypothesis space. As the number of nodes (or depth) increases, the hypothesis space grows. There is a preference bias: the simplest consistent explanation is the best. Therefore, the smallest decision tree that correctly classifies all of the training examples is best. Finding the provably smallest decision tree is NP-hard, so instead of constructing the absolute smallest tree consistent with the training examples, construct one that is pretty small. In 1952 MIT student David Huffman devised, in the course of doing a homework assignment, an elegant coding scheme that is optimal in the case where all symbols' probabilities are integral powers of ½. A Huffman code can be built in the following manner: (1) rank all symbols in order of probability of occurrence, (2) successively combine the two symbols of the lowest probability to form a new composite symbol; eventually we will build a binary tree where each node is the probability of all nodes beneath it, and (3) trace a path to each leaf, noticing direction at each node. We want to determine which attribute in a given set of training feature vectors is most useful for discriminating between the classes to be learned. Information gain tells us how important a given attribute of the feature vectors is. We will use it to decide the ordering of attributes in the nodes of a decision tree. Use information gain to build decision tree. Many case studies have shown that decision trees are at least as accurate as human experts.

Human Associative Memory (HAM): This model was developed by John Anderson and Gordon Bower (1973) [935]. It is a model of memory. This memory is organized as a network of propositional binary trees. When an informant asserts a statement to HAM, the system parses the sentence and builds a binary tree representation. As HAM is informed of new sentences, they are parsed and formed into new tree-like structures with existing ones. When HAM is posed with a query it is formed into a tree structure called a probe. This structure is then matched against memory structures for the best match. The structure with the closest match is used to formulate an answer to the query. Matching is accomplished by first locating the leaf nodes in memory that match leaf nodes in the probe. The corresponding links are then checked to see if they have the same labels and in the same order. The search process is constrained by searching only node groups that have the same relation links. Access to nodes in HAM is accomplished through word indexing in LISP.

Learning decision tree is a tree-structured plan of a set of attributes to test in order to predict the output. To decide which attribute should be tested first, simply find the one with the highest information gain.

- Then recurse. Take the Original Dataset, and partition it according to the value of the attribute. Build tree from these records. Base case if all records in current data subset have the same output then don't recurse. Base case if all records have exactly the same set of input attributes then don't recurse.

To build, if all output values are the same in *DataSet*, return a leaf node that says "predict this unique output":

- If all input values are the same, return a leaf node that says "predict the majority output"
- Else find attribute X with the highest Info Gain
- Suppose X has n_X distinct values (i.e., X has arity n_X).
  - Create and return a non-leaf node with n_X children.
  - The ith child should be built by calling.

BuildTree (DS_i,*Output*) where DS_i built consists of all those records in DataSet for which X=ith distinct value of X. For a mobile robot, integrating Soar has demonstrated autonomous navigation and SLAM technologies. **SOAR** (stands for **S**tate, **O**perator, **A**nd **R**esult; however, it is no longer regarded as an acronym) is a cognitive architecture, originally created by John Laird, Allen Newell, and Paul Rosenbloom at Carnegie Mellon University. The goal of the Soar project is to develop the fixed computational building blocks necessary for general intelligent agents; agents that can perform a wide range of tasks and encode, use, and learn all types of knowledge to realize the full range of **cognitive capabilities** found in humans, such as **decision-making**, **problem-solving**, **planning**, and **natural language understanding**. It is both a theory of what cognition is and a computational implementation of that theory. Soar embodies multiple hypotheses about the computational structures underlying general intelligence, many of which are shared with other cognitive architectures, such as the ACT-R by John R. Anderson. The original theory of cognition underlying Soar is the **Problem Space Hypothesis**, which is described in Allen Newell's book (**Unified Theories of Cognition**) and dates back to one of the first AI systems created by Newell, Herb Simon, and J. Cliff Shaw in the **Logic Theorist**, first presented in 1955. Soar is an implementation of the Problem Space Computational Model. There is knowledge search from multiple sources. The Problem Space Hypothesis contends that all goal-oriented behavior can be cast as **search** through a space of possible states (a *problem space*) while attempting to achieve a goal. At each step, a single operator is selected and then applied to the agent's current state, which can lead to internal changes, such as retrieval of knowledge from long-term memory or modifications or external actions in the world. Inherent to the Problem Space Hypothesis is that all behavior, even a complex activity such as planning, is decomposable into a sequence of selection and the application of primitive operators, which when mapped onto human behavior, takes ~ 50 ms (Figure 5.24).

Problem-solving in **SOAR** equates to search through a state space. Chunking, a simple uniform learning mechanism, coverts a solution from weak methods (e.g., hill climbing and means end analysis) into a rule.

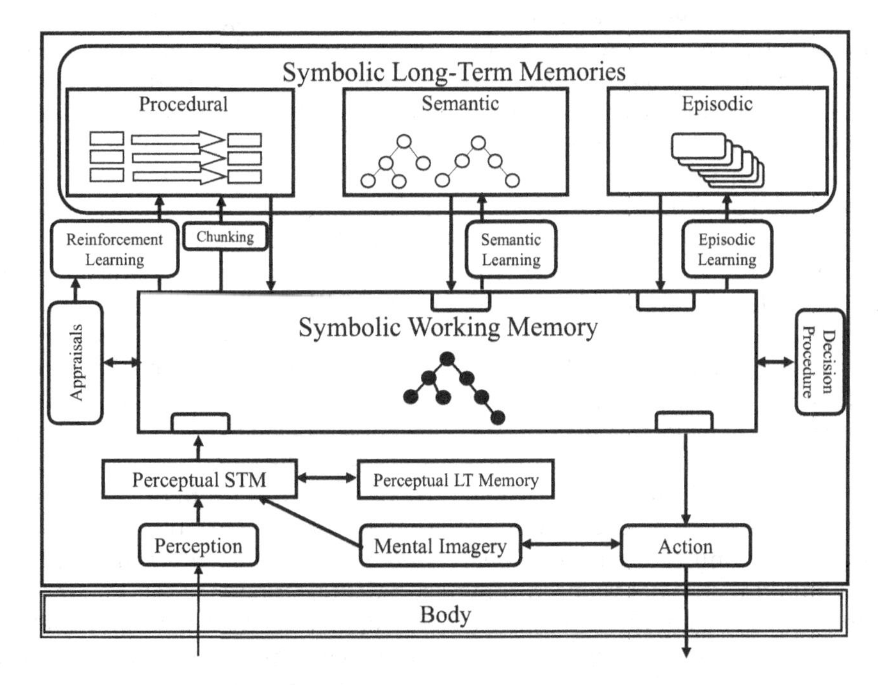

FIGURE 5.24
SOAR cognitive architecture.

General intelligent entities are distinguished by their ability to pursue a wide variety of goals embedded in many different problem spaces and to use large bodies of different types of knowledge in many ways: to assess the current situation in the environment, to react to changes in the environment, to deliberately select actions in order to pursue goals, to plan future actions, to predict future states, to reflect on past behavior in order to improve future performance, and to adapt to regularities in the environment, all in real time.

A Markov network or MRF (Markov Random Field) is a model that represents the relationship between a set of variables by using their joint distributions. It is similar to a Bayesian network in its representation of dependencies; however, the differences are that Bayesian networks are directed and acyclic, whereas Markov networks are undirected and may be cyclic. Approaches such as Markov Decision Processes (MDPs) work well for specific tasks where there are limited and predictable numbers of features; however, they do not scale to complex behavior and planning; nor do they address how an agent efficiently manages its memory of situations, events, and the structured knowledge it acquires through experience. SOAR interfaces to a variety of robotic simulators and a simple mobile robot. The SOAR team has made significant extensions to SOAR that add new memories and new non-symbolic reasoning to SOAR's original symbolic processing, which improves SOAR's abilities to control robots. These extensions include mental imagery, episodic and semantic memory, reinforcement learning, and continuous model learning. A small mobile robot was used with the SOAR cognitive architecture to create a flexible but adaptive cognitive robotic agent. It incorporates both symbolic and non-symbolic processing and has multiple learning mechanisms. In addition, one obvious place for future work is the integration of the hardware with the software. Another place for future research is to push further on the interaction between low-level perception and high-level cognition. SVS (Spatial Visual System) provides one level of processing where these come together, but as of yet, this soar integration does not have a general theory (or implementation) of adaptive low-level perception that translates noisy pixels into object descriptions, categories, features, and spatial relations.

Decision-making and language understanding using abstracted concepts are verified using a real robot. The mechanism behind high-level cognitive functions, such as action planning, language understanding, and logical thinking, has not yet been fully implemented in robotics. Intelligent robots require a framework for the simultaneous comprehension of concepts, actions, and language as a first step toward this goal. This can be achieved by integrating various cognitive modules and leveraging mainly multimodal categorization by using multilayered multimodal latent Dirichlet allocation (mMLDA). The integration of reinforcement learning and mMLDA enables actions based on deeper understanding. Furthermore, the mMLDA, in conjunction with grammar learning and based on the Bayesian hidden Markov model (BHMM), allows the robot to verbalize its own actions and understand user utterances.

However, the realization of flexible and versatile intelligence, such as that evident in humans, remains a difficult problem. In particular, it is fair to say that robots that can use language and appropriately plan and perform various actions are not yet a reality. Several cognitive functionalities, such as perception, language, and decision-making, are intertwined in a complex manner to realize such intelligence. Based on such a premise, the research questions seem to be two-fold; how do individual cognitive models relate each to other, and how do they develop each other? In other words, the appropriate connection of all modules should be studied, observing how each module learns in the entire structure. One of the promising research directions toward AGI is the use of generative probabilistic modeling. In fact, some studies have shown the usefulness of the probabilistic models to acquire knowledge by self-organizing multimodal information that the robot obtains through its own experience. Because such knowledge is abstracted and linked to language, it can be reused for various tasks.

One option for an integrated intelligent system is to examine the acquisition of multiple cognitive functions such as concept formation, decision-making, and language learning by robots through the integration of multiple cognitive modules, centered on mMLDA, as well as to verify the effect on integration of multiple modules. The robot observes multimodal information through various sensors by acting in a real environment. Learning can be applied to the various modules in the form of language learning, policy learning, and action learning. If we understand the underlying principles and mechanisms of the development of natural cognition in human babies through social interaction, we can use this knowledge to inform the design of cognitive capabilities in artificial agents such as robots. Such principles and mechanisms can be implemented in the cognitive architecture of robots and tested through developmental experiments with robots. This is the aim of developmental robotics and will explore the current achievements and challenges in the design of autonomous mental development via social interaction in robots and the benefit of a mutual interaction between developmental psychologists and developmental roboticists. The next figure is another example of a process model for the mind, interfacing with its environment and sandwiched between sensing/perception and action (Figure 5.25).

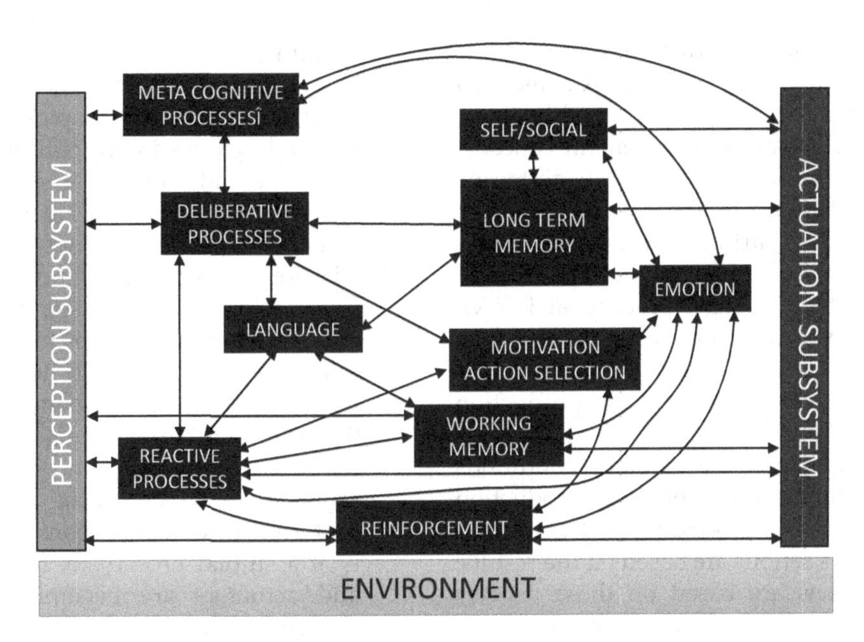

FIGURE 5.25
High-level architecture of the mind.

Human cognitive architecture can be summarized as follows: we have a limited working memory that deals with all conscious activities and an effectively unlimited long-term memory that can be used to store schemas of varying degrees of automaticity. Intellectual skill comes from the construction of large numbers of increasingly sophisticated schemas with high degrees of automaticity. Schemas bring together multiple elements that can be treated as a single element and allow us to ignore myriads of irrelevant elements. Working memory capacity is freed, allowing processes to occur that otherwise would overburden working memory. Automated schemas both allow fluid performance on familiar aspects of tasks and by freeing working memory capacity, permitting levels of performance on unfamiliar aspects that otherwise might be quite impossible.

Storing Knowledge: Episodic memory holds a history of previous states, while semantic memory contains previously known facts. One common tradition distinguishes declarative from procedural representations. Declarative encodings of knowledge can be manipulated by cognitive mechanisms independent of their content. For instance, a notation for describing devices might support design, diagnosis, and control. First-order logic is a classic example of such a representation. Production rules are a common means of representing procedural knowledge. In general, procedural representations let an agent apply knowledge efficiently, but typically in a flexible manner. Another important set of properties concerns the manner in which a cognitive architecture organizes knowledge in its memory. One choice that arises here is whether knowledge is stored in some "at" scheme or in a more structured manner (Figure 5.26).

Production systems and first-order logic are two examples of frameworks, in that the stored memory elements make no direct reference to each other. In contrast, stored elements in structured frameworks make direct reference to other elements. One such approach involves a task hierarchy, in which one plan or skill calls directly on component tasks, much as in subroutine calls. Similarly, a part of hierarchy

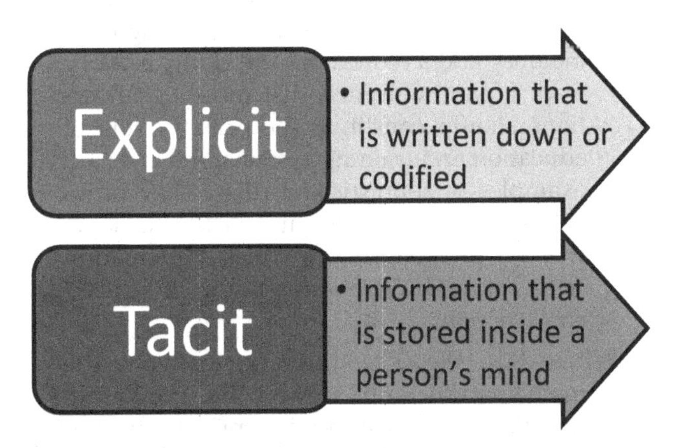

FIGURE 5.26
Two types of knowledge – explicit and tacit.

describes a complex object or situation in terms of its components and relations among them. A somewhat different organization occurs with an is-a hierarchy, in which a category refers to more general concepts (its parents) and more specialized ones (its children). Most architectures commit to either a "at" or structured scheme, but task, part-of, and is-a hierarchies are complementary rather than mutually exclusive. The second type of knowledge, procedural knowledge, is made up of production rules representing procedural skills that manipulate declarative knowledge as well as the environment. Each production rule is essentially a condition-action rule that generates the specified actions if the specified conditions are satisfied. Chunking is represented as a production rule that summarizes the processing that generated the results. A chunk's actions are based on the results, whereas its conditions are based on those aspects of the goals above the subgoal that were relevant to determining the results. Once the agent has learned a chunk, it fires in new situations that are similar along relevant dimensions, often giving the required results directly and thus avoiding the impasse that led to its formation.

Working Memory: Working memory can be equated with consciousness. Working memory is capable of holding only about seven items or elements of information at a time. According to schema theory, knowledge is stored in long-term memory in the form of schemas. A schema categorizes elements of information according to the manner in which they will be used. Schemas are stored in long-term memory. One of their obvious functions is to provide a mechanism for knowledge organization and storage. It is not their only function. Schemas also reduce working memory load. In summary, schema construction has two functions: the storage and organization of information in long-term memory and a reduction of working memory load. It can be argued that these two functions should constitute the primary role of education and training systems. Schemas are examples of sophisticated rules. They probably become automated in exactly the same way as problem-solving rules. Automation is therefore an important factor in schema construction.

Where Are We: The difference between the mathematical mind {esprit de geometrie} and the perceptive mind {esprit de finesse} and the reason that mathematicians are not perceptive is that they do not see what is before them, and that, accustomed to the exact and plain principles of mathematics, and not reasoning till they have well inspected and arranged their principles, they are lost in matters of perception where the principles do not allow for such arrangement. Distributed robotics is now where robotics was in the 1970s; algorithms are developed in an architecture-dependent way on a task-by-task basis, with no formal foundations. In the early 1980s, the realization that uncertainty was the fundamental obstacle in robot control led to the invention of configuration space, which caused a qualitative leap in the advancement of robot science. Robotics is at a similar crossroads today. As sensors and actuators are becoming smaller and cheaper, teams of robots working together are becoming more pervasive. Cooperative robotics has the potential of expanding greatly the application domains of robots but remains a largely unexplored field. In addition to all the challenges of single robot systems, cooperative robotics has the added difficulties of cooperation and communication, and of combining discrete and continuous systems. This has created a great opportunity for the development of a mathematical basis for distributed robotics that is grounded in engineering issues. In computer science, one often learns a lot about the structure of an algorithmic problem by parallelizing it; a similar methodology may be useful in robotics (Figure 5.27).

So where are we with robotics and AI. The main idea of converging AI and Robotics is to try to optimize its level of autonomy through learning. This level of intelligence can be measured as the capacity of predicting the future, either in planning a task, or in interacting (either by manipulating or by navigating) with the world. Robots with intelligence have been attempted many times. System AI models are in chained form; however, feasible robot trajectories are in closed-form parameterization. Steering control is closed-form, piecewise constant solution (polynomial steering) and collision avoidance is explicitly condition based on its geometry and time. Robots make decisions in intelligent environments. The objectives for decision-making are to optimize inhabitant productivity, minimize operating costs, and maximize inhabitant comfort if applicable. The decision process has to be safe. Decisions made can never endanger inhabitants or cause damage, and decisions should be within the range accepted by the inhabitants. In the

FIGURE 5.27
Nao robot.

connectionist (Subsymbolic) Hypothesis, Smolensky stated "The intuitive processor is a sub conceptual connectionist dynamical system that does not admit a complete, formal and precise conceptual-level description". The inner workings of an ANN are difficult to interpret, but are they substantially different to a symbolic system? Newell and Simon proclaimed "A physical symbol system has the necessary and sufficient means for intelligent action" in 1976. But basically, the Physical Symbol System Hypothesis states:

- a system, embodied physically, that is engaged in the manipulation of symbols
- an entity is potentially intelligent if and only if it instantiates a physical symbol system
- symbols must *designate*
- symbols must be *atomic*
- symbols may combine to form *expressions.*

Complex tasks are often hard to learn since they involve long sequences of actions that must be correct for reward to be obtained. There is a theory of scaling up; there is a theory where the intelligent robot can learn complex tasks from simpler tasks. Complex tasks can be learned as shorter sequences of simpler tasks such as:

- Control strategies that are expressed in terms of subgoals are more compact and simpler
- Fewer conditions must be considered if simpler tasks are already solved
- New tasks can be learned faster
- Hierarchical Reinforcement Learning with learning with abstract actions and acquisition of abstract task knowledge.

The different decision approaches for autonomous robots are (1) Preprogrammed decisions using timer-based automation, (2) Reactive decision-making systems where decisions are based on condition-action rules and are driven by the available facts, (3) Goal-based decision-making systems when decisions are made in order to achieve a particular outcome, and (4) Utility-based decision-making systems are made in order to maximize a given performance measure. The different AI decision-making techniques that support the previously described decision approaches are given by this table (Table 5.4).

Autonomous and intelligent mobile robots and their environments are extremely complex physical systems. Robot workspaces are dynamic and inherently unpredictable, while robotic sensors and actuators have limited capabilities and are prone to noise. In order to reduce the complexity and to provide robustness to noise, adoption of appropriate geometric constrains in combination with effective probabilistic modeling of related information is required. The achieved abstraction not only makes robotic problems computationally feasible but also provides robustness in the presence of noise, or other, often unpredictable, factors. A probabilistic model must include the robot's state, the environment, the robot's actions and observations as encoded by its sensors, and of course must be able to handle noise. A learning algorithm recognizes patterns, generates patterns, recognizes anomalies, and makes predictions (Figure 5.28).

A robot must learn from what it perceives. Its initial configuration reflects prior knowledge of the environment and the robot gains experience so that knowledge of its environment may be modified/augmented. Autonomy compensates for partial or incorrect prior

TABLE 5.4
Table of decision-making approaches

DECISION-MAKING APPROACHES	AI - TECHNIQUES
Reactive Decision Making	Rule-based Expert Systems
Goal-based Decision Making	Planning
Decision-Theoretic Decision Making	Belief Networks
	Markov Decision Process
Learning Techniques	Neural Networks
	Reinforcement Learning

FIGURE 5.28
Stereo camera pair.

knowledge and will start with initial built-in knowledge (a priori) and having the ability to learn new information. The robot will learn to interpret sensor information but recognizing objects in the environment is difficult, sensors provide prohibitively large amounts of data, and programming of all required objects is generally not possible. By learning new strategies and tasks because new tasks must be learned on-line in the laboratory and different inhabitants require new strategies even for existing tasks. This requires the adaptation of existing control policies where user preferences can change dynamically and the changes in the environment have to be reflected. We can formulate two kinds of tasks with particular interest in robotic applications. Localization is an inference task that computes the probability that the robot is at pose z at time t given all observations up to time t (forward recursions only). Map building (learning task) determines the map m that maximizes the probability of the observation sequence. Learning is needed when the information about the environment is incomplete, when the environment is changing, when we use a sequence of percepts to estimate the missing details, and when it is hard for us to articulate the knowledge needed to building AI systems, for example, try writing a program to recognize visual input like various types of flowers.

Herb Simon describes learning as "Learning denotes changes in the system that are adaptive in the sense that they enable the system to do the tasks drawn from the same population more efficiently and more effectively the next time". More generically, he defines learning as: "Learning is any process by which a system improves performance from experience". An adaptive system is a set of interacting or interdependent entities, real or abstract, forming an integrated whole that together can respond to environmental changes or changes in the interacting parts. Feedback loops represent a key feature of adaptive systems, allowing the response to changes; examples of adaptive systems include natural ecosystems, individual organisms, human communities, human organizations, and human families. Some artificial systems can be adaptive as well; for instance, robots employ control systems that utilize feedback loops to sense new conditions in their environment and adapt accordingly. Learning is a hallmark of intelligence; many would argue that a system that cannot learn is not intelligent. Without learning, everything is new; a system that cannot learn is not efficient because it re-derives each solution and repeatedly makes the same mistakes. But why do we believe we have the license to predict the future? The different learning approaches for robot systems are:

- Supervised learning by teaching: Robots can learn from direct feedback from the user that indicates the correct strategy. (The robot learns the exact strategy provided by the user).
- Learning from demonstration (imitation): Robots learn by observing a human or a robot perform the required task. (The robot must be able to "understand" what it observes and map it onto its own capabilities).
- Learning by exploration: Robots can learn autonomously by trying different actions and observing their results. (The robot learns a strategy that optimizes a reward).

The robot can learn sensory patterns. For example, we need to learn to identify objects. So, how can a particular object be recognized? Programming recognition strategies are difficult because we do not fully understand how we perform recognition. Learning techniques permit the robot system to form its own recognition strategy. Supervised learning can be used by giving the robot a set of pictures and the corresponding classification using neural networks and decision trees. An autonomous robots must be able to learn new tasks even without input from the user. One approach is to learn to perform a task in order to optimize the reward the robot obtains (via Reinforcement Learning). Reward must be provided either by the user or the environment using intermittent user feedback and generic rewards indicating unsafe or inconvenient actions or occurrences. The robot has to explore its actions to determine what their effects are, such as actions that change the state of the environment and actions achieve different amounts of reward. During learning, the robot has to maintain

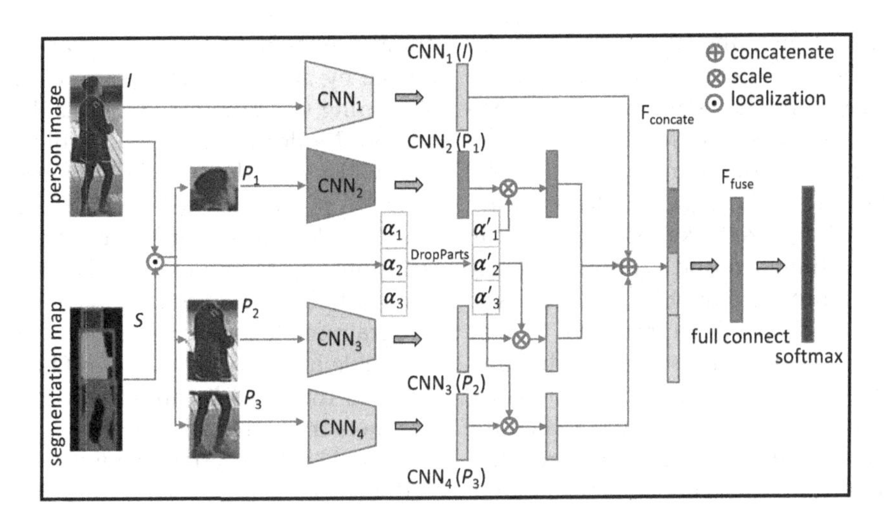

FIGURE 5.29
Sensor learning model.

a level of safety. Learning problems are classified as: (1) Supervised learning – Given a set of input/output pairs, learn to predict the output if faced with a new input, (2) Unsupervised Learning – Learning patterns in the input when no specific output values are supplied, and (3) Reinforcement Learning – Learn to interact with the world from the reinforcement you get (Figure 5.29).

Given a sample set of inputs and corresponding outputs, find a function to express this relationship: Pronunciation=Function from letters to sound, Bowling=Function from target location (or trajectory?) to joint torques, and Diagnosis=Function from lab results to disease categories. Aspects of Function Learning include memory, averaging, and generalization. Learning with explanation (case to rules) are discovery and data mining, while learning with no explanation are neural networks and CBR. What is rationality? The purpose of action is to reach the goal, given the information/knowledge possessed by the robot. The notion of rationality does not necessarily include success of the actions chosen. Everything seems okay so far. Computational constraints can we possibly specify EXACTLY the domain the AI will work in? A look-up table of reactions to percepts is far too big. However, most things that could happen, don't. Rule-based systems use logic languages, knowledge bases, and inference engines. Ability areas for AI include computer vision, Natural language recognition, Natural language generation, Speech recognition, Speech generation, of course Robotics, and Games/Entertainment.

Today's AI-enabled systems based on symbolic logic are "brittle". They can often achieve super-human performance in narrow domains. When pushed outside the boundaries of their programming, however, they can fail, and fail badly. They can go from super smart to super dumb in an instant. Unlike humans, machines cannot flexibly adapt to novel situations. They will do precisely what they are programmed to do. Is Machine Learning a Solution? Learning systems do not need to follow rigid rules. However, machine learning with giant datasets and huge, inscrutable black-box DNNs can lead to some surprises. Robots are inherently uncertain. This uncertainty arises from four major factors:

1. Environment stochastic, unpredictable
2. Robot stochastic
3. Sensor limited, noisy
4. Models are inaccurate.

So, what does all this mean? Norbert Wiener and cybernetics dealt with closed systems that are deterministic. Cybernetics uses a Newtonian Approach for steersmanship (control) and stability using feedback and transformations. The system is singular, stable (invariant), and has complete information. As a result, these systems can utilize an *automatic* processes that are fast using correlations. But that is not the real world. The real world has both structured and unstructured components. At the other extreme of being deterministic is a stochastic or random environment that is an open system that is indeterminate. There is complexity to handle with incomplete information (having uncertainty). This is a slower AI process. This problem requires a statistical approach where there are several possible options versus individual values. The models utilized are plural, meaning there are options. As a result, this can be *autonomous* processes using Markov Chains to handle the uncertainty and AI techniques for decision-making. It's been easier to mechanize many of the high-level cognitive tasks we usually associate with "intelligence" in people, for example, symbolic integration, proving theorems, playing chess, and some aspect of medical diagnosis. However, it's been very hard to mechanize tasks that animals can easily do such as walking around without running into things, catching prey and avoiding predators or interpreting complex sensory information (visual, aural, etc.), modeling the internal states of other animals from their behavior, or working as a team (ants, bees). Is there a fundamental difference between the two categories? Why are some complex problems (e.g., solving differential equations, database operations) not subjects of AI?

Moravec's Paradox is the observation made by AI and robotic researchers that, contrary to traditional assumptions, reasoning requires very little computation, but sensorimotor (sensors actuators) skills require enormous computational resources. What is easy for the ancient, neural "technology" inherent in people but difficult for the modern, digital technology of computers (and vice versa) in this Paradox. Hans Moravec wrote: "It is comparatively easy to make computers exhibit adult level performance on intelligence tests or playing checkers, and difficult or impossible to give them the skills of a one-year-old when it comes to perception and mobility". Deep learning simulates a network resembling the layered neural networks of our brain. Based on large quantities of data, the network learns to recognize patterns and links to a high level of accuracy and then connect them to courses of action without knowing the underlying causal links. This implies that it is difficult to provide deep-learning AI with some kind of transparency in how or why it has made a particular choice by, for example, by expressing an intelligible reasoning (for humans) about its decision process, as we do. Modern machines combine powerful multicore CPUs with dedicated hardware designed to solve parallel processing. GPU and FPGA are the most popular dedicated hardware commonly available in computer systems developing AI systems. FPGA is a reconfigurable digital logic containing an array of programmable logic blocks and a hierarchy of reconfigurable interconnections. An FPGA is not a processor and therefore it cannot run a program stored in the memory (Table 5.5).

Most AI people agree that without parallelism there will ultimately be a barrier to further progress due to the lack of computing power. Commercial processor advancements are motivated by the observation that "Current" computers are extremely weak in basic functions for processing speech, text, graphics, picture images, and other non-numeric data, and for AI type processing such as inference, association, and learning. The centrality of (1) Problem-solving and inference, (2) Knowledge-base management, and (3) Intelligent interfaces requires enormous computing power. Thus, a new type of computing power is required, one more symbolic and inferential in character than conventional systems. Also, the system was explicitly assumed to rely on very large knowledge bases and to provide specialized capabilities for knowledge and data base management. For example, "... A database machine with 100 to 1,000 Gb capacity" and able "... to retrieve the knowledge bases required for answering a question within a few seconds". "The intention of software for the knowledge base management function will be to establish knowledge information processing technology where the targets will

TABLE 5.5
Cybernetics versus Stochastic Systems

CYBERNETICS	STOCHASTIC (RANDOM)
Closed-system	Open-system
Deterministic Systems	Indeterminate Systems
Newtonian Approach	Statistical Approach
Steersmanship & Stability	Decision Making
Feedback & Transformations	Markov Chains
Correlations	Possible options versus individual values
Complete Information	Incomplete Information (having uncertainty)
Singular	Plural
Stable (invariant)	Complexity
Fast processes	Slower processes
Automatic Processes	Autonomous Processes

Moravec's Paradox	
Mind-numbingly Easy	Insanely Hard
• Language Translation	• Walking, Running
• Playing Chess	• Seeing & Recognizing
• Calculus	• Smelling

FIGURE 5.30
Moravec's Paradox.

be development of knowledge representation systems, knowledge base design and maintenance support systems, large-scale knowledge base systems, knowledge acquisition experimental systems, and distributed knowledge management systems" (Figure 5.30).

It is speculated that "the intelligent interface function will have to be capable of handling communication with the computer in natural language, speech, graphics, and picture images so that information can be exchanged in a ways natural to man". Ultimately the system will cover a basic vocabulary (excluding specialist terms) of up to 100,000 words and up to 2,000 grammatical rules, with a 99% accuracy in syntactic analysis. "The object speech inputs will be continuous speech in Japanese standard pronunciation by multiple speakers (from 5th Generation Computer Systems Project), and the aims here will be a vocabulary of 50,000 words, a 95% recognition rate for individual words, and recognition of processing within 3 times the real time of speech". "The system should be capable of storing roughly 10,000 pieces of graphic and image information and utilizing them for knowledge information processing". Despite these new AI solutions, the aforementioned hardware limitations were a big restriction during "learning" training. With recent hardware advances, such as the parallelization using GPU, the cloud computing and the multicore processing finally led to the present stage of AI. In this stage, deep neural nets have made tremendous progress in terms of accuracy and they can now recognize complex images and perform voice translation in real time. However, researchers are still dealing with issues relating to the overfitting of the networks, since large datasets are often required and not always available. For this reason, more sophisticated training procedures have recently been proposed. For example, in 2006, Geoffrey Hinton introduced the idea of unsupervised pretraining and Deep Belief Nets. This approach has each pair of consecutive layers trained separately, using an unsupervised model similar to the one used in the Restricted Boltzmann Machine; then the obtained parameters are frozen, and a new pair of layers are trained and stacked on top of the previous ones. This procedure can be repeated many times leading to the development of a deeper architecture with respect to the traditional neural nets (Figure 5.31).

Visual recognition was largely accepted to be something easy for humans but difficult for machines. But now, with the emergence of deep learning, humans will not be able to claim that as an advantage for much longer. There are tens of thousands of machine-learning algorithms, including hundreds of new ones every year. Every ML algorithm has three components: Representation, Optimization, and Evaluation. There have been some key lessons learned such as learning can be viewed as using direct or indirect experience to approximate a chosen target function, function approximation can be viewed as a search through a space of hypotheses (representations of functions) for one that best fits a set of training data, and different learning methods assume different hypothesis spaces (representation languages) and/or employ different search techniques. **Boltzmann Machines** represent a type of neural network model by using stochastic units with a specific distribution (for example Gaussian). It is a network of symmetrically coupled stochastic binary units {0,1}. Learning procedure involves several steps called Gibbs sampling, which gradually adjust the individual weights to minimize the reconstruction errors. They are useful if it is required to model probabilistic relationships between variables. A variant of this machine is the Restricted Boltzmann Machines where the visible and hidden units are restricted to form a bipartite graph that allows implementation of more efficient training algorithms.

Reinforcement Learning starts with given a sequence of states and actions with (delayed) rewards, output a policy which is a mapping from states → actions that tells you what to do in a given state. When designing a learning system, (1) choose the training experience, (2) choose exactly what is to be learned, that is, the target function, (3) choose how to represent the target function, and (4) choose a learning algorithm to infer the target function from the experience. More specifically, deep learning is considered an evolution of machine learning. It uses a programmable neural network that enables a machine to make accurate decisions without help from humans. More specifically, deep learning is considered an evolution of machine learning. Deep learning is a subfield of machine learning that structures algorithms in layers to create an "artificial neural network" that can learn and make intelligent decisions on its own. In fact, deep learning is machine learning and functions in a similar way (hence why the terms are sometimes loosely interchanged) to Machine Learning. However,

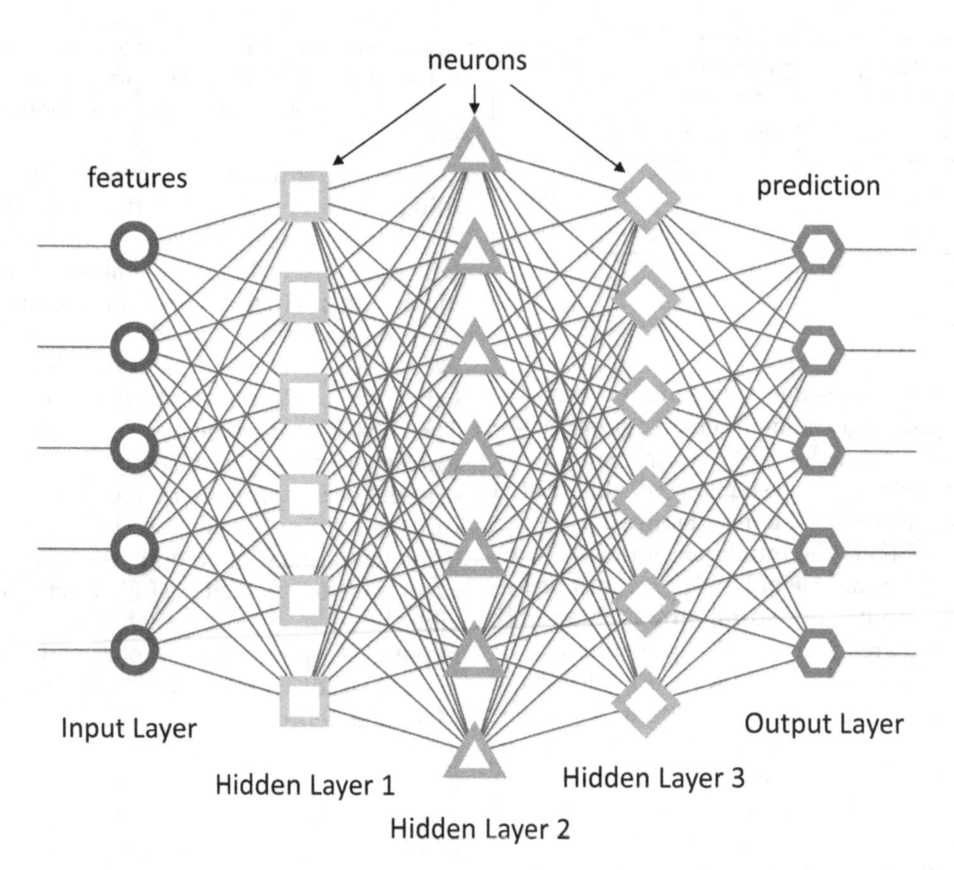

FIGURE 5.31
Deep learning approach with multiple layers.

its capabilities are different. A deep-learning model is designed to continually analyze data with a logic structure similar to how a human would draw conclusions (Figure 5.32).

The design of an ANN is inspired by the biological neural network of the human brain, leading to a process of learning that's far more capable than that of standard machine-learning models. To achieve its performance in tasks such as speech perception or object recognition, the brain extracts multiple levels of representation from the sensory input. Backpropagation was the first computationally efficient model of how neural networks could learn multiple layers of representation, but it required labeled training data and it did not work well in deep networks. The limitations of backpropagation learning can now be overcome by using multilayer neural networks that contain top-down connections and training them to generate sensory data rather than to classify it. Learning multilayer generative models might seem difficult, but a recent discovery by Hinton makes it easy to learn nonlinear distributed representations one layer at a time. "Because of the 'all-or-none' character of nervous activity, neural events and the relations among them can be treated by means of propositional logic". It's a tricky prospect to ensure that a deep-learning model doesn't draw incorrect conclusions, like other examples of AI, it requires lots of training to get the learning processes correct. But when it works as it's intended to, functional deep learning is often received as a scientific marvel that many consider being the new backbone of true AI.

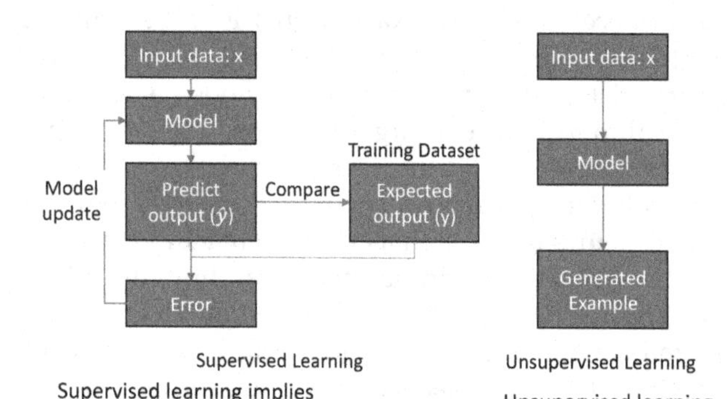

FIGURE 5.32
Reinforcement learning model.

Deep Learning is essentially a neural network with three or more layers (a multi-later perceptron).

It is a new area of machine-learning research, which has been introduced with the objective of moving machine learning closer to AI. These neural networks attempt to simulate the behavior of the human brain, albeit far from matching its ability, allowing it to "learn" from large amounts of data. The adjective "deep" in deep learning refers to the use of multiple layers in the network. Early work showed that a linear perception cannot be a universal classifier, but that a network with a nonpolynomial activation function with one hidden layer of unbounded width can. Deep learning is a modern variation which is concerned with an unbounded number of layers of bounded size, which permits practical application and optimized implementation, while retaining theoretical universality under mild conditions. In deep learning the layers are also permitted to be heterogeneous and to deviate widely from biologically informed connectionist models, for the sake of efficiency, trainability, and understandability, whence the "structured" part. In convolutional layers, each of the various types of deep-learning models is made by stacking multiple layers of regression models. Within these models, different types of layers have evolved for various purposes. One type of layer that warrants particular mention is convolutional layers. Unlike traditional fully connected layers, convolutional layers use the same weights to operate all across the input space. The pioneering works in neural networks with convolutional layers (CNNs) applied them to the task of image recognition. Now, CNNs have become well established as a highly effective deep-learning model for a diversity of image-based applications.

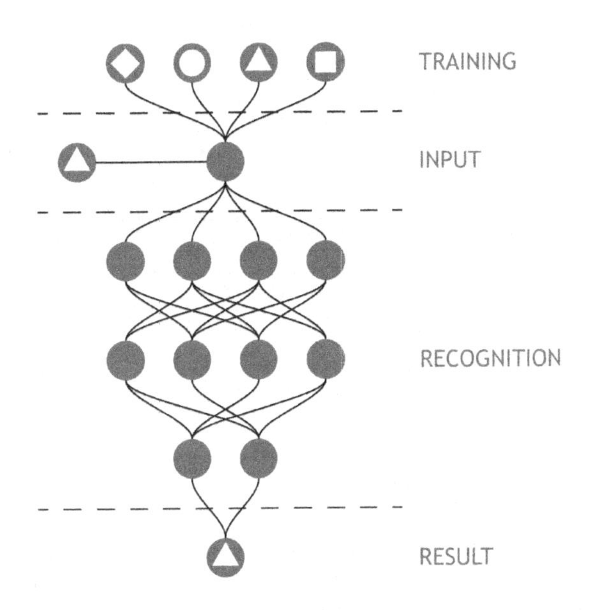

FIGURE 5.33
Connectionist Approach, starting with training and followed by analysis, leading to.

The DNN Mathematical Function is given as:

$$f : R^n \rightarrow R^m \tag{5.2}$$

This equation has high input and output dimensionality, in the order of:

$$n \in \left[10^2, 10^6\right] m \in \left[2, 10^5\right]$$

104–107 parameters to learn and 106–109 connections (operations)/image. As a result, learning the parameters requires quadrillions of floating point operations (Figure 5.33).

Backpropagation is now the most widely used tool in the field of ANNs. At the core of backpropagation is a method for calculating derivatives exactly and efficiently in any large system made up of elementary subsystems or calculations that are represented by known, differentiable functions; thus, backpropagation has many applications that do not involve neural networks as such. Backpropagation is a simple method that is now being widely used in areas like pattern recognition and fault diagnosis. It presents the basic equations for backpropagation through time and discusses applications to areas like pattern recognition involving dynamic systems, systems identification, and control. Further extensions of this method, to deal with systems other than neural networks, are systems involving simultaneous equations or true recurrent networks, and other practical issues which arise with this method. Pseudocode can be provided to clarify the algorithms. The chain rule for ordered derivatives and the theorem that underlies backpropagation is key to this approach (Figure 5.34).

Deep-learning architectures such as DNNs, deep belief networks, deep belief networks, deep reinforcement learning, recurrent neural network, and convolutional neural networks have been applied to fields including computer vision, speech recognition, natural language processing, machine translation, material inspection and board game programs, where they have produced results comparable to and in some cases surpassing human expert performance. Again, a DNN is an ANN with multiple layers between the input and output layers. There are different types of neural networks but they always consist of the same components: neurons, synapses, weights, biases, and functions. These components function similarly to the human brain and can be trained like any other ML algorithm. For all of its benefits, deep-learning does pose some drawbacks. Perhaps most significant is the volume of training data required, which is particularly problematic in robotics because generating

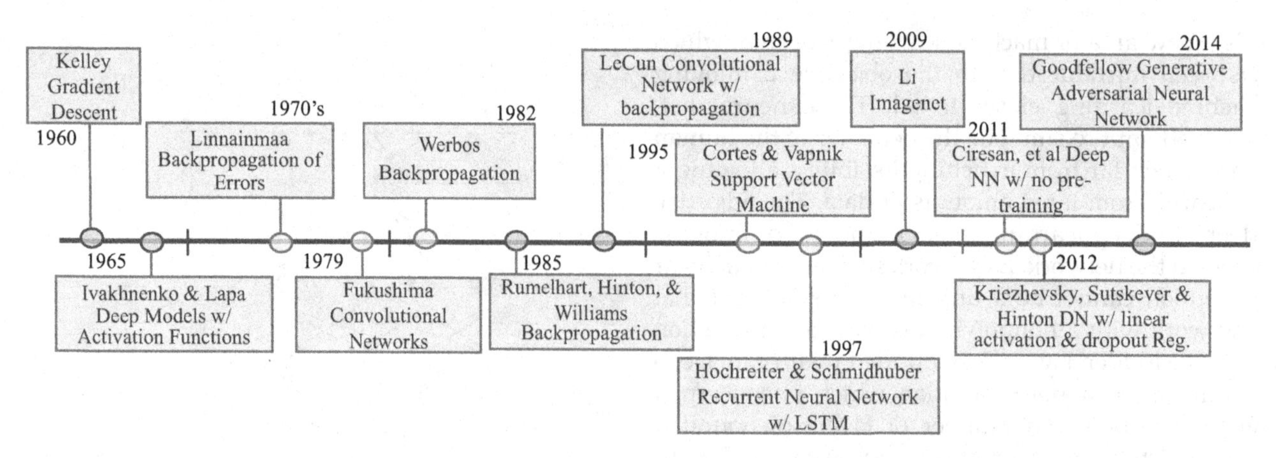

FIGURE 5.34
History of deep learning.

training data on physical systems can be expensive and time consuming. Training time is another challenge associated with the sheer size of DNNs. Typical models involve up to millions of parameters and can take days to train on parallel hardware, which is practical only for frequently repeated tasks that provide adequate payback on training time invested. Deep learning has shown promise in significant sensing, cognition, and action problems, and even the potential to combine these normally separate functions into a single system. Deep learning has shown promise in significant sensing, cognition, and action problems, and even the potential to combine these normally separate functions into a single system. Deep reinforcement learning has also been used to control dynamic systems from video, without direct access to state information. Visuomotor control requires an even closer integration between object perception and grasping, mapping image data directly to actuator control signals. ML and deep learning are applied to robotics in the areas of navigation, speech recognition, facial recognition, scene labeling, inference, manipulation (object and grasping), robotic vision, object recognition, control policies, novel dynamics, anticipating human actions, scene understanding and sensor fusion, and high-level task planning.

Another important field in robotics is the optimization of controllers. Currently, robots are often treated as a black box in this optimization process, which is the reason why derivative-free optimization methods such as evolutionary algorithms or reinforcement learning are omnipresent. When gradient-based methods are used, models are kept small or rely on finite difference approximations for the Jacobian. This method quickly grows expensive with increasing numbers of parameters, such as those found in deep learning. Recently, significant progress has been made by combining advances in deep learning for sensory processing. An obvious approach to adapting deep reinforcement learning methods such as Deep Q-Network (DQN) to continuous domains is to simply discretize the action space. However, this has many limitations, most notably, the curse of dimensionality. Considering a standard reinforcement learning setup that consists of an AI/robot interaction with its environment, E in discrete timesteps. At each timestep t, the agent receives an observation x_t, takes an action at, and receives a scalar reward r_t. It is not possible to straightforwardly apply Q-learning to continuous action spaces because in continuous spaces, finding the greedy policy requires an optimization of a_t every timestep; this optimization is too slow to be practical with large, unconstrained function approximators and nontrivial action spaces. This work combines insights from recent advances in deep learning and reinforcement learning, resulting in an algorithm that robustly solves challenging problems across a variety of domains with continuous action spaces, even when using raw pixels for observations. As with most reinforcement learning algorithms, the use of nonlinear function approximators nullifies any convergence guarantees; however, experimental results have demonstrated stable learning without the need for any modifications between environments.

Evolutionary Computation: The evolutionary approach to AI is based on the computational models of natural selection and genetics. Evolutionary computation (EC) consists of machine-learning optimization and classification paradigms that are roughly based on evolution mechanisms such as biological genetics and natural selection. The EC field comprises four main areas: genetic algorithms (GAs), evolutionary programming, evolution strategies, and genetic programming. EC consists of machine-learning optimization and classification paradigms that are roughly based on evolution mechanisms such as biological genetics and natural selection. Search is not random and is directed toward regions that are likely to have higher fitness values. Different EC paradigms make different uses of stochasticity. A **GA** is a search-based algorithm used for solving optimization problems in machine learning. This algorithm is important because it solves difficult problems that would take a long time to solve. A GA is a heuristic search algorithm used to solve search and optimization-type problems. This algorithm is a subset of evolutionary algorithms, which are used in computation. GAs employ the concept of genetics and natural selection to provide solutions to problems. GAs are characterized as:

- Reflect in a primitive way some of the natural processes of evolution
- Gas perform highly efficient search of the problem hyperspace
- Gas work with a population of individuals (chromosomes)
- Number of elements in each individual equals the number of parameters in the optimization problem
- If w is the number of parameters, and we have b bits per parameter, then the search space is 2wb

The variables being optimized comprise the phenotype space. The binary strings upon which the operators work comprise the genotype space. Genetic programming (GP) evolves hierarchical computer programs. Programs are represented as tree structure populations (most work originally done in LISP). These structures may vary in size and complexity and the goal is to obtain the desired output for a given set of inputs, within the search domain. Population members of GPs are executable structures (generally, computer programs) rather than strings of bits and/or variables. The fitness of an individual population member in a GP is measured by executing it. Generic Gas measure of fitness depends on the problem being solved. The hierarchical structure of GPs includes representation and preparation and terminal set and function set. For Representation and Preparation, each program is represented as a parse tree, functions are at internal tree points, variables and constants are at external (leaf) points and preliminary steps are to specify – Terminal set (variables and constants), Function set, Fitness measure, System control parameters, and Termination conditions. The Terminal Set and Function Set: Consists of system state variables and constants relevant to problem, Functions selected are limited only by programming language implementation, Math functions (sin, exp,...), Arithmetic functions, Conditional operators (if... then, etc.), and Boolean operators (AND, NOT,...).

Finally, Machine Learning allows robots to learn from mistakes and adapt. People get smarter through experience. Through technology such as machine learning, robotics applications may have the same ability. When that happens, they might not need continual time-intensive training from humans. Machine learning has had a significant impact on robot vision because "robots seeing" involves more than just computer algorithms; engineers and roboticists also have to account for camera hardware that allows robots to process physical data. Imitation learning is closely related to observational learning, a behavior exhibited by infants and toddlers using imitation learning.

Bayesian or probabilistic models are a common feature of this machine-learning approach. Self-supervised learning approaches enable robots to generate their own training examples in order to improve performance; this includes using a priori training and data captured close range to interpret "long-range ambiguous sensor data". Autonomous learning, which is a variant of self-supervised learning involving deep-learning and unsupervised methods, has also been applied to robot and control tasks. Coordination and negotiation are key components of multi-agent learning, which involves machine learning–based robots (or agents, this technique has been widely applied to games) that are able to adapt to a shifting landscape of other robots/agents and finding "equilibrium strategies". Challenges for deep learning are a number of robotics-specific learning, reasoning, and embodiment. There is a need for better evaluation metrics that highlight the importance and unique challenges for deep robotic learning in simulation and explore the spectrum between purely data-driven and model-driven approaches (Figure 5.35).

As envisaged by John McCarthy, logic has turned out to be one of the key knowledge representations for AI. Logic, in the form of the First-Order Predicate Calculus, provides a representation for knowledge that has clear semantics, together with well-studied rules of inference. This has led to the development of the subject of Logic Programming (LP), which can be viewed as a key part of Logic-Based AI. The subtopic of LP concerned with Machine Learning is known as "Inductive Logic Programming", which again can be broadened to Logic-Based Machine Learning (LBML)

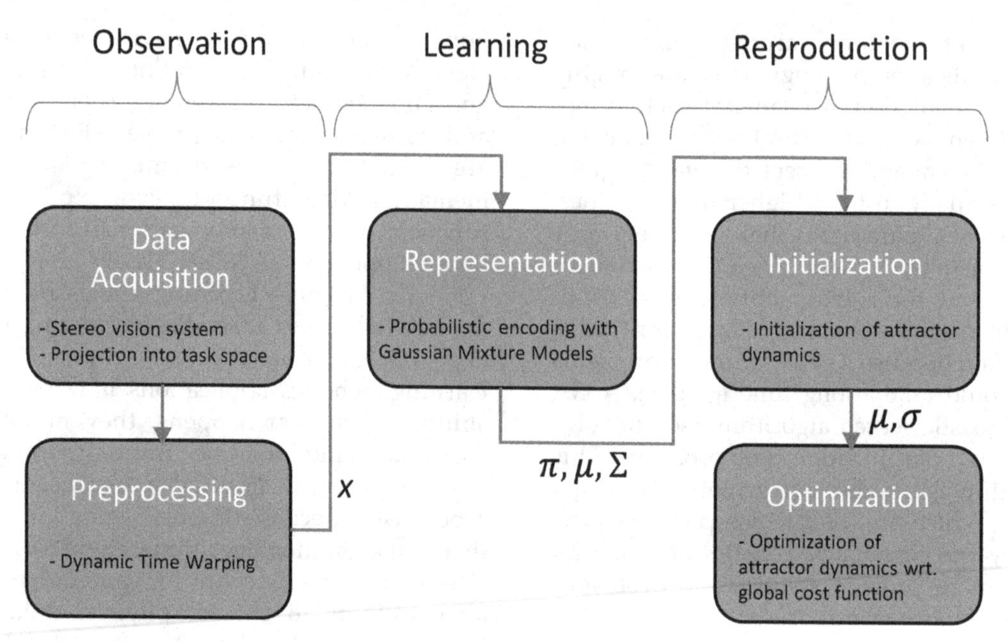

FIGURE 5.35
Machine learning: observation to learning to reproduction.

by dropping Horn clause restrictions. The modern study of LBML has largely been involved with the study of the learning of logic programs, or ILP. ILP is a general form of Machine Learning that involves the construction of logic programs from examples and background knowledge.

$$\mathbf{LP} = \mathbf{Logic} + \mathbf{Control} \tag{5.3}$$

The equation emphasizes the role of the programmer in providing sequencing control when writing Prolog programs.

$$\begin{aligned}\mathbf{ILP} = \mathbf{Logic} + \mathbf{Statistics} \\ + \mathbf{Computational\ Control}\end{aligned} \tag{5.4}$$

The logical part of ILP is related to the formation of hypotheses while the statistical part is related to evaluating their degree of belief. As in LP, the Computational Control part is related to the sequencing of search carried out when exploring the space of hypotheses.

In terms of software, there is a trend of committing to LP. The requirements clearly suggested the use of a rich, symbolic programming language that is capable of supporting a broad spectrum of programming styles. The symbolic processing language plays the role of a broad spectrum "Kernel language" for the computing machine, spanning the range from low-level operating system details up to application software. Researchers need to decide to explore an "And-Parallel" approach to parallel LP. This means that the subgoals of a clause are explored in parallel with shared variable bindings being the means of communication. This requires:

1. Databases and knowledge base support,
2. Constraint LP,
3. Parallel theorem proving, and
4. Natural language understanding a parallel database system called Kappa-P. This is a "nested relational" database system.

It is a constraint LP language with object-orientation features such as object identity, complex objects described by the decomposition into attributes and values, encapsulation, type hierarchy, and methods. A sequential constraint LP language commercial KBS tools have been recoded in C, which includes algebraic, Boolean, set, and linear constraint solvers. In the United States, the AI research community continues to use LISP as a vehicle for the rapid development of research insights. The most difficult problems in AI are informally known as **AI-complete** or **AI-hard**. This implies that the difficulty of these computational problems is equivalent to that of solving the central AI problem of Strong AI. To call a problem AI-complete reflects on an attitude that it could not be solved by a simple specific algorithm. AI-complete problems are hypothesized to integrate many AI tasks for a robot such as (1) perception (vision, sound, haptics), (2) reasoning (search, route planning, action planning), and (3) learning (recognition of objects/locations and exploration).

Others would include a **Bongard problem** (kind of puzzle) where the objective is to spot the differences between the two sides of relatively simple diagrams (a pattern recognition problem), and natural language understanding. Current AI systems can solve very simple and/or restricted versions of AI-complete problems, but not in its full generality. When AI researchers attempt to "scale up" their systems to handle more complicated, real-world situations, the programs tend to become excessively brittle, and without commonsense knowledge or a rudimentary understanding of the situation, they fail as unexpected circumstances outside of its original problem context begin to appear. Common sense is **the knowledge that all humans have** inherently. It consists of facts about the everyday world, such as "Lemons are sour", that all humans are expected to know and is currently an unsolved problem in Artificial General Intelligence. Such knowledge is unspoken and unwritten, and humans take it for granted.

Today, the U.S. Army pursues a Robotic and Autonomous System strategy to maintain overmatch to win in a complex world. This strategy is intended to serve as a catalyst to drive the necessary changes in technological innovation, acquisition priorities, doctrine, and the nature of how humans and machines should collaborate. The doctrine pursued is to increase SA, lighten the soldier's physical and cognitive workload, sustain the force with improved throughput and efficiency, facilitate movement and maneuver, and ultimately protect the force.

6

Artificial Intelligence: Today

The foundations for developing Artificial Intelligence and Robotics begin with the *science* of cognition with the brain and mind, the *engineering* of artificial intelligence (AI) and robotic systems, and *mathematics* as the tools to develop these systems. **Science** is a systematic endeavor that builds and organizes knowledge in the form of testable explanations and predictions about the universe. Scientists in neuroscience and cognitive psychology (Latin: cognoscere; Greek: gignoskein to know, perceive) are interested in the "problem" of the mind and brain. **Engineering** is the use of those previously mentioned scientific principles to design and build machines, structures, and other items, including robots, autonomous vehicles, manipulators, drones, and many other "smart" systems. Engineers are interested in the "solution" rather than the "problem", the domain of science. **Mathematics** is an area of knowledge that includes the topics of numbers, formulas, and related structures, shapes and the spaces in which they are contained, and quantities and their changes. Mathematics is essential in the natural sciences, engineering, computer sciences, and social sciences. For empirical sciences, natural sciences use tools from the formal sciences, such as mathematics and logic, and information about nature into measurements which can be explained as clear statements of the "Laws of Nature", developed from data and can be further matured through mathematics. The polemical literature in AI usually contains attacks on something called dualism, but what most critics fail to see is that they themselves display dualism in a strong form, for unless one accepts the idea that the mind is completely independent of the brain or of any other physically specific system, one could not possibly hope to create minds just by designing programs. The mind according to strong AI is independent of the brain. It is a computer program, and as such, has no essential connection to any specific hardware. An important point is that simulation is not the same as duplication and that facts hold as much importance for thinking about arithmetic as it does for feeling angst (Figure 6.1).

FIGURE 6.1
AI on chip icon.

Science

There is an assumption of Aristotle's view that the brain is part of the anatomy of vertebrate animals. And, historically, the way we thought of the brain and nervous system was in the light of vertebrate organisms (the approach of neuroscience). A brain was defined as an organ acting as the control hub of the nervous system in animals displaying cephalization. More generally, the brain was defined as an organ acting as the control hub of the nervous system in animals displaying cephalization. The brain controls all functions of the human body, interprets information from the outside world, and embodies the essence of the mind and soul. Intelligence, creativity, emotion, and memory are a few of the many things governed by the brain. The brain is composed of the cerebrum, cerebellum, and brainstem. And communication theory must be one of the fundamental disciplines that will help us understand what brains are and how they work, including having criteria like intelligence, thinking, reasoning, and other so-called cerebral higher functions. The brain integrates sensory information and directs motor responses; and in higher vertebrates, it is also the center of learning. Junctions between neurons, known as synapses, enable electrical and chemical messages to be transmitted from one neuron to the next in the brain, a process that underlies basic sensory functions and that is critical to learning, memory and thought formation, and other cognitive activities.

The human brain is the substrate for human intelligence. By simulating the human brain, AI builds

DOI: 10.1201/9781032673134-6

computational models that have learning capabilities and perform intelligent tasks approaching the human level. AI is the field of computer science that attempts to build machines that can think, learn, and act intelligently, similar to a human. AI systems are powered by machine-learning algorithms and neural networks (NNs), which allow them to learn directly from massive amounts of data. The appearance of language is a large leap in human history that accelerates the learning process of humans about this world because it helps humans better describe the world as special data (***X*** and ***Y***). With language, samples are generated by tagging sensory inputs with words or lingual logics as outputs, or even language represents inputs themselves. Language plays an important role in training the brains of human individuals in a supervised way (Figure 6.2).

The development of human intelligence occurs through three phases: (1) natural selection, in which the environment determines the cost function and minimizes errors by eliminating unfit behaviors while favoring fit behaviors; (2) mimicking, in which older generations define the actual outputs and younger generations learn from observing and imitating older behaviors, in which emotion facilitates signaling the actual outputs and prediction errors; and (3) language, in which actual outputs are defined by language (words, sentences, stories) composed of other people and prediction errors are minimized by agreeing on influential people.

J.C.R. Licklider, who had a background in psychology and computer science, became interested in computers in the late 1950s and was interested in how these new machines could amplify humanity's collective intelligence. When he reviewed the existing literature, he found that programmers aimed to "teach" these machines how to perform pre-existing human activities, such as playing chess or language translation, with greater aptitude and efficiency than humans. The problem was that the existing paradigm saw humans and machines as being intellectually equivalent beings. Licklider (ARPA Information Processing Techniques Office's first director) believed that humans and machines were fundamentally different in their cognitive capacities and strengths, and humans were good at certain intellectual activities like being creative and exercising judgment. He believed that computers were good at other cognitive tasks, like remembering data and processing it quickly. Instead of having computers imitate human intellectual activities, Lick proposed an approach in which humans and machines would collaborate, each making use of their particular advantages. Licklider believed that humans would 1 day interact seamlessly with computers. He suggested that this strategy would shift the focus from competition (like in playing games) and facilitate previously unimaginable forms of intelligent activity through collaboration (symbiosis). "Symbiotic means" complementary as in a symbiotic relationship between computers and humans as a period before AI. For Licklider, a promising existing example of this symbiosis was a system of computers, networking equipment, and human operators. And when at the Advanced Research Project Agency (later renamed to DARPA), he had the opportunity to put some of his ideas into practice. This later idea evolved after leaving his office as the ARPANET, the predecessor to the Internet. Today, many great leaps forward in machine-learning applications are underpinned by collaborative networks of humans and machines.

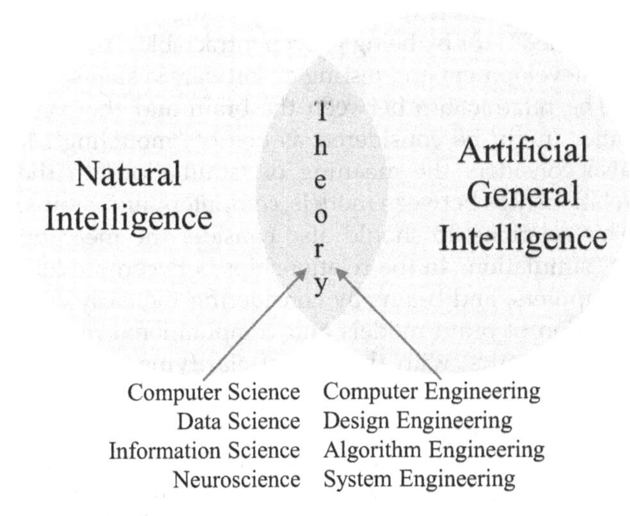

FIGURE 6.2
Natural intelligence versus artificial general intelligence.

Also in the early 1950s, as calculating machines were coming into their own, a few pioneer thinkers began to realize that digital computers could be more than number crunchers. At that point, two opposed visions of what computers could be, each with its correlated research program, emerged and struggled for recognition. One faction saw computers as (1) a system for manipulating mental symbols; the other, (2) as a medium for modeling the brain. One sought to use computers to instantiate a formal representation of the world; while the other, to simulate the interactions of neurons. One took problem-solving as its paradigm of intelligence; and the other, learning. One utilized logic, the other statistics. One school was the heir to the rationalist, reductionist tradition in philosophy; and the other viewed itself as idealized, holistic neuroscience. Among AI researchers, the idea of the computer and mind being equal in one or more fundamental aspects, as mirroring each other's operations and contents, has come to be known as the "strong AI" thesis. According to this way of thinking, the value of computer programs and models is not that they provide one predictive or falsifiable account of mental phenomena among others. Instead, they reproduce, make transparent, or provide an "existence proof" (Gardner) for the contents and operations of the mind. Programs and algorithms in this sense can serve both as description and as proof for the kinds of "mathematical laws of reasoning" that Leibniz and his predecessors had only dreamed of bringing to light. These mental entities can be construed in a variety of ways – as algorithmic and procedural, or as "connectionist" or relational. But in each case, as John Searle explains, strong AI assigns its computer programs or simulations a very special significance: "programs are not mere tools that enable us to test psychological explanations; rather, the programs are *themselves* the explanations" (Searle) (Figure 6.3).

John Searle put its best when asked is the brain's mind a computer program? His answer was "no" because a program merely manipulates symbols whereas a brain attaches meaning to them. The real question is "Could a machine think just by virtue of implementing a computer program? And is the program by itself constitutive of thinking?" This is a completely different question because it is not about the physical, causal properties of actual or possible physical systems but rather about the abstract, computational properties of formal computer programs that can be implemented in any sort of substance at all, provided only that the substance is able to carry the program. Thus, many AI researchers believe that by designing the right programs with the right inputs

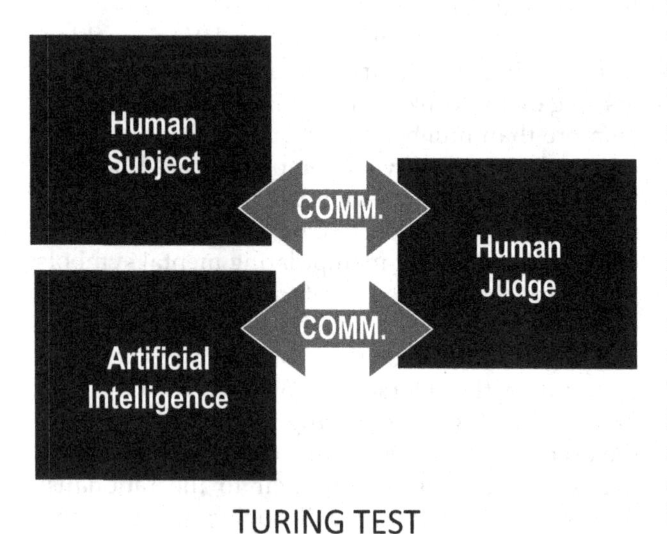

FIGURE 6.3
Turing test configuration model.

and outputs, they are literally creating minds. Some believe furthermore that AI has a scientific check in the Turing test for determining success or failure. The Turing test, as currently understood, is simply this: if a computer can perform in such a way that an expert cannot distinguish its performance from that of a human who has a certain cognitive ability-say, the ability to do addition or to understand Chinese, then the computer also has that ability. So, the goal is to design programs that will simulate human cognition in such a way as to pass the Turing test. What is more, such a program would not merely be a model of the mind; it would literally be a mind, in the same sense that a human mind is a mind. Strong AI claims that thinking is merely the manipulation of formal symbols, and that is exactly what the computer does, manipulate formal symbols. This view is often summarized by saying, "The mind is to the brain as the program is to the hardware". The parallel, "brain-like" character of the processing, however, is irrelevant to the purely computational aspects of the process.

This basic idea of the reducibility of thought to formalized logical, mathematical, or mechanical rules and processes lies at the heart of AI. It is already explicit in the first definition for this field: AI is proposed as "a study" proceeding from "the conjecture that every aspect of learning or any other feature of intelligence can in principle be so precisely described that a machine can be made to simulate it". Simply put, this hypothesis is as follows: if thinking is indeed a form of mechanical calculation, then development of calculating machines (i.e., computers) has the potential to unlock the "mathematical laws of human reasoning" posited in the West's rationalistic tradition. If thinking is information processing, in other words, then hypotheses about how we think can either be proven or falsified by being successfully "modeled" – or by being proven intractable – through the development and testing of software systems.

The relationship between the brain and the computer might be considered as one of "modeling". It also considers the meaning of "simulation" in the relationships between models, computers, and brains. This relationship should also consider the meaning of "simulation" in the relationships between models, computers, and brains by considering the early convolution of brain models and computational models in cybernetics, with the aim of clarifying their significance for more current debates in the cognitive sciences. "Model" is a challenging concept, in part because it is both a noun and a verb, an object, and a practice, and it takes many prepositional forms: *X* models *Y*, *X* is a model of *Y*, *X* is a model for *Y*, *Y* is modeled on *X*, *Y* is modeled after *X*, and even *Y* models

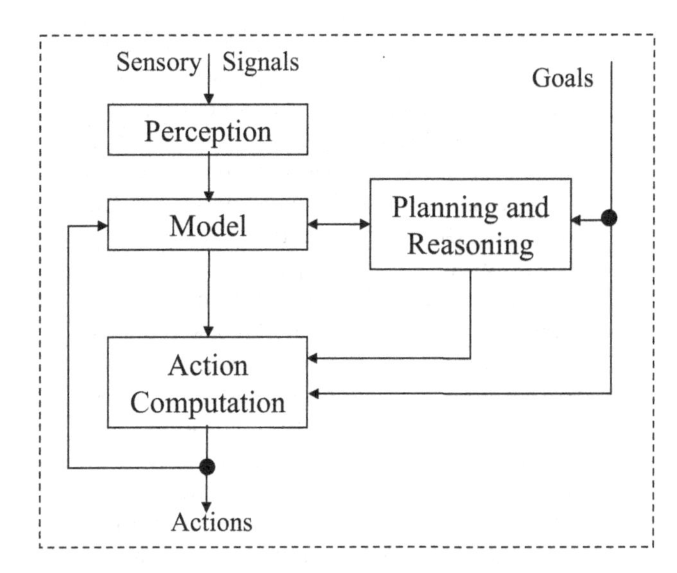

FIGURE 6.4
An AI/robot system.

for Z. Thus, we could say that the brain was a model for the structure of a computer (or the computer was modeled on the brain) in the sense that the designers of the early computers, such as John von Neumann, treated the biological brain like an artist's model, and crafted the computer in its image (Figure 6.4).

Of course, once completed, we might be inclined to think of the artist's sculpture as a sort of model of the subject it is based upon and so we might think of the computer as a model of the brain. This would seem to be the case for certain electronic devices that were models of the performance and behavior of the brain, such as W. Ross Ashby's Homeostats which were meant to be models of the adaptive properties of the brain, or W. Grey Walter's Tortoises which were meant to be models of the dynamic drives of the living brain. In another sense, the first computers were seen by some, such as Alan Turing, as models of the fundamental structure of the brain in the sense that the digital computer was an engineered device that worked on the same principles as the brain and could be used to test theories about how the higher level functions of the brain might operate. For each of these cases, the actual process of modeling and the construction of each device were unique and complex and involved further models of the brain and behavior, both concrete and abstract. Asaro has considered the nature of working models as represented by the Homeostats and Tortoises elsewhere. This discussion will add to both theoretical and working models in another class, the "simulation", which is a term used in nearly as many ways as a "model". For many, a "simulation" is simply synonymous with a "model", while for others it is what Asaro calls a "working model", because its behavior can become the subject of empirical inquiry. In its more precise usage, "simulation" generally refers to a special class of models defined by their use; specifically, computational models used to approximate the behavior of a system of equations that are too difficult to solve by analytical techniques. In their most general case, PDP (parallel distributed processing) systems are simulations of NNs found in the brain. They consist of large numbers of individual processing units connected together in varying degrees of complexity. Individual units typically perform simple computations; they process information by sending excitatory or inhibitory signals to other units in varying degrees of intensity, dependent entirely upon the signals of the units simultaneously connected to them. A PDP network is not programmed with explicit rules nor does it create representations of the world to manipulate. Instead, a network is repeatedly exposed to "input" information concerning the world and "output" expected responses. By adjusting its internal connections, the network learns to associate the expected response to the situation.

Classical conditioning (also known as Pavlovian or respondent conditioning) is learning through association and was discovered by Ivan Pavlov, a Russian physiologist. In simple terms, two stimuli are linked together to produce a new learned response in a person or animal. John B. Watson proposed that the process of classical conditioning (based on Pavlov's observations) was able to explain all aspects of human psychology. Everything from speech to emotional responses was simply patterns of stimulus and response. B.F. (Burrhus Frederic) Skinner believed that we do have such a thing as a mind, but that it is simply more productive to study observable behavior rather than internal mental events. The work of Skinner was rooted in a view that classical conditioning was far too simplistic to be a complete explanation of complex human behavior. He believed that the best way to understand behavior is to look at the causes of an action and its consequences. He called this approach operant conditioning. Skinner's theory of operant conditioning was based on the work of Edward Thorndike (1905). Thorndike studied learning in animals using a puzzle box to propose the theory known as the "Law of Effect". B.F. Skinner coined the term operant conditioning; it means roughly changing of behavior by the use of reinforcement which is given after the desired response. Rudolf Carnap suggested his work on inductive logic could possibly provide a "neutral" framework to help clarify the problem of estimation in theoretical statistics.

If you pair a neutral stimulus (NS) with an unconditioned stimulus (US) that already triggers an unconditioned response (UR) that NS will become a conditioned stimulus (CS), triggering a conditioned response (CR) is similar to the original unconditioned response.

Jaakko Hintikka was a pioneer of possible-worlds semantics, epistemic logic, inductive logic, game-theoretical semantics, the interrogative approach to inquiry, and independence-friendly logic. It is often said that there are two kinds of arguments: deductive and inductive, where the premises of deductive arguments are intended to guarantee the truth of their conclusions, while inductive arguments involve some risk of their conclusions being false even if all of their premises are true. Hintikka is known as the main architect of game-theoretical semantics and of the interrogative approach to inquiry, and also as one of the architects of distributive normal forms, possible-worlds semantics, tree methods, infinitely deep logics, and the present-day theory of inductive generalization. Heidegger refutes the traditional representational theory of mind, which holds that we form meaningful mental representations of the world and manipulate them when thinking. Martin Heidegger does not deny the possibility of mental phenomena: he does, however, reject the idea that such phenomena create "internal meanings" of the world. Formalizing "its understanding" in order to gain commonsense knowledge is at an impasse because Heidegger's commonsense understanding of everyday know-how does not consist of procedural rules, but rather an unformalizable knowing-what-to-do in everyday situations. A proper explanation of the blank spaces on the map of science (can) only be made by a team of scientists, each a specialist in his own field but each possessing a thoroughly sound and trained acquaintance with the fields of his neighbors.

Consciousness: Consciousness is everything we experience, from a city scene teeming with shapes, colors, and noises to darkness and silence. It has been defined as what abandons us every night when we fall into a dreamless sleep and returns the next morning when we wake up. Many philosophers have claimed that understanding how the brain produces subjective experience may lie forever beyond the realm of scientific explanation. The first problem has to do with the necessary and sufficient conditions that determine whether consciousness is present or not. Solving the first problem means that we would know whether a physical system can generate consciousness and to what extent – a level of consciousness. The second problem has to do with the necessary and sufficient conditions that determine the specific nature of consciousness. Solving the second problem means that we would know what kind of consciousness is generated, the content of consciousness. The point is simply this: every time we experience a particular conscious state out of such a huge repertoire of possible conscious states, we gain access to a correspondingly large amount of information. This conclusion is in line with the classical definition of information as reduction of uncertainty among a number of alternatives.

There is a claim that *Consciousness* is an integrated form of information theory where:

- Information: reduction in uncertainty,
- Uncertainty: formalized as information entropy (Shannon),
- Integrated Information (Φ): Information generated by a system over and above information generated by its parts taken separately,
- Quale/Qualia: Specific individual instances of subjective conscious experience,
- Qualia Space (Q): a multidimensional space, defining all possible states of the system, within which, any particular shape specifies a unique quale, and
- Claim (II): The quantity of consciousness in a system is its Φ and the quality of consciousness is its shape in Q.

This aspect of the theory is that the information associated with the occurrence of a conscious state is information from the perspective of an integrated system. When each of us experiences a particular conscious state, that conscious state is experienced as an integrated whole; it cannot be subdivided into independent components, that is, components that are experienced independently. If consciousness is integrated information, then a physical system should be able to generate consciousness to the extent that it can rapidly enter any of a large number of available states (information); yet it cannot be decomposed into a collection of causally independent subsystems (integration). According to this theoretical framework, consciousness comes about when the parameters of a physical system are such that it can integrate a large amount of information over a short period.

Alan Turing argued that a properly programmed computer could in principle exhibit intelligent behavior. The argument rests on Turing's own discovery of the existence of a Universal Turing Machine (UTM), an abstract automaton that can imitate any other formally specifiable computer. Newell stated at the most abstract level, the class of mechanisms called computers are the only known mechanisms that are sufficiently plastic in their behavior to match the plasticity of human cognition. The human brain is famously flexible, or "plastic", because neurons can do new things by forging new or stronger connections with other neurons. But if some connections strengthen, neuroscientists have reasoned, neurons must compensate lest they become overwhelmed with input. They are also the only known mechanism capable of producing behavior that can be described as "knowledge dependent". Because of such properties, computing remains the primary candidate for meeting the dual needs of (1) explaining cognition in mechanistic terms and (2) accounting for certain otherwise problematic aspects of cognition, in particular the fact that behavior can be systematically influenced by inducing differences in beliefs or goals. This notion of mechanism arose in conjunction with attempts to develop a completely formal, content-free foundation for mathematics. The "Hilbert program" was one of the most ambitious attempts to build up mathematics by purely formal means, without regard to questions of what the formalism was about. Some of this enterprise succeeded based on the work of Frege, and of Russell and Whitehead. On the other hand, one of the greatest intellectual achievements of our age was the demonstration by purely formal means that the ultimate goal of complete formalization was in principle not achievable (this was done originally by Gödel and subsequently by Turing, Church, Post, and others).

Theory of Multiple Intelligence: This theory suggests that traditional psychometric views of intelligence are too limited. Gardner suggested that all people have different kinds of "intelligences." He proposed that there are eight intelligences and has suggested the possible addition of a ninth known as "existentialist intelligence." The theory claims that human beings have different ways in which they process data, each being independent.

The problem that we must deal with is the question of the kinds of knowledge available to an understander. Every theory of processing must also be a theory of memory. To put this in another way, if psychologists show that recognition confusions occur between two entities in memory, this will have to be taken as evidence against a theory that said those two entities existed and were processed entirely separately. The dominant paradigm, which is referred to as Traditional AI, has focused on formalizing the process of thinking into rules, symbols, and representations of the world. As such, its roots can be found in the philosophical traditions of reductionism and rationalism. The second paradigm, which is referred to as Parallel Distributed Processing (PDP), has focused on using computers to emulate the neurological structure of the brain. Concerned less with formalization than its underlying computational structure, this approach was developed from the neurosciences, Gestalt Theory, and work in perception. Traditional AI has approached the formalization by postulating mental representations, which both Dreyfus and Heidegger reject. The computational paradigm used by Traditional AI approaches has been described by Newell and Simon as a physical symbol system. It can be characterized by its use of abstract symbols to represent salient features in a "**microworld**" – an artificially constructed problem domain that simulates a subset of the real world. Syntactic rules manipulate these symbols to reflect the processes and relations that occur in the microworld. The technique is powerful: symbols and rules are capable of representing every fact and process that can occur within the constraints of the explicitly defined problem domain.

Considering the efforts made in AI during the 1970s and beyond, it is clear that the criticisms concerning microworlds and the representation of knowing-how have become influential in Traditional AI theorizing. New proposals, such as Minsky's **frame system** or Schank's **scripts**, used complex representational structures in an attempt to address these issues. Scripts attempt to enrich microworlds by representing human-world interactions and attempt to capture the kind of commonsense know-how humans use in everyday situations. Minsky's approach of decomposing the commonsense knowledge of, say, "how to use a spoon" is characteristic of AI's information processing model: the use of a spoon is a conglomeration of a huge number of actions and rules-the degree of tension the fingers must use to hold the spoon, the proper angle to hold the spoon so that food will not slide off and so on.

The hope was that the restricted and isolated "microworlds" could be gradually made more realistic and combined so as to approach real-world understanding.

But researchers confused two domains which, following Heidegger, we shall distinguish as universe and world. A set of interrelated facts may constitute a universe, like the physical universe, but it does not constitute a world. The latter, like the world of business, the world of theater, or the world of the physicist, is an organized body of objects, purposes, skills, and practices on the basis of which human activities have meaning or make sense. To see the difference, one can contrast the meaningless physical universe with the meaningful world of the discipline of physics. The world of physics, the business world, and the theater world make sense only against a background of common human concerns. They are local elaborations of the one commonsense world we all share. That is, sub-worlds are not related like isolable physical systems to larger systems they compose, but are rather, local elaborations of a whole, which they presuppose. Microworlds were not worlds but isolated meaningless domains, and it has gradually become clear that there was no way they could be combined and extended to arrive at the world of everyday life.

AI has been wrestling with what has come to be called the **commonsense knowledge** problem. The representation of knowledge was always a central problem for work in AI, but the two earlier periods, cognitive simulation and microworlds, were characterized by an attempt to avoid the problem of commonsense knowledge by seeing how much could be done with as little knowledge as possible. In his recent work on computational simulations, Winsberg considers three traditional attempts to account for computational simulations in physics: as metaphors, as experiments, and as a third middle mode. The view of simulations as metaphors, while perhaps never expressed clearly as such, nonetheless holds that simulations are essentially just brute-force number-crunching procedures, used when analytic techniques are impossible – in other words, a degenerate form of theorizing. The view that simulations are experiments, and the computer is an experimental target, holds that there is some mimetic relation between the simulation and the simulated, such that the simulation can mimic the real and act as a stand-in. The third mode holds that simulations are an entirely different kind of thing, lying somewhere between theorizing and experiment.

The difficulty of learning large models is a severe limitation. Modern graphics processors far surpass the computational capabilities of multicore CPUs and have the potential to revolutionize the applicability of deep unsupervised learning methods. We develop general principles for massively parallelizing unsupervised learning tasks using graphics processors. Modern graphics processors far surpass the computational capabilities of multicore CPUs and have the potential to revolutionize the applicability of deep unsupervised learning methods. Stanford develops general principles for massively parallelizing unsupervised learning tasks using graphics processors. With the invention of increasingly efficient learning algorithms over the past decade, these models have been applied to several machine-learning applications, including computer vision, text modeling, and collaborative filtering, among others. These models are especially well suited to problems with high dimensional inputs, over which they can learn rich models with many latent variables or layers. When applied to images, these models can easily have tens of millions of free parameters, and ideally, we would want to use millions of unlabeled training examples to richly cover the input space.

Data Science: studies data and how to extract meaning from it, using a series of methods, algorithms, systems, and tools to extract insights from both structured and unstructured data. Machine learning automates the process of data analysis and goes further to make predictions based on collecting and analyzing large amounts of data on certain populations. Thus, data science is a field that studies data and how to extract meaning from it, whereas machine learning is a field devoted to understanding and building methods that utilize data to improve performance or inform predictions. The data scientist uses data to understand and explain its phenomena through analyses. The scientist analyzes, gathers, cleans, and studies the datasets for input into machine learning. Data engineers transform the raw data and prepare the data for training within a NN or deep NN. The dataset is stored and managed in databases within its infrastructure. The models allow machine-learning engineers to conduct statistical analysis to understand patterns in the data. Data Scientists often incorporate machine-learning techniques in their work, while machine-learning engineers are well-versed in data structures, algorithms, and software architectures. Both disciplines require knowledge of software engineering and software design (Figure 6.5).

Information Science: is an academic field that is primarily concerned with the science and practice dealing with the effective collection, storage, retrieval, and use of information. It is concerned with recordable information and knowledge. Information science involves

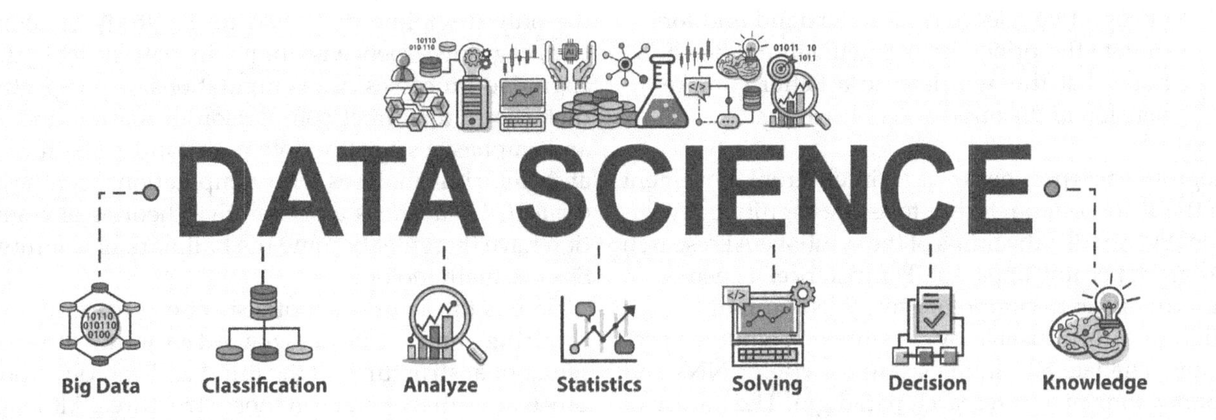

FIGURE 6.5
Data science taxonomy figure.

several disciplines, including behavioral science, social science, business, and computer science. It deals with various forms of information and the communication, processing, evaluation, analysis, and storage of information by machines and people. Information science involves working with computers and enormous quantities of information. Information science can be divided into three major areas: (1) Information systems including operations research and computer science, (2) Computer–human interaction (cognitive studies, communication, and psychology), and (3) Studying computing from a social studies perspective relative to economics, science, technology, and law.

Information theory, also known as the **mathematical theory of communication**, is an approach that studies **data processing and measurement in the transmission of information.** Information theory is the mathematical treatment of the concepts, parameters, and rules governing the transmission of messages through communication systems. The mathematical theory of information was proposed in **1949** by mathematician and engineer **Claude Shannon** and biologist **Warren Weaver**. However, it was the result of research initiated almost 30 years prior by scientists such as **Andrei Markov** and **Ralph Hartley**. **However, it was Alan Turing** who created the blueprint for a machine capable of processing pieces of data through the emission of symbols, which was the last precedent for the development, culmination, and consecration of what would be called the *Mathematical Theory of Communication*. The techniques used in information theory are probabilistic and some view information theory as a branch of probability theory. In a given set of possible events, the information of a message describing one of these events quantifies the symbols needed to encode the event optimally. "Optimal" means that the obtained code word will determine the event unambiguously, isolating it from all others in the set, and will have minimal length, that is, it will consist of a minimal number of symbols. Information theory also provides methodologies to separate real information from noise and to determine the channel capacity required for optimal transmission conditioned on the transmission rate.

Cognitive Science: is the interdisciplinary, scientific study of the mind and its processes with input from linguistics, psychology, neuroscience, philosophy, computer science/AI, and anthropology. It examines the nature, the tasks, and the functions of cognition. Similar to AI, Cognitive Science is an interdisciplinary field that studies the fundamental workings of the mind. It investigates perception, action, language, knowledge, development, and thinking from multiple perspectives: theoretical, experimental, and computational, with the aim of gaining a better understanding of human cognition and the nature of intelligent systems. Near the end of the 19th century, Santiago Ramon y Cajal discovered that the brain consists of discrete cells. These neurons signal each other through contacts at specialized points called synapses. The term *cognitive science* was coined by Christopher Longuet-Higgins in his 1973 commentary on the Lighthill report, which concerned the then-current state of AI research and the goal

> of cognitive science is to understand and formulate the principles of intelligence with the hope that this will lead to a better comprehension of the mind and of learning.

Cognitive Sciences began as an intellectual movement in the 1950s, often referred to as the cognitive revolution. Eventually, the limits of the symbolic AI research program became apparent. For instance, it seemed to be unrealistic to comprehensively list human knowledge in a form usable by a symbolic computer program. The late 80s and 90s saw the rise of NNs and connectionism as a research paradigm. The nature of mental representations and their acquisition and use are important themes, as are the comparison between human and AI, and the relation between human cognition and its biological foundations. One accomplishment that has eluded cognitive science is a unified theory that explains the full range of psychological phenomena, in the way that evolutionary and genetic theory unify biological phenomena, and relativity and quantum theory unify physical theory.

The cognitive revolution in psychology was a counter-revolution. The first revolution occurred much earlier when a group of experimental psychologists, influenced by Pavlov and other physiologists, proposed to redefine psychology as the science of behavior. They argued that mental events are not publicly observable. Thus, the only objective evidence available is, and must be, behavioral. By changing the subject to the study of behavior, psychology could become an objective science based on scientific laws of behavior. Behaviorism was an exciting adventure for experimental psychology but by the mid-1950s, it had become apparent that it could not succeed fully. Alberto Greco described a meta-theoretic system, suggesting how cooperation between cognitive disciplines may have a true explanatory value. In this system, a single commonsense "fact" is described as a different "state" from the perspective of different disciplines (as a physical state, or a state of the body, of the brain, of consciousness, etc.). Such descriptions include new states resulting from changes of state ("events"), disposed along a time sequence (called "flow"). A parallel representation of different flows, describing from various disciplinary standpoints, the same events occurring in a certain time course (called a "flow-chain") allows us to establish the nature of correspondences and the links between events in the same or different flows. Greco argues that a multidisciplinary exchange is really needed for explanation when a cognitive phenomenon includes events that are correlated but cannot be causally linked inside a single flow, that is, using a set of descriptions belonging to a single discipline. By no means the only discipline dedicated to the study of cognition, cognitive science is unique in its basic tenet that cognitive processes are computations, a perspective that allows for direct comparison of natural and AI and emphasizes a methodology that integrates formal and empirical analyses with computational synthesis. *Computer simulations* as generative theories of cognition have therefore become the hallmark of Cognitive Science methodology.

Here is the central hypothesis of cognitive science: **thinking** can best be understood in terms of representational structures in the mind and computational procedures that operate on those structures. Although there is much disagreement about the nature of the representations and computations that constitute thinking, the central hypothesis is general enough to encompass the current range of thinking in cognitive science, including connectionist theories. Computer models are often very useful for theoretical investigation of mental processes. Comprehension of cognitive science models requires noting the distinctions and the connections among four crucial elements: theory, model, program, and platform. A computational model makes these structures and processes more precise by interpreting them by analogy with computer programs that consist of data structures and algorithms. An algorithm is simply a recipe for doing something. It contains a sequence of steps that, if followed correctly, produce the intended results. Software algorithms are methods for making the computers produce some result. All software applications use algorithms. AI often revolves around the use of algorithms. It is a set of instructions that a mechanical computer can execute. A complex algorithm is often built on top of another and simpler one, and a common way to visualize it is with a tree design. Vague ideas about representations can be supplemented by precise computational ideas about data structures, and mental processes that can be defined algorithmically. To test the models, it must be implemented in programming languages. Although formal logic has not been the most influential psychological approach to mental representation, there are several reasons for beginning our survey with it. First, many basic ideas about representation and computation have grown out of the logical tradition. Second, many philosophers and AI researchers today take "logic" as central to work on **reasoning**. Third, logic has substantial representational power that must be matched by other approaches to mental representation that may have more computational efficiency and psychological plausibility.

In cognitive science, Computational Processing and Information Processing are viewed differently. On one side stands mainstream cognitivism, which is based on the manipulation of representations, and on the other side, there are non-cognitivist approaches such as behaviorism and Gibsonian psychology, which reject mental representations. However, most cognitive scientists agree that knowledge in the mind consists of mental representations. As stored-program computing took off, experimental psychologists were facing a growing pile of anomalies that motivated looking inside the behaviorist's black box. George Miller showed that short-term memory capacity stayed constant at around 7 "chunks" of information because items could be recoded into new "chunks": for example, a 10-digit number is more easily remembered by being recoded into 3 chunks (e.g., 123-456-7890). This showed that internal cognitive machinery was needed to explain memory. Noam Chomsky argued that children's linguistic output was governed by grammatical rules (or violated those rules in regular ways) that were underdetermined by the speech they heard as stimulus. This evidence of the "poverty of the stimulus" (its inadequacy to explain the output) showed that internal operations were needed to explain language. Memory allows us to store information for later retrieval. Memory is often thought of as consisting of both a long-term and short-term store. Long-term memory allows us to store information over prolonged periods (days, weeks, and years). Unfortunately, we do not yet know the practical limit of long-term memory capacity, while short-term memory allows us to store information over short time scales (seconds or minutes). Memory is often grouped into declarative and procedural forms.

Computational models require a mathematically and logically formal representation of a problem. Computer models are used in the simulation and experimental verification of different specific and general properties of intelligence. Computational modeling can help us understand the functional organization of a particular cognitive phenomenon. Approaches to cognitive modeling can be categorized as:

1. symbolic, on abstract mental functions of an intelligent mind by means of symbols.
2. subsymbolic, on the neural and associative properties of the human brain; and
3. across the symbolic – subsymbolic border, including hybrid.

The stored-program computer of von Neumann's design was the machine for which these first psychologically interpreted internal state transitions were developed. They were specified in the form of software programs written in high-level programming languages. The Computational-Representational Understanding of Mind has contributed to much theoretical understanding and practical application. But no single approach has emerged as the clearly most powerful explanation of human cognitive capacities. Different approaches have different representational and computational advantages and disadvantages.

For example, decision-making is a "high-level" cognitive process that is clearly distinguishable from other processes in at least two ways: it builds on more basic cognitive processes such as perception, memory, and attention, and it is uniquely identified by its essential element: the process of choice. Choice is the act of selecting among alternatives, whether they are present at the same time or they develop over time. The process of choice is highly influenced by the cognitive processes that occur before a choice is made (e.g., perception, recognition, and judgment) and those that occur after a choice is made (e.g., feedback and learning). Notably, the judgment process, which precedes choice, involves evaluating the merits of and preferences for different alternatives. These two processes, judgment and choice, have been the main focus of the study of decision-making, a field often referred to as "judgment and decision-making".

A decision maker perceives information from its environment and transforms that information to find and create alternatives, build preferences, and evaluate options that lead to a choice. An action is executed and it naturally results in changes in the environment. Then, feedback (i.e., the knowledge of outcomes from actions taken) must be processed in order to reinforce or not past decisions (i.e., learn from past choices). This is referred to as a closed-loop process view of decision-making, and it is the principal flow of events behind most models of dynamic decision-making and models of decisions from experience in dynamic situations. A recent development in decision sciences has expanded on the initial view of choice as a learning process. It has great potential to expand our understanding and provide insights into the dynamic decision-making process. The development involves a shift of attention to how decisions are made based on experience (i.e., decisions from experience), rather than based on explicit descriptions of options (Figure 6.6).

In computer science, AI research is defined as the study of "intelligent agents": any device that perceives its environment and takes actions that maximize its chance of successfully achieving its goals. Cognitive Science researchers use experimental paradigms that involve repeated decisions rather than one-shot decisions, the estimation of possible outcomes and probabilities are based on the observed outcomes rather than from a written description and learning from feedback. All these are natural processes for making

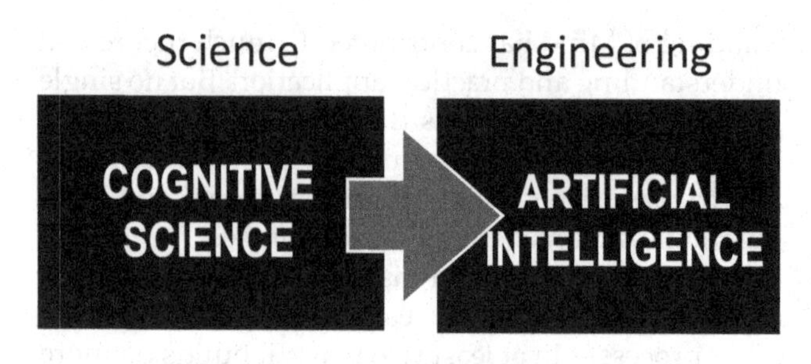

FIGURE 6.6
Relationship between cognitive science and AI model.

decisions in many real-world situations where alternatives, outcomes, and probabilities are unknown. The experimental paradigm often involves two alternatives, represented as two unlabeled buttons, each representing a probability distribution of outcomes unknown to participants. However, Cognitive Science still lacks a comprehensive theory of language learning and use, although different approaches have shed considerable light on different aspects of language. Some aspects of grammar and pronunciation, for example, are plausibly described in terms of rules, but rule-based approaches have helped little with understanding the nature of the lexicon or the role of metaphor in language production and comprehension. Under this point of view, the mind could be characterized as a set of complex associations, represented as a layered network. Some critics argue that there are some phenomena that are better captured by symbolic models, and that connectionist models are often so complex as to have little explanatory power. Recently symbolic and connectionist models have been combined, making it possible to take advantage of both forms of explanation. Cognitive Science aims to explain agency in material terms and in particular, in mathematical terms that bridge logic (mind) and engineering (matter). Finally, although study of the brain largely disappeared from cognitive science in the 1960s and 1970s, partly because research on brain processes seemed too remote to contribute to understanding cognitive operations, the late 1980s and 1990s saw the emergence of cognitive neuroscience (Figure 6.7).

Bottomline, artificial intelligence differs from conventional computer systems in:

1. Being Non-Algorithmic, and
2. Being the only systems that discovers the solution and then executes it. (Other computer systems have the solution designed by the programmer and only execute the solution.)

Computer Science is concerned with theory and fundamentals, while software engineering is concerned with the practicalities of developing and delivering useful software. For example, branch and bound is the core algorithm behind many mixed integer programming (MIP) solvers and is particularly useful for smaller problems or when the problem has numerous constraints. Additionally, its straightforward nature makes it accessible, and no hard math formulas are needed. However, computer science theories are still insufficient to act as a complete underpinning for software engineering. System engineering is concerned with all aspects of computer-based systems development including hardware, software, and process engineering. A structured approach to software engineering is based on developing software that includes system models, notations, rules, design advice, and process guidance.

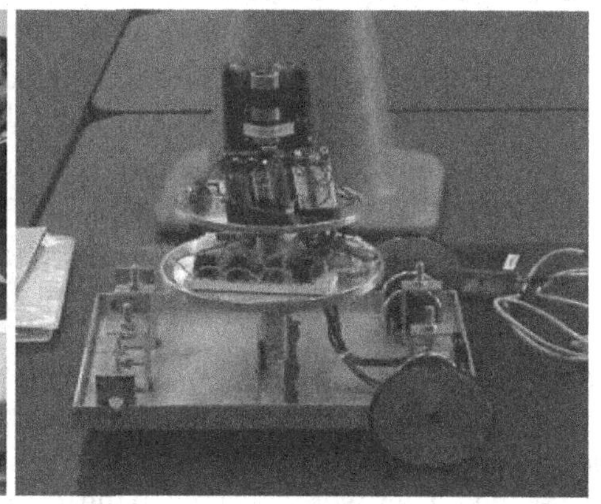

FIGURE 6.7
Students studying robotics.

The *Mythical Man-Month* by author Fred Brooks is a collection of essays on software projects that demonstrated to programmers that they, as a group, are very poor at estimating the costs that are required to complete a software project, where the costs are measured in terms of time, effort, and money. The central theme is that adding manpower to a software project that is behind schedule delays it even longer. This observation led to the emerging discipline of software engineering.

Logic is the science of the correctness or incorrectness of reasoning, or the study of the evaluation of arguments. A statement is a declarative sentence, or part of a sentence, that can be true or false. A proposition is what is meant by a statement (the idea or notion it expresses and this might be the same for different sentences). In logic, an argument is a collection of statements or propositions, some of which are intended to provide support or evidence in favor of one of the others, and premises are those statements or propositions in an argument that are intended to provide the support or evidence required. The conclusion is the statement or proposition for which the premises are intended to provide support. Two things are required of a good argument: (1) its premises have to be true (factually correct), and (2) the premises have to provide support for the conclusion (validation). The theory of correct reasoning is also called the theory of logical inference, the theory of proof, or the theory of deduction. It is often said that the most important critical test of any scientific theory is its usefulness and accuracy in predicting phenomena before the phenomena are observed. Any such prediction must involve the application of the principles of logical inference. Prediction is a major premise behind of AI. The most important defect in the classical tradition of logic was the failure to relate logic as the theory of inference to the kind of deductive reasonings that are continually used in mathematics. Leibniz had some insight into the necessity of making this connection, but not until the latter part of the 19th century and the beginning of this century were systematic relations between logic and mathematics established, primarily through the work of Frege, Peano, and Russell. But only in recent years has there been formulated a completely explicit theory of inference adequate to deal with all the standard examples of deductive reasoning in mathematics and the empirical sciences (Figure 6.8).

FIGURE 6.8
Logic in computer science.

The number of people who have contributed to these developments is extensive, but perhaps most prominent are Kurt Godel, David Hilbert, and Alfred Tarski. In the 1930s, Kurt Gödel founded modern theoretical computer science and introduced a *universal coding language* that was based on the integers. He allowed for formalizing the operations of any digital computer in axiomatic form. Gödel used it to represent data (such as axioms and theorems) and programs (such as proof-generating sequences of operations on the data). He famously constructed formal statements that talk about the computation of other formal statements, especially self-referential statements which imply that they are not decidable. Given a computational theorem prover that systematically enumerates all possible theorems from an enumerable set of axioms, he identified fundamental limits of algorithmic theorem proving, computing, and any computation-based AI. A correct piece of reasoning, whether in mathematics or physics, is valid by virtue of its logical form, and one of the most powerful methods for eliminating conceptual vagueness is to isolate a small number of concepts basic to the subject at hand and then to define the other concepts of the discipline in terms of the basic set.

By the early 1960s, Church in collaboration with Alan Turing defined the logical notion of type with programming language features (previously called "modes") that distinguish between integer and floating-point arithmetic. The ubiquity and utility of types in computer programming may be one reason why computer scientists tend to prefer elaborately typed formalisms. Logic Programming is the name of a programming paradigm that was developed in the 1970s. Rather than viewing a computer program as a step-by-step description of an algorithm, the program is conceived as a logical theory, and a procedure call is viewed as a theorem of which the truth needs to be established. Thus, executing a program means searching for a proof. In traditional (imperative) programming languages, the program is a procedural specification of how a problem needs to be solved. In contrast, a logic program concentrates on a declarative specification of what the problem is. So, logic is the study of formal representations of objects in the real world, and the formal statements that are true about them. Computational Logic looks at the computational aspects of logic. The basic idea is that logical statements can be executed on a machine such as a computer. This has far-reaching consequences that ultimately led to: logic programming, deduction systems for mathematics and engineering, logical design and verification of computer software and hardware, deductive databases and software synthesis as well as logical techniques for analysis in the field of mechanical engineering. Computer Science has utilized many subfields of Logic in areas such as program verification, semantics of programming languages, automated theorem proving, and logic programming. If we want to prove theorems in a formal logic, computer power is necessary and ideally, we would like the computer to prove our theorems automatically. And as a general rule, greater automation requires the choice of a less expressive formalism.

A proof plays a central role in this work because there is a belief with most mathematicians that proofs are essential for genuine understanding. Proofs also play a growing role in computer science; they are used to certify that both software and hardware will always behave correctly, something that no amount of testing can do. Simply put, proof is a method of establishing truth. The goal of computational mathematics, put simply, is to find or develop algorithms that solve mathematical problems computationally (i.e., using computers). In particular, we desire that any algorithm we develop fulfills four primary properties:

- Accuracy. An accurate algorithm is able to return a result that is numerically very close to the correct, or analytical, result.
- Efficiency. An efficient algorithm is able to quickly solve the mathematical problem with reasonable computational resources.
- Robustness. A robust algorithm works for a wide variety of inputs x.
- Stability. A stable algorithm is not sensitive to small changes in the input x.

Logic, itself, is used in natural language processing, theorem proving, game theory, automatic answering systems, ontologies, control systems, and graphical user interfaces.

Computational Logic: is the use of logic to perform or reason about computation. It bears a similar relationship to computer science and engineering in the same way mathematical logic bears to mathematics and as philosophical logic bears to philosophy. Thus, it is synonymous with "logic in computer science". The term was introduced by John A. Robinson. This expression claims that "computational logic" is "surely a better phrase than 'theorem proving', for the branch of artificial intelligence which deals with how to make machines do deduction efficiently". Computational logic has also come to be associated with logic programming, because much of the early work in logic programming in the early 1970s also took place in the Department of Computational Logic at the University of Edinburgh. Some of the key areas of logic that are particularly significant are computability theory (formerly called recursion theory), modal logic, and category theory. The theory of computation is based on concepts defined by logicians and mathematicians such as Alonzo Church and Alan Turing. Church first showed the existence of algorithmically unsolvable problems using his notion of lambda-definability. Turing gave the first compelling analysis of what can be called a mechanical procedure and Kurt Gōdel asserted that he found Turing's analysis "perfect". In addition, some other major areas of theoretical overlap between logic and computer science are: (1) Gōdel's incompleteness theorem, (2) frame problem, (3) Curry-Howard correspondence, and (4) Category theory. From the beginning of the field, it was realized that technology to automate logical inferences could have great potential to solve problems and draw conclusions from facts.

PROLOG (PROgramming in LOGic) is a language designed to enable programmers to build a database of facts and rules, and then to have the system answer questions by a process of logical deduction using the facts and rules in the database. The name PROLOG was originally intended as the name for the programming language developed by Alain Colmerauer and Phillipe Roussel in the summer of 1972. The name was suggested by Roussel's wife, Jacqueline, as an abbreviation for programmation en logique. In time, however, this abbreviation has been used to refer to the concept of logic programming in general. It is a confusing notion, as claims made for the general concept of logic programming do not always hold for the programming language, PROLOG, and vice versa. Logic programming shares with mechanical theorem proving the use of logic to represent knowledge and the use of deduction to solve problems by deriving logical consequences. However, it differs from mechanical theorem proving in two distinct but complementary ways: (1) It exploits the fact that logic can be used to express definitions of computable functions and procedures; and (2) it exploits the use of proof procedures that perform deductions in a goal-directed manner, to run such definitions as programs. A consequence of using logic to represent knowledge is that such knowledge can be understood declaratively. A consequence of using deduction to derive consequences in a computational manner is that the same knowledge can also be understood procedurally. Thus, logic programming allows us to view the same knowledge both declaratively and procedurally. **Symbolic Artificial Intelligence** is the term for the collection of all methods in AI research that are based on high-level symbolic (human-readable) representations of problems, logic, and search. Computational semantics is a relatively new discipline that combines insights from formal semantics, computational linguistics, and automated reasoning. The aim of computational semantics is to find techniques for automatically constructing semantic representations for expressions of human language, representations that can be used to perform inference.

Mathematics

The Greek root of the word "mathematics" is *mathēma*, which means "what is learned". For argument's sake, **mathematics** provides a model of **knowledge**. It is also eternal, that is, what is true now will always be true. Numbers seem to be the content of objective truth and objective truths are about objects. For example, a number is an object, following Plato's theory of Forms and is not explicitly part of space and time. The **theory of Forms** or **theory of Ideas** is a philosophical theory that the physical world is not as real or true as timeless, absolute, unchangeable ideas. Plato believed that true knowledge/intelligence is the ability to grasp the world of **Forms** with one's mind. A Form is *aspatial* (transcendent to space), *atemporal* (transcendent to time) and is an objective "blueprint" of perfection. It is assumed that it is possible to gain knowledge on forms, and mathematics from thinking. Together, **Geometry** is used to organize space in three-dimensions and **Time** is a sequence of numbers in one-dimension, describing knowledge both spatially and temporally. Thus, **logic** is a branch of **mathematics** (logicism) and **computers** are a machine for logic.

Logicism is the thesis that all of mathematics, or core parts of it, can be reduced to logic. This is an initial, rough characterization, since it leaves open, among others, what "logic" is meant to encompass and how it is to be characterized. Theoretical computer science was developed out of logic, the theory of computation (if this is to be considered a different subject from logic), and some related areas of mathematics. So theoretically, minded computer scientists are well informed about logic even when they aren't logicians. Computer scientists in general are familiar with the idea that logic provides techniques for analyzing the inferential properties of languages, and with the distinction between a high-level logical analysis of a reasoning problem and its implementations. Logic, for instance, can provide a specification for a programming language by characterizing a mapping from programs to the computations that they license. A compiler that implements the language can be incomplete, or even unsound, as long as in some sense it approximates the logical specification. This makes it possible for the involvement of logic in AI applications to vary from relatively weak uses in which the logic informs the implementation process with analytic insights, to strong uses in which the implementation algorithm can be shown to be sound and complete. In some cases, a working system is inspired by ideas from logic but acquires features that at first seem logically problematic but can later be explained by developing new ideas in logical theory. This sort of thing has happened, for instance, in logic programming.

McCarthy's methodological position has not changed substantially since it was first articulated in (McCarthy) and elaborated and amended in (McCarthy and Hayes). The motivation for using logic is that even if the eventual implementations do not directly and simply use logical reasoning techniques like theorem proving, a logical formalization helps us to understand the reasoning problem itself. The claim is that without an understanding of what the reasoning problems are, it will not be possible to implement their solutions. Plausible as this Platonic argument may seem, it is in fact controversial in the context of AI; an alternative methodology would seek to learn or evolve the desired behaviors. The representations and reasoning that this methodology would produce might well be too complex to characterize or to understand at a conceptual level. Time and temporal reasoning have been associated with logic since the origins of scientific logic with Aristotle. The idea of a logic of tense in the modern sense has been familiar since at least the work of Jan Lukasiewicz, but the shape of what is commonly known as tense logic was standardized by Arthur Prior's work in the 1950s and 1960s. As the topic was developed in philosophical logic, tense logic proved to be a species of modal logic; Prior's work was heavily influenced by both Hintikka and Kripke, and by the idea that the truth of tense logical formulas is relative to world-states or temporal stages of the world; these are the tense-theoretic analogs of the timeless possible worlds of ordinary modal logic. Thus, the central logical problems and techniques of tense logic were borrowed from modal logic. For instance, it became a research theme to work out the relations between axiomatic systems and the corresponding model-theoretic constraints on temporal orderings.

The purely logical frame problem can be solved using monotonic logic, by simply writing explicit axioms stating what does not change when an action is performed. This technique can be successfully applied to quite complex formalization problems. But non-monotonic solutions to the framework have been extensively investigated and deployed, these lead to new and interesting lines of logical development. Over the last 25 years or so, many profound relations have emerged between logic and grammar. Computational linguistics (or natural language processing) is a branch of AI, and it is fairly natural to classify some of these developments under logic and AI. But many of them also belong to an independent tradition in logical foundations of linguistics; and in many cases, it is hard (and pointless) to attempt a classification. Eventually, researchers began to realize that the most important problem in **natural language processing** was **inference**. The single most important fact about the primitive ACTS was that they helped to organize the inference problem. At this point, we began to take seriously the problem of codifying the kinds of causal relations that there were. This work was crucial to the inference problem since we had come to believe that the major inferences were (forward) consequences and (backward) reasons. Thus, the primary task of the inference process was to fill in causal chains. We identified four kinds of causal links: RESULT, REASON, INITIATE, and ENABLE. The last 4 years have found us developing the system of plans, goals, themes, and scripts for use in understanding systems.

It was the development of the notion of universality of formal mechanism, first introduced in the work on foundations of mathematics in the 1930s, which provided the initial impetus for viewing mind as a symbol processing system. Universality implies that a formal symbol processing mechanism can produce any arbitrary input–output function that we can specify insufficient detail. Put in more familiar terms, a Universal Machine can be programmed to compute any formally specified function. This extreme plasticity in behavior is one of the reasons why computers have from the very beginning been viewed as artifacts that might be capable of exhibiting intelligence. Many people who were not familiar with this basic idea have misunderstood the capacity of machines. Arthur Norman Prior was the main founder of temporal logic and the creator of tense logic. Overall, its work can be broken down into three major moment movements: (1) an anachronistic encounter between Diodorus Chronos and St. Thomas Aquinas; (2) a mathematization of temporal logic; and (3) the encoding of tense logic. Prior's study of the *Master Argument* focuses on the inter-definability between modal and temporal notions.

The same work that provided demonstrations of particular in-principle limitations of formalization also provided demonstrations of its universality. Thus, Alan Turing, Emil Post, and Alonzo Church independently developed distinct formalisms which they showed were complete in the sense that they were powerful enough to formally (i.e., "mechanically") generate all sequences of expressions that could be interpreted as proofs, and hence could generate all provable theorems of logic. In Turing's case, this took the form of showing that there exists a universal mechanism (a particular Turing machine called the Universal Machine "UM") that could simulate any mechanism describable in its formalism. It does this by accepting a description of the mechanism to be simulated, and then carrying out a procedure whose input–output behavior is identical to that which would have been generated by the machine whose description it was given (Figure 6.9).

Statistics provides us with a way to describe how new data can be best used to update our beliefs, and in this way there are deep links between statistics and psychology. In fact, many theories of human and animal learning from psychology are closely aligned with ideas from the new field of machine learning. Machine learning is a field at the interface of statistics and computer science that focuses on how to build computer algorithms that can learn from experience. While statistics and machine learning often try to solve the same problems, researchers from these fields often take very different approaches; the famous statistician Leo Breiman once referred to them as "The Two Cultures" to reflect how different their approaches can be [113]. In this book, we will try to blend the two cultures together because both approaches provide useful tools for thinking about data.

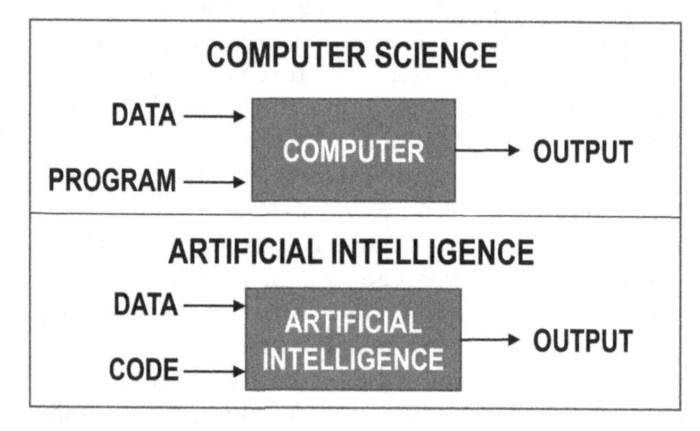

FIGURE 6.9
Computer science vs. artificial intelligence model.

An AI **model** is a mathematical model used to make predictions or decisions. Some of the common types of AI models include:

- Linear regression
- Logistic regression
- Classification
- Segmentation
- Decision trees
- Neural networks/network analysis.

Presupposing cognition as basis of behavior, among the most prominent tools in the modeling of behavior are computational-logic systems, connectionist models of cognition, and models of uncertainty.

Engineering

Computers are relevant to cognition in many ways. Allen Newell had discussed a range of views of the possible relation between computing and cognition. At the most abstract level, the class of mechanisms called computers are the only known mechanisms that are sufficiently plastic in their behavior to match the plasticity of human cognition. They are also the only known mechanism capable of producing behavior that can be described as "knowledge dependent". Because of such properties, computing remains the primary candidate for meeting the dual needs of (1) explaining cognition in mechanistic terms and (2) accounting for certain otherwise problematic aspects of cognition; in particular the fact that behavior can be systematically influenced by inducing differences in beliefs or goals. At a more concrete level, computers provide a way to deal with a number of problems that plague the attempt to understand cognition. Among them are the complexity of the processes underlying cognition, and the need for a theory that bridges the gap from internal processing to actual instances of behavior. Such a theory is sometimes said to meet the "sufficiency condition" (Figure 6.10).

A key issue is "what is the role of computers in computer science". Zenon Pylyshyn examined a variety of such roles from the instrumental use of computers to express theories, through its role as a source of ideas, and to the bold empirical claim that cognition is quite literally a species of computing. In 1951, Turing argued that a properly programmed computer could in principle exhibit intelligent behavior. The argument rests on Turing's own discovery of the existence of a UTM,

FIGURE 6.10
Engineering design.

an abstract automaton that can imitate any other formally specifiable computer. Computers are relevant to cognition in many ways per Allen Newell. For example, at the most abstract level, the class of mechanisms called computers are the only known mechanisms that are sufficiently plastic in their behavior to match the plasticity of human cognition. They are also the only known mechanism capable of producing behavior that can be described as "knowledge dependent". Because of such properties, computing remains the primary candidate for meeting the dual needs of (1) explaining cognition in mechanistic terms and (2) accounting for certain otherwise problematic aspects of cognition; in particular the fact that behavior can be systematically influenced by inducing differences in beliefs or goals. At a more concrete level, computers provide a way to deal with a number of problems that plague the attempt to understand cognition. Among them are the complexity of the processes underlying cognition, and the need for a theory that bridges the gap from internal processing to actual instances of behavior. Such a theory is sometimes said to meet the "sufficiency condition".

In general, mathematicians and computer scientists attempt to determine whether a problem is solvable by trying them out on an imaginary machine. In Turing's original theoretical machine and in every real digital computer, a distinction is made between the processor and the memory. The processor "writes" symbolic expressions into memory, alters them, and "reads" them. Reading certain of these symbols causes specified actions to occur which may change other symbols. The memory may consist of a tape, a set of registers, or any form of working storage. The expressions that are written are complex symbols that are made up of simpler symbols, just the way sentences are complex symbols made up of simpler symbols in a systematic way. The processor (or, in the case of logic, the rules of inference) then transforms the expressions into new expressions in a special kind of systematic way. As has already been mentioned, symbolic expressions have a semantics, that is, they are codes for something, or they mean something. Therefore, the transformations of the expressions are designed to coherently maintain this meaning, or to ensure that the expressions continue to make sense when semantically interpreted in a consistent way. If you can arrange for the computer to transform them systematically in the appropriate way, the transformations can correspond to useful mathematical operations, such as addition or multiplication (Figure 6.11).

In scientific computing, as well as in the history of computer applications up to the 1970s, the most frequently encountered domain of representation was doubtlessly that of numbers, and consequently, the most common transformations over expressions were those that mirror mathematical functions over numbers. But if the symbolic expressions were codes for propositions or beliefs or knowledge, as they might be if they were expressions in some symbolic logic, then the computer might transform them in ways corresponding to proofs or inferences, or perhaps to a sequence of "thoughts" that occur during commonsense reasoning. The important thing is that, according to the Classical View, certain kinds of systems, which include both minds and computers, operate on representations that take the form of symbolic codes. There is one more important property that such symbolic codes must have, according to the Classical View. In the Classical View, symbol systems display the meanings of a complex expression or depend in a systematic way on the meaning of its parts (or constituents). This is the way ordinary language, formal logic, and even the number system work, and there are good reasons believing that they must work that way in both practical computing and modeling cognition. The idea that the brain thinks by writing symbols and reading them sounds absurd to many. It suggests to some people that we have been influenced too much by the way current electronic computers work. The basic source of uneasiness seems to come from the fact that we do not have the subjective experience that we are manipulating symbols. The content of knowledge is related to the state of a system by a semantic relation, which is quite a different relation from the ones that appear in natural laws (for one thing, the object of the relation need not exist). It says that knowledge is encoded by a system of symbolic codes, which themselves are physically realized, and that it is the physical properties of the codes that cause the behaviors in question. It must be stressed that at present there exists no alternative to what Allen Newell has called the "physical symbol system" assumption for dealing with reasoning in a mechanical way, even though there are many speculative discussions of how one might eventually be able to do without symbols.

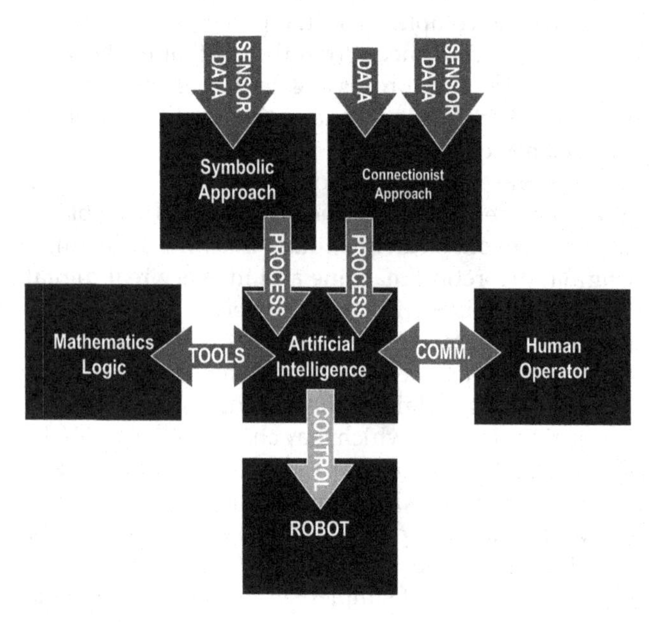

FIGURE 6.11
AI/robotics model.

It may be *premature* to some to consider combining the two main approaches to AI, since so far neither has accomplished enough to be on solid ground. However, NN modeling may simply be getting a deserved chance to fail as did the symbolic approach. Still there is an important difference to remember as each research program struggles on. The physical symbol system approach seems to be failing because it is simply false to assume that there must be a theory of every domain. NN modeling, however, is not committed to this or any other philosophical assumption. However, simply building an interactive net sufficiently similar to the one our brain has evolved to may be just too hard. Indeed, the commonsense knowledge problem, which has blocked the progress of symbolic representation techniques for 20 years, may be looming on the neural net horizon, although connectionists may not yet recognize it. All NN modelers agree that for a net to be intelligent, it must be able to generalize, that is, given sufficient examples of inputs associated with one particular output, it should associate further inputs of the same type with that same output. The Connectionist (Subsymbolic) Hypothesis states: "The intuitive processor is a subconceptual connectionist dynamical system that does not admit a complete, formal and precise conceptual-level description" (Figure 6.12). However, the inner workings of an artificial neural network (ANN) are difficult to interpret, but the issue to ponder are they substantially different to a symbolic system?

As Zenon Pylyshyn suggested, computers can enter into the detailed process of constructing models of cognitive processes at several levels. By working within the AI tradition, we have been more concerned with explaining the general abilities or capacities in question, while postponing the detailed empirical validation of the mechanisms and algorithms actually used in the model. David Marr has proposed that there are three levels at which cognitive processes may be studied. He referred to these as the level of the computation, the level of the algorithm, and the level of the mechanism. The area of Cognitive Science that is sometimes known as "Information Processing Psychology" has been dominated by the empirical validation of mini-models. The commitment to the construction of a model meeting the sufficiency condition, that is, which actually generates token behaviors, forces one to confront the problem of how, and under what conditions, the internal representations and the rules are invoked in the course of generating actions. The most common idea has been that of control moving from point to point, or from instruction to instruction, in a largely predetermined way. Such sequencing of instructions makes the notion of flow of control quite naturally, and branch instructions make it equally natural to think of passing or sending control to another locus. When control passing is combined with a primitive message-passing facility (for passing arguments), subroutines become possible. And since subroutines can be nested, that is, subroutines can themselves send control to still lower subroutines with the assurance that it will eventually find its way back, the notion of a hierarchy of control also emerges. Cognitive Science is ultimately interested in whether the computational model is empirically valid whether it corresponds to human cognitive processes or not. The notion of an algorithm is somewhat better established in computer science than is the notion of method. The general assumption in cognitive science has been that the appropriate level of comparison corresponds roughly to the intuitive notion of an algorithm. And every algorithm is different in theory, but many algorithms use similar approaches. For example, if you can divide a problem into pieces and able to solve the pieces separately, a divide-and-conquer approach may work well. Or if you can define a problem as being made up of smaller problems, a recursive method may be useful.

In response to the need to design this declarative component, a subfield of AI known as knowledge representation emerged during the 1980s. Knowledge representation deals primarily with the representational and reasoning challenges of this separate component. Theoretical work in logical AI and in philosophical logic overlap to a large extent. Both are interested in developing non-metamathematical applications of logic, and the core topics are very similar. This overlap is due not only to commonality of interest but to direct influence of philosophical logic on logical AI; there is ample evidence, as we will see, that the first generation at least of AI logicists read and were influenced by the literature in philosophical logic. The importance of applications in logical AI, and the scale of these applications, represent a new methodology for logic; one that would have been impossible without mechanized reasoning.

A **model** is a program that outputs the data. AI practitioners model the data, because they need to do something with that model, that is, making predictions. Figure 6.12 depicts a young engineer building his own robot. In computer science, "A model of x is a Turing Machine that outputs x and then Halts". When they update the model, they use Bayesian theorem. There is a probability distribution over all possible models, and to get a new data point, the distribution is updated with this observation, using

$$P(H \mid O) = P(O \mid H) \quad P(H), P(O) \tag{6.1}$$

referred to as the normalizing constant.

So, we have our observed string x and all programs that compute x (and more). Technically, for all minimal programs, removing any bits from the end will cause it not to compute x anymore:

- a probability distribution on that set,
- and we update that distribution with each new observation using Bayes law.

Bayesian model updating aims to estimate uncertain model parameters by minimizing the discrepancies between measured and predicted responses.

FIGURE 6.12
Young engineer building a robot.

Cognitive algorithms, the central concept in computational psychology, are understood to be executed by the cognitive architecture. According to the strong realism view that many of us have advocated, a valid cognitive model must execute the same algorithm as that carried out by subjects. In order for an algorithm to serve as a model of a cognitive process, it must be presented in some standard or canonical form or notation, for example, as a program in some programming language. Although this machine is universal, in the sense that it can be programmed to compute any computable function, anyone who has tried to write procedures for it will have noticed that most computations are extremely complex. More importantly, however, the complexity of the sequence of operations it must go through varies with such things as the task and the nature of the input in ways that are quite different from that of machines with a more conventional architecture. The distinction between directly executing an algorithm and executing it by first emulating some other functional architecture is crucial to cognitive science. It bears on the central question of which aspects of the computation can be taken literally as part of the model and which aspects are to be considered as mere implementation details. From the point of view of cognitive science, it is important to be explicit about why a model works the way it does and to independently justify the crucial assumptions about the cognitive architecture. That is, it is important for the use of computational models as part of an explanation, rather than merely in order to mimic some performance, that we not take certain architectural features for granted simply because they happen to be available in our computer language. Connectionists claim that symbol systems are not needed or there are very good reasons for maintaining that reasoning and other knowledge-dependent or rational processes requiring symbol processing and, moreover, that these processes are extremely pervasive in the phenomena that have been studied in cognitive science.

The problems of heuristic programming, of making computers solve really difficult problems, are divided into five main areas: Search, Pattern Recognition, Learning, Planning, and Induction. A computer can do, in a sense, only what it is told to do. But even when we do not know how to solve a certain problem, we may program a machine (computer) to Search through some large space of solution attempts. Unfortunately, this usually leads to an enormously inefficient process. With Pattern-Recognition techniques, efficiency can often be improved, by restricting the application of the machine's methods to appropriate problems. Pattern Recognition, together with Learning, can be used to exploit generalizations based on accumulated experience, further reducing the search. By analyzing the situation and using Planning methods, we may obtain a fundamental improvement by replacing the given search with a much smaller, more appropriate exploration. To manage broad classes of problems, machines will need to construct models of their environments, using some scheme for Induction.

In refuting one's inability to articulate knowing "how", as evidence of a qualitatively different kind of knowledge, Fodor offers a distinction between mental competences or skilled abilities and mental traits like intelligence or sensibility. Knowing "how" to do something is evidence of a competency, but not necessarily a trait like intelligence. Moreover, traits like intelligence are not dependent on competencies. Richer knowledge representations require advances in robotic movement, vision, and interaction to learn from its environment; yet such advances in robotics first require advances in knowledge representations in such fundamental areas such as representing the robot's own body, the solidity of objects, the effects of movement on perspective and more. Dreyfus does not consider enriched representational schemes any kind of advance toward machine understanding at all. The problem lies in an unjustified belief concerning human ability in the world: why would one consider, as Minsky and Fodor do, the explicit representation of human practices to be formalizable? This makes sense only in the context of the highly constrained microworld in which a program operates and reveals serious discrepancies between microworlds and the real world of human experiences. The attempt to gain machine understanding through enriched representational schemes of the world has, so far, met with failure.

NN (Neural Network): performs "shallow learning", as opposed to "deep learning" which utilizes multiple layers. The NNs is based on the method of least squares and linear regression. Linear regression is a type of regression model that assumes a linear relationship between the target and features, while least squares regression is a method used to find the optimal parameters for a linear regression model. In an NN, the forward pass computes the output of the network given the input data, while a backward pass computes the output error for the expected output and then goes backward into the network and updates the weights using gradient descent. Backpropagation is the combination of the forward pass and the backward press where the idea is to calculate the difference between prediction and actual values to fit the hyperparameters of the method used.

> Backpropagation applies forward and backward passes, sequentially and repeatedly while trying to minimize the errors. The backpropagation algorithm is widely used to train deep NNs and is just Leibniz's chain rule of differential calculus. Backpropagation is essentially an efficient way of implementing the chain rule for deep networks. Gradient descent teaches an NN to translate input patterns from a training set into desired output patterns. Thus, gradient descent is the process of using gradients to find the minimum value of the cost function, while backpropagation is calculating those gradients by moving in a backward direction in the NN. Augustin Louis-Cauchy's gradient descent uses the incremental weakening of certain NN connections and strengthens others in the course of many trials, such that the NN behaves more and more like some teachers.

A NN is also referred to as a **biological neural network** and is based on thoughts and body activities that result from interactions among neurons within the brain. The preliminary theoretical underpinning for NNs was proposed by Alexander Bain (logic) in 1873 and William James (psychology) in 1890. Bain postulated that for each human activity, certain sets of neurons would fire and when repeated, the connections between those neurons would strengthen. Similarly, James came to the same conclusion but also suggested that memories and actions were the result of electrical currents flowing among the neurons in the brain. In 1898, C.S. Sherrington conducted experiments to test James' Theories. In 1943, McCullough and Pitts created a computational model of NNs based on mathematics and algorithms and called the model "threshold logic". In 1986, Rumelhart and McClelland provided a full exposition on the use of connectionism in computers to simulate neural processes. The brain learns by changing the strengths or weights of the connections, which determine how strongly neurons influence each other, and which seem to encode all your lifelong experience.

ELM (Extreme Learning Machines) are feedforward NNs. A recurrent NN (RNN) is characterized by the direction of the flow of information between its layers, in contrast to a uni-directional feedforward NN. It is a bi-directional ANN, meaning that it allows the output from some nodes to affect subsequent input to the same nodes. RNNs have feedback connections, such that one can follow directed connections from certain internal nodes to others and eventually end up where one started. The most popular RNN is the Long Short-Term Memory (LSTM) and the most popular feedforward NN is a version of the LSTM-based Highway Net called ResNet. Successful learning in *deep* feedforward network architectures started in 1965 in Ukraine when Alexey Ivakhnenko and Valentin Lapa introduced the first general, working learning algorithms for deep MLPs with arbitrarily many hidden layers (already containing the now popular multiplicative gates). Seppo Linnainmaa publishes what's now known as backpropagation, the famous algorithm for credit assignment in networks of differentiable nodes, also known as "reverse mode of automatic differentiation". Backpropagation is now the foundation of widely used NN software.

The field of machine learning has used statistical modeling, including generative models, to model and predict data. Beginning in the late 2000s, the emergence of deep learning drove progress and research in image and video processing, text analysis, speech recognition, and other tasks. ANNs can be most adequately characterized as computational models with particular properties such as the ability to adapt or learn to generalize or to cluster or organize data and which operation is based on parallel processing. However, most deep NNs were trained as discriminative models, performing classification tasks such as convolutional NNs-based image classification. **NNs** (also known as connectionist models or PDP) are the functional unit of *Deep Learning* and are known to mimic the behavior of the human brain to solve complex data-driven problems. It consists of a pool of simple processing units that communicate by sending signals to each other over a large number of weighted connections. The input data is processed through different layers of artificial neurons stacked together to produce the desired output.

The prominent NN architectures include:

- Feedforward NNs, including perceptrons and radial basis function networks, transforming patterns from input to output. They are the archetypical NN, having layers that consist of either input, hidden, or output nodes.
- Recurrent networks are feedforward networks with connections within layers.
- LSTMs provide a resolution for the vanishing and exploding gradient problems, by introducing gates and explicitly defined memory cells.
- Autoencoders compress (encode) and regenerate (decode) information by transforming it through a smaller hidden layer with symmetrical surrounding layers.
- In Hopfield networks, each neuron is connected to all other neurons, and all neurons

are both input and output nodes. (Restricted) Boltzmann machines are similar (considered an extension to the Hopfield Network) to the extent that only some neurons are input neurons, while others are hidden.

- Convolutional networks are deep learning architectures that typically contain convolutional and pooling layers, used for approximate scanning of patterns that are often spatially correlated.
- Generative adversarial networks or GANs actually consist of two networks: one tasked with generating data (the generator), and the other with predicting whether the data have been generated or not (the discriminator) (Figure 6.13).

In 2014, advancements such as the variational autoencoder and generative adversarial network produced the first practical deep NNs capable of learning generative, rather than discriminative, models of complex data such as images. **Generative artificial intelligence** or **generative AI** is a type of AI system capable of generating text, images, or other media in response to prompts. Generative AI models learn the patterns and structure of their input training data and then generate new data that has similar characteristics. These deep generative models were the first able to output not only class labels for images, but to output entire images. In 2017, the Transformer network enabled advancements in generative models, leading to the first generative pre-trained transformer in 2018. This was followed in 2019 by *GPT-2* which demonstrated the ability to generalize unsupervised to many different tasks as a *Foundational model*. A foundation model is a large AI model trained on a vast quantity of data at scale (often by self-supervised learning or semi-supervised learning) resulting in a model that can be adapted to a wide range of downstream tasks. In 2021, the release of *DALL-E*, a transformer-based pixel generative model, followed by *Midjourney* and *Stable diffusion* marked the emergence of practical high-quality AI art from natural language prompts. In 2023, *GPT-4* was released. A team from Microsoft Research concluded that "it could reasonably be viewed as an early (yet still incomplete) version of an artificial general intelligence (AGI) system". Generative AI has potential applications across a wide range of industries, including art, writing, software development, healthcare, finance, gaming, marketing, and fashion. Recent progress in generative AI has been driven by four main factors: computing power, earlier innovations in model architecture, having the ability to "pre-train" using large amounts of unlabeled data, and making refinements in training techniques. However, there are also concerns about the potential misuse of generative AI, such as in creating *fake news* or *deepfakes*, which can be used to deceive or manipulate people. ChatGPT, Bing, and Bard excel at making stuff up, even if you just want facts (Figure 6.14).

ANNs are helpful for solving complex problems, while CNNs (Convolution Neural Networks) are best for solving Computer Vision (CV)-related problems. RNNs (Recurrent Neural Networks) are proficient in natural language processing. In general, CNN tends to be a more powerful and accurate way of solving classification problems and ANN is still dominant for problems where datasets are limited, and image inputs

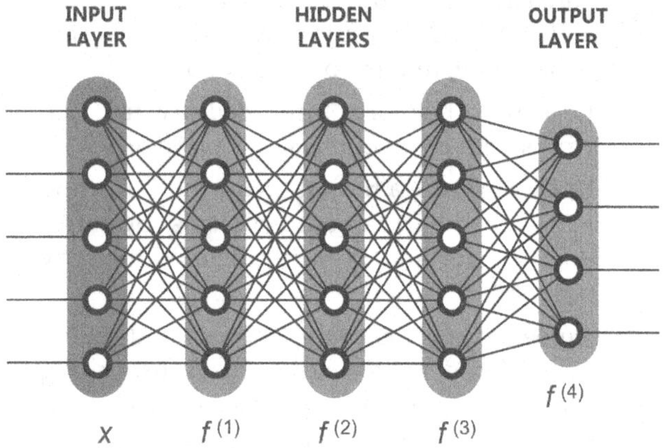

INPUT LAYER HIDDEN LAYERS OUTPUT LAYER

x $f^{(1)}$ $f^{(2)}$ $f^{(3)}$ $f^{(4)}$

FIGURE 6.13
Deep neural network model.

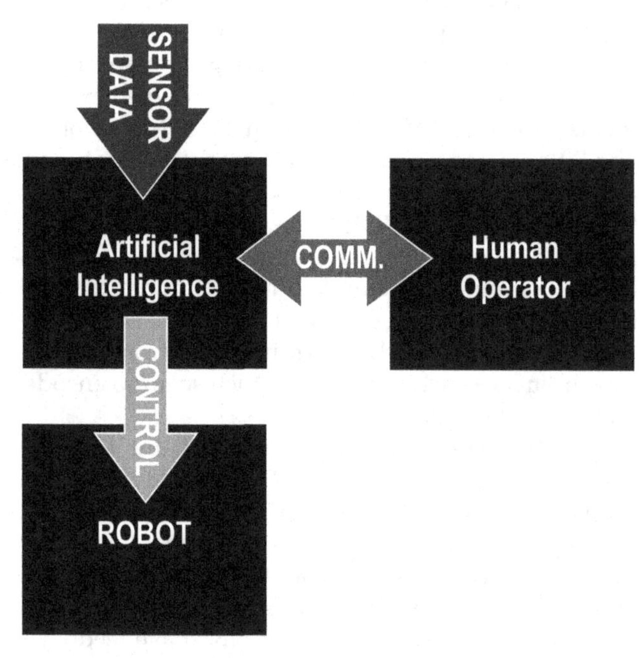

FIGURE 6.14
Sense-Plan-Act robot paradigm.

are not necessary. As in robots, Stuart Russell equates intelligence with the ability to act **rationally**; that is, to choose actions that can be expected to achieve one's objectives. For example, reaching one's goal or objective requires reasoning, planning, decision-making, perception, and learning. Control theorists would minimize cost functions; operations researchers maximize rewards; statisticians minimize an expected loss function; and economists, of course, maximize the utility of individuals, the welfare of groups, or the profit of corporations. Of great importance is the plasticity of human preferences, which brings up both the philosophical problem of how to decide on behalf of a human whose preferences change over time (Pettigrew) and the practical problem of how to ensure that AI systems are not incentivized to change human preferences in order to make them easier to satisfy. In addition to being used for feature learning, NNs can also be applied to reinforcement learning as a policy search method, by representing the controller as a NN and optimizing the controller's parameters using a genetic algorithm.

Discussion on Symbolic and Connectionist AI

In the philosophy of artificial intelligence, GOFAI ("Good old fashioned artificial intelligence") is classical symbolic AI, as opposed to other approaches, such as neural networks, situated robotics, narrow symbolic AI or neuro-symbolic AI. "Representationalists" hold that postulating representational (or "intentional" or "semantic") states are essential to a theory of cognition; and according to Representationalists, there are states of mind that function to encode states of the world. In contrast, "Eliminativists" think that psychological theories can dispense with such semantic notions as **representation**. According to Eliminativists, the appropriate vocabulary for psychological theorizing is neurological (or perhaps behavioral or possibly syntactic); but in any event, there is not a vocabulary that characterizes mental states in terms of what they represent. Connectionists are on the Representationalist side of this debate and as referenced by Rumelhart and McClelland: "PDPs are explicitly concerned with the problem of internal representation". It is asserted that the specification of what the states of a network represent is an essential part of Connectionist's model.

Representationalism claims that the "cognitive level" is dispensable in favor of a more precise and biologically motivated level of theory. For example, there are a lot of discussions in the Connectionist literature about processes that are "subsymbolic", and therefore presumably not representational. This is misleading because connectionist modeling is consistently representational in practice, and representationalism is generally endorsed by every theorist who also likes the idea of cognition "emerging from the subsymbolic". Thus, Rumelhart insisted that PDP models are "...strongly committed to the study of representation and process". Similarly, Smolensky takes Connectionism to articulate regularities at the "subsymbolic level" of analysis, and it turns out that subsymbolic states do have semantics even though it's not the semantics of representations at the "conceptual level".

To avoid confusion and misunderstandings, we adopt the terms *symbolic*, *subsymbolic*, and *analogical* with the following meanings. "Symbolic" stands for a representation system in which the *atomic* constituents of representations are, in their turn, representations. Such a representation system has a compositional syntax and semantics. The typical case of a symbolic system is an interpreted logical theory. We call a representation "subsymbolic" if it is made by constituent entities that are not representations in their turn, for example, pixels, sound images as perceived by the ear, and signal samples; subsymbolic units in NNs can be considered particular cases of this category. The term "analogical" refers to a representation in which the constituents and their relations are one-to-one with the represented reality. In this category, we include mental models defined by Johnson-Laird, mental imagery, and diagrammatic representations. Analogical representations play a crucial role in multimedia knowledge representation. Note that, in this sense, analogical is not opposite to digital; in the hybrid model, we are developing analogical components that are implemented by computer programs (Table 6.1).

TABLE 6.1

Table of symbolic vs. subsymbolic comparison

Symbolic AI	Subsymbolic AI
- Explicit symbolic programming - Inference, search algorithms - AI programming languages - Rules, ontologies, plans, goals, ... - Introspection more useful for coding - Easier to Debug - Easier to explain - Easier to control - Small Data - More useful for explaining people's thoughts - Better for abstract problems	- Bayesian learning - Deep learning - Connectionism - Neural Nets/Backpropagation - LDA, SVM, HMM, PMF, alphabet soup - Robust against noise - Better performance - Less apriori knowledge upfront - Easier to scale up - Big Data - More useful for connecting neuroscience - Better for perceptual problems

Connectionism refers to the explanation of cognition as arising from the interactions between simple (subsymbolic) processing elements. It has close links to cybernetics, which focuses on the development of control structures from which intelligent behavior emerges. Connectionism came to be equated with the use of ANNs that abstract away from the details of biological NNs. An ANN is a computational model that is loosely inspired by the human brain as it consists of an interconnected network of simple processing units (artificial neurons) that learn from experience by modifying their connections. By connecting multiple neurons, one obtains a NN that implements some nonlinear function $y=f(x)$, where the f_i are nonlinear transformations and stands for the network parameters (i.e., weight vectors). After training a NN, representations become encoded in a distributed manner as a pattern that manifests itself across all its neurons. Alan Turing was one of the first to propose the construction of computing machinery out of trainable networks consisting of neuron-like elements (Copeland and Proudfoot). Marvin Minsky, one of the founding fathers of AI, is credited for building the first trainable ANN, called SNARC.

A completely different approach to the explanation of cognition as emerging from bottom-up principles is the view that cognition should be understood in terms of formal symbol manipulation. This computationalist view is associated with the cognitivist program which arose in response to earlier behaviorist theories. It embraces the notion that to understand natural intelligence, one should study internal mental processes rather than just externally observable events. That is, cognitivism asserts that cognition should be defined in terms of formal symbol manipulation, where reasoning involves the manipulation of symbolic representations that refer to information about the world as acquired by perception. This view is formalized by the physical symbol system hypothesis (Newell and Simon), which states that "a physical symbol system has the necessary and sufficient means for intelligent action". This hypothesis implies that artificial agents when equipped with the appropriate symbol manipulation algorithms, will be capable of displaying intelligent behavior.

We investigate AI and robotics using a structured systems engineering approach, where we investigate the robot's operating environment (terrestrial, space, underwater, etc.) and AI's operating system, and try to understand the various interfaces (software to software and software to hardware) within each system/subsystem. After understanding the problem to be solved, requirements, constraints, functional analysis, and having a concept of operation, the next phase will be to develop the robot's architecture, both functionally and physically. This results in a robot hardware architecture and a software (including algorithms) architecture. The key phase in this engineering design process is integration and verification, including testing as the primary form of verification. Now getting back to AI and in reality, we know very little about the human brain. We know about the connections, but we don't know how information is processed. Learning, for example, doesn't just require good memory but also depends on speed, creativity, attention, focus, and, most importantly, flexibility. It's not that we use 10% of our brains, but merely that we only understand about 10% of how it works. The brain is the most complex part of the human body. It is the seat of intelligence, interpreter of the senses, initiator of body movement, and controller of behavior. What seems apparent about the wonders of the brain is that it is more efficient than the analogies that we are using with software and CPU/GPU hardware.

There are two major pathways for AI development: (1) symbolic AI and (2) the connectionist approach to AI. Symbolic AI is commonly known as rule-based AI, good old-fashioned AI (GOFAI), and classic AI. Classical AI uses explicit and logical rules arranged in a hierarchy to manipulate symbols in a serial manner. Early AI development research is based on symbolic AI, which relied on inserting human behavior and knowledge in the form of computer codes. Humans have used symbols to drive meaning from things and events in the environment around us. Thus, symbols become the building block for cognition. Any application made with symbolic AI has a combination of characters signifying real-world concepts or entities through a series of symbols. These symbols can easily be arranged through networks and lists, arranged hierarchically. Such arrangements tell the AI algorithms how each symbol is related to each other in totality. Information in symbolic AI is processed through something that is referred to as an **expert** system. It is where the if/then pairing directs the algorithm to the parameters on which it can behave. These expert systems are man-made **knowledge bases**, and the **inference** engine is a term given to a component that refers to the knowledge base and selects rules to apply to given symbols.

Symbolic artificial intelligence is the term for the collection of all methods in AI research that are based on high-level symbolic representations of **problems**, **logic**, and **search**. Symbolic AI uses tools such as logic programming, production rules, semantic nets, and frames and is used to develop applications such as knowledge-based systems (especially, expert systems), symbolic mathematics, automated theorem provers, ontologies, the semantic web, and automated **planning** and **scheduling** systems. The symbolic AI paradigm led to seminal ideas in search, symbolic programming languages, agents, multi-agent systems, and the strengths and limitations of formal **knowledge** and **reasoning** systems. Researchers in the 1960s and the 1970s were convinced that symbolic approaches would eventually succeed in creating a machine with artificial general intelligence and considered the ultimate goal of this field. Critics, however, argued that it didn't actually channel heuristic thinking, which includes guesswork and shortcuts, and instead showed precise trial-and-error problem-solving. In other words, it could approximate the workings of the human mind but not the spontaneity of its thoughts.

Pros and Cons of Symbolic AI: Symbolic AI is well suited for applications that are based on crystal clear rules and goals. If you want this AI to beat a human in the game of chess, then we need to teach the algorithm the specifics of chess. This framework acts like a boundary that helps it to operate properly. However, symbolic AI falls short when it is required to encounter variations. Taking an example of machine vision, which might look at a product from all the possible angles. It would be tedious and time-consuming to create rules for all the possible combinations. The real world contains huge amounts of data and numerous variations. It is difficult to anticipate all the possible alterations in a given environment. Finally, symbolic AI suffers from the exponential explosion needed for timely search problems and fails the *Commonsense Knowledge Problem*. The **commonsense knowledge problem** is to create a database that contains the general knowledge most individuals are expected to have, represented in an accessible way to AI programs that use natural language.

In general, information travels from the input layer to the output layer in an artificial cognition model. The publication of *Parallel Distributed Processing* in 1986 signaled the beginning of the connectionist revolution in cognitive science. Interest in connectionist style of models all but died in 1969 when Marvin Minsky and Seymour Papert published *Perceptrons*, a book that showed the severe limitations of the state of the art at the time where perceptrons were not capable of calculating a simple exclusive OR function. In general, perceptrons operate on a simple input–output model, similar to the behaviorist stimulus-response paradigm. The Connectionist AI group showed that changing the structure of the networks can free the resulting systems from the limitations shown previously by Minsky and Papert. These changes came by adding additional levels to the network architecture (see next figure). These models (which are called "feedforward" because information always travels from the input layer toward the output layer) in conjunction with a learning rule called "backpropagation", have been used in countless models of cognition and are what most people think of when discussing NNs. The additional layer of the feedforward architecture allows for more complex processing and therefore the feedforward model can model a wider range of behaviors.

Connectionism is a theory of information processing. Connectionism relies on the parallel processing of sub-symbols, using statistical properties instead of logical rules to transform information. Connectionism is often mistaken for associationism, but connectionism has borrowed concepts from associationism such as distributed representations, hidden units, and supervised learning. **Connectionism artificial intelligence** is an approach to modeling perception and cognition that explicitly employs some of the mechanisms and styles of the processing that is believed to occur in the human brain. Basically, we are mimicking our physical brains through machine-learning algorithms and NNs, rather than simulating how we reason. In particular, connectionist models usually take the form of NNs, which are composed of a large number of very simple components wired together. NN models were inspired by and resemble the anatomy and physiology of the human nervous system. Key aspects of many NN models are that they are able to learn and their behavior improves with training or experience. **Connectionism** is an approach to AI that was developed out of attempts to understand how the human brain works at the neural level and, in particular, how people learn and remember (For that reason, this approach is sometimes referred to as neuron-like computing). This AI is based on how a human mind functions and its neural interconnections based on the way the human brain is built and this prevailing connectionist approach was originally known as PDP. It is an ANN approach that stresses the parallel nature of neural processing, as well as the distributed nature of neural representation. This technique of AI software development is also sometimes called a perceptron to signify a single neuron. An application built with Connectionist AI tends to get more intelligent as we keep on feeding data and learning patterns, and the relations associated with the environment and with itself (the notion of learning). On the other hand, symbolic AI gets hand-coded (Figure 6.15).

To understand Connectionist AI, let's take the example of an ANN. Each one is made up of hundreds of single units processing elements and artificial neurons. They are a layered format with weights forming connections with this structure where weights are adjustable parameters. In Connectionist AI, all the processing

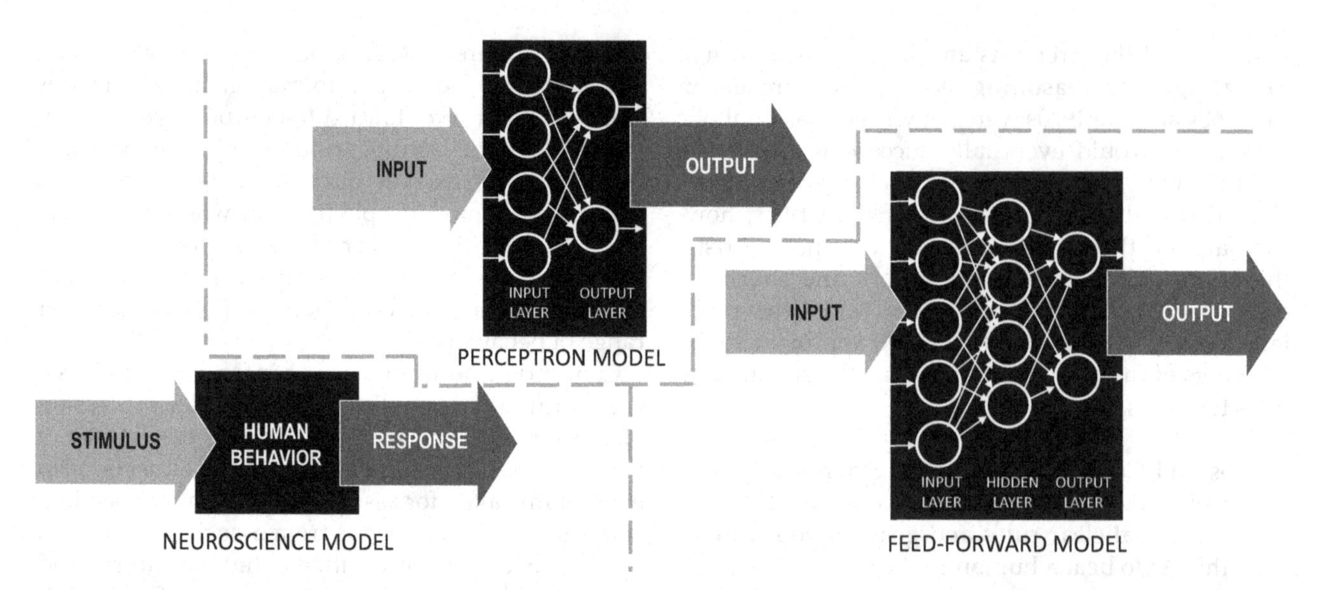

FIGURE 6.15
Neuroscience vs. perceptron vs. feedforward model.

elements have weighted units, output, and a transfer function. The transfer function assesses multiple inputs and then combines them into a single output value. Each weight in the algorithm efficiently evaluates directionality and importance and eventually the weighted sum is the component that activates the neuron. When all is done, then the activated signal passes through the transfer function and produces one output.

Connectionist AI has eight essential properties:

1. A set of processing units
2. A state of activation
3. An output function for each unit
4. A pattern of connectivity among units
5. A propagation rule for propagating patterns of activities through the network of connectivity's
6. An activation rule for combining the inputs impinging on a unit with the current state of that unit to produce a new level of activation for the unit
7. A ***learning rule*** whereby patterns of connectivity are modified by experience
8. An environment within which the system must operate.

These eight properties are mapped onto the six functional properties of the neuron.

Pros and Cons of Connectionist AI: When you have high-quality training data, Connectionist AI is a good option to be fed with that data. Even though this AI model gets smarter as data is fed into it, it still needs the support of accurate information (cleansing the data) to start the whole learning process. This approach is commonly used when there is a large amount of images/data that are required to be verified by humans for correctness and assign annotations for contexts. With all the pros, this AI often cannot explain how it reaches its solution. Thus, it is advised not to select this AI as the primary or the sole choice as the conclusions drawn by it cannot be explained and would require the help of a third party. This approach is non-deterministic, which does not explain how it reached the conclusion that it did. Early researchers focused on the symbolic-type of AI however, while today, the **Connectionist AI** is more popular.

Connectionist work, in general, does not need to be biologically realistic and therefore suffers from a lack of neuroscientific plausibility. However, the structure of NNs is derived from that of biological neurons, and this parallels low-level AI structure, and is often argued to be an advantage of connectionism in modeling cognitive structures compared with other approaches. One area where connectionist models are thought to be biologically implausible is with respect to error-propagation networks that are needed to support learning, but error propagation can explain some of the biologically generated electrical activity seen at the scalp in event-related such as the **N400** (a component of time-locked Electroencephalography (EEG)

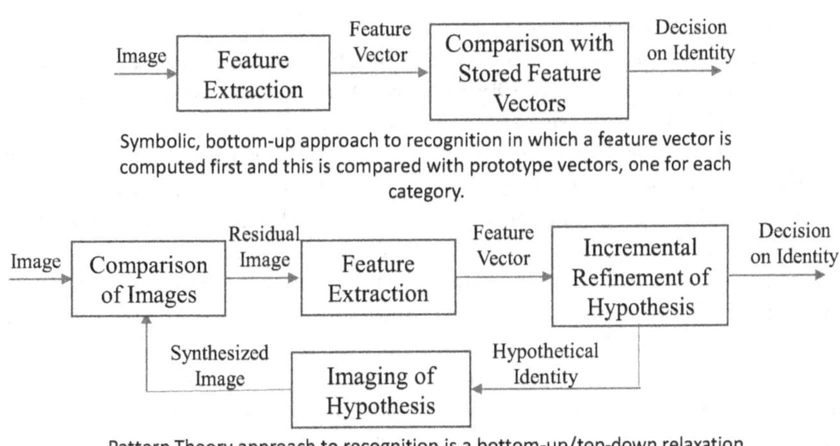

FIGURE 6.16
Bottoms-up recognition vs. bottom-up, top-down recognition.

signals known as event-related potentials (ERP)) and **P600** (an ERP component, or peak in electrical brain activity measured by EEG). It is a language-relevant ERP component and is thought to be elicited by hearing or reading grammatical errors and other syntactic anomalies, and this provides some biological support for one of the key assumptions of connectionist learning procedures.

The basic problem of object recognition is to determine the nature of the isomorphic relation between the connectional hierarchy in the sensory brain and the hierarchical representation of perceptual processing that leads to object recognition. ANNs have been trained to perform object recognition. In psychology, ANNs are called connectionist models. Common features of connectionist models:

1. they assume the distribution of knowledge in assemblies of units, neurons, or nodes that represent the component elements of knowledge.
2. the nodes are interconnected in networks by synapses.
3. the networks are layered.
4. some connections between layers are reciprocal, supporting re-entrant processing.
5. layers are connected by parallel, convergent, and divergent connections.
6. networks learn by modification of synaptic weights.
7. they learn by unsupervised learning: synapses are strengthened (weights increase) by temporal coincidence of pre- and post-synaptic activity.

Object recognition as interaction of sensation and memory in cortex. Object recognition implies that learning of an object has previously occurred. Therefore, the sensory analysis used in perception is thought to be guided by perceptual knowledge stored as perceptual network memories. And there is a difference between passive and active perception. For object recognition to be guided implies that it is an active, not a passive, process. But this conclusion is not universally accepted. For the case of vision, we next consider the competing views of perception as being either passive or active. The difference between passive and active perception may be expressed in computational terms: passive perception is strictly feedforward (top figure), whereas active perception involves feedback (bottom figure) from higher "downstream" stages to lower "upstream" stages (Figure 6.16).

ART (Adaptive Resonance Theory) models carry out hypothesis testing, search, and incremental fast or slow, self-stabilizing learning, recognition, and prediction in response to large nonstationary databases (big data). Perception is seen to consist of the classification of sensory items by the binding of features according to Gestalt grouping principles. Sensory information arriving at the cortex is organized according to sets of spatial and temporal relations between elementary sensory features. These relations define the informational structure that becomes stored in memory through perception. Visual perception involves the recognition of objects (figure) as distinct from their backgrounds (ground). Objects appear to "stand out" from the background. Figure-ground perception in vision usually depends on edge assignment and how that effects shape perception. It may be bistable, meaning that either of two (stable) figures may be perceived. This may occur when a visual pattern is too ambiguous for the visual system to recognize it with a single unique interpretation. Object recognition is seen as the categorization of sensory information according to the memory structure that has been built up by prior experience. Memory structure is viewed as a spatial pattern of linked perceptual memory networks.

The history of both approaches is illustrated in the next figure where the research activities have cycled from a connectionist approach to a symbolic approach and back to the connectionist approach. There has been a complete paradigm shift away from expert systems. The current AI approach has become inductive; it is no longer a question of coding rules as for expert systems, but of letting computers discover them alone by correlation and classification, based on a massive amount of data. The trend looks to continue with the connectionist approach, but to develop an intelligent system, there is a probability that this will require both approaches in a hybrid form such as neuro-symbolic AI. NNs, a subsymbolic approach, had been pursued from the early days (Figure 6.17).

Early examples in Connectionist AI are Rosenblatt's perceptron learning work, the backpropagation work of Rumelhart, Hinton and Williams, and recent work in convolutional NNs by LeCun et al. However, NNs were not viewed as successful until about 2012 when big data became commonplace, the general consensus in the Al community was that the so-called neural-network approach was hopeless. Systems at that time just didn't work that well, compared to other methods. A revolution came in 2012, when a number of people, including a team of researchers working with Hinton, worked out a way to use the power of GPUs to enormously increase the power of NNs. Over the next several years, deep learning had spectacular success in handling **vision, speech recognition, speech synthesis, image generation**, and **machine translation**. However, since 2020, as inherent difficulties with bias, explanation, comprehensibility, and robustness became more apparent with deep learning approaches; an increasing number of AI researchers have called for combining the best of both the symbolic and NN approaches and addressing areas that both approaches have difficulty with such as commonsense reasoning. The following table highlights the strength of each approach (Table 6.2).

Classicists and connectionists all assign semantic content to something. Connectionists assign semantic content to "nodes" (i.e., to units or aggregates of units; whereas Classicists assign semantic content to expressions, i.e., to the sorts of things that get written on the tapes of Turing machines and stored at addresses in von Neumann machines). But Classical theories disagree with Connectionist theories about what primitive relations hold among these content-bearing entities. Connectionist theories acknowledge only casual connectedness as a primitive relation among nodes; when you know how activation and inhibition flow among them, you know everything there is to know about how the nodes in a network are related. By contrast, Classical theories acknowledge not only causal relations among the semantically evaluable objects that they posit but also a range of structural relations, of which constituency is paradigmatic. Neural-symbolic computation aims at integrating robust connectionist learning algorithms with sound symbolic reasoning. The recent impact of neural learning, in particular of deep networks, has led to the creation of new representations that have, so far, not been used for reasoning. For now, Smolensky determined that connectionist models may well offer an opportunity to escape the brittleness of symbolic AI systems,

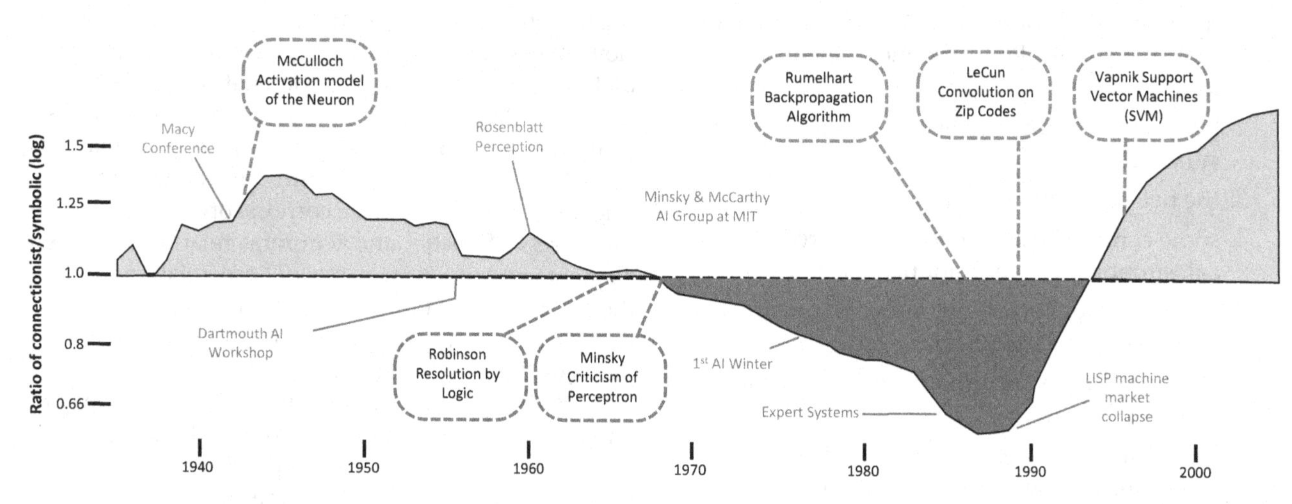

FIGURE 6.17
History of connectionist and symbolic AI.

TABLE 6.2
Table of weak AI technologies – symbolic or connectionist

Weak-AI Area	Best Approach	Miscellaneous
Pattern Recognition	Connectionist	Based on data analytics
Expert Systems	Symbolic	Based on rules, knowledge base, information
Search	Symbolic	Known techniques
Learning	Connectionist	Backpropagation, convolution, SVM
Logic	Symbolic	Based on mathematics and proofs
Planning & scheduling	Symbolic	Needed to reach a goal
Perception or Machine Vision	Connectionist	Object classification and identification
Natural Language Understanding	Connectionist	Language Translation and to communicate with a human
Speech	Connectionist	Communicate with a human
Diagnostics	Connectionist	Needed to recover from faults and extended to predict end of useful life
Reasoning	Symbolic	Aid in decision making
Inference	Symbolic	Needed for Reasoning and predict future state
Decision Making	-	Basis for Autonomy
Anomaly Detection	Connectionist	Based on data analytics for health management

a chance to develop more human-like intelligent systems, but only if we can find ways of naturally instantiating the sources of power of symbolic computation within fully connectionist systems.

Neuro-symbolic artificial intelligence refers to a field of research and applications that combines machine-learning methods based on ANNs, such as deep learning, with symbolic approaches to computing and AI, as can be found for example in the AI subfield of knowledge representation and reasoning. Earlier attempts to integrate logic and connectionist systems have mainly been restricted to propositional logic, or first-order logic without function symbols. To the best of our knowledge, all rule extraction techniques for connectionist networks are propositional in the sense that they only generate propositional rules. We would like to see a theory where in various layers of increased expressiveness logics, their corresponding connectionist models, their time and space complexities, their properties concerning learning and rule extraction as well as learning and rule extraction algorithms are specified.

We can understand ANNs as an abstraction of the physical workings of the brain, while we can understand formal logic as an abstraction of what we perceive, through introspection, when contemplating explicit cognitive reasoning. A natural question to ask is how these two abstractions can be related or even unified, or how symbol manipulation can arise from a neural substrate. There is an observation that neural and symbolic approaches to AI can complement each other with respect to their strengths and weaknesses. For example, deep learning systems are trainable from raw data and are robust against outliers or errors in the base data, while symbolic systems are brittle with respect to outliers and data errors and are far less trainable. Symbolic systems, on the other hand, can make explicit use of expert knowledge, and are to a high extent self-explanatory, as their algorithms can be inspected and understood in detail by a human, while neural learning systems cannot readily take advantage of available coded expert knowledge, and are black boxes that make understanding their decision-making processes very hard. Their coupling may be through different methods, including the calling of deep learning systems within a symbolic algorithm, or the acquisition of symbolic rules during training. Neuro-symbolic AI is an emerging subfield of AI that promises to combine knowledge representation and deep learning in order to improve deep learning and explain outputs of deep learning-based systems (Figure 6.18).

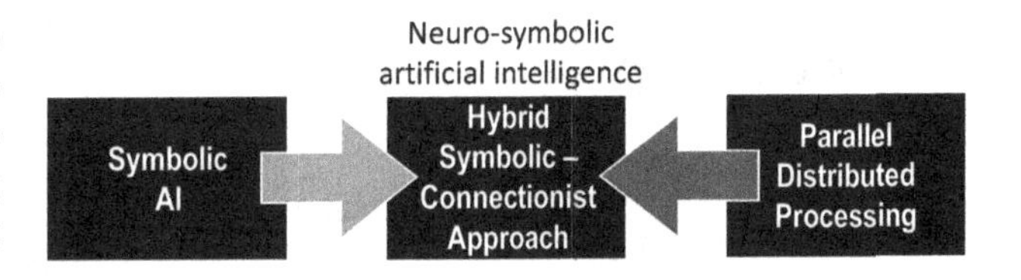

FIGURE 6.18
Future AI paradigm?

In AI and robotics, scene understanding is the task of identifying and reasoning about entities, that is, objects and events, which are bundled together by spatial, temporal, functional, and semantic relations. René Descartes regarded thoughts themselves as symbolic representations and perception as an internal process. For example, we use NNs to recognize the color and shape of an object. And as opposed to pure NN-based models, the hybrid AI can learn new tasks with less data and is now explainable. The symbolic component is used to represent and reason with abstract knowledge. By combining AI's statistical foundation with its knowledge foundation, researchers get the most effective cognitive analytics results with the least number of problems and less spending. It has been postulated that a combination of deep learning with the high-level reasoning capabilities present in the symbolic, logic-based approaches is necessary to progress toward more general AI systems. The core idea is that an intelligent agent receives percepts from the external world in the form of formulae in some logical system (e.g., first-order logic), and infers, on the basis of these percepts and its knowledge base, what actions should be performed to secure the agent's goals. Koller has investigated the marriage between probability theory and logic. And, in general, the very recent arrival of so-called *human-level* AI is being led by theorists seeking to genuinely integrate the three paradigms set out above.

A combination of symbolic and subsymbolic AI approaches is referred to as an in-between method or a hybrid method. Apart from the core symbolic or subsymbolic methods, nowadays there are symbolic applications with subsymbolic characteristics and vice versa. The main differences between these two AI fields are as follows:

a. Symbolic approaches produce logical conclusions, whereas subsymbolic approaches provide associative results.
b. The human intervention is common in the symbolic methods, while the subsymbolic learns and adapts to the given data.
c. The symbolic methods perform best when dealing with relatively small and precise data, while the subsymbolic ones are able to handle large and noisy datasets.

The symbolic techniques are defined by explicit symbolic methods, such as formal methods and programming languages, and are usually used for deductive knowledge. They consist of first-order logic rules, while other methods include rules, ontologies, decision trees, planning, and reasoning. These symbolic techniques are defined by explicit symbolic methods, such as formal methods and programming languages, and are usually used for deductive knowledge. In terms of applications, the symbolic methods work best on well-defined and static problems, and on manipulating and modeling abstractions (Figure 6.19).

In contrast, subsymbolic methods establish correlations between input and output variables. Such relations have high complexity and are often formalized by functions that map the input to the output data or the target variables. Subsymbolic AI includes statistical learning methods, such as Bayesian learning, deep learning, backpropagation, and genetic algorithms. Bayesian learning methods can be used to learn representations of probabilistic functions, particularly belief networks. The subsymbolic methods are more robust against noisy and missing data and generally have high computing performance. They are easier to scale up, therefore, they are well suitable for big datasets and large knowledge graphs. Moreover, they are better for perceptual problems and the subsymbolic methods are more robust against noisy and missing data, and generally have high computing performance. Most common applications of subsymbolic methods include prediction, clustering, pattern classification and recognition of objects, and natural language processing (NLP) tasks. The goal is to create a comprehensive cognitive model which integrates statistical learning with logical reasoning but can also perform well with noisy data. In practice, these learning systems can combine logical rules with data, and at the same time fine-tuning the knowledge based on

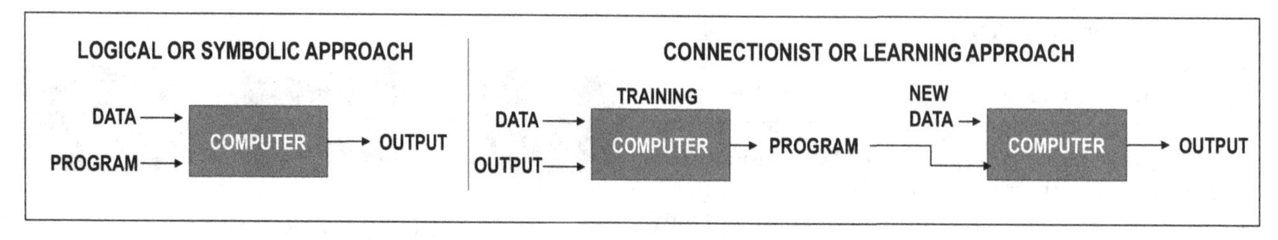

FIGURE 6.19
Symbolic approach to learning approach to computation.

the input data. However, logical reasoning alone does not suffice for real-world decision-making.

It is assumed that it is possible to combine learning and reasoning parts according to a specific problem. However, the existing hybrid models so far are non-generalizable and cannot be applied in multiple domains; thus, every time the model has to be developed to answer a specific question. At this point, there is no guide deciding the combinations of symbolic and subsymbolic parts for computation and representation. Some hybrid methods proposed so far include:

- connectionist expert systems
- multi-agent systems
- hybrid representations
- neural-fuzzy
- neural-symbolic computing
- learning and reasoning.

However, as imagined by Bengio, such a direct neural-symbolic correspondence was insurmountably limited to the aforementioned *propositional* logic setting. It lacks the ability to model complex real-life problems involving abstract knowledge with relational logic representations, the research in propositional neural-symbolic integration remains a small niche. This inability of classic NNs to capture logic reasoning beyond the propositional expressiveness was then often referred to as *propositional fixation* problem, coined by John McCarthy (Figure 6.20).

Hybrid representation can be used to promote an effective integration path. Such description differs from others since its entities aim at representing AI models in general, allowing to describe both non-symbolic and symbolic knowledge, the integration between them and their corresponding processors. Moreover, the entities also support representing workflows, leveraging traceability to keep track of every change applied to models and their related entities (e.g., data or concepts) throughout the lifecycle of the models. By integrating these two approaches (neural-symbolic integration), researchers hope to suppress their individual shortcomings. Hilario splits the various approaches to neural-symbolic integration

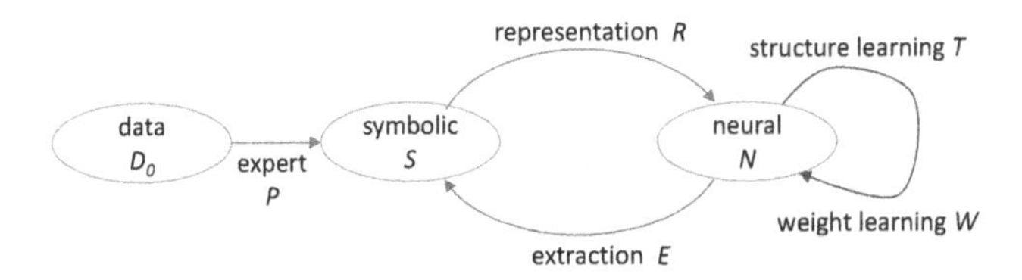

FIGURE 6.20
The concept of a neural-symbolic learning cycle.

into two types of strategy: unified and hybrid. The former are strategies that aim at combining NNs with symbolic capabilities that are part of the neural structures and processes alone. The latter are strategies that rely on a synergistic combination of neural and symbolic models. According to Hilario, this hybrid combination can be either translational or functional. Translational hybrids are systems where NNs act as processors, like the ones in the unified strategies, but there are translations between the symbolic structures and the NNs. The translational models translate representations between NNs and symbolic structures. For example, functional hybrids are comprised of symbolic structures, NNs, and their corresponding processors. Hilario has also identified four integration modes for the functional hybrids, which vary according to how symbolic and neural subsystems cooperate. In the chain processing mode, one of the subsystems (neural or symbolic) is the central processor, while the other handles the preprocessing or post-processing tasks. In the sub-processing mode, one of the two subsystems is embedded and subordinate to the other, which acts as the primary problem solver. In the meta-processing mode, one subsystem is the first-level problem solver, and the other is a mid-level function compared to the first. In the co-processing mode, the subsystems are equal partners in the problem-solving process (Figure 6.21).

Another approach maps a rule-based expert system into a neural architecture in its structural and behavioral aspects. Others have looked into building models that combine learning and reasoning but have to reconcile the inherent differences between statistics and logic. The neural-symbolic uses the NN that provides machinery for parallel computation and learning, while logic provides explanations for the network models that facilitate interaction with the world and other systems. The challenges for neural-symbolic integration today emerge from the goal of effective integration, expressive reasoning, and robust learning. The combination of a NN and a rule-based approach is proposed as a general framework for pattern recognition. This approach enables supervised and unsupervised learning, while providing the probability estimates of the output classes as in identifying textures.

For example, a combined unsupervised and supervised learning approach is used to analyze texture, the fundamental properties of the visible surface. The first stage performs feature extraction and transforms the image space into an array of N-dimensional feature vectors, each vector corresponding to a local window in the original image. The second stage analyzes the N-dimensional feature space and finds clusters in it, thus quantizing the continuous input features.

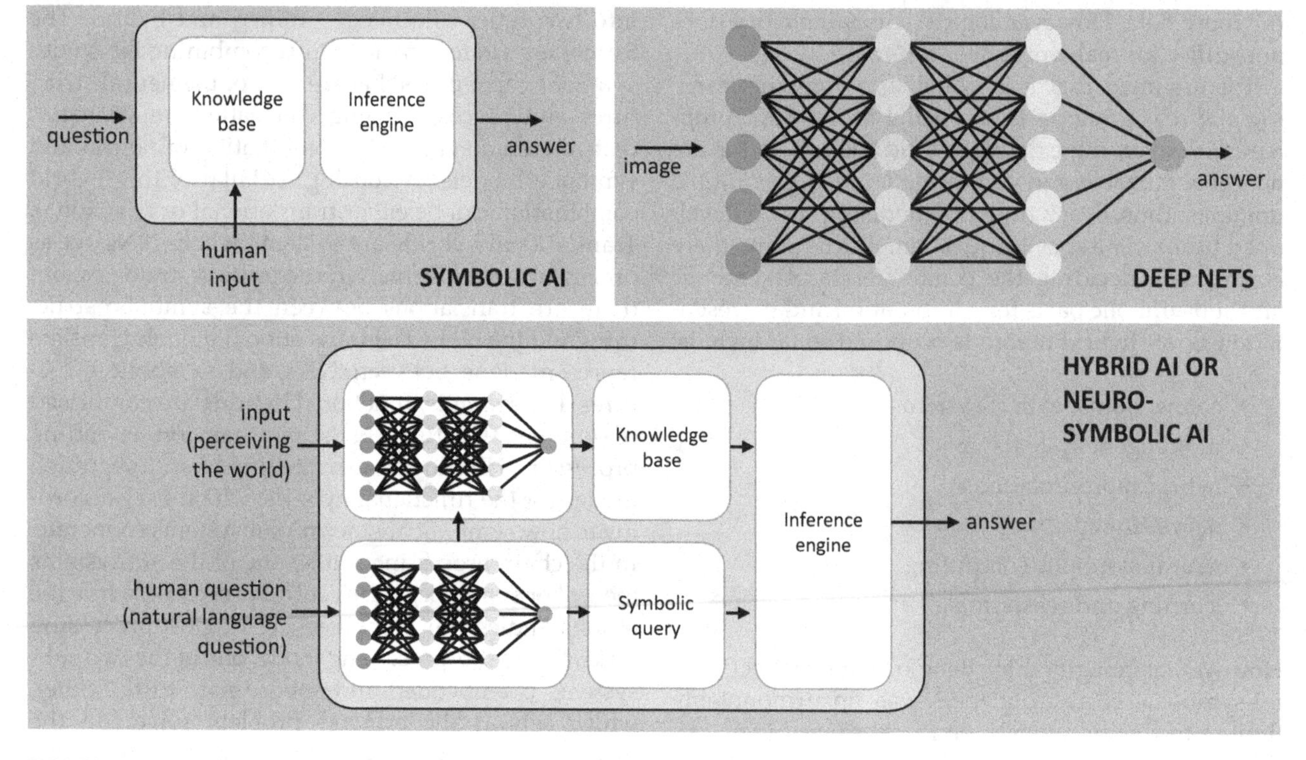

FIGURE 6.21
Hybrid AI or neuro-symbolic AI.

The quantized feature vectors are the input to the third stage of analysis in which categorization of the input into one of the pre-learned texture classes is achieved. Relational learning is the union of inductive logic programming, statistical relational learning, and multi-relational data mining; it constitutes a general class of techniques and methodology for learning from structured data such as graphs, networks, relational databases, and background knowledge. This is a step toward explainable-AI. Others have proposed network architecture that combines a rule-based approach with that of the NN paradigm. The primary motivation for this approach is to ensure that the knowledge embodied in the network is explicitly encoded in the form of understandable rules. This enables the network's decision to be understood and provides an audit trail of how that decision was arrived at. Another generalized approach uses the idea of neural-symbolic learning cycle, that is, the idea of a learner repeatedly alternating between symbolic and subsymbolic representations. And finally, a method that can extract expert-comprehensible rules from trained NNs (Figure 6.22).

Another method for the extraction of expert-comprehensible rules from trained NNs is using the KBANN (Knowledge-Based Artificial Neural Network) algorithm, which translates symbolic domain knowledge into NNs; defining the topology and connection weights of the networks it creates. KBANN is a hybrid learning system built on top of connectionist learning techniques that maps, in the presented spirit, problem-specific "domain theories", represented by propositional logic programs, into feedforward NNs, and then refines this reformulated knowledge using backpropagation. It uses a knowledge base of domain-specific inference rules to define what is initially known about a topic. The first link inserts domain knowledge, which needs to be neither complete nor correct, into an NN using KBANN. The second link trains the KNN (Networks created

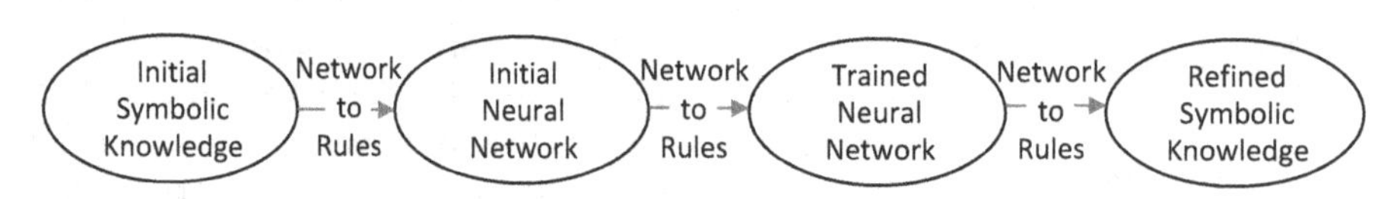

FIGURE 6.22
Integration specification and execution engine on top of integration modes.

using KBANN are called KNNs) using a set of classified training examples and standard neural learning methods. The final link extracts rules from trained KNNs. Rule extraction is an extremely difficult task for arbitrarily configured networks but is somewhat less daunting for KNNs due to their initial comprehensibility. This idea has also been later extended by providing corresponding algorithms for symbolic knowledge extraction back from the learned network, completing what is known in the NSI community as the "neural-symbolic learning cycle".

However, in the meantime, a new stream of neural architectures based on **dynamic computational graphs** has become popular in modern deep learning to tackle structured data in the (non-propositional) form of various sequences, sets, and trees. Most recently, an extension to arbitrary (irregular) graphs then became extremely popular as Graph Neural Networks (GNNs). These dynamic models finally enable to skip the preprocessing step of tuning the relational representations, such as interpretations of a relational logic program into the fixed-size vector (tensor) format. They do so by effectively reflecting the variations in the input data structures into variations in the structure of the neural model itself, constrained by some shared parameterization (symmetry) scheme reflecting the respective model prior. And while this particular dynamic deep learning models introduced so far, such as the GNNs, are still just somewhat specific graph-propagation heuristics rather than universal (logic) reasoners, the paradigm of dynamic neural computation finally opens the door to properly reflect relational logic reasoning in NNs. We also find that data movement poses a potential bottleneck, as it does in many machine-learning workloads. Neuro-Symbolic Integration (or Neural-Symbolic Integration) is concerned with the combination of ANNs with symbolic methods, for example, from logic-based knowledge representation and reasoning in AI.

As an example, the probabilistic approach also provides a basis for making optimal decisions under uncertainty. This is realized by extending probability theory with decision theory. The position that ANN are sufficient for modeling all of cognition may seem exceedingly naive. For example, state-of-the-art question-answering systems such as IBM's Watson use ANN technology as a minor component within a larger largely symbolic framework, and the AlphaGo system, which plays the game of Go at an expert level, combines NNs with Monte Carlo tree search.

Deep Learning and Traditional Computer Vision: Computer vision is an area of machine learning dedicated to interpreting and understanding images and video. Many of the CV techniques invented in the past 20 years have become irrelevant in recent years because of Deep Learning (DL). The CV models are designed to translate visual data, based on features and contextual information identified during training. This enables models to interpret images and video and apply those interpretations to predictive or decision-making tasks. DL is used in digital image processing to solve difficult problems (e.g., image colorization, classification, segmentation and detection). DL methods such as CNNs mostly improve prediction performance using big data and plentiful computing resources and have pushed the boundaries of what was possible with super-human accuracy. However, DL is not going to solve all CV problems. There are some problems where traditional techniques with global features are a better solution. However, the advent of DL may open many new doors to do something with traditional techniques to overcome the many challenges DL brings (e.g., computing power, time, accuracy, characteristics, and quantity of inputs, among others). DL is all about learning or "credit assignment" across many layers of an NN accurately, efficiently, and without supervision and is of recent interest due to enabling advancements in processing hardware (Figure 6.23).

The development of CNNs has had a tremendous impact in the field of CV and is responsible for a big jump in the ability to **recognize objects**. CNNs make use of kernels (also known as filters), to detect features (such as edges) throughout an image. A kernel is just a matrix of values, called weights, which are trained to detect specific features. As their name indicates, the main idea behind the CNNs is to spatially convolve the kernel on a given input image to check if the feature it is meant to detect is present. It provides a value representing how confident it is that a specific feature is present, a convolution operation is carried out by computing the dot product of the kernel and the input area where the kernel is overlapped (the area of the original image the kernel is looking at is known as the receptive field). To facilitate the learning of kernel weights, the convolution layer's output is summed with a bias term and then fed to a nonlinear activation function. Depending on the nature of data and classification tasks, these activation functions are selected accordingly. And to speed up the training process and reduce the amount of memory consumed by the network, the convolutional layer is often followed by

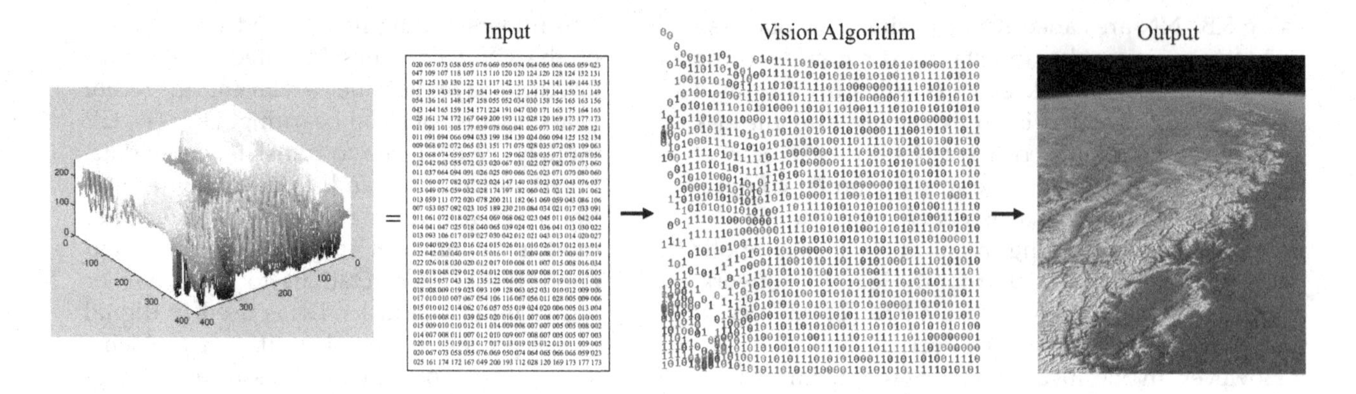

FIGURE 6.23
Classical computer vision.

a pooling layer to remove redundancy present in the input feature.

Deep CNNs work by dividing the tasks in order to understand an image. For example, the first layer is used to detect edges, the next layers could combine the detected edges to simpler shapes, the next hidden layers could solve illumination problems and the final layers will do the prediction and matching to the input images. Traditional feature-based approaches include: Scale Invariant Feature Transform (SIFT), Speeded Up Robust Features (SURF), Features from Accelerated Segment Test (FAST), Hough transforms, and Geometric hashing. Feature descriptors such as SIFT and SURF are generally combined with traditional machine-learning classification algorithms such as Support Vector Machines and K-Nearest Neighbors to solve the aforementioned problems. DL is sometimes an overkill as often traditional CV techniques can solve a problem much more efficiently and with a fewer lines of code. Today, traditional CV techniques are used when the problem can be simplified so that they can be deployed on low-cost microcontrollers or to limit the problem for DL techniques by highlighting certain features in the data, augmenting the data, or aiding in the dataset annotation. Classic ("classic" is synonymous with "traditional") CV algorithms are well-established, transparent, and optimized for performance and power efficiency, while DL offers greater accuracy and versatility at the cost of large amounts of computing resources (Figure 6.24).

The fusion of machine-learning metrics and Deep Network has become very popular due to the simple fact that it can generate better models. DL needs big data and often millions of data records are required. However, training a DNN takes a very long time depending on the computing hardware availability, where training can take a matter of hours or days. Subsymbolic CV algorithms are based on CNNs and have provided a dramatic improvement in performance compared to the traditional, symbolic image processing algorithms. CNNs are NNs with a

FIGURE 6.24
Deep learning for computer vision and image processing.

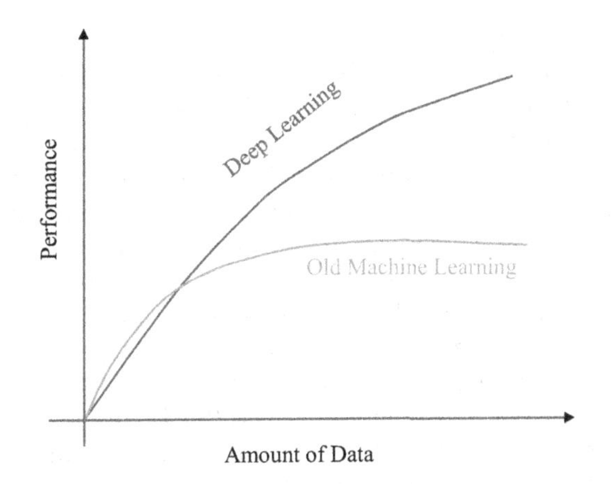

FIGURE 6.25
Performance of deep learning concerning the amount of data required.

multi-layered architecture that is used to gradually reduce the data and calculations to the most relevant set. This set is then compared against known data to identify or classify the data input (Figure 6.25).

When an image is processed by a CNN, each base color used in the image (e.g., red, green, and blue) is represented as a matrix of values. These values are evaluated and condensed into 3D tensors (tensors are multidimensional arrays with a uniform type, in this case of color images), which are collections of stacks of feature maps tied to a section of the image. These tensors are created by passing the image through a series of convolutional and pooling layers, which are used to extract the most relevant data from an image segment and condense it into a smaller, representative matrix. This process is repeated numerous times (depending on the number of convolutional layers in the architecture). The final features extracted by the convolutional process are sent to a fully connected layer, which generates the predictions. As these technologies improve, the incorporation of CV applications to robotic tasks is becoming more useful.

Localization and object detection in an image can be used to identify multiple objects in complex scenes. CV algorithms are highly compute-intensive and may require multiple GPUs to run at production scale. The strength of the Deep Convolutional NNs (DCNN) is in its layering architecture. It uses a three-dimensional NN to process each of the red, green, and blue elements of the image at the same time. This considerably reduces the number of artificial neurons required to process an image, as compared to traditional feedforward NNs. Again, the DCNN receives images as an input and uses them to train the classifier. This network employs special mathematical operations called a "convolution", instead of matrix multiplication. The pooling layers gradually reduce the size of the image, keeping only the most important information. Finally, the softmax function is applied at the end to the outputs of the fully connected layers, giving the probability of a class that the image belongs to. Recently, Visual Prompting is a method in which the user can label just a few areas of an image (supervised learning), resulting in a much faster and easier approach than conventional labeling, which typically requires completely labeling every image in the training set. This new approach accelerates the building process as only a few simple visual prompts are required.

Today, DDN for visual pattern recognition are used for image classification, detection, or segmentation problems. Pre-NN research has advanced slowly, being hampered by high processing power requirements. This process is highly nonlinear and very complex. According to Dr. Dan Cireşan, DNNs would have high input and output dimensionality: 10^4–10^7 parameters to learn, 10^6–10^9 connections (or operations) per image, and learning the parameters requires quadrillions of floating-point operations. The CNN uses a hierarchical architecture designed for image processing that is loosely inspired by biology. Training is computationally intensive with the feature extraction layer composed of 321,800 weights and 26.37 million connections, and the classification layer consisting of 1,881,255 weights having 1.88 million connections. By training on 1,121,749 training samples, it took 27 hours to forward propagate one epoch and 14 months to train 30 epochs. An **epoch** is a fixed date and time used as a reference from which a computer measures) system time. By incorporating a GPU processor configuration, a single-threaded CPU version of a CNN took **1** day to train on a GPU as opposed to **2 months** on a CPU. By training a DNN on a GPU, it takes 3.54 hours/epoch. Dr. Cireşan recorded a 1-year training time for 30 epochs on a CPU and 1 week on a GPU. As another example, it took **5 months** of training time for up to seven epochs on a CPU as opposed to **3 days** on a GPU. The bottom line, a GPU configuration makes a tremendous difference in training DNNs and is already better and much faster than humans on many difficult problems. DNNs using a combination of CNNs are now the state of the art on many image classification, detection, and segmentation tasks.

There are many more challenging problems in CV such as in robotics, augmented reality, automatic panorama stitching, virtual reality, 3D modeling, motion estimation, video stabilization, motion capture, video processing, and scene understanding which cannot simply be easily implemented in a differentiable manner with DL but need to be solved using the other "traditional" techniques. DL excels at solving closed-end

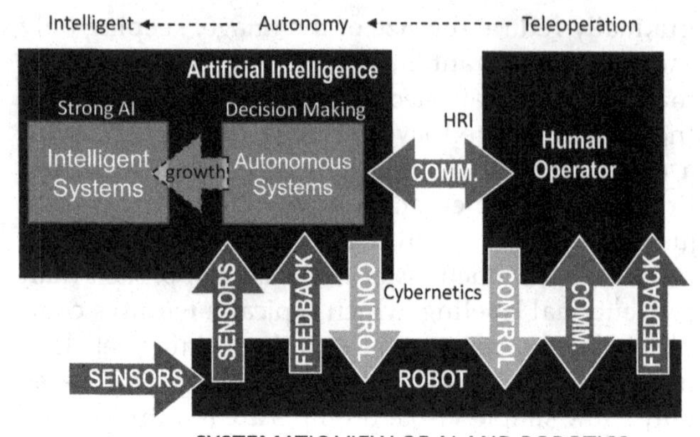

FIGURE 6.26
Systematic view of AI and robotics.

classification problems, in which a wide range of potential signals must be mapped onto a limited number of categories, given that there is enough data available and the test set closely resembles the training set.

Today, No-code AI platforms may make it possible for anyone, regardless of technical background, to unlock the potential of AI with just a few clicks and without writing code. No-code AI tools offer pre-trained machine-learning models, eliminating the need for model development and allowing for faster development and deployment times. Affective Computing is an area of AI research and technology development that zeroes in on human emotions. Affective Computing as a field strives to develop systems capable of recognizing, interpreting, and simulating human emotions (Figure 6.26).

From the system engineering perspective, there are multiple methods for controlling a robot: teleoperation (or man-in-the-loop), autonomous, or intelligent control, and based on AI technologies available such as by developing select weak AI modules. Robotics and hardware engineers typically focus on controls or what the robot has to do. Depicted in the previous figure is a holistic view of the relationship of AI with robotics. The two key pieces in addition to robotics and AI are the environment in which the robot operates and by understanding the role of the human operator as planned for understanding the concept of operations. Past experience has indicated that designers have had technical issues associated with sliding autonomy, and it was discovered it to be easier to adapt the ends of the Sheridan autonomy spectrum, with a bias toward teleoperation for its ease of implementation. The domain of the military and aerospace desires autonomous system for unmanned reconnaissance, surveillance, offensive/defensive maneuvers, and exploration. The weak point of a remotely operated vehicle system is its communication subsystems.

Autonomy with unmanned vehicles will take the soldier out of harm's way and does not require the full complement of modern AI technologies. The vision for military missions is to function without the presence of human errors, which can affect performance. The reduction in the number of human operators will enable new capabilities (the weakness on some missions is the human body itself, e.g., a human pilot) and a reduction in the human labor cost. Intelligent systems (as well as intelligent robots) will require more AI technologies in an integrateable form factor to do useful work and potentially pass the Turing test. The user's desire is to implement intelligent systems that could predict its future state with accurate models (Figure 6.27).

Sensors are used to model the state of its external environment, but its internal state is just as important to model. For a robot to perform as anticipated, it requires a healthy vehicle that is "happy". If the robot is not healthy (somehow injured), it may have to adjust its planned action. Decision-making on the robot considers both the utility of its resources (e.g., onboard power and resources available to the robot) and accomplishing the goal of its mission. The use of autonomy is highly dependent on the nature of its environment such as being structured or highly unstructured, which involves many uncertainties. If we know what we want the robot to do, we need to integrate all the miscellaneous technologies into a monolithic vehicle that we call a robot. The human body and mind are a very complex system which we are trying to replicate with engineering technologies. For example, the human muscle is better than any commercial actuator. And if we can better understand the workings of the human brain, we may be better able to replicate its functions.

FIGURE 6.27
Mapping a robot's environment with a drone.

AI-Complete implies that these are difficult computational problems, assuming that intelligence is computational, and is equivalent to that of solving the central AI problem that can make computers as intelligent as people (or strong AI). To label a problem as AI-complete reflects an attitude that it would not be solved by a simple specific algorithm. Volume 2 of Foundations for AI and Robotics goes into detail into the history of the various technologies, including early researchers from the turn of the century and other researchers later when the topical area was officially formalized. It will include some groundbreaking events, a little work history at Martin Marietta from the 1980s until the early 2000s, some further technology developments, and what to expect in the coming future.

7

Epilogue

The expectations of smart machines have been high since the writings of Isaac Asimov. The public was promised machines that could think with human intelligence. Artificial intelligence (AI) is arguably the most exciting field in robotics. In the Near Future, this capability would include the ability to learn just about anything, the ability to reason, the ability to use language, and the ability to formulate original ideas. Roboticists are nowhere near achieving this level of AI today, but we, as a society, have made a lot of progress with more limited AI. When we consider robotics, we must understand the big picture where robotics and AI are tightly coupled and should be viewed as synonymous. However, robotics has failed completely to live up to the predictions of the 1950s. Once we understand how the mind operates, we will be able to "program" detailed descriptions of these principles into inexpensive computers. In pursuit of intelligent robots, waves of researchers have grown disheartened and scores of start-up companies have gone out of business. It is not the mechanical "body" that is unattainable, but the computer-based artificial brain that is still well below the level of sophistication needed to build a human-like robot (Figure 7.1).

Early attempts at strong AI or artificial general intelligence (AGI) brought to light the daunting complexity of trying to emulate human intelligence. However, during the last few decades, work on weak AI, intelligence targeted to very specific domains or tasks, has met with considerably more success. As a result, today AI permeates our daily lives, playing a role in everything from anti-lock braking systems to warehouse stocking to electronic trading on stock exchanges. Language translation and chatbots are in many homes. Machine learning is used on video surveillance systems, especially in China. Little by little, AI has taken on roles previously performed by people and bested them in ways once unimaginable. For example, computer phone attendants are capable of routing hundreds of calls a minute. Robot-operated warehouses have sped up the delivery of items to packers in seconds. Pattern-matching algorithms pick out the correct image from among thousands in a matter of moments. When reflecting on the great potential benefits of general intelligence, we tend to consider narrow AI applications as separate entities that can very well be outperformed by a broader AGI that presumably can deal with everything.

FIGURE 7.1
Ingenuity, the Mars helicopter.

And on the robotic side, already 30,000 patients with Parkinson's disease have neural implants. Google, Apple, Tesla, and Aurora are experimenting with computers that can drive cars. There are more than 2,000 robots (aerial and ground) fighting in Afghanistan alongside the human troops. In a game show that is a once-in-a-life-time event in the history of AI, but this time the computer will be the guest, an IBM super-computer nicknamed Watson competed on *Jeopardy!* Watson runs on 90 servers and takes up an entire room, and in a practice match, it finished ahead of two former champions, Ken Jennings and Brad Rutter. It got every question it answered right, but much more importantly, it didn't need help understanding the questions (or, strictly speaking, the answers), which were phrased in plain English. Watson isn't strong AI, but if strong AI happens, it will arrive gradually, bit by bit, and this will have been one of the bits. But until now, nothing could compete with a human being when it came to general knowledge about the world, but maybe Watson on Jeopardy can change that perception. In 2016, over 200 million people watched online as AlphaGo emerged a surprise 4-1 victor of The Google DeepMind Challenge match in Seoul, South Korea. Robots are moving away from the factory floors of the 20th century to today with animatronics in entertainment, robot toys, personal services such as hotel servants and restaurant waiters, expert systems to detect cancer and da Vinci for

DOI: 10.1201/9781032673134-7

FIGURE 7.2
iGART robot.

robotic surgery, agricultural drones for mapping and applying pesticides, agricultural harvesting, operations in hazardous environments requiring people in Personal Protection Equipment (PPE) such as radiation, space, underwater, and underground. So, we will have both the hardware and software to achieve human-level intelligence in a machine by 2029 by experts in the field. And Moravec is convinced that the decades-old dream of a useful, general-purpose autonomous robot will be realized in the not-too-distant future. Automation (together with robotics and AI) is changing the future (Figure 7.2).

Over the past 20 years, machine learning has evolved from a field of laboratory demonstrations to a field of significant commercial value. Although we do not yet know how to make computers learn nearly as well as people learn, many successful machine-learning applications have been developed, ranging from data-mining programs that learn to detect fraudulent credit card transactions to information-filtering systems that learn users' reading preferences to autonomous vehicles that learn to drive on public highways. Data mining involves applying machine-learning methods to historical data to improve future decisions. For example, given such time-series data, one is often interested in learning to predict features that occur late in the time series based on features that are known earlier. There is a new robust set of first-generation machine-learning algorithms that are appropriate for many problems including decision tree – learning algorithms. Another application niche for machine learning lies in constructing computer programs that are simply too difficult to program by hand. Many sensor-interpretation problems fall into this category such as for today's top speech-recognition systems all use machine-learning methods to some degree to improve their accuracy. An example is image-classification problems, such as the face-recognition task. During the coming years, a shift is seen to take place in which software developers eventually abandon the goal of manually hand crafting increasingly complex software systems and shift to a software development methodology of manually designing the global structure of the system but using machine-learning methods to automatically train individual software components that cannot be found in a software library and to optimize global performance. Prof. Tom Mitchell at CMU envisions major improvements with machine learning for the incorporation of prior knowledge with training data, lifelong learning, machine learning embedded in programming language, and machine learning for natural language. In the future, computer will write their own programs using learning. Learning algorithms, also known as learners, are algorithms that will make other algorithms. Finally, with machine learning, computers could write their own programs, so we don't have to (Figure 7.3).

The controversy extends beyond whether machines can "think" like humans. The "thinking" machine, if it becomes a reality, would undoubtedly have profound social implications and cause irrevocable changes in

FIGURE 7.3
Autonomous tractor for agriculture.

FIGURE 7.4
Computer vision used for surveillance.

the very foundation of our society. Today's Trends and the Fourth Industrial Revolution are accelerating robotics. We should expect to see improvements in power (currently a limiting factor in robotics), new materials, and continued advancements in computing. Robotics have been pervasive in manufacturing, starting in the 1960s. The cost of hardware has been getting lower and cheaper. There has been a renewed focus on software and algorithms, with the ROS framework setting standards to build on past experience and aid in integrations. It is already possible to train a language model AI (Generative Pre-trained Transformer3 or GPT-3) with a gigantic dataset and then have it learn various tasks based on a handful of examples – one or few-shot learning. GPT-3 (developed by OpenAI) can do this with language-related tasks, but there is no reason why this should not be possible with image and sound, or with combinations of these three. And most recently, there has been an infusion of venture capitalist funding to push the public sector in advancing the technologies. Core technologies from other domains (for example, vision, biology, cognitive science, or biomechanics) are becoming crucial components in more and more modern robotic systems (Figure 7.4). But what happens if AI is smarter than a person?

President Obama has noticed this race for technology superiority in AI and robotics. Here are the findings of "A Roadmap for U.S. Robotics: From Internet to Robotics". Some of the major findings include the introduction of 600,000 new jobs in manufacturing over the last 5–6 years. Sensors and computing power have become cheap enough to be easily adopted for robot applications, which is resulting in a revolution in control and flexibility of systems. Logistics is seeing major growth, for example, e-commerce has seen 40% growth with new methods with enterprises such as Amazon Express and Uber Food. The core challenge in manufacturing and assembly is the flexible integration with human operators and collaborators. As new robots are introduced into society, it is essential that the United States also consider the training of the workforce to ensure efficient utilization of these new technologies. The next step is how do we prepare for the future with robotics and AI. It is important to see optimism about AI and machine learning in improving people's lives by helping to solve the world's greatest challenges and inefficiencies. Federal R&D has identified some areas of opportunity in robotics and AI and recommends ways to coordinate R&D to maximize benefits and build a highly trained workforce. Growth in this field has dramatically increased the need for people with relevant skills to support advancing the field, but also having a data-literate citizenry that is able to read, use, interpret, and communicate with data and matters that are affected by AI. There is a risk in automation potentially creating economic inequality within society, but a goal would be workers that should be retrained and able to succeed in occupations that are complimentary to, rather than competing with, automation. There is a challenge in building systems that can transition from the laboratory to the real world and adapting gracefully to unforeseen situations is difficult and necessary for safe operations.

> In Turing's Own Words: The original question, "Can machines think?", *I believe to be too meaningless to deserve discussion. Nevertheless, I believe that at the end of the century the use of words and general educated opinion will have altered so much that one will be able to speak of machines thinking without expecting to be contradicted.*
>
> – *Alan Turing (1950)*

Glossary of Terms

A*(pronounced "A-star"): a graph traversal and path planning algorithm, which is used in many fields of computer science due to its completeness, optimality, and optimal efficiency.

Abduction: is a reasoning process that tries to form plausible explanations for abnormal observations (reasons from effects to causes).

Accelerated Processing Unit (APU): combines the CPU and GPU on a single chip (to form a combined processing unit) to improve efficiency, and to reduce physical size and manufacturing costs as manufacturers combine electronic components into single chips, also known as a System-on-a-Chip (SoC).

Acceleration: a change in velocity (that is changing speed and/or direction of travel).

Accuracy: is the measurement of the deviation between the command characteristic and the attained characteristic or the precision with which a computed or calculated robot position can be attained. Accuracy is normally worse than the arm's repeatability. Accuracy is not constant over the workspace, due to the effect of link kinematics.

Agent: an agent is anything that can be viewed as perceiving its environment through sensors and acting upon that environment through effectors.

AGI: artificial general intelligence or "strong AI" exhibits intelligence in a wide range of contexts and problem spaces, rather than only in specific niches.

Algorithm: a formula given to a computer in order for it to complete a task (i.e., a set of rules for a computer).

Algorithm Theoretical Basis Documents (ATBDs): describe the physical theories, mathematical procedures, and model assumptions applied in the calculation of?

ANN: artificial neural networks process data to make decisions in a way that is inspired by the structure and functionality of the human brain.

API (Application Program Interface): a software component that allows you to integrate someone else's program into your own application without needing to understand the underlying code.

Architecture: is an engineering discipline that deals with the technological aspects and multi-disciplinary approach to planning, designing, building, and operating unmanned machines (hardware) and functions (software).

Architecture's style: refers to the computational structure that defines communication between components within the architecture.

Arithmetic Logic Unit (ALU): is a combinational digital circuit that performs arithmetic and bitwise operations on integer binary numbers.

Arm or Manipulator: an interconnected set of links and powered joints comprising a robot manipulator that supports and/or moves a wrist and hand or end-effector through space. The arm itself does not include the end-effector.

Articulated Manipulator: a manipulator with an arm that is broken into sections (links) by one or more joints. Each of the joints represents a degree of freedom in the manipulator system and allows translation and rotary motion.

Articulation: describes a jointed device, such as a jointed manipulator. The joints provide rotation about a vertical axis, and elevation out of the horizontal plane. This allows a robot to be capable of reaching into confined spaces.

Artificial Intelligence (AI): a subset of computer science that deals with computer systems performing tasks with similar, equal, or superior intelligence to that of a human (e.g., decision-making, object classification and detection, speech recognition, and translation).

AI-Complete (or AI-hard): implies that these are difficult computational problems, assuming that intelligence is computational, and is equivalent to that of solving the central artificial intelligence problem that can make computers as intelligent as people (or strong AI).

Artificial Life: a field of study wherein researchers examine systems related to natural life, its processes, and its evolution, through the use of simulations with computer models, robotics, and biochemistry.

Artificial Neural Networks: software that roughly emulates the way neurons operate in the brain.

Artificial General Intelligence (AGI): refers to a true thinking machine. AGI is typically considered to be more or less synonymous with the terms HUMAN-LEVEL AI or STRONG AI.

Artificial Super Intelligence: a system with intellectual powers beyond those of humans across a comprehensive range of categories and fields of endeavor and is a hypothetical state of AI.

Augmented Reality (AR): is an interactive experience that combines the real world and computer-generated content. The content can span multiple sensory modalities, including visual, auditory, haptic, somatosensory and olfactory.

Augmented Transition Network (ATN): type of graph theoretic structure used in the operational definition of formal languages, used especially in parsing relatively complex natural language.

Autoencoder: a type of artificial neural network used to learn efficient coding of unlabeled data (unsupervised learning).

Automated Reasoning: to use the stored information to answer questions and draw new conclusions (relates to inference).

Automation: a process is performed by using programmable machines. The process is not only supported by machines but these machines can work in accordance with a program that regulates the behavior of the machine. (Common Misspellings: automaition).

Automaton (plural: automata or automatons): is a relatively self-operating machine, or control mechanism designed to automatically follow a predetermined sequence of operations or respond to predetermined instructions.

Autonomous agency theory (AAT): is *a viable system theory* (VST) that models autonomous social *complex adaptive systems.*

Autonomous Systems: can independently plan and decide sequences of steps to achieve a specified goal without micro-management.

Backpropagation: shorthand for "backward propagation of errors", is a method of training neural networks where the system's *initial* output is compared to the *desired* output, then adjusted until the difference (between outputs) becomes minimal.

Backward chaining: strategy of working backward for Reason/Cause of a problem.

Bagging (Bootstrap Aggregation): improves the accuracy by taking repeated samples from the data and combining results from all individual decision trees.

Bayesian Classification Model: is a probabilistic approach to learning and inference based on a different view of what it means to learn from data, in which probability is used to represent uncertainty about the relationship being learnt.

Bayesian Hierarchical Modeling: a statistical model written in multiple levels (hierarchical form) that estimates the parameters of the posterior distribution using the Bayesian method. The sub-models combine to form the hierarchical model, and Bayes' theorem is used to integrate them with the observed data and account for all the uncertainty that is present. The result of this integration is the posterior distribution, also known as the updated probability estimate, as additional evidence on the prior distribution is acquired.

Bayesian Networks: also known as Bayes network, Bayes model, belief network, and decision network, is a graph-based model representing a set of variables and their dependencies.

Behavior Tree (BT): is a way to structure the switching between different task bins for an autonomous agent.

Bias: AI bias refers to the tendency of a model to make certain predictions more often than others. Bias can be caused due to the training data of a model or its inherent assumptions.

Big Data: big data is a term that describes datasets that are too large or too complex to process using traditional methods. It involves analyzing vast sets of information to extract valuable insights and patterns to improve decision-making.

Blackboard: it is the memory inside computer, which is used for communication between the cooperating expert systems.

Body: provides the structure of the robot.

Boltzmann Machine: a stochastic spin-glass model with an external field, i.e., a Sherrington-Kirkpatrick model, that is a stochastic Ising model. It is a statistical physics technique applied in the context of cognitive science. It is also classified as a Markov Random field.

Boosting: a method used in machine learning to reduce errors in predictive data analysis. Data scientists train machine learning software, called machine learning models, on labeled data to make guesses about unlabeled data.

Brain-computer interface: a direct connection between a brain and a computer, allowing the brain to command the computer or the computer to monitor the brain.

Cartesian Coordinates: is a type of coordinate system that specifies the location of a point in two-dimensional space by a pair of numerical numbers, which further specify the distance to fixed axes that are perpendicular to each other. In simple terms, an XY graph represents a two-dimensional Cartesian Coordinate System. When a point is specified in a three-dimensional space (XYZ graph), it constitutes a three-dimensional Cartesian coordinate system. A robot's TCP position is specified in a Cartesian Coordinate.

Case-Based Reasoning: using old experiences to understand and solve new problems. In case-based reasoning, a reasoner remembers a previous situation similar to the current one and uses that to solve the new problem.

Cellular Automaton or Automata (CA): is a collection of cells arranged in a grid of specified shape, such that each cell changes state as a function of time, according to a defined set of rules driven by the states of neighboring cells.

Central Processing Unit (CPU): is the main circuit board and processor of the Controller System.

Chatbots: a chat robot that can converse with a human user through text or voice commands. Utilized by e-commerce, education, health, and business industries for ease of communication and to answer user questions.

Chain Rule: provides us a technique for finding the derivative of composite functions, with the number of functions that make up the composition determining how many differentiation steps are necessary

Closed-loop: control achieved by a robot manipulator by means of feedback information.

Cloud Robotics: a field of robotics that attempts to invoke cloud technologies such as cloud computing, cloud storage, and other internet technologies centered on the benefits of converged infrastructure and shared services. When connected to the cloud, robots can benefit from the powerful computation, storage, and communication resources of modern data center in the cloud, which can process and share information from various robots or agent.

Clustering: algorithm technique that allows machines to group similar data into larger data categories.

Cognition: the collection of mental processes and activities used in perceiving, learning, remembering, thinking, and understanding, and the act of using those processes.

Cognitive Computing: computerized model that mimics human thought processes by data mining, natural language processing, and pattern recognition.

Cognitive Science: the study of the human mind and brain, focusing on how the mind represents and manipulates knowledge and how mental representations and processes are realized in the brain.

Commonsense Reasoning: is a human-like ability to make presumptions about the type and essence of ordinary situations humans encounter every day.

Complexity: used to characterize something with many parts where those parts interact with each other in multiple ways, culminating in a higher order of emergence greater than the sum of its parts.

Complexity of a Pattern: is measured by the number of attributes and the relationships between these attributes

Computational Complexity Theory: imposes the theoretical bounds on the inherent complexity of different learning tasks measured in terms of computational effort, number of training examples, number of mistakes, etc.

Computational Learning Theory: a branch of artificial intelligence that studies algorithms and mathematical models of machine learning. It focuses on the theoretical foundations of learning to understand how machines can acquire knowledge, make predictions, and improve their performance.

Computational Lithography: is the set of mathematical and algorithmic approaches designed to improve the resolution attainable through photolithography. It uses algorithmic models of the manufacturing process, calibrated with key data from our machines and from test wafers.

Computationalism (in the philosophy of mind): is the view that mental processes, including perceptual processes, are computational (relating to or using computers, and relating to the process of mathematical calculation).

Computers: invented to "compute", to solve "complex mathematical problems".

Computer Vision (or Machine Vision): when a machine processes visual input from image files (JPEGs) or camera feeds or process visual information.

Continuous Path: describes the process where by a robot is controlled over the entire path traversed, as opposed to a point-to-point method of traversal.

Control Algorithm: a monitor used to detect trajectory deviations in which sensors detect such deviations and torque applications are computed for the actuators.

Control Mode: the means by which instructions are communicated to the robot.

Control theory: forms the theoretical foundation of procedures that learn to control processes in order to optimize predefined objectives and to predict the next state of the process they are controlling.

Controller: a device that forms part of a control system – often taking the error between the desired state and the actual state and generating data used to affect an actuator.

Convolutional Neural Network: known as Shift Invariant or Space Invariant Artificial Neural Networks (SIANN), based on the shared-weight architecture of the convolution kernels or filters that slide along input features and provide translation-equivalent responses known as feature maps, most commonly applied to analyze visual imagery.

Convolutional Neural Network (CNN): is a regularized type of feed-forward neural network that learns feature engineering by itself via filters (or kernel) optimization.

Covariance Matrix: a square matrix giving the covariance between each pair of elements of a given random vector. Any covariance matrix is symmetric and positive semi-definite and its main diagonal contains variances.

Cybernetics: the science of control and communication (in the animal and the machine), the science of feedback systems, incorporating control, learning, and interaction.

Coordinate System (or Frame): defines a reference position and orientation from which a robot position can be measured.

Data Mining: the process of sorting through large sets of data in order to identify recurring patterns while establishing problem-solving relationships (automating the process of searching for patterns in the data).

Data Processing Unit (DPU): channel controller, a programmable specialized electronic circuit with hardware acceleration of data processing for data-centric computing.

Data Science: involves extracting insights from data using scientific methods, algorithms, and systems.

Decision-Making: making choices by identifying a decision, gathering information, and assessing alternative resolutions.

Decision tree (or a classification tree): a tree in which each internal (non-leaf) node is labeled with an input feature.

Decision-tree Learning: is supervised learning approach used in statistics, data mining, and machine learning. The goal of using a Decision Tree is to create a training model that can be used to predict the class or value of the target variable by learning simple decision rules inferred from prior data (training data).

Deduction: the action of deducting or subtracting something (reasons from causes to effects).

Deductive Reasoning: a logical approach where you progress from general ideas to specific conclusions.

Deep Belief Network: a generative graphical model, or alternatively a class of deep neural network, composed of multiple layers of latent variables ("hidden units"), with connections between the layers but not between units within each layer.

Deepfakes: synthetic media in which a person in an existing image or video is replaced with someone else's likeness. While the act of creating fake content is not new, deepfakes leverage powerful techniques from artificial intelligence and machine learning to manipulate or generate visual and audio content that can more easily deceive.

Deep Learning: a machine learning technique that teaches computers how to learn by rote (i.e., machines mimic learning as a human mind would, by using classification techniques). Deep learning is a type of machine learning that uses deep (or many layered) layers.

Degrees of Freedom: the number of independent directions or joints of the robot, which would allow the robot to move its end-effector through the required sequence of motions. For arbitrary positioning, six degrees of freedom are needed: three for position (left-right, forward-backward and up-down), and three for orientation (yaw, pitch, and roll).

Denavit–Hartenberg: for robot manipulators, the Denavit–Hartenberg parameters (also called DH parameters) are the four parameters associated with a particular convention for attaching coordinate reference frames to the links of spatial kinematic chains.

Dexterity: a measure of the robot's ability to follow complex paths.

Diagnostics: analyze vast amounts of data and identify patterns and relationships.

Diffusion models: generate new data by reversing a diffusion process, i.e., information loss due to noise intervention. The main idea here is to add random noise to data and then undo the process to get the original data distribution from the noisy data.

Diffusion models: these work by adding random data (known as "noise") to the data its learning about, then figuring out how to remove it while preserving the original data, thus learning what's important and what can be discarded.

Digital Computer: a computer that performs calculations and logical operations with quantities represented as digits, usually in the binary number system.

Digital Ecosystem: several software platforms or cloud services that work in tandem across a network.

Digital Twin: involves the invention of a mirror world made up of large-scale, very accurate digital models of people and objects in virtual environments to create a lifelike digital reproduction of the actual world.

Drone: also known as an unmanned aerial vehicle (UAV) is an aircraft without any human pilot, crew or passengers on board.

Dynamics: the study of motion, the forces that cause the motion, and the forces due to motion. The dynamics of a robot arm are very complicated as they result from the kinematical behavior of all masses within the arm's structure.

Echo State Networks: a type of reservoir computer that uses a recurrent neural network with a sparsely connected hidden layer (with typically 1% connectivity). The connectivity and weights of hidden neurons are fixed and randomly assigned.

Electronics: the science of the flow and control of electrons from devices such as batteries through electric components such as resistors and motors.

Embedded system: a built-in processor, providing real-time control as a component of a larger system, often with no direct user interface.

Emotional Intelligence: the ability to recognize, understand, and manage our emotions and those of others.

End-effector: an accessory device or tool, specifically designed for attachment to the robot wrist or tool mounting plate to enable the robot to perform its intended task. (Examples may include: gripper, spot weld gun, arc weld gun, spray point gun, or any other application tools.)

Energy Source: any electrical, mechanical, hydraulic, pneumatic, chemical, thermal, or other source.

Entropy: is a scientific concept as well as a measurable physical property that is most commonly associated with a state of disorder, randomness, or uncertainty.

Environment: it is the part of real or computational world inhabited by the agent or robot.

Epistemology: study of the kinds of knowledge that are required for solving problems in the world. It investigates what distinguishes justified belief from opinion.

Epoch: is a fixed date and time used as a reference from which a computer measures) system time.

Error Ellipse: a graphical tool used to illustrate the pair-wise correlation that exists between computed values

Error Function: the error function assigns a single value that represents the difference between the desired and actual values of one or several dependent variables.

Expert System: a computer program that uses artificial intelligence (AI) technologies to simulate the judgment and behavior of a human or an organization that has expertise and experience in a particular field.

Explainability: in AI, explainability refers to how easily humans can understand and explain the inner working of algorithms, either for a specific decision or in terms of their overall logic.

Facial recognition: includes facial verification (checking one face against data about one face, e.g., to check the identity of a phone user), identification (analyzing many faces against data about many faces to identify individuals). Can also include facial classification to estimate age, gender, or other categories such as mood or personality.

Feedback: the return of information from a manipulator or sensor to the processor of the robot to provide self-correcting control of the manipulator.

Feed Forward Neural Network: is the first and simplest type of artificial neural network devised. In this network, the information moves in only one direction, forward, from the input nodes, through the hidden nodes (if any), and to the output nodes. There are no cycles or loops in the network.

Fixed Automation (Hard Automation): automated, electronically controlled system for simple, straight, or circular motion. These systems are mainly used for large production runs where little flexibility is required.

Force: the strength associated with action or movement.

Force Feedback: a sensing technique using electrical signals to control a robot end-effector during the task of the end-effector. Information is fed from the force sensors of the end-effector to the robot control unit during the particular task to enable enhanced operation of the end-effector.

Force–Torque Sensing: for a manipulator, an electronic device that is designed to monitor, detect, record and regulate linear and rotational forces exerted upon it.

Formal Reasoning: is characterized by rules of logic and mathematics, with fixed and unchanging premises and is concerned only with the forms of arguments.

Forward Chaining: strategy of working forward for conclusion/solution of a problem.

Forward Kinematics: computational procedures determine where the end-effector of a robot is located in space. The procedures use mathematical algorithms along with joint sensors to determine its location.

Frame problem: a problem of representing the real-word situations to be computationally tractable.

Fuzzy Logic: a form of logic which deals with approximate values, as opposed to logic signals which are true or false only.

Gain: the size of the output of an element divided by its input – in a Proportional Controller, its output is the error * the gain.

Generative Adversarial Networks (GANs): are a class of machine learning frameworks where two neural networks contest with each other in a game (in the form of a zero-sum game, where one agent's gain is another agent's loss).

Generative AI (GenAI): can generate all kinds of data, including audio, code, images, text, simulations, 3D objects, videos, and so forth. It takes inspiration from existing data, but also generates new and unexpected outputs, breaking new ground in the world of product design, art, and many more.

Generative Pretrained Transformers (GPT): a generative AI Tool used to develop large language models and to power chatbots, for example, ChatGPT.

Generative Query Network: a framework within which machines learn to perceive their surroundings by training only on data obtained by themselves as they move around scenes.

Genetic algorithm: an algorithm based on principles of genetics that is used to find solutions efficiently and quickly to difficult problems.

Global Navigation Satellite System (GNSS): a general term describing any satellite constellation that provides positioning, navigation, and timing (PNT) services on a global or regional basis.

GOFAI: good old-fashioned artificial intelligence refers to first-generation "symbolic AI" systems that predate and are often more explainable, than machine learning.

Gradient Descent: a first-order iterative optimization algorithm for finding a local minimum of a differentiable function. The idea is to take repeated steps in the opposite direction of the gradient (or approximate gradient) of the function at the current point because this is the direction of steepest descent. GD is an iterative optimization process that searches for an objective function's optimum value (Minimum/ Maximum).

Graph Neural Network (GNN): a class of artificial neural network for processing data that can be represented as graphs.

Graphical Processing Unit (GPU): a specialized electronic circuit designed to manipulate and alter memory to accelerate the creation of images in a frame buffer intended for output to a display device.

Gripper: an end-effector that is designed for seizing and holding (ISO 8373) and "grips" or grabs an object. It is attached to the last link of the arm. It may hold an object using several different methods, such as: applying pressure between its "fingers", or may use magnetization or vacuum to hold the object, etc.

Hallucination or Artificial Hallucination: a confident response by an AI that does not seem to be justified by its training data.

Haptics: interaction involving the sense of touch.

Haar-like Features: are digital image features used in object recognition similar to Haar wavelets instead of using image intensities.

Heuristic: a computer science technique designed for quick, optimal, solution-based problem-solving.

Hidden Markov Model (HMM): a statistical approach that is frequently used for modeling biological sequences. In applying it, a sequence is modeled as an output of a discrete stochastic process, which progresses through a series of states that are "hidden" from the observer.

Histogram of Oriented Gradients (HOG): is a feature descriptor used in computer vision and image processing for the purpose of object detection. The technique counts occurrences of gradient orientation in localized portions of an image.

Holonomic: all degrees of freedom are controllable.

Hopfield Network: a form of recurrent artificial neural network and serves as content-addressable ("associative") memory systems with binary threshold nodes, or with continuous variables.

Human-centered AI: AI that seeks to augment the abilities of, address the societal needs of, and draw inspiration from human beings.

Human–Robot Interface: study of interactions between humans and robots.

Humanoid: a non-human entity with human form or characteristics.

Hyperparameters: hyperparameters are settings that define how an algorithm or a machine learning model learns and behaves. Hyperparameters include learning rate, regularization strength, and the number of hidden layers in the network. You can tinker with these parameters to fine-tune the model's performance according to your needs.

Image recognition: the process of identifying or detecting an object or feature of an object in an image or video.

Induction: the process or action of bringing about or giving rise to something (reasons from specific cases to general rules).

Inference: a conclusion reached on the basis of evidence and reasoning.

Information theory: measures of entropy and optimal codes are germane and central to the issue of delimiting optimal training sequences for encoding a hypothesis.

Instruction: a piece of a program commanding the computer to do something.

Integral control: a method used in Feedback Control where the output of the controller depends on its input (typically the error) and previous values of its output.

Intelligence: the ability to acquire and apply skills and knowledge.

Intelligent robot: a robot whose actions are at least in part determined by the robot.

Interface: a boundary between the robot and machines, transfer lines, or parts outside immediate environment.

Inverse Optimal Control: also known as inverse reinforcement learning, it's the problem of recovering an unknown reward function in a Markov decision process from expert demonstrations of the optimal policy.

Industrial Automation: also referred to as automation, uses numerical control during the use of control systems (e.g., computers) to control industrial machinery and processes, replacing human operators.

Industrial Robot: a re-programmable multifunctional manipulator designed to move material, parts, tools, or specialized devices through variable programmed motions for the performance of a variety of tasks (R15.06). The principle components are: one or more arms that can move in several directions, a manipulator, and a computer controller that gives detailed movement instructions.

Joint Space: (1) Joint Space (or Joint Coordinates) is just a method of defining the position of the robot in terms of the value of each axis instead of as a TCP position. For example, the Home Position of a robot is often defined in Joint Space as each axis being at 0°. (2) The set of joint positions.

Kalman Filter: also known as linear quadratic estimation (LQE), is an algorithm that uses a series of measurements observed over time, including statistical noise and other inaccuracies, and produces estimates of unknown variables that tend to be more accurate than those based on a single measurement alone, by estimating a joint probability distribution over the variables for each timeframe.

Karnaugh Map (KM or K-map): is a method of simplifying Boolean algebra expressions where it reduces the need for extensive calculations by taking advantage of humans' pattern-recognition capability.

Kinematics: the relationship between the motion of the endpoint of a robot and the motion of the joints. For a Cartesian Robot, this is a set of simple linear functions (linear tracks that may be arranged in *X*, *Y*, *Z* directions), for a revolute topology (joints that rotate) however, the kinematics are much more complicated involving complicated combinations of trigonometry functions. The kinematics of an arm is normally split into forward and inverse solutions.

Knowledge: facts, information, and skills acquired through experience or education, the theoretical or practical understanding of a subject.

Knowledge Base: set of sentences in a formal language representing facts about the world.

Knowledge Engineering: acquiring knowledge from human experts and other resources.

Knowledge Representation: to store and manipulate information (logical and probabilistic representations).

Labeled Data: data that are accompanied with information about the data, for example, a picture of cat that is labeled as containing a cat.

Language: is a system of signs having meaning by convention. In this sense, language need not be confined to the spoken word.

Language of Thought Hypothesis (LOTH): claim that mental representation has a linguistic structure. A representational system has a linguistic structure if it employs both a *combinatorial syntax* and a *compositional semantics.*

Large Language Model (LLM): is a language model consisting of a neural network with many parameters (typically billions of weights or more), trained on large quantities of unlabeled text using self-supervised learning.

Limited Memory: systems with short-term memory limited to a given timeframe.

Liquid State Machine: a type of reservoir computer that uses a spiking neural network. An LSM consists of a large collection of units (called *nodes*, or *neurons*). Each node receives time-varying input from external sources (the inputs) as well as from other nodes.

Localization: in robotics, it is about determining where a robot is.

Logic: a branch of mathematics concerned with signals that can be only true or false, and which form the basis of modern computers.

Logical calculus: is a calculus whose deductive system is aimed at capturing logical inferences among the class of expressions allowed by the calculus' grammar.

Logical Deduction: mental process of drawing deductive inferences where an inference is deductively valid if its conclusion follows logically from its premises, that is, it is impossible for the premises to be true and the conclusion to be false.

Logicism: refers to the doctrine that mathematics is a part of (deductive) logic and mathematics is an extension of logic, and therefore that some or all mathematics is reducible to logic. It effectively holds that mathematical theorems and truths are logically necessary.

Long Short-Term Memory (LSTM): unlike standard feedforward neural networks, LSTM has feedback connections. Such a recurrent neural network (RNN) can process not only single data points (such as images) but also entire sequences of data (such as speech or video).

Machine Learning: focuses on developing programs that access and use data on their own, leading machines to learn for themselves and improve from learned experiences; intelligence from data; to adapt to new circumstances; and to detect and extrapolate patterns.

Machine Translation: an application of natural language processing used for language translation (human-to-human) in text- and speech-based conversation.

Manipulator: a machine or robotic mechanism of which usually consists of a series of segments (jointed or sliding relative to one another) for the purpose of grasping and/or moving objects (pieces or tools), usually in several degrees of freedom. The control of the manipulator may be by an operator, a programmable electronic controller, or any logic system (for example a cam device and wired).

Markovian: statistical processes that exhibit the Markov property, in which the conditional probability distribution of future states of the process, given the present state and all past states, depends only upon the present state and not on any past states.

Mathematical Logic: is the study of logic within mathematics.

Memory: the mental processes of acquiring and retaining information for later retrieval, and the mental storage system in which these processes operate.

Mental Models: are any sort of concept, framework, or worldview that a person carries around in his/her mind (e.g. people construct microcosmic representations of the world in their minds to help them understand and process their experiences).

Micro Aerial Vehicle (MAV): is a class of unmanned aerial vehicles that has a size restriction and may be autonomous. Modern craft can be as small as 15 centimeters.

Microcontroller: a small computer system, typically on a single integrated circuit, comprising the microprocessor, memory, and peripherals.

Mind: with consciousness and self-awareness.

Mission Planning: has the task of assigning missions (goal sets) to each robot in the fleet by communicating with the logistics planner which models world state information. It speeds up the programming process when the robot is in a complex environment. Instead of planning every single move ourselves, the motion planner can automatically create one or more good routes for the robot to follow.

Mobile Robot: a self-propelled and self-contained robot that is capable of moving over a mechanically unconstrained terrain.

Model: a computer program that's been trained by a machine learning algorithm to perform a specific task.

Model Predictive Control (MPC): is an optimal control strategy that uses the model of the system to obtain an optimal control sequence by minimizing an objective function.

Motion Planning: a computational problem to find a sequence of valid configurations that moves the object from the source to destination.

Moore's Law: the 1965 prediction that the number of transistors on a computer chip would double every two years. It still holds true, despite recently falling slightly behind schedule. Today, Moore's law is used more generally to refer to the trend for computer power to increase exponentially.

Narrow AI: or "weak AI" refers to the current paradigm of AI tools that exhibit intelligence only in specific niches such as playing chess or recognizing cats.

Natural Language: a language that has developed naturally in use (as *contrasted* with an artificial language or computer code).

Natural Language Generation (NLG): natural language generation deals with the creation of human-readable text from structured data.

Natural Language Processing (NLP): helps computers process, interpret, and analyze human language and its characteristics by using natural language data (to communicate with the machine).

Neural Networks: networks of nerve cells in the brains of humans and animals.

Neural Ordinary Differential Equations: instead of specifying a discrete sequence of hidden layers, we parameterize the derivative of the hidden state using a neural network. The output of the network is computed using a blackbox differential equation solver. These continuous-depth models have constant memory cost, adapt their evaluation strategy to each input, and can explicitly trade numerical precision for speed.

Neural Processing Unit (NPU): is a specialized accelerator for AI workloads that do not run the operating system nor process graphics. Still, it can easily assist in doing both when those workloads are accelerated using neural networks.

Neural Radiance Field (NeRF): a fully connected neural network that can generate novel views of complex 3D scenes, based on a partial set of 2D images.

Neurobiology: information processing found in biological organisms motivated Artificial Neural Network models of learning.

Neuromorphic Processor: an approach to computing that is inspired by the structure and function of the human brain.

No-Code AI: creating AI programs without writing code.

Non-Holonomic: some degrees of freedom are not directly controllable.

Object Detection: the task of identifying a given object in an image.

Object-Oriented Programming (OOP): is a programming paradigm based on the concept of "objects", which can contain data and code: data in the form of fields (often known as attributes or *properties*), and code, in the form of procedures (often known as methods).

Ontology: is concerned with identifying, in the most general terms, the kinds of things that actually exist.

Operant Conditioning: changing behavior by the use of reinforcement which is given after the desired response.

Operator: the person designated to start, monitor, and stop the intended productive operation of a robot or robot system. An operator may also interface with a robot for productive purposes.

Oracle (AI): is a hypothetical AI designed to answer questions but is prevented from gaining any goals and subgoals that involve modifying the world beyond its limited environment.

Parameters: parameters are the settings and weights that control how each "neuron" or node processes and transforms input data.

Path: the continuous locus of positions (or points in three-dimensional space) traversed by the tool center point and described in a specified coordinate system.

Path Planning: lets a robot find the shortest and most obstacle-free path from a start to goal state. The path can be a set of states (position and orientation) or waypoints. Path planning requires a map of the environment along with start and goal states as input.

Pattern: is a set of repeating, significant attributes.

Pattern Recognition: automated recognition of patterns found in data.

Perception: as an internal process and the ability to see, hear, or become aware of something through the senses.

Perceptrons: an algorithm for supervised learning of binary classifiers and is a neural network unit that does certain computations to detect features or business intelligence in the input data.

Percepts: it is the format in which the agent obtains information about the environment.

Philosophy: argumentations like "the simplest hypothesis is the best" underlie the reasoning process of machine learning algorithms.

Physical System: is a collection of physical objects. Everything outside the system is known as the environment.

Pick-and-Place Task: a repetitive part transfer task composed of a picking action followed by a placing action.

Planning: a sequence of steps, and is a combination of an algorithm (search) and a set of representations.

Pose: alternative term for robot configuration, which describes the linear and angular position. The linear position includes the azimuth, elevation, and range of the object. The angular position includes the roll, pitch, and yaw of the object.

Position: the definition of an object's location in 3D space, usually defined by a 3D coordinate system using X, Y, and Z coordinates.

Precision: the amount of information that is conveyed by a number in terms of its digits and shows the closeness of two or more measurements to each other.

Problem-solving: basically is a state-space search, i.e., primarily an algorithm and not a representation.

Process Mining: applies data science to discover, validate, and improve workflows.

Production system: is a computer program typically used to provide some form of artificial intelligence, which consists primarily of a set of rules about behavior but it also includes the mechanism necessary to follow those rules as the system responds to states of the world.

Programmable Logical Controller (PLC): a solid-state control system, which has a user-programmable memory for storage of instructions to implement specific functions such as: I/O control logic, timing, counting arithmetic, and data manipulation. A PLC consists of a central processor, input/output interface, memory, and programming device, which typically uses relay equivalent symbols.

Programmable Matter by Folding: an example of a system capable of achieving multiple shapes for multiple functions.

Programmable Robot: a feature that allows a robot to be instructed to perform a sequence of steps and then to perform this sequence in a repetitive manner. It can then be reprogrammed to perform a different sequence of steps if desired.

Prompt: short text.

Prompt engineering: is the process of structuring text that can be interpreted and understood by a generative AI model. A prompt is natural language text describing the task that an AI should perform.

Proportional-Integral-Derivative (PID) controller: is a three-term controller that employs feedback to read a sensor, then compute the desired actuator output by calculating proportional, integral, and derivative responses and summing those three components to compute this output.

Pruning: overriding unnecessary and irrelevant considerations in AI systems.

Psychology: a general term encompassing various mental abilities, including the ability to remember and use what one has learned, in order to solve problems, adapt to new situations, and understand and manipulate one's reality.

Pushdown Automata: are computational models, for example, theoretical computer-like machines that can do more than a finite state machine, but less than a Turing machine.

Q-Learning: a model-free reinforcement learning algorithm to learn the value of an action in a particular state. It does not require a model of the environment, and it can handle problems with stochastic transitions and rewards without requiring adaptation.

Qualitative Reasoning: a means to express conceptual knowledge such as the physical system structure, causality, start and end of processes, assumptions, and conditions under which facts are true, qualitatively distinct.

Quantum Computing: an area of computer science that uses the principles of quantum theory, explaining the behavior of energy and materials on the atomic and subatomic levels. Quantum computing uses subatomic particles such as electrons or photons.

Qubit: is a short for "quantum bit" and is the basic unit of information in quantum computing and counterpart to the bit (binary digit) in classical computing.

Radial Basis Function: are means to approximate multivariable (also called *multivariate*) functions by linear combinations of terms based on a single univariate function (the radial basis function).

Random Decision Forest: an ensemble learning method for classification, regression and other tasks that operates by constructing a multitude of decision trees at training time.

Reach: the volume of space (envelope), which a robot's end-effector can reach in at least one orientation.

Reactive Machines: can analyze, perceive, and make predictions about experiences, but do not store data; they react to situations and act based on the given moment.

Real-Time Processing: handling or processing information at the time events occur or when the information is first created (see also Real-Time System).

Rectified Linear Unit (ReLU): is a non-linear activation function used for deep neural networks in machine learning.

Recurrent Neural Network (RNN): a type of neural network that makes sense of and creates outputs based on sequential information and pattern recognition.

Recurrent Neural Network (RNN): is one of the two broad types of artificial neural network, characterized by direction of the flow of information between its layers. In contrast to the uni-directional feedforward neural network, it is a bi-directional artificial neural network, meaning that it allows the output from some nodes to affect subsequent input to the same nodes.

Recursive Neural Network: is a kind of deep neural network created by applying the same set of weights recursively over a structured input, to produce a structured prediction over variable-size input structures, or a scalar prediction on it, by traversing a given structure in topological order.

Reinforcement Learning: a machine learning method where the reinforcement algorithm learns by interacting with its environment, and is then penalized or rewarded based off of decisions it makes.

Real-Time Kinematic Positioning (RTK): the application of surveying to correct for common errors in current satellite navigation (GNSS) systems.

Real-Time System: a computer system in which the computer is required to perform its tasks within the time restraints of some process simultaneously with the system it is assisting. The computer processes system data (input) from the sensors for the purpose of monitoring and computing system control parameters (outputs) required for the correct operation of a system or process.

Reasoning: the action of thinking about something in a logical, sensible way and is also computation.

Regression: regression algorithms, used in statistical modeling and in artificial intelligence, are a set of statistical processes for estimating the relationships between a dependent variable and one or more independent variables.

Reinforcement Learning: is a machine learning training method based on rewarding desired behaviors and/or punishing undersized ones.

Remote Control: is a component of an electronics device used for operating the television device wirelessly from a short line-of-sight distance.

Remotely Piloted Vehicle: an unmanned vehicle capable of being controlled from a distant location through a communication link. It is normally designed to be recoverable.

Repeatability: a measure of how close an arm can repeatedly obtain a taught position.

Retrieval-Augmented Generation (RAG): is an AI framework for retrieving facts from an external knowledge base to ground large language models (LLMs) on the most accurate, up-to-date information and to give users insight into LLMs' generative process.

Retrieval-Augmented Generation (RAG): is a design (or architecture) approach for Generative AI that combines information retrieval techniques with generative language models (LLMs) to enhance the quality and relevance of generated text.

Robot: is a machine, especially one programmable by a computer, capable of carrying out a complex series of actions teleoperated, automatically, or autonomously.

Robotic Process Automation (RPA): uses software with artificial intelligence and machine learning capabilities to perform repetitive tasks once completed by humans.

Rough Sets: are approximation sets that deviate from the idea of ordinary sets and are a technique that may be utilized to discover previously unknown structural connections when working with incomplete or noisy data.

Rovers: machines that drive, either by human or robotic control, on planetary surfaces.

Rule: it is a format of representing knowledge base in Expert System. It is in the form of IF-THEN-ELSE.

SCADA: stands for *Supervisory Control and Data Acquisition* and is a category of software applications for controlling industrial processes, which is the gathering of data in real time from remote locations in order to control equipment and conditions.

SCARA Robot: a cylindrical robot having two parallel rotary joints (horizontally articulated) and provides compliance in one selected plane.

Search: the platform learns from data on users to automatically generate the most accurate and relevant search experiences.

Self-Awareness: is the ability to focus on yourself and how your actions, thoughts, or emotions do or don't align with your internal standards.

Semantic Network: is a simple representation scheme that uses a graph of labeled nodes and labeled, directed arcs to encode knowledge.

Semiotics: is the systematic study of sign processes and meaning making. Semiosis is any activity, conduct, or process that involves signs, where a sign is defined as anything that communicates something, usually called a meaning, to the sign's interpreter.

Sensor: instruments used as input devices for robots, which enable it to determine aspects regarding the robot's environment, as well as the robot's own positioning.

Sentience: the capacity of a being to experience feelings and sensations.

Sequence to Sequence Learning: about training models to convert sequences from one domain (e.g., sentences in English) to sequences in another domain (e.g., the same sentences translated to French).

Servo-System: a system in which the controller issues commands to the motors, the motors drive the arm, and an encoder sensor measures the motor rotary motions and signals the amount of the motion back to the controller. This process is continued many times per second until the arm is repositioned to the point requested.

Shell: a software that helps in designing inference engine, knowledge base, and user interface of an expert system.

Scale-Invariant Feature Transform (SIFT): a computer vision algorithm to detect, describe, and match local features in images.

Simulated Annealing: a probabilistic technique for approximating the global optimum a given function and is a metaheuristic to approximate global optimization in a large search space for an optimization problem.

Singularity: a configuration where two joints of the robot arm become co-axial (aligned along a common axis). In a singular configuration, smooth path following is normally impossible and the robot may lose control.

Situation Awareness (SA): the perception of the elements in the environment within a volume of time and space, the comprehension of their meaning, and the projection of their status in near future.

SLAM (Simultaneous, Localization and Mapping): is the computational problem of constructing or updating a map of an unknown environment while simultaneously keeping track of a robot's location within it.

Sparse Coding: a class of unsupervised methods for learning sets of over-complete bases to represent data efficiently. The aim of sparse coding is to find a set of basis vectors ϕ_i such that we can represent an input vector x as a linear combination of these basis vectors: $x = k\Sigma\mathrm{I} = \mathrm{I} a_i \phi_i$.

Speech Recognition: an interdisciplinary subfield of computer science and computational linguistics that develops methodologies and technologies that enable the recognition and translation of spoken language into text by computers with the main benefit of searchability.

Spin Glass: a magnetic state characterized by randomness, besides cooperative behavior in freezing of spins at a temperature called "freezing temperature".

Stable Diffusion: is a deep learning, text-to-image model and is primarily used to generate detailed images conditioned on text descriptions. It can also be applied to other tasks such as inpainting, outpainting, and generating image-to-image translations guided by a text prompt.

Statistics: involves the collection, description, analysis, and inference of conclusions from quantitative data.

Stochastic Gradient Descent (SGD): is a variant of the Gradient Descent algorithm that is used for optimizing machine learning models.

Strong AI: see AGI.

Structured Data: clearly defined data with easily searchable patterns.

Structured Query Language (SQL): is a domain-specific, programming language for storing data and processing information in a relational database.

Subsumption: a control architecture that was proposed in opposition to traditional AI, or GOFAI. Instead of guiding behavior by symbolic mental representations of the world, subsumption architecture couples sensory information to action selection in an intimate and bottom-up fashion.

Superintelligence: is defined as "An intellect that is much smarter than the best human brains in practically every field, including scientific creativity, general wisdom, and social skills".

Supervised Learning: a type of machine learning where output datasets teach machines to generate desired outcomes or algorithms (akin to a teacher-student relationship).

Support Vector Mechanism (SVM): Also called support vector networks, SVMs are supervised learning models with associated learning algorithms that analyze data used for classification and regression analysis.

Swarm intelligence (SI): is one of the computational intelligence techniques which are used to solve complex problem. SI involves collective study of the individuals behavior of population interact with one another locally.

Syllogism: is a kind of logical argument that applies deductive reasoning to arrive at a conclusion based on two propositions that are asserted or assumed to be true.

Symbolic AI (or Classical AI): represents human knowledge in a declarative form (i.e., facts and rules).

Symbolic Integration: in calculus, the problem of finding a formula for the antiderivative, or *indefinite integral*, of a given function $f(x)$.

Tactile Sensing: a device that measures information arising from physical interaction with its environment. Tactile sensors are generally modeled after the biological sense of cutaneous touch which is capable of detecting stimuli resulting from mechanical stimulation, temperature, and pain (although pain sensing is not common in artificial tactile sensors).

Target Drone: is an unmanned, remote-controlled aerial vehicle, usually used in the training of anti-aircraft crews. In their simplest form, target drones often resemble radio controlled model aircraft. More advanced drones are made from large, old anti-ship missiles which had their warheads removed.

Task: It is the goal the agent is trying to accomplish.

Technological Singularity: is a hypothetical future point in time at which technological growth becomes uncontrollable and irreversible, resulting in unforeseeable changes to human civilization.

Teleoperation: refers to the operation of a system or machine at a distance.

Telepresence: refers to a set of technologies that allow a person to feel as if they were present, to give the appearance or effect of being present via telerobotics, at a place other than their true location.

Telerobotics: area of robotics concerned with the control of semi-autonomous robots from a distance, chiefly using television, wireless networks, or tethered connections.

Tensor Flow: where computations are expressed as stateful dataflow graphs.

Thinking: is the manipulation of symbols.

Thoughts: themselves as symbolic representations.

Tokenization: Tokenization is the process of splitting a text document into smaller units called tokens. These tokens can represent words, numbers, phrases, symbols, or any elements in text that a program can work with. The purpose of tokenization is to make the most sense out of unstructured data without processing the entire text as a single string, which is computationally inefficient and difficult to model.

Tool: A term used loosely to define a working apparatus mounted to the end of the robot arm, such as a hand, gripper, welding torch, and screw driver.

Training Data: sets of data that are processed by machine learning algorithms to improve their functionality.

Transfer Learning: a system that uses previously learned data and applies it to a new set of tasks.

Transformer: is a deep learning model that adopts the mechanism of self-attention, differentially weighting the significance of each part of the input data.

Transformer Neural Network: A transformer is a deep learning architecture developed by Google and based on the multi-head attention mechanism. It has no recurrent units, and thus requires less training time than previous recurrent neural architectures, such as long short-term memory (LSTM), and its later variation has been prevalently adopted for training large language models (LLM) on large (language) datasets (text is converted to numerical representations called tokens).

Translation: communication of the meaning of a source-language text by means of an equivalent target-language text.

Tree Search: a tree data structure used for locating specific keys from within a set. In order for a tree to function as a search tree, the key for each node must be greater than any keys in subtrees on the left, and less than any keys in subtrees on the right.

Trusted AI: is AI capability that can provide reasonable confidence that it has satisfied user-defined objectives in a proper and interpretable way over its lifetime.

Trusted Autonomy: trusted autonomy encompasses self-monitoring, self-adjustment, and learning to accomplish safe and high-performance complex task execution in time-varying, complex, and unknown environments.

Turing Test: a test created by computer scientist Alan Turing to see if machines could exhibit intelligence equal to or indistinguishable from that of a human.

Turing Machine: a mathematical definition of algorithmic computation.

Unmanned Aerial Vehicle: is an aircraft with no pilot on board. A UAV is defined as being capable of controlled, sustained-level flight and powered by a jet or reciprocating engine.

Unstructured Data: data without easily searchable patterns (e.g., audio, video, social media content).

Unsupervised Learning: a type of machine learning where an algorithm is trained with information that is neither classified nor labeled, thus allowing the algorithm to act without guidance (or supervision).

Variational Autoencoders: this is a type of model that learns how data is constructed by encoding it in a simple way that captures its essential characteristics and then figuring out how to reconstruct it.

Vector Database: is a collection of data stored as mathematical representations, which makes it easier for machine learning models to remember previous inputs.

Virtual Reality (VR): the use of computer modeling and simulation that enables a person to interact with an artificial three-dimensional (3-D) visual or another sensory environment.

Vision Sensor: a sensor that identifies the shape, location, orientation, or dimensions of an object through visual feedback, such as a television camera.

YOLO (You Only Look Once): a neural network model mainly used for the detection of objects in images and in real-time videos.

Waldo: a means of capturing motion data. Waldos were originally developed within the film industry in order to control robotic or animated character.

Waypoint Navigation: waypoints are sets of coordinates that identify a point in physical space. These coordinates can include longitude, latitude, and altitude. A waypoint is a predetermined geographical position that is defined in terms of latitude/longitude coordinates (altitude optional). Waypoints may be a simple named point in space or may be associated with existing navigational aids, intersections, or fixes.

Weak AI: see "narrow AI".

Work Envelope: The set of all points that a manipulator can reach without intrusion. Sometimes the shape of the work space, and the position of the manipulator itself can restrict the work envelope.

Bibliography

I Artificial Intelligence Background

Mind

[1] Descartes, R., *Treatise on Man*, translated from *De l'homme et de la formation du foetus*, Prometheus Publishing, Paris, 1664.

[2] Aristotle, *Prior Analytics*, Translation by A.J. Jenkinson with Minor Emendations by Daniel Kolak, MIT Classics, Cambridge, MA, 350 BC

[3] Descartes, R., *Rules for the Direction of the Mind*, Bobbs-Merrill Company, Inc., written in English, Indianapolis, 1962.

[4] Singer, P.N., "Galen's Pathological Soul: Diagnosis and Therapy in Ethical and Medical Texts and Contexts", In *Mental Illness in Ancient Medicine*, C. Thumiger and P.N. Singer (Eds.), Brill, Leiden, Netherlands, Jan. 2018, pp. 381–420.

[5] Helvetius, M., *A Treatise on Man; his Intellectual Faculties and his Education* (in French by W. Hooper), Albion Press, Printed for James Cundee, IVT-Lane, Vernor, Hood & Sharpe, 31, Poultry, 1810.

[6] Hobbes, T., *Leviathan or the Matter, Forme, & Power of a Common-Wealth Ecclesiasticall and Civill.*, Andrew Crooke, at the Green Dragon in St. Pauls Church-yard, 1651.

[7] de La Mettrie, J.O., *Man a Machine* (French-English), The Open Court Publishing Co., Chicago, IL, 1912.

[8] Hofstadter, D., Gdel, Escher, *Bach: An Eternal Golden Braid*, Penguin Books, New York, NY, 1994.

[9] Chalmers, D., "Facing up to the Problem of Consciousness", *Journal of Consciousness Studies*, Vol. 2, No. 3, 1995, pp. 200–219.

[10] Husserl, E., *Ideas Pertaining to a Pure Phenomenology and to a Phenomenological Philosophy*, Martinus Nijhoff Publishers, Part of Kluwer Academic Publishers, Leiden, Netherlands,1982.

[11] Ogden, C. and Richards, I., *The Meaning of Meaning*, A Harvest Book, Harcourt, Brace & World, Inc., New York, NY, 1923.

[12] Vygotsky, L., *Mind and Society*, Harvard University Press, Cambridge, MA, 1930.

[13] Bergmann, M., Moor, J., and Nelson, J., *The Logic Book*, McGraw Hill Companies, New York, NY, 1990.

[14] Licklider, J.C., "Man-Computer Symbiosis", *IRE Transactions of Human Factors in Electronics*, Vol. HFE-1, Mar. 1960, pp. 4–11.

[15] Jaynes, E.T., *Probability Theory: The Logic of Science*, Cambridge University Press, New York, NY, 2003.

[16] Bateson, G., *Steps to an Ecology of Mind*, Jason Aronson Inc, Northvale, NJ, 1972.

[17] Bateson, G., *Mind and Nature: A Necessary Unity*, E.P. Dutton, New York, NY, 1979.

[18] Dennet, D., Brainchildren*: Essays on Designing Minds*, A Bradford Book, MIT Press, Cambridge, MA, 1998.

[19] Albus, J. and Meystel, A., *Engineering of Mind: An Introduction to the Science of Intelligent Systems*, John Wiley & Sons, Inc., New York, NY, 2001.

[20] Albus, J., "The Engineering of Mind", *Information Sciences*, Vol. 117, 1999, pp 1–18.

[21] Albus, J., "Toward a Computational Theory of Mind", *Journal of Behavioral and Brain Sciences/Journal of Mind Theory*, Vol. 1, No. 1, 2006, pp. 1–38.

[22] Penrose, R., *The Emperor's New Mind: Concerning Computers, Minds, and the Laws of Physics*, Penguin Books, New York, NY, Nov. 1989.

[23] Chalmers, D., *The Conscious Mind: In Search of a Fundamental Theory*, Oxford University Press, Oxford, UK, 1996.

[24] Pollock, J., *Cognitive Carpentry: A Blueprint for How to Build a Person*, MIT Press, Cambridge, MA, 1995.

[25] Berkeley, E., *Giant Brains or Machines That Think*, Science Editions, Inc., New York, NY, 1961.

[26] Penrose, R., *Shadows of the Mind: A Search for the Missing Science of Consciousness*, Oxford University Press, Oxford, UK, 1994.

[27] Penrose, R., *The Road to Reality: A Complete Guide to the Laws of the Universe*, Jonathan Cape, London, UK, 2004.

[28] Hubel, D. and Wiesel, T., "Shape and Arrangement of Columns in Cat's Striate Cortex", *Journal of Physiology*, Vol. 165, 1963, pp. 559–568.

[29] Hubel, D. and Wiesel, T., "Receptive Fields of Single Neurones in the Cat's Striate Cortex", *Journal of Physiology*, Vol. 148, 1959, p. 574.

[30] Engelbart, D., "Augmenting Human Intellect: A Conceptual Framework", AFOSR-3223, SRI Project No. 3578, Summary Report for the Air Force Office of Scientific Research, Oct. 1962.

[31] Lucas, J., "Minds, Machines, Godel", *Philosophy*, Vol. 36, No. 137, Jul. 1961, pp. 112–127.

[32] Maslin, K.T., *An Introduction to the Philosophy of Mind*, Polity Press, Blackwell Publishers, Maiden, MA, 2001.

[33] Lowe, E.J., *An Introduction to the Philosophy of Mind*, Cambridge University Press, Cambridge, UK, 2000.

[34] Feser, E., *Philosophy of Mind: A Short Introduction*, Oneworld Publications, Oxford, England, 2006.

[35] Rapaport, W., "How Minds can be Computational Systems", *Journal of Experimental & Theoretical Artificial Intelligence*, Vol. 10, 1998, pp. 403–419.

[36] Putnam, H., "Minds and Machines", In *Dimensions of Mind*, S. Hook (Ed.), New York University Press, New York, 1960 (and in *Journal of the ACM*,Vol. 7, No. 3, 1960, pp. 201–215).

[37] McDermott, D., *Mind and Mechanism*, MIT Press, Cambridge, MA, 2001.

[38] Aristotle, *The Complete Works of Aristotle*, J. Barnes (Ed.), Princeton University Press, Princeton, NJ, 1991.

[39] Hegel, G.W.F., *Philosophy of Mind: Part Three of the Encyclopedia of the Philosophical Sciences*, Translated by W. Wallace, Clarendon Press, Oxford, Great Britain, 1830.
[40] Wallace, W., *Hegel's Philosophy of Mind*, Clarendon Press, Oxford Great Britain, 1894.
[41] Haugeland, J., "Semantic Engines: An Introduction to Mind Designs", In *Mind Design*, J. Haugeland (Ed.), MIT Press, Cambridge, MA, 1981.
[42] Mind Design II*: Philosophy, Psychology, and Artificial Intelligence*, J. Haugland (Ed.), A Bradford Book, MIT Press, Cambridge, MA, 1997.
[43] Dennett, D., *Darwin's Dangerous Idea: Evaluation and the Meaning of Life*, Penguin Books, London, England, 1995.
[44] Dennett, D., *Kinds of Minds: Toward an Understanding of Consciousness*, Basic Books, New York, NY, 1996.
[45] Frye, D., Zelazo, P., and Palfai, T., "Theory of Mind and Rule-based Reasoning", *Cognitive Development*, Vol. 10, No. 4, Oct.-Dec. 1995, pp. 467–649.
[46] Hull, C., *Principles of Behavior: An Introduction to Behavior Theory*, Appleton-Century-Crofts, Inc. New York, NY, 1943.
[47] Skinner, B.F., *Beyond Freedom and Dignity*, A Penguin Book Ltd, Middlesex, England, 1971.
[48] Lashley, K.S., "Basic Neural Mechanisms in Behavior", *Psychological Review*, Vol. 37, 1930, pp. 1–24.
[49] Clark, R., Manns, J., and Squire, L., "Classical Conditioning, Awareness, and Brain Systems", *Trends in Cognitive Sciences*, Vol. 6, No. 12, Dec. 2002, pp. 524–531.
[50] Clark, R. and Squire, L., "Classical Conditioning and Brain Systems: The Role of Awareness", *Science*, Vol. 280, No. 5360, Apr. 1998, pp. 77–81.
[51] Kurzweil, R., *How to Create a Mind: The Secret of Human Thought Revealed*, Penguin Books, London, UK, Aug. 2013.
[52] Mach, E., *The Analysis of Sensations and the Relation of the Physical to the Psychical*, The Open Court Publishing Co., Great Britain, 1914.
[53] Mach, E., *The Science of Mechanics: A Critical and Historical Account of its Development*, The Open Court Publishing Co., Chicago, IL, 1919.
[54] Crick, F., *The Astonishing Hypothesis: The Scientific Search for the Soul*, Charles Scribner's Sons, Macmillan Publishing Company, New York, NY, 1994.
[55] *The Mechanical Mind in History*, P. Husbands, O. Holland, and M. Wheeler (Eds.), A Bradford Book, MIT Press, Cambridge, MA, 2008.
[56] Baynes, T.S., *Logic, or The Art of Thinking: Being the Port-Royal Logic*, Simpkin, Marshall and Co., London, UK, 1850.
[57] La Mettrie, J., *Man a Machine*, Open Court Publishing Co., Chicago, IL, 1912.
[58] Merrill, N. and Chuang, J., "Models of Minds: Reading the Mind Beyond the Brain", *ACM CHI* 2019, Glasgow, Scotland, UK, May 2019.
[59] *Open Mind*, T. Metzinger and J. Windt (Eds.), MIND Group, Frankfurt am Main, 2015.
[60] Ryle, G., *The Concept of Mind*, Hutchinson Publishing Co., London, UK, 1949.
[61] Fodor, J., "The Mind-Body Problem", *Scientific American*, Vol. 244, No. 1, Jan. 1981, pp. 114–123.
[62] Haugeland, J., "What is Mind Design?", In *Mind Design II: Philosophy, Psychology, and Artificial Intelligence*, J. Haugeland (Ed.), MIT Press, Cambridge, MA, 1997.
[63] Goertzel, B., *The Hidden Pattern: A Patternist Philosophy of Mind*, Brown Walker Press, Irvine, CA, Jun. 2006.
[64] Craik, K., *The Nature of Explanation*, Cambridge at the University Press, Cambridge, Great Britain, 1952.
[65] Bruner, J., Goodnow, J., and Austin, G., *A Study of Thinking*, John Wiley & Sons, Inc., New York, NY, 1956.
[66] Dyson, G., *Darwin among the Machines: The Evolution of Global Intelligence*, Helix Books, Addison-Wesley Publishing Co., Boston, MA, 1997.
[67] Grenander, U., *Probabilities on Algebraic Structures*, John Wiley & Sons, Inc., New York, NY, 1963.
[68] Crane, T., *The Mechanical Mind: A Philosophical Introduction to Minds, Machines and mental Representation*, Routledge, Abingdon, Oxon, 2016.
[69] Babbage, C., *Passages from The Life of a Philosopher*, Longman, Green, Longman, Roberts, & Green, London, UK, 1864.
[70] Cordeschi, R., "The Discover of the Artificial. Some Protocybernetic Developments 1930-1940", *AI & Society*, Vol. 5, 1991, pp. 218–238.
[71] Babbage, C., *On the Economy of Machinery and Manufacturing*, Charles Knight, Pall Mall East, 1832.
[72] Russell, B., *The Analysis of Mind*, Pennsylvania State University, Hazelton, PA, 1921.
[73] McLuhan, M. and Fiore, Q., *The Medium is the Massage: An Inventory of Effects*, Harmondsworth, Allen Lane/Penguin Books (Originally the University of Michigan), London, UK, 1967.
[74] Kahneman, D., Sibony, O., and Sunstein, C., *Noise: A Flaw in Human Judgment*, William Collins, London, Great Britain, 2021.
[75] Weber, A., *History of Philosophy*, Translation by F. Thilly, Charles Scribner's Sons, New York, NY, 1907.
[76] Bennett, M., *A Brief History if Intelligence*, Harper Collins Publisher, New York, NY, Oct. 2023.
[77] Grenander, U., *A Calculus of Ideas: A Mathematical Study of Human Thought*, World Scientific Publishing, Singapore, Sept. 2012.
[78] Jefferson, G., "The Mind of Mechanical Man", *British Medical Journal*, Vol. 1, Jun. 1949, p. 1105.
[79] Freeman, W.J., "Nonlinear Brain Dynamics and Intention According to Aquinas", *Mind & Matter*, Vol. 6, No. 2, 2008, pp. 207–234.
[80] Albus, J., "A Theory of Cerebellar Function", *Mathematical Biosciences*, Vol. 10, 1971, pp. 25–61.
[81] Borges, J.L., "Ramon Llull's Thinking Machines", Viking Press, New York, NY, 1937, pp. 155–159.
[82] Walter, W.G., *The Living Brain*, Penguin Book, Middlesex, Great Britain, 1961.
[83] de Spinoza, B., *Ethics Preceded by On the Improvement of the Understanding*, Hafner Publishing Co. New York, NY, 1949.
[84] Spinoza, B., *Spinoza Complete Works*, Translation by S. Shirley, Hackett Publishing, Co., Indianapolis, IN, 2002.

[85] Quetelet, M.A., *A Treatise on Man and the Development of his Faculties*, Published by William and Robert Chambers, Edinburgh, 1842.
[86] Craik, K., "Hypothesis on the Nature of Thought", In *The Nature of Explanation*, W. Barnes (Ed.), Cambridge University Press, Cambridge, 1943.
[87] Gdel, K, *On Formally Undecidable Propositions of Principia Mathematica and Related Systems*, Dover Publishing, Inc., New York, NY, 1931 (Also In Formal Unentscheidbare Stze der "Principia Mathematica" Und Verwandter Systeme I, Monatshefte fr Mathematik und Physik", No. 38, 1931, pp. 173–198).
[88] Wilkes, M.V., "Can Machines Think?", *Proceedings of the I.R.E.*, Vol. 41, Oct. 1953, pp. 1231–1234.
[89] Eddington, A., *The Philosophy of Physical Science*, Cambridge at the University Press, Cambridge, Great Britain, 1939.
[90] Bain, A., *Mind and Body: The Theories of their Relation*, D. Appleton & Company, New York, NY, 1873.
[91] Thompson, D.W., *On Growth and Form*, Cambridge University Press, London, England, 1917.
[92] Gadamer, H.-G., *Truth and Method*, The Continuum Books, London, Great Britain, 1975.
[93] Oakley, B., *A Mind for Numbers: How to Excel at Math and Science (Even If You Flunked Algebra)*, Jeremy P. Tarcher/Penguin, Penguin Group, New York, NY, 2014.
[94] Marcus, G., *The Algebraic Mind: Integrating Connectionism and Cognitive Science*, The MIT Press, Cambridge, MA, 2001.
[95] Tononi, G., *Phi-A Voyage from the Brain to the Soul*, Pantheon Books, 2012.
[96] Hintikka, J., "The Unambiguity of Aristotelian Being", *The Society for Ancient Greek Philosophy Newsletter*, No. 238, 1981.
[97] Asaro, P., "Computers as Models of the Mind: On Simulations, Brains and the Design of Early Computers", In *The Search for a Theory of Cognition*, S. Franchi and F. Bianchini (Eds.), Value Inquiry Book Series, Vol. 238, 2011, pp. 89–114.
[98] Craik, K., "Chapter 5 – Hypothesis on the Nature of Thought", In *The Nature of Explanation*, Cambridge University Press, Cambridge, UK, 1943.
[99] Searle, J., "Is the Brain's Mind a Computer Program?", *Scientific American, Nature Portfolio*, Springer Nature, Berlin, Germany, Jan. 1990, pp. 26–31.
[100] Katz, M., "Jerry Fodor and the Representational Theory of Mind", In *Philosophy of Mind: The Key Thinkers*, A. Bailey (Ed.), Bloomsbury Publishing, London, UK, 2014.
[101] Fodor, J., "The Mind-Body Problem", *Scientific American, Nature Portfolio*, Springer Nature, Berlin, Germany, 1981.
[102] Fodor, J., *The Modularity of Mind*, MIT Press, Cambridge, MA, 1983.
[103] Fodor, J., *The Language of Thought*, Thomas Y. Crowell Co., Inc., New York, NY, 1975.
[104] Wittgenstein, L., *Philosophical Investigations*, Translated by G.E.M. Anscombe, Basil Blackwell & Mott, Ltd., Oxford, Great Britain, 1953.
[105] Vygotsky, L., *Mind and Society*, Transcribed by A. Blunden and N. Schmolze, Harvard University Press, Cambridge, MA, 1930.
[106] Nasar, S., *A Beautiful Mind*, Simon & Schuster Paperbacks, New York, NY, 1998.
[107] Gamble, D., "Critical Notice", *Australasian Journal of Philosophy*, Vol. 70, No. 3, London, UK, 2006, pp. 343–357.
[108] Sowa, J.F., "Architectures for Intelligent Systems", *IBM System Journal*, Vol. 41, No. 3, 2002, pp. 331–349.

Cognitive Neuroscience

[109] Cajal, S.R., *Histologie du Systeme NerVeux: De L'Homme & Des Vertebres*, Generalites, Moelle, Ganglions, Rachidiens, Bulbe & Protuberance, Paris, France, 1909.
[110] Golgi, C., "The Neuron Doctrine - Theory and Facts", In *Nobel Lecture*, Dec. 1906.
[111] Malcom, N., "The Conceivability of Mechanism", *The Philosophical Review*, Vol. 77, No. 1, Jan. 1968, pp. 45–72.
[112] Penfield, W., *Mystery of the Mind: A Critical Study of Consciousness and the Human Brain*, Princeton University Press, Princeton, NJ, 1975.
[113] Traub, R. and Miles, R., *Neuronal Networks of the Hippocampus*, Cambridge University Press, Cambridge, UK, May 1991.
[114] Traub, R., Jefferys, J., and Whittington, M., *Fast Oscillations in Cortical Circuits*, The MIT Press, Cambridge, MA, May 1999.
[115] Markram, H., "The Blue Brain Project", *Nature Reviews: Neuroscience*, Vol. 7, Feb. 2006, pp. 153–160.
[116] Richards, R., "Christian Wolff's Prolegomena to Empirical and Rational Psychology: Translation and Commentary", *Proceedings of the American Philosophical Society*, Vol. 124, No. 3, Jun. 1980, pp. 227–239.
[117] Eccles, J. and Gibson, W., *Sherrington: His Life and Thought*, Springer-Verlag, Berlin, Heidelberg, Germany, 1979.
[118] Jackson, J.H., "The Croonian Lectures on Evolution and Dissolution of the Nervous System", *British Medical Journal*, Vol. 1, No. 1215, 1884, p. 703.
[119] Wilson, S., "Knowledge Growth in an Artificial Animal", In *Adaptive and Learning Systems*, K. Narendra (Ed.), Springer, Boston, MA, 1986.
[120] Edelman, S., *Computing the Mind: How the Mind Really Works*, Oxford University Press, Oxford, UK, Sept. 2008.
[121] Menninger, K., *The Human Mind*, Garden City Publishing Co., Inc., Garden City, NY, 1930.
[122] Ramachandran, V.S. and Blakeslee, S., *Phantoms in the Brain*, William Morrow and Company, Harper Collins, UK, 1998.

[123] Sejnowski, T., Koch, C., and Churchland, P., "Computational Neuroscience", *Science*, Vol. 241, No. 4871, Sept. 1988, pp. 1299–1306.
[124] Edelman, G., "Neural Darwinism: Selection and Reentrant Signaling in Higher Brain Function", *Neuron*, Vol. 10, Feb. 1993, pp. 115–125.
[125] Stanley-Jones, D. and Stanley-Jones, K., *The Kybernetics of Natural Systems: A Study in Patterns of Control*, Pergamon Press Ltd., Oxford, Great Britain, Jan. 1960.
[126] Dickens, W. and Cohen, J., "Instinct and Choice", In *Nature and Nurture: The Complex Interplay of Genetic and Environmental Influences on Human Behavior and Development*, C. Coll (Ed.), Lawrence Erlbaum Assoc., Mahwah, NJ, 2003.
[127] *Neuroscience and Philosophy*, F. Brigard and W. Sinnott-Armstrong (Eds.), MIT Press, Cambridge, MA, 2022.
[128] Breiman, L., "Statistical Modeling: The Two Cultures", *Statistical Science*, Vol. 16, No. 3, 2001, pp. 199–231.
[129] Lepore, E. and Pylyshyn, Z., *What is Cognitive Science?*, Wiley-Blackwell, Hoboken, NJ, Oct. 1999.
[130] Fodor, J., *Concepts: Where Cognitive Science Went Wrong*, Clarendon Press, Oxford, Great Britain, 1998.
[131] Boden, M., *The Creative Mind: Myths and Mechanisms*, George Weidenfeld and Nicolson Ltd., London, UK, 1990.
[132] Boden, M., *Minds as Machine: A History of Cognitive Science, Vol. 1 & 2*, Clarendon Press, Oxford, UK, 2006.
[133] Gleitman, L. and Papafragou, A., "Language and Thought", In *The Cambridge Handbook of Thinking and Reasoning*, K. Holyoak and R. Morrison (Eds.), Cambridge University Press, Cambridge, UK, Apr. 2005.
[134] Ryskin, R., Yoon, S.O., and Brown-Schmidt, S., "Language, Perspective, and Memory", In *Emerging Trends in the Social and Behavioral Sciences: An Interdisciplinary, Searchable, and Linkable Resources*, R. Scott and S. Kosslyn (Eds.), John Wiley & Sons, Hoboken, NJ, May 2015.
[135] Pagn, O., "The Brain: A Concept in Flux", *Philosophical Transactions of the Royal Society B*, Vol. 374, No. 1774, Jun. 2019, p. 20180383.
[136] Lytton, W., *From Computer to Brain: Foundations of Computational Neuroscience*, Springer-Verlag, New York, NY, 2002.
[137] Ermentrout, B. and Terman, D., *Foundations of Mathematical Neuroscience*, Springer, New York, NY, 2010.
[138] Sacaleaunu, V., Mohan, A., Covache-Busuioc, R., Costin, H., and Ciurea, A., "Wilhem von Waldeyer: Important Steps in Neural Theory, Anatomy and Citology", *Brain Sciences*, Vol. 12, No. 224, 2022, p. 224.
[139] Fodstad, H., Kondziolka, D., and Lotbiniere, A., "The Neuron Doctrine, the Mind, and the Arctic", *Neurosurgery*, Vol. 47, No. 6, Dec. 2000 pp. 1381–1389.
[140] Guillery, R., "Observations of Synaptic Structures: Origins of the Neuron Doctrine and its Current Status", *Philosophical Transactions of the Royal Society B*, Vol. 360, 2005, pp. 1281–1307.
[141] Thorpe, S., "Spike Arrival Times: A Highly Efficient Coding Scheme for Neural Networks", In *Parallel Processing in Neural Systems and Computers*, R. Eckmiller, G. Hartmann, and G. Hauske (Eds.), North-Holland Elsevier, 1990, pp. 91–94.
[142] Helmholtz, H.V., *Handbuch der Physiologischen Optik*, Allgemeine Encyklopdie der Physik, Leipzig, Leopold Voss, 1867.
[143] *Databasing the Brain: From Data to Knowledge (Neuroinformatics)*, S. Koslow and S. Subramaniam (Eds.), Wiley, New York, NY, Mar. 2005.
[144] Fodor, J. and Pylshyn, Z., "Connectionism and Cognitive Architecture: A Critical Analysis", In *Mind Design II: Philosophy, Psychology, Artificial Intelligence*, J. Haugeland (Ed.), A Bradford Book, The MIT Press, Cambridge, MA, 1988 (Also in *Cognition*, Vol. 28, 1988, pp. 3–71).
[145] Craik, F. and Lockhart, R., "Levels of Processing: A Framework for Memory Research", *Journal of Verbal Learning and Verbal Behavior*, Vol. 11, 1972, pp. 671–684.
[146] Hodgkin, A. and Huxley, A., "Currents Carried by Sodium and Potassium Ions through the Membrane of the Giant Axon of Loligo", *Journal of Physiology*, Vol. 116, 1952, pp. 449–472.
[147] Edelman, G., *Neural Darwinism: The Theory of Neuronal Group Selection*, Basic Books, New York, NY, 1987.
[148] Hempel, C., *Aspects of Scientific Explanation and Other Essays in the Philosophy of Science*, The Free Press, New York, NY, 1965.
[149] Varela, F., Thompson, E., and Rosch, E., *The Embodied Mind: Cognitive Science and Human Experience*, The MIT Press, Cambridge, MA, 1991.
[150] Adrian, E.D., *The Basis of Sensation: The Action of the Sense Organs*, Hafner Publishing Co., New York, NY, 1964.
[151] Sherrington, C., *The Integrative Action of the Nervous System*, Yale University Press, New Haven, CT, Oct. 1906.
[152] Rosenfeld, I., *The Invention of Memory: A New View of the Brain*, Basic Books, Inc, Publishers, New York, NY, 1988.
[153] Mahpraja, Y., *The Mysteries of Mind*, Today & Tomorrow's Printers and Publishers, New Delhi, India, 1982.
[154] Hodgkin, A. and Huxley, A., "A Quantitative Description of Membrane Current and its Application to Conduction and Excitation in Nerve", *Journal of Physiology*, Vol. 117, 1952, pp. 500–544.
[155] Sterratt, D., Graham, B., Gillies, A., and Willshaw, D., *Principles of Computational Modelling in Neuroscience*, Cambridge University Press, Cambridge, UK, 2011.

Psychology of Cognition

[156] Pope, A., An Essay on Man: *Moral Essays and Satires*, Cassell & Company, Ltd., London, England, 1891.
[157] James, W., *The Principles of Psychology, Vol. I*, Henry Holt and Company, New York, NY, 1890.
[158] Spencer, H., *The Principles of Psychology, Vol. I*, Williams and Norgate, London, England, 1890.
[159] James, W., "Habit", In *The Principles of Psychology*, Chapter 4, by Author W. James, Henry Holt and Company, Dover Publications, New York, NY, 1914.
[160] Shewhart, W., *Economic Control of Quality of Manufactured Product*, D. Van Nostrand, Inc., New York, NY, 1923.
[161] Sternberg, R. and Sternberg, K., *Cognitive Psychology*, Wadsworth, Cengage Learning, Belmont, CA, 2009.
[162] *Documents of Gestalt Psychology*, M. Henle (Ed.), University of California Press, Berkeley and Los Angeles, CA, 1961.
[163] Atkinson, R. and Shiffrin, R., "Human Memory: A Proposed System and its Control Processes", *Psychology of Learning and Motivation*, Vol. 2, 1968, pp. 89–195.
[164] Lashley, K.S., "The Problem of Serial Order in Behavior", In *Cerebral Mechanisms in Behavior; The Hixon Symposium*, L.A. Jeffress (Ed.), Wiley, 1951, pp. 112–146.
[165] *Cognitive Psychology*, N. Braisby and A. Gellatly (Eds.), Oxford University Press, Oxford, UK, 2005.
[166] Bialek, W., "Thinking About the Brain", In *Physics of Biomolecules and Cells*, H. Flyvbjerg, F. Jlicher, P. Ormos, and F. David (Eds.), Springer, Berlin, 2002.
[167] Neisser, U., *Cognitive Psychology*, Meredith Publishing CO., Des Moines, IA, 1967.
[168] Thorndike, E., *The Elements of Psychology*, A.G. Seiler, The Mason Press, Syracuse, NY, 1905.
[169] Hull, C., "The Conflicting Psychologies of Learning - A Way Out", *Psychological Review*, Vol. 42, 1935, pp. 491–516.
[170] Malcom, N., "Dreaming", In *Studies in Philosophical Psychology*, R. Holland (Ed.), Routledge & Kegan Paul, Humanities Press, London, Great Britain, 1959.
[171] Piaget, J., *The Origins of Intelligence in Children*, International Universities Press, Inc., New York, NY, 1952.
[172] Pylyshyn, Z., "Computing in Cognitive Science", In *Foundations of Cognitive Science*, M. Posner (Ed.), MIT Press, Cambridge, MA, 2009, pp. 51–91.
[173] Computation, *Cognition, and Pylyshyn*, D. Dedrick and L. Trick (Eds.), MIT Press, Cambridge, MA, 1989.
[174] *Cognitive Psychology: Key Readings in Cognition*, D. Balota and E. Marsh (Eds.), Psychology Press, London, UK, Sep. 2004.
[175] Plunkett, K., Karmiloff-Smith, A., Bates, E., Elman, J., and Johnson, M., "Connectionism and Developmental Psychology", *Journal of Child Psychology, Psychiatry*, Vol. 38, No. 1, 1997, pp. 53–80.
[176] Thorndike, E., *The Fundamentals of Learning*, Bureau of Publications, Teachers College, Columbia University, New York, NY, 1932.
[177] Anderson, J.R., *Cognitive Psychology and Its Implications*, Worth Publishers, New York, NY, 2009.
[178] Anderson, J., *Cognitive Psychology and Its Implications*, Worth Publishers, New York, NY, 2000.
[179] *William James: Selected Unpublished Correspondence: 1885-1910*, F. Down Scott (Ed.), Ohio State University Press, Columbus, OH, 1986.
[180] Spoehr, K. and Lehmkuhle, S., *Visual Information Processing*, W H Freeman & Co., New York, NY, Mar. 1982.
[181] Fischer, J. and Whitney, D., "Serial Dependence in Visual Perception", Nature Neuroscience, Vol. 17, No. 5, Mar. 2014, pp. 738–743.
[182] Marr, D. and Nishihara, H.K., "Representation and Recognition of the Spatial Organization of Three-Dimensional Shapes", *Proceedings of the Royal Society of London. Series B. Biological Sciences*, Vol. 200, 1978, pp. 269–294.
[183] Marr, D. and Hildreth, E., "Theory of Edge Detection", *Proceedings of the Royal Society of London. Series B. Biological Sciences*, Vol. 207, 1980, pp. 187–217.
[184] Broadbent, D.E., *Perception and Communication*, Pergamon Press Ltd., Oxford, UK, 1958.
[185] Tulving, E., "Episodic Memory and Common Sense: How Far Apart", *Philosophical Transactions of the Royal Society of London. Series B: Biological Sciences*, Vol. 356, 2001, pp. 1505–1515.
[186] Tulving, E., "Memory and Consciousness", *Canadian Psychology*, Vol. 26, No. 1, 1985, p. 1.
[187] Skinner, B.F., *Verbal Behavior*, Prentice-Hall, Inc., Hoboken, NJ, 1957.
[188] Broomhead, D.S. and Lowe, D., "Multivariable Functional Interpolation and Adaptive Networks", *Complex Systems*, Vol. 2, 1988, pp. 321–355.
[189] Vanderplas, J. and Garvin, E., "The Association Value of Random Shapes", *Journal of Experimental Psychology*, Vol. 57, No. 3, 1959, p. 147.
[190] Friedenberg, J. and Silverman, G., *Cognitive Science: An Introduction to the Study of Mind*, Sage Publications, Inc., Thousand Oaks, CA, 2006.
[191] Titchener, E., *A Text-Book of Psychology*, The MacMillan Company, New York, NY, 1928.
[192] Miller, G., Galanter, E., and Pribram, K., *Plans and Structure of Behavior*, Holt, Rinehart and Winston, Inc., Austin, TX, 1960.
[193] Jaynes, E.T., "Information Theory and Statistical Mechanics", *Physical Review*, Vol. 106, No. 4, May 1957, p. 620.
[194] Hayes, P., "The Second Naïve Physics Manifesto", In *Formal Theories of the Commonsense World*, J. Hobbs and B. Moore (Eds.), Ablex Publishing Corp., New York, NY, 1985, pp. 1–36.
[195] Hayes, P., "The Naïve Physics Manifesto", In *Expert Systems in the Micro-electronic Age*, D. Michie (Ed.), Edinburgh University Press, Edinburgh, Scotland, UK, 1978.
[196] Sloman, A., "The Structure of the Space of Possible Minds", In *The Mind and the Machine: Philosophical Aspects of Artificial Intelligence*, S. Torrance (Ed.), Halstead Press, Australia, Jan. 1984, pp. 173–182.

[197] Rescorla, M., "Cognitive Maps and the Language of Thought", *The British Journal of the Philosophy of Science*, Vol. 60, No. 2, 2009, pp. 377–407.

[198] Raaijmakers, J. and Shiffrin, R., "Models of Memory", In *Steven's Handbook of Experimental Psychology*, Third Edition, H. Pashler and D. Medin (Eds.), Vol. 2: Memory and Cognitive Processes, John Wiley & Sons, New York, NY, 1988.

[199] Skinner, B.F., "The Technology of Teaching", *Proceedings of the Royal Society of London, Series B, Biological Sciences*, Vol. 162, No. 989, Jul. 1965, pp. 427–443.

[200] Dennett, D., *Brainstorms: Philosophical Essays on Mind and Psychology*, The MIT Press, Cambridge, MA, Jul. 1981.

[201] Tulving, E. and Thomson, D., "Encoding Specificity and Retrieval Processes in Episodic Memory", *Psychological Review*, Vol. 80, No. 5, 1973, pp. 352–373.

[202] Maier, N.R.F., "Reasoning in humans II: The solution of a problem and its appearance in consciousness", *Journal of Comparative Psychology*, Vol. 12, No. 2, 1931, pp. 181–194.

[203] Searle, J., "Is the Brain's Mind a Computer Program?", *Scientific American*, Vol. 262, No. 1, Jan. 1990 pp. 25–31.

[204] Boden, M., "Computer Models of Creativity", *AI Magazine*, Vol. 30, Jul. 2009, pp. 23–34

[205] Rosen, R., *Fundamentals of Measurement and Representation of Natural Systems Research Series*, First Edition, Vol. 1, Elsevier Science Ltd., Amsterdam, Netherlands, May 1978.

[206] Rumelhart, D. and McClelland, J., "Levels Indeed! A Response to Broadbent", *Journal of Experimental Psychology: General*, Vol. 114, No. 2, 1985, pp. 193–197.

[207] Dretske, F., "Précis of Knowledge and the Flow of Information", *The Behavioral and Brain Sciences*, Vol. 6, 1983, pp. 55–90.

[208] Varela, F., *Principles of Biological Autonomy*, Elsevier North Holland, New York, NY, 1979.

[209] Skinner, B.F., "Teaching Machines", *Science*, Vol. 128, No. 3330, Oct. 1958, pp. 969–977.

[210] Block, N., "Chapter 11: The Mind as the Software of the Brain", In *Thinking: An Invitation to Cognitive Science*, E. Smith and D. Osherson (Eds.), The MIT Press, Cambridge, MA, 1995.

[211] Pylyshyn, Z., "Computation and Cognition: Issues in the Foundations of Cognitive Science", *The Behavioral and Brain Sciences*, Vol. 3, No. 1, 1980, pp. 111–132.

[212] Vygotsky, L., *Thinking and Speech*, Cambridge University Press, Cambridge, MA, 1934.

[213] Caligiore, D. and Fischer, M., "Vision, Action and Language", *Psychological Research*, Vol. 77, 2013, pp. 1–6.

[214] Cherry, E.C., "A History of the Theory of Information", *Symposium on Information Theory*, Royal Society, London, UK, Sept. 1950.

[215] Asaro, P., "Computers as Models of the Mind: On Simulations, Brains and the Design of Early Computers", In *The Search for a Theory of Cognition*, S. Franchi & F. Bianchini (Eds.), Vol. 238, Brill Publishing, Leiden, Netherlands, Jan. 2011, pp. 89–114.

[216] Pylyshyn, Z., "Return of the Mental Image: Are there Really Pictures in the Brain?", *Trends in Cognitive Sciences*, Vol. 7, No. 3, Mar. 2003, pp. 113–118.

[217] Gupta, S. and Huang, D.-Y., "On Detecting Influential Data and Selecting Regression Variables", Technical Report #89-28C, Purdue University, W. Lafayette, IN, 1990.

[218] James, W., *The Principles of Psychology, Vol. II*, Henry Holt and Company, New York, NY, 1890.

[219] Tulving, E., "Episodic Memory: From Mind to Brain", *Annual Review of Psychology*, Vol. 53, 2002, pp. 1–25.

[220] Tulving, E., "How Many Memory Systems are There?", *American Psychologist*, Vol. 40, No. 4, Apr. 1985, pp. 385–398.

[221] Freund, R., Loftus, G., and Atkinson, R., "Applications of Multiprocess Models for Memory to Continuous Recognition Tasks", *Journal of Mathematical Psychology*, Vol. 6, No. 3, 1969, pp. 576–594.

[222] Craik, F.I.M. and Tulving, E., "Depth of Processing and the Retention of Words in Episodic Memory", *Journal of Experimental Psychology: General*, Vol. 104, No. 3, 1975, pp. 268–294.

[223] Atkinson, R. and Shiffrin, R., "The Control Processes of Short-Term Memory", Stanford Psychology Series, Tech. Report 173, Stanford University, Stanford, CA, Apr. 1971.

[224] Titchener, E.B., *Lectures on the Experimental Psychology of the Thought-Process*, The MacMillan Company, Syracuse, NY, Nov. 1909.

[225] Tononi, G. and Edelman, G., "Consciousness and Complexity", *Science's Compass, Science Magazine*, Vol. 282, Dec. 1998, pp. 1846–1851.

[226] Boyd, J., *Destruction and Creation*, U.S. Army Command and General Staff College, Fort Leavenworth, KS, 1987.

[227] Shallice, T., *From Neuropsychology to Mental Structure*, Cambridge University Press, Cambridge, UK, 1988.

[228] Spencer, H., *The Principles of Psychology*, Vol. 1, Williams and Norgate, London, UK, 1890.

[229] Koffka, K., *Principles of Gestalt Psychology*, Kegan Paul, Trench, Trubner & Co., Ltd., London, England, 1935.

Logicism & Computability

[230] Schrder, E., *Vorlesungen ber die Algebra der Logik. 1.* Band, in German, Druck und Verlag von B.G. Teubner, Leipzig, 1890.

[231] Russell, B., *The Principles of Mathematics*, Vol. 1, Cambridge University Press, Cambridge, UK, 1903.

[232] Jevons, W.S., *Pure Logic or the Logic of Quality Apart from Quantity: with Remarks on Boole's System and on the Relation of Logic and Mathematics*, Edward Stanford, London, UK, 1864.

[233] Venn, J., *Symbolic Logic*, MacMillan and Co., London, UK, 1881.

[234] Hintikka, J., *Knowledge and Belief: An Introduction to the Logic of the Two Notions*, Cornell University Press, Ithaca, NY, 1962.

[235] Curry, H., "Functionality in Combinatory Logic", *Proceedings of the National Academy of Sciences*, Vol. 20, Sept. 1934, pp. 584–590.

[236] Tarski, A., "The Concept of Truth in Formalized Languages", In *Logic, Semantics, Metamathematics*, A. Tarski (Ed.), Clarendon Press, Oxford, UK, 1936, pp 152–278.

[237] Quine, W., *A System of Logistic*, Harvard University Press, Cambridge, MA, 1934.

[238] Boole, G., *An Investigation of the Laws of Thought*, Walton and Maberly, London, Great Britain, 1854.

[239] Peirce, C., "How to Make Our Ideas Clear", *Popular Science Monthly*, Vol. 12, Jan. 1878, pp. 286–302.

[240] Gilmore, P.C., "A Proof Method for Quantification Theory: Its Justification and Realization", *IBM Journal of Research and Development*, Vol. 4, No. 1, 1960, pp. 28–35.

[241] Carnap, R., *Philosophy and Logical Syntax*, Psyche Miniatures, General Series No. 70, Kegan Paul, Trench, Trubner & Co., Ltd, Broadway House Carter Lane, London, UK, 1935.

[242] Carnap, R., *Logical Syntax of Language*, Kegan Paul, Trench, Trubner & Co., Ltd., London, UK, 1937.

[243] Davis, M., *The Universal Computer: The Road from Leibniz to Turing*, AK Peters Book, CRC Press, Boca Raton, FL, 2012.

[244] Hilbert, D. and Ackermann, W., *GrundzÄugen der theoretischen Logik (Principles of Theoretical Logic)*, Springer-Verlag, Berlin-Heidelberg, Germany, 1928.

[245] Carnap, R., *The Logical Structure of the World 'I': Pseudoproblems in Philosophy*, Translated by Rolf George, University of California Press, Berkeley and Los Angeles, 1967.

[246] McLaughlin, T., "On an Extension of a Theorem of Friedberg", *Notre Dame Journal of Formal Logic*, Vol. III, No. 4, Oct. 1962, pp. 270–273.

[247] Dedekind, R., *Essays on the Theory of Numbers*, The Open Court Publishing Co., Chicago, IL, 1901.

[248] Tarski, A., *Introduction to Logic and to the Methodology of the Deductive Sciences*, Oxford University Press, New York, NY, 1941.

[249] Mendelson, E., *Introduction to Mathematical Logic*, Sixth Edition, CRC Press, Taylor and Francis Group, Boca Raton, FL, 2015.

[250] Carnap, R., *Meaning and Necessity*, University of Chicago Press, Chicago, IL, 1947.

[251] Quine, W., *Mathematical Logic*, Harvard University Press, Cambridge, MA, 1940.

[252] Martin Davis on Computability, *Computational Logic, and Mathematical Foundations*, E.G. Omodeo and A. Policriti (Eds.), Springer International Publishing, Switzerland, 2016.

[253] Davis, M., "One Equation to Rule Them All", Memorandum RM-5494-PR, Prepared for the United States Air Force Project Rand, The Rand Corporation, Santa Monica, CA, Feb. 1968.

[254] Wittgenstein, L., *Tractatus Logico-Philosophicus*, Translated by C.K. Ogden, Kegan Paul, Trench, Trubner & Co, Ltd., London, UK, 1922.

[255] Whitehead, A.N. and Russell, B., *Principia Mathematica*, Vol. 1, University Press, Cambridge, UK, 1910.

[256] Hilbert, D., *The Foundations of Geometry*, The Open Court Publishing Company. La Salle, IL, 1950.

[257] Church, A., *Introduction to Mathematical Logic*, Princeton University Press, Princeton, NJ, 1956.

[258] Church, A., *The Calculi of Lambda-Conversion*, Annals of Mathematics Studies, No. 6, Princeton University Press, Princeton, NJ, 1941.

[259] Huth, M. and Ryan, M., *Logic in Computer Science: Modelling and Reasoning about Systems*, Cambridge University Press, Cambridge, UK, 2004.

[260] Presburger, M., "Uber die vollstandigkeit eines gewissen systems der arithmetik ganzer zahlen, in welchem die addition als einzige operation hervorstritt," *Sprawozdanie z I Kongresu Matematikow Krajow Slowcanskich*, Warszawa, 1929, pp. 92–101.

[261] Davis, M. and Putnam, H., "A Computing Procedure for Quantification Theory", *Journal of the ACM*, Vol. 7, No. 3, 1963, pp. 201–215

[262] Robinson, J.A., "Automatic Deduction with Hyper-Resolution", *International Journal of Computer Mathematics*, Vol. 1, 1965, pp. 227–234.

[263] Fitting, M., *First-Order Logic and Automated Theorem Proving*, Springer-Verlag, Berlin, Heidelberg, 1996.

[264] Boole, G., *The Mathematical Analysis of Logic: Being an Essay Towards a Calculus of Deductive Reasoning*, Macmillan, Barclay and Macmillan, Cambridge, 1847.

[265] Frege, G., *Begriffsschrift, eine der arithmetischen nachgebildete Formelsprache des reinen Denkens* (in French), Verlag von Louis Nebert, Halle, 1879.

[266] *Hilary Putnam on Logic and Mathematics*, G. Hellman and R. Cook (Eds.), Springer Nature Switzerland, Cham, Switzerland, 2018.

[267] Kowalski, R., *Computational Logic and Human Thinking: How to be Artificially Intelligent*, Cambridge University Press, Cambridge, UK, Aug. 2011

[268] Tarski, A., "Truth and Proof", *Scientific American*, Vol. 220, No. 6, Jun. 1969, pp. 63–77.

[269] Sipser, M., *Introduction to the Theory of Computation*, Third Edition, Cengage Learning, Boston, MA, 2013.

[270] Rapaport, W., "Cognitive Science", In *Encyclopedia of Computer Science*, Third Edition, A. Ralston and E. Reilly (Eds.), Van Nostrand Reinhold Publishing, New York, NY, 1993, pp. 185–189.

[271] Critchlow, C. and Eck, D., *Foundations of Computation*, Carol Critchlow and David Eck Publishers, Open Textbook Library, University of Minnesota, 2011.

[272] Lehman, E., Leighton, F.T., and Meyer, A., *Mathematics for Computer Science*, Samurai Media Ltd., 2017.

[273] Wigderson, A., *Mathematics and Computation: A Theory Revolutionizing Technology and Science*, Princeton University Press, Princeton, NJ, 2019.

[274] Hilbert, D. and Bernays, P., *Grundlagen Der Mathematik*, Verlag Von Julius Springer, Berlin, Germany, 1934.

[275] Carroll, L., *Symbolic Logic: Part 1 Elementary*, MacMillan and Co., Ltd., London, UK, 1896.

[276] Snapper, E., "The Three Crises in Mathematics: Logicism, Intuitionism and Formalism", *Mathematics Magazine*, Vol. 52, No. 4, Sept. 1979, pp. 207–216.

[277] Nilsson, N., "The Physical Symbol System Hypothesis: Status and Prospects", *50 Years of Artificial Intelligence, Lecture Notes in Computer Science (LNCS)*, Vol. 4850, Springer Nature, 2007.

[278] Wang, H., "Proving Theorems by Pattern Recognition I", *Communications of the ACM, Association for Computing Machinery*, Vol. 3, No. 4, Apr. 1960, pp. 220–234.

[279] Shoenfield, J., *Mathematical Logic*, Addison-Wesley Publishing Co., Reading, MA, 1967.

[280] Horty, J., "Defaults with Priorities", *Journal of Philosophical Logic*, Vol. 36, 2007, pp. 367–413.

[281] Searle, J., "Minds, Brains, and Programs," *Behavioral and Brain Sciences*, Vol. 3, No. 3, Sept. 1980, pp. 417–458.

[282] Boolos, G., *Logic, Logic, and Logic*, Harvard University Press, Cambridge, MA, 1998.

[283] Wos, L., Robinson, G., and Carson, D., "Efficiency and Completeness of the Set of Support Strategy in Theorem Proving", *Journal of the Association for Computing Machinery*, Vol. 12, No. 4, Oct 1965, pp. 536–541.

[284] Stefferud, E., "The Logic Theory Machine: A Model Heuristic Program," Memorandum RM-3731-CC, The Rand Corporation, Prepared for Carnegie Corporation, Jun. 1963.

[285] Husserl, E., *Logical Investigations, Vol. 1 & II*, Routledge & Kegan Paul, Ltd., Taylor & Francis Group, London, UK, 2001.

[286] Mill, J.S., A *System of Logic, Ratiocinative, and Inductive: Being a Connected View of the Principles of Evidence, and the Methods of Scientific Investigation*, John Parker, West Strand, London, UK.

[287] Huntington, E., "A New Set of Independent Postulates for the Algebra of Logic, with Special Reference to Whitehead and Russell's Principia Mathematica", *Proceedings of the National Academy of Sciences*, Vol. 18, No. 2, Feb. 1932, pp. 179–180.

[288] Huntington, E., "Sets of Independent Postulates for the Algebra of Logic", *Transactions of the American Mathematical Society*, Vol. 5, Apr. 1904, pp. 288–309.

[289] Robinson, J.A., "Theorem-Proving on the Computer", *Journal of the ACM, Association for Computing Machinery*, Vol. 10, No. 2, Apr. 1963, pp. 163–174.

[290] Allaire, P. and Bradley, R., "Symbolical Algebra as a Foundation for Calculus: D.F. Gregory's Contribution", *Historia Mathematica*, Vol. 29, 2002, pp. 395–426.

[291] Kripke, S., *Naming and Necessity*, Harvard University Press, Cambridge, MA, 1972.

[292] Kneale, W. and Kneale, M., *The Development of Logic*, Oxford at the Clarendon Press, Oxford University Press, London, Great Britain, 1962.

[293] Cellucci, C., *The Theory of Gdel*, Springer Nature, Cham, Switzerland, 2022.

[294] *Logical Reasoning with Diagrams*, G. Allwein and J. Barwise (Eds.), Oxford University Press, New York, NY, 1996.

[295] Putnam, H., *Representation and Reality*, MIT Press, Cambridge, MA, 1988.

[296] *Dimensions of Mind: A Symposium*, S. Hook (Ed.), New York University Press, New York, NY, 1960.

[297] Pribram, K., "Mind, Brain, Consciousness: The Organization of Competence and Conduct", In *The Psychobiology of Consciousness*, J.M. Davidson and R.J. Davidson (Eds.), Springer, Boston, MA, 1980.

[298] Verburgt, L., "The Venn-MacColl Dispute in Nature", *History and Philosophy of Logic*, Vol. 41, No. 3, 2020, pp. 244–251.

[299] Whitehead, A.N., *Symbolism: Its Meaning and Effect*, MacMillan Co., New York, NY, 1927.

[300] Gardner, M., *Logic Machines & Diagrams*, McGraw-Hill Book Company, Inc., New York, NY, 1958.

[301] Venn, J., "On the Diagrammatic and Mechanical Representation of Pro-positions and Reasoning", *Philosophical Magazine and Journal of Science*, Vol. 9, No. 59, Jul. 1880, pp. 1–18.

[302] Gödel, K., "Diskussion zur Grundlegung der Mathematik, Erkenntnis 2", *Monatshefte Fur Mathematik Und Physik*, Vol. 32, No. 1, 1931, pp. 147–148.

[303] Gödel, K., "Über formal unentscheidbare Sätze der Principia Mathematica und verwandter Systeme I", *Monatshefte Fur Mathematik Und Physik*, Vol. 38, No. 1, 1931, pp. 173–198.

[304] Deleuze, G., *Francis Bacon: the Logic of Sensation, Continuum*, London, England, 1981 (in French).

[305] Turner, H., "Representing Actions in Logic Programming and Default Theories: A Situation Calculus Approach", *The Journal of Logic Programming*, Vol. 31, No. 1–3, 1997, pp. 245–298.

[306] Hoare, C.A.R., "An Axiomatic Basis for Computer Programming", *Communications of the ACM*, Vol. 12, No. 10, Oct. 1969, pp. 576–580.

[307] Dijkstra, E., "Go To Statement Considered Harmful", *Communications of the ACM*, Vol. 11, No. 3, New York, NY Mar. 1968.

[308] von Wright, G., *An Essay in Modal Logic*, North-Holland Publishing Co., Amsterdam, the Netherlands, 1951.

[309] *Church's Thesis: Logic, Mind and Nature*, A. Olszewski, B. Brozek, and P. Urbanczyk, (Eds.), Copernicus Center Press, Krakow, Poland, 2014.

[310] Ivancevic, V. and Ivancevic, T., *Computational Mind: A Complex Dynamics Perspective*, Springer-Verlag, Berlin, Heidelberg, 2007.

[311] Pylyshyn, Z., "Computing in Cognitive Science", In *Foundations of Cognitive Science*, M. Posner (Ed.), A Bradford Book, MIT Press, Cambridge, MA, 1989.

[312] Aumann, R., *Lectures on Game Theory*, Westview Press, Inc., Boulder, CO, 1989

[313] *Peirce: Selected Philosophical Writings,* Peirce Edition Project (Ed.), Vol. 2, Indiana University Press, Bloomington, IN, 1998.

[314] Hoare, G.T.Q., "1936: Post, Turing and 'A Kind of Miracle' in Mathematical Logic", *The Mathematical Gazette*, Vo. 88, No. 511, The Mathematical Association, Mar. 2004, pp. 2–15.

[315] *The Logical Legacy of Nikolai Vasiliev and Modern Logic,* V. Markin and D. Zaitsev (Eds.), Vol. 387, Studies in Epistemology, Logic, Methodology and Philosophy of Science, Springer International Publishing, Cham, Switzerland, 2017.

[316] Copi, I., Cohen, C., and McMahon, K., *Introduction to Logic, Pearson*, Essex, England, 2014.

[317] Tarski, A., *Introduction to Logic and to the Methodology of Deductive Science,* Oxford University Press, New York, NY, 1941.

[318] Corpina, F., "The Ancient Master Argument and Some Examples of Tense Logic", *Argumenta, Issue 1, No. 2, Journal of Analytic Philosophy*, Roma, Sassari, 2016, pp. 245–258.

[319] Blackburn, P., "Arthur Prior and Hybrid Logic", *Synthese*, Vol. 150, No. 3, *The Logic of Time and Modality*, Springer Nature, Berlin, Germany, Jun. 2006, pp. 329–372.

[320] Harrison, J., Urban, J., and Wiedijk, F., "History of Interactive Theorem Proving", in *Handbook of the History of Logic, Computational Logic*, J. Siekmann (Ed.), Vol. 9, 2014, pp. 135–214.

[321] Fodor, J., *Psychosemantics, the Problem of Meaning in the Philosophy of Mind, A Bradford Book*, The MIT Press, Cambridge, MA, 1987.

[322] Blackburn, P. and Jørgensen, K., *"Reichenbach, Prior and Hybrid Tense Logic", Synthese, An International Journal of Epistemology, Methodology and Philosophy of Science*, Vol. 193, Springer, Nov. 2016, pp. 3677–3689.

[323] Stefferud, E., "The Logic Theory Machine: A Model Heuristic Program", Memorandum RM-3731–CC, The Rand Corporation, Santa Monica, CA, Jun. 1963.

[324] Beziau, J.-Y., "Is Modern Logic Non-Aristotelian?", in *The Logical Legacy of Nikolai Vasiliev and Modern Logic,* V. Markin and D. Zaitsev (Eds.), Springer-Verlag, Cham, Switzerland, 2017.

[325] Kowalski, R., *Computational Logic and Human Thinking: How to be Artificially Intelligent*, Cambridge University Press, Cambridge, UK, 2011.

[326] *The Aftermath of Syllogism: Aristotelian Logical Argument from Avicenna to Hagel*, M. Sgarbi and M. Cosci (Eds.), Bloomsbury Academic, London, UK, 2018.

[327] Kneale, W. and Kneale, M., *The Development of Logic*, Oxford University Press, London, Great Britain, 1962.

[328] Additional References on Logicism continues in Vol. 2.

II Robotics

General Robotics

[329] Gevartar, W., "An Overview of Artificial Intelligence and Robotics," Vol. II - Robotics, Report No. NBSIR 82-2479, Mar. 1982.

[330] Paul, R., *The Early Stages of Robotics*, IFAC (International Federation of Automatic Control) Real Time Digital Control Applications, Guadalajara, Mexico, 1983.

[331] Springer *Handbook of Robotics*, B. Siciliano and O. Khatib (Eds.), Marcel Dekker, Inc., Springer Verlag, Berlin, 2000.

[332] Culbertson, J., *The Minds of Robots: Sense Data, Memory Images, and Behavior in Conscious Automata*, University of Illinois, Champaign, IL, Jan. 1965.

[333] *A Roadmap for US Robotics: From Internet to Robotics,* H. Christensen, et al. (Organizer and Sponsor) San Diego, CA, Sept. 2020.

[334] Siciliano, B., Sciavicco, L., Villani, L., and Oriolo, G., *Robotics: Modelling, Planning and Control*, Springer-Verlag London, Limited, Berlin, Germany 2010.

[335] *Principles of Robot & Artificial Intelligence,* D. Franceschetti, Salem Press, Grey House Publishing, Ipswich, MA, 2018.

[336] *Artificial Intelligence and Robotics*, H. Lu and X. Xu (Eds.), Springer International Publishing, Cham, Switzerland, 2018.

[337] Lynch, K. and Park, F., *Modern Robotics: Mechanics, Planning and Control*, Cambridge University Press, Cambridge, UK, 2017.

[338] Putnam, H., "Robots: Machines or Artificially Created Life?", *The Journal of Philosophy*, Vol. 61, No. 21, Nov. 1964, 668–691.

[339] Churchland, P. and Sejnowski, T., The *Computational Brain,* A Bradford Book, MIT Press, Cambridge, MA, 1992.

[340] Todd, D.J., *Fundamentals of Robot Technology,* Kogan Page, Ltd., London, UK, 1986.

[341] Hewitt, C., "Description and Theoretical Analysis (Using Schemata) of Planner: A Language and Manipulating Models in a Robot", Artificial Intelligence Laboratory, Massachusetts Institute of Technology, AI-TR-258 Report, Cambridge, MA, Apr. 1972.

[342] Fahimi, F., *Autonomous Robots: Modeling, Path Planning, Control*, Springer Science + Business Media, New York, NY, 2009.

[343] Vepa, R., *Nonlinear Control of Robots and Unmanned Aerial Vehicles: An Integrated Approach*, CRC Press, Taylor & Francis Group, Boca Raton, FL, 2017.

[344] McKinnon, P., *Robotics: Everything you Need to Know about Robotics from Beginner to Expert*, CreateSpace Independent Publishing Platform, Scotts Valley, CA, Jan. 2016.

[345] Cai, Z., *Robotics: From Manipulator to Mobilebot*, World Scientific Publishing Co., Singapore, Singapore, Aug. 2022.

[346] Niku, S., *Introduction to Robotics: Analysis, Control, Applications*, Pearson Education, Inc., London, UK, 2011.

[347] Popovic, M., *Biomechanics and Robotics*, Jenny Stanford Publishing, Taylor & Francis, Boca Raton, FL, 2013.

[348] *Foundations of Robotics: A Multi-disciplinary Approach with Python and ROS*, D. Herath and D. St-Onge (Eds.), Springer Nature Singapore Pte Ltd, Singapore, Singapore, 2022.

[349] Encyclopedia of Robotics, M. Ang, O. Khatib, and B. Siciliano (Eds.), Springer, Berlin, Heidelberg, Germany, 2021.

[350] Kyrarini, M., Lygerakis, F., Rajavenkatanarayanan, A., Sevastopoulos, C., Nambiappan, H., Chaitanya, K., Babu, A., Mathew, J., and Makedon, F., *A Survey of Robots in Healthcare*, MDPI, Basel, Switzerland, 2021.

[351] Corke, P., *Robotics, Vision and Control: Fundamental Algorithms in MATLAB*, Springer International Publishing, Cham, Switzerland 2011.

[352] Koren, Y., *Robotics for Engineers*, McGraw-Hill, New York, NY, 1985.

[353] Husbands, P., *Robots: What Everyone Needs to Know*, Oxford University Press, Oxford, UK, 2021.

[354] Weng, J., McClelland, J., Pentland, A., Sporns, O., Stockman, I., Sur, M., Thelen, E., "Autonomous Mental Development by Robots and Animals", *Science Magazine*, Vol. 291, No. 5504, *American Association for the Advancement of Science (AAAS)*, Washington, D.C., Jan. 2001, pp. 599–600.

[355] Norman, D. and Shallice, T., "Attention to Action: Willed and Automatic Control of Behavior", In *Consciousness and Self-Regulation: Researchers in Research and Theory*, Vol. 4, R. Davidson, G. Schwartz, D. Shapiro (Eds.), Springer Science + Business Media, LLC, New York, NY, 1986, pp. 1–18.

[356] Nise, N., *Control Systems Engineering*, Wiley, Hoboken, NJ. 1999.

[357] Maekawa, A., Kume, A., Yoshida, H., Hatori, J., Naradowsky, J., and Saito, S., "Improvised Robotic Design with Found Objects", 32nd Conference on Neural Information Processing Systems (NIPS 2018), Montréal, Canada, Dec. 2018.

[358] Russell, S., *Human Compatible: Artificial Intelligence and the Problem of Control*, Viking Penguin, New York, NY, 2019.

[359] *Cognitive Robotics*, A. Cangelosi and M. Asada (Eds.), The MIT Press, Cambridge, MA, 2022.

[360] Yim, J., Nadan, P., Zhu, J., Stutt, A., Payne, J.J., Pavlov, C., and Johnson, C., "Double-Anonymous Review for Robotics", *arViv*:2406.10059v1, Jun. 2024.

Cybernetics

[361] Ashby, W.R., *An Introduction to Cybernetics*, Chapman & Hall, Inc., London, Great Britain, 1957.

[362] Maltz, M., *Psycho-Cybernetics*, A Kangaroo Book, Published by Pocket Books, New York, NY, 1960.

[363] Kline, R., The *Cybernetics Moment: Or Why We Call Our Age the Information Age*, John Hopkins University Press, Baltimore, MD, 2015.

[364] Heims, S.J., *The Cybernetics Group*, The MIT Press, Cambridge, MA, 1991.

[365] *Cybernetic*, Vol. 1, No. 1, Summer-Fall, American Society of Cybernetics, Washington, DC, 1985.

[366] Heylighen, F., "Principles of Systems and Cybernetics: An Evolutionary Perspective", In *Cybernetics and Systems*, R. Trappl (Ed.), World Science, Singapore, 1992, pp. 3–10.

[367] Cordeschi, R., *The Discovery of the Artificial: Behavior, Mind and Machines Before and Beyond Cybernetics*, Kluwer Academic Publishers, Amsterdam, 2002.

[368] "Cybernetics: Circular Causal and Feedback Mechanisms in Biological and Social Systems", *Transactions of the Eight Conference on Cybernetics*, H. von Foerster (Ed.), Josiah Macy, Jr. Foundation, New York, NY, Mar. 1951.

[369] Makarieva, A., "Systems Ecology: Cybernetics", In *Systems Ecology*, S. Jorgensen and B. Fath (Eds.), Vol. 1 of Encyclopedia of Ecology, Elsevier, Oxford, UK, 2008, pp. 806–812.

[370] Rid, T., *Rise of the Machines: A Cybernetic History*, W.W. Norton & Co., New York, NY, Jun. 2016.

[371] Mindell, D., *Between Human and Machine: Feedback, Control, and Computing before Cybernetics*, John Hopkins University Press, Baltimore, MD, 2002.

[372] Wiener, N., *Norbert Wiener: A Life in Cybernetics (Ex-Prodigy: My Childhood and Youth and I am a Mathematician: The Later Life of a Prodigy)*, The MIT Press, Cambridge, MA, 2017.

[373] Rosenblueth, A., Wiener, N., and Bigelow, J., "Behavior, Purpose and Teleology", *Philosophy of Science*, Vol. 10, No. 1, Jan. 1943, pp. 18–24.

[374] *Cybernetics: The Macy Conferences (1946-1953), The Complete Transactions*, C. Pias (Ed.), The University of Chicago Press, Chicago, IL, 2016.

[375] *Heinz von Foerster,* 1911-2002: Cybernetics *&* Human Knowing, S. Brier and R. Glanville (Eds.), Imprint Academic, Exeter, Mar. 2004.

[376] Ivakhnenko, A.G. and Lapa, V.G., "Cybernetics and Forecasting Techniques", In *Modern Analytic and Computational Methods in Science and Mathematics: A Group of Monographs and Advanced Textbooks*, R. Bellman (Ed.), American Elsevier Publishing Co, New York, NY, Jan. 1967.

[377] von Foerster, H., "On Self-Organizing Systems and their Environments", In *Self-Organizing Systems*, M. Yovits and S. Cameron (Eds.), Pergamon Press, London, 1960, pp. 31–50.

[378] Segal, L., *The Dream of Reality: Heinz von Foerster's Constructivism*, WW Norton & Company, Penguin Books Canada Ltd., Ontario, Canada, 1986.

[379] Heylighen, F. and Joslyn, C., "Cybernetics and Second-order Cybernetics", In *Encyclopedia of Physical Science and Technology*, R. Meyers (Ed.), Vol. 4, Academic Press, New York, NY, 2003, pp. 155–170.

[380] Kassel, S., "Soviet Cybernetics Research: A Preliminary Study of Organizations and Personalities," A Rand Report prepared for the Advanced Research Projects Agency, Report R-909-ARPA, Dec. 1971.

[381] Edwards, J.R., "A Cybernetic Theory of Stress, Coping, and Well-Being in Organizations", *Academy of Management Review*, Vol. 17, No. 2, 1992, pp. 238–274.

[382] Arbib, M., "From Cybernetics to Brain Theory, and More: A Memoir", *Cognitive Systems Research*, Vol. 50, 2018, pp. 83–145.

[383] Francois, C., "Systemics and Cybernetics in a Historical Perspective", *Systems Research and Behavioral Science System Research*, Vol. 16, 1999, pp. 203–219.

[384] Ivakhnenko, A.G., "Heuristic Self-Organization in Problems of Engineering Cybernetics", *Automatica*, Vol. 6, Pergamon Press, 1970, pp. 207–219.

Industrial Robots

[385] Snyder, W., *Industrial Robots: Computer Interfacing and Control*, Prentice Hall Inc., Englewood Cliffs, NJ, 1985.

[386] Gruenke, D., Programming Fanuc Robots for Industry Application, American Technical Publishers, Orland Park, IL, Apr. 2021.

[387] *Handbook of Industrial Automation*, R. Shell and E. Hall (Eds.), Marcel Dekker, Inc., New York, NY, 2000.

[388] *Handbook of Industrial Robotics*, S. Nof (Ed.), Wiley & Sons, Hoboken, NJ, 1998.

[389] Gupta, A.K., Arora, S.K., and Westcott, J.R., *Industrial Automation and Robotics*, Mercury Learning and Information, Dulles, VA, 2017.

[390] Fazlollahtabar, H. and Saidi-Mehrabad, M., *Autonomous Guided Vehicle: Methods and Models for Optimal Path Planning*, Springer International Publishing, Switzerland, 2015.

[391] Sandler, B.-Z., *Robotics: Designing the Mechanisms for Automated Machinery*, Academic Press, San Diego, CA, 1999.

[392] Ross, L., Fardo, S., and Walach, M., *Industrial Robotics Fundamentals*, Goodheart-Willcox, Tinley Park, IL, Aug. 2021.

Mechatronics

[393] *The Mechatronics Handbook*, R. Bishop (Ed.), CRC Press LLC, 2002.

[394] de Silva, C., *Mechatronics: A Foundation Course*, CRC Press, Boca Raton, FL, Jun. 2010.

[395] Bolton, W., *Mechatronics: Electronic Control System in Mechanical and Electrical Engineering*, Pearson Higher Education, London, England, Dec. 2018.

[396] Aström, K. and Murray, R., *Feedback Systems: An Introduction for Scientists and Engineers*, Princeton University Press, Princeton, NJ, 2020.

Systems & Systems Theory

[397] Von Bertalanffy, L., *General Systems Theory: Foundations, Development, Applications*, George Braziller, Inc., New York, NY, Feb. 1969.

[398] Forrester, J., "Counterintuitive Behavior of Social Systems", *Technology Review*, Vol. 73, No. 3, Jan. 1971, pp. 52–68.

[399] Muscettola, N., Nayak, P., Pell, B., and Williams, B., "Remote Agent: To Boldy Go Where No AI System has Gone Before", *Artificial Intelligence*, Vol. 103, No. 1–2, Aug. 1998, pp. 5–47.

[400] Jones, M.T., *Artificial Intelligence: A Systems Approach*, Infinity Science Press, LLC, Hingham, MA, 2008.

[401] Khatib, O., Yokoi, K., Brock, O., Chang, K., and Casal, A., "Robots in Human Environments: Basic Autonomous Capabilities", *The International Journal of Robotics Research*, Vol. 18, No. 7, Jul. 1999, pp. 684–696.

[402] Wilson, M., *Implementation of Robot Systems: An Introduction to Robotics, Automation, and Successful Systems Integration in Manufacturing*, Butterworth-Heinemann, Elsevier, London, UK, 2015.

[403] Becht, G., "Systems Theory, the Key to Holism and Reductionism," *BioScience*, Vol. 24, No. 10, Oct. 1974, pp. 569–579.

[404] Gleick, J., *Chaos: Making a New Science*, Viking Penguin Inc, New York, NY, 1987.

[405] Alligood, K., Sauer, T., and Yorke, J., *Chaos: An Introduction to Dynamical Systems*, Springer-Verlag, New York, NY, 1996.

[406] Ivakhnenko, A.G., "Polynomial Theory of Complex Systems", *IEEE Transactions on Systems, Man, and Cybernetics*, Vol. SMC-1, No. 4, Oct. 1971, pp. 364–378.

[407] Downey, A., *Think Complexity*, Green Tea Press, Needham, MA, 2016.

[408] Afzal, A., Goues, C., Hilton, M., and Timperley, C., "A Study on Challenges of Testing Robotic Systems", *2020 IEEE 13th International Conference on Software Testing, Validation and Verification (ICST)*, Oct. 2020, pp. 96–107.

[409] *Robotics: Science and Systems V*, J. Trinkle, Y. Matsuoka, and J. Castellanos (Eds.), Penguin Random House, The MIT Press, Cambridge, MA, Jul. 2010.

[410] Bellingham, J. and Rajan, K., "Robotics in Remote and Hostile Environments", *Science Magazine*, Vol. 318, Dec. 2007, pp. 1098–1102.

[411] Thurner, S., Hanel, R., Klimek, P., *Introduction to the Theory of Complex Systems*, Oxford University Press, Oxford, UK, 2018.

[412] Sowa, J.F., "Architectures for Intelligent Systems", *IBM Systems Journal*, Vol. 41, No. 3, 2002.

Human-Robot Theory & Interface

[413] Hartley, R.V.L., "Transmission of Information", *Proceedings of the International Congress of Telegraphy and Telephony*, Lake Como, Italy, Sept. 1927.

[414] Card, S., Moran, T., and Newell, A., *The Psychology of Human-Computer Interaction*, Lawrence Erlbaum Associates, Inc., Mahwah, NJ, 1983.

[415] McLuhan, M., *Understanding Media: The Extension on Man*, Gingko Press, Berkeley, CA, 1944.

[416] Shneiderman, B., *Human Factors of Interactive Software*, Department of Computer Science, University of Maryland, College Park, MD, 1983.

[417] Lawrence, D., "Stability and Transparency in Bilateral Teleoperation", *IEEE Transactions on Robotics and Automation*, Vol. 9, No. 5, Oct. 1993, pp. 624–637.

[418] Walker, J., Colonnese, N., and Okamura, A., "Noise, But Not Uncoupled Stability, Reduces Realism and Likeability of Bilateral Teleoperation", *IEEE Robotics and Automation Letters*, Vol. 1, No. 1, Jan. 2016, pp. 562–569.

[419] Breazeal, C., Edsinger, A., Fitzpatrick, P., and Scassellati, B., "Active Vision for Sociable Robots", *IEEE Transactions on Man, Cybernetics and Systems*, Vol. 31, No. 5, Sept. 2001, pp. 443–453.

[420] Yan, H., Ang Jr., M., and Poo, A., "A Survey on Perception Methods for Human-Robot Interaction in Social Robots", *International Journal of Social Robotics*, Vol. 6, Jan. 2014, pp. 85–119.

[421] Wilkins, D., Lee, T., and Berry, P., "Interactive Execution Monitoring of Agent teams", *Journal of Artificial Intelligence Research*, Vol. 18, 2003, pp. 217–261.

[422] Corliss, W. and Johnsen, E., *Teleoperator Controls: An AEC-NASA Technology Survey*, NASA Office of Technology Utilization, Washington, DC, Dec. 1968.

[423] Vallee, J., Passport to Magic*: On UFOs, Folklore, and Parallel Worlds*, Contemporary Books, Chicago, IL, Jan. 1993.

[424] Shneiderman, B., "Direct Manipulation: A Step Beyond Programming Languages", *Computer*, Vol. 16, Aug. 1983, pp. 57–69.

[425] Sproull, L., Subramani, M., Kiesler, S., Walker, J., and Waters, K., "When the Interface is a Face", *Human-Computer Interaction*, Vol. 11, No. 2, 1996, pp. 97–124.

[426] Horvitz, E., "Principles of Mixed-Initiative User Interface", *Proceedings of the SIGCHI Conference on Human Factors in Computing Systems (CHI '99)*, Pittsburgh, PA, Association for Computing Machinery, May 1999, pp. 159–166.

[427] Horvitz, E., Jacobs, A., and Hovel, D., "Attentive-Sensitive Alerting", *Proceedings of the Fifteenth Conference on Uncertainty in Artificial Intelligence (UAI1999), Report No. UAI-P-1999-PG-305-313*, Stockholm, Sweden, Morgan Kaufmann, 1999.

[428] Kelley, C., "Manual Control: Theory and Applications," AD No. AD449586, Defense Documentation Center, Prepared for Office of Naval Research, Jun. 1964.

[429] Veloso, M., "The Increasingly Fascinating Opportunity for Human-Robot-AI Interaction: The CoBot Mobile Service Robots", *ACM Transactions on Human-Robot Interaction*, Vol. 7, No. 1, May 2018, p. 5.

[430] Teilhard de Chardin, P., *The Phenomenon of Man*, Harper Perennial Modern Thought, Harper Collins, New York, NY, 2008.

[431] Pepper, R., "Human Factors in Remote Vehicle Control", *Proceedings of the Human Factors Society, 30th Annual Meeting*, Dayton, OH, 1986.

[432] Martin, J., *Design of Man-Computer Dialogues*, Prentice-Hall, Inc., Englewood Cliffs, NJ, 1973.

[433] *The Human-Computer Interaction Handbook: Fundamentals, Evolving Technologies and Emerging Applications*, A. Sears and J. Jacko (Eds.), Lawrence Erlbaum Association, Taylor & Francis, New York, NY, 2008.

[434] Sheridan, T. and Ferrell, W., *Man-Machine Systems*, MIT Press, Cambridge, MA, 1974.

[435] Breazeal, C. and Scassellati, B., "How to Build Robots that Make Friends and Influence People", *Proceedings of the 1999 IEEE/RSJ International Workshop on Intelligent Robots and Systems (IROS)*, Kyongju, Korea, Oct. 1999.

[436] Vertut, J. and Coiffert, P., *Teleoperation & Robotics: Applications & Technology*, Englewood Cliffs, NJ, Prentice Hall, 1985.

[437] Sheridan, T., *Telerobotics, Automation, and Human Supervisory Control*, MIT Press, Boston, MA, 1992.

[438] Pepper, R.L. and Cole, R.E., *Display System Variables Affecting Operator Performance in Undersea Vehicles and Work Systems*, NOSC Technical Report 269, Naval Ocean Systems Center, San Diego, CA, 92152, Jun. 1978.

[439] Norman, K., *Cyberpsychology: An Introduction to Human-Computer Interaction*, Cambridge University Press, Cambridge, UK, 2008.

[440] Rich, E., Knight, K., and Nair, S., *Artificial Intelligence*, Third Edition, Tata McGraw Hill Education, New Delhi, India, 2009.

[441] Patterson, D., *Introduction to Artificial Intelligence and Expert Systems*, Prentice-Hall, Hoboken, NJ, Jan. 1990.

[442] *Encyclopedia of Artificial Intelligence: The Past, Present, and Future of AI*, P. Frana and M. Klein (Eds.), ABC-CLIO, Llc., Santa Barbara, CA, 2021.

Manipulators/Arms

[443] Pollard, W., 1942, *Position Controlling Apparatus*, US Patent 2,286,571.

[444] Goertz, R.C., Mar. 1953, *Remote-Control Manipulator*, US Patent 2,632,571.

[445] Goertz, R., Grove, D., Grimson, J., Park, V., and Kohut, F., *Manipulator for Slave Robot*, United States Patent 2,978,118, Patented on Apr. 4, 1961.

[446] Goertz, R.C. "Mechanical Master-Slave Manipulator", *Nucleonics*, Vol. 12, No. 11, 1954, pp. 45–46.

[447] Duffy, J., *Statics and Kinematics with Applications to Robotics*, Cambridge University Press, Cambridge, 1996.

[448] Goertz, R.C. and Bevilacqua, F., "A force-reflecting positional servomechanism, *Nucleonics*, Vol. 10, No. 11, 1952, pp. 43–45.

[449] Slotine, J.-J. and Li, W., *Applied Non-Linear Control*, Prentice Hall, Inc., Englewood Cliffs, NJ, 1991.

[450] Slotine, J.-J. and Li, W., "Adaptive Manipulator Control: A Case Study", *IEEE Transactions on Automatic Control*, Vol. 33, No. 11, Nov. 1988, pp. 995–1003.

[451] Lewis, F., Dawson, D., and Abdallah, C., *Robot Manipulator Control: Theory and Practice*, Marcel Dekker, Inc., New York, NY, 2004.

[452] Kurdila, A. and Ben-Tzvi, P., *Dynamics and Control of Robotic Systems*, John Wiley & Sons Ltd, Hoboken, NJ, 2020.

[453] Kersten, L., "The Lemma Concept: A New Manipulator", *Mechanical and Machine Theory*, Vol. 12, 1977, pp. 77–84.

[454] Kelly, R., Santibez, V., and Lora, A, *Control of Robot Manipulators in Joint Space*, Springer-Verlag London Limited, London, UK, 2005.

[455] Angeles, J., *Fundamentals of Robotic Mechanical Systems: Theory, Methods, and Algorithms*, Second Edition, Springer-Verlag, Inc., New York, NY, 2003.

[456] Peshkin, M., *Robotic Manipulation Strategies*, Prentice Hall, Englewood Cliffs, NJ, 1990.

[457] Rimon, E. and Burdick, J., *The Mechanics of Robot Grasping*, Cambridge University Press, Cambridge, UK, 2019.

[458] Grasping in Robotics, G. Carbone (Ed.), Springer-Verlag, London, UK, 2013.

[459] Craig, J., *Introduction to Robotics: Mechanics and Control*, Pearson Prentice Hall, Hoboken, NJ, 2005.

[460] Murray, R., Li, X., and Sastry, S., *A Mathematical Introduction to Robotic Manipulation*, CRC Press, Boca Raton, FL, 1994.

[461] Spong, M., Hutchinson, S., and Vidyasagar, M., *Robot Modeling and Control*, John Wiley & Sons, Inc., Hoboken, NJ 2020.

[462] Crane, C. and Duffy, J., *Kinematic Analysis of Robot Manipulators*, Cambridge University Press, Cambridge, UK 1998.

[463] Riven, E., *Mechanical Design of Robots*, McGraw-Hill, New York, NY, 1988

[464] Tsai, L.W., *Robot Analysis: The Mechanics of Serial and Parallel Manipulators*, Wiley, New York, NY, 1999.

[465] Asada, H. and Slotine, J.-J., *Robot Analysis and Control*, Wiley-Interscience, New York, NY, 1991.

[466] Fu, K., Gonzalez, R., and Lee, C.S.G., Robotics: Control, Sensing, Vision, *and Intelligence*, McGraw-Hill, New York, NY, 1987.

[467] Schilling, R., *Fundamentals of Robotics: Analysis & Control*, Prentice-Hall, Hoboken, NJ, 1990.

[468] Allen, J., Karchak, A., and Bontrager, E., "Design and Fabricate a Pair of Rancho Anthropomorphic Manipulator Arms", Final Project Report, Contract No. NAS 8-28361, Period: 2/17/72 thru 12/31/72.

[469] Hogan, N., "Impedance Control: An Approach to Manipulation: Part I – Theory", *Journal of Dynamic Systems, Measurement, and Control*, Vol. 107, Mar. 1987, pp. 1–7.

[470] Pang, T., Suh, H.J.T., Yang, L., and Tedrake, R., "Global Planning for Contact-Rich Manipulation via Local Smoothing of Quasi-Dynamic Contact Models", *IEEE Transactions on Robotics*, Vol. 39, No. 6, Aug. 2023, pp. 4691–4711.

[471] Devol Jr., G.C., 1961, *Programmed Article Transfer*, US Patent 2,988,237.

[472] Inoue, H., "Computer Controlled Bilateral Manipulator", *Bulletin of the JSME (Japanese Society of Mechanical Engineer)*, Vol. 14, No. 69, 1971, pp. 199–207.

[473] Hogan, N., Impedance Control: An Approach to Manipulation: Part II - Implementation, *Transaction of the ASME*, Vol. 107, No. 1, Mar. 1985, pp. 8–16

[474] *Robot Motion: Planning and Control*, M. Brady, J. Hollerbach, T. Johnson, T. Lozano-Prez, and M. Mason (Eds.), MIT Press, Cambridge, MA, Mar., 1983.

[475] Moran, M., "Evolution of Robotic Arm", *Journal of Robotic Surgery*, Vol. 1, 2007, pp. 103–111.

[476] Simeon, T., Cortes, J., Sahbani, A., and Laumond, J.P., "A General Manipulation Task Planner", In *Algorithmic Foundations of Robotics* V, J.D. Boissonnat, J. Burdick, K. Goldberg, and S. Hutchinson (Eds.), Springer Tracks in Advanced Robotics, Vol. 7, Springer, Berlin, Heidelberg, 2004.

[477] Rahman, T., Sample, W., Seliktar, R., and Alexander, M., "A Body-powered Functional Upper Limb Orthosis", *Journal of Rehabilitation Research and Development*, Vol. 37, No. 6, Nov./Dec. 2000, pp. 675–680.

[478] Das, H., Slotine, J.-J., and Sheridan, T., "Inverse Kinematic Algorithms for Redundant Systems", *Proceedings of the 1988 IEEE International Conference on Robotics and Automation (ICRA)*, Vol. 1, Philadelphia, PA, Apr. 1988, pp. 43–48.

[479] Khatib, O., "Unified Approach for Motion and Force Control of Robotic Manipulators: The Operational Space Formulation", *IEEE Journal of Robotics and Automation*, Vol. RA-3, No. 1, Feb., 1987.

[480] Bejczy, A., "Sensors, Controls, and Man-Machine Interface for Advanced Teleoperation", *Science*, Vol. 208, No. 4450, Jun. 1980, pp. 1327–1335.

[481] Paul, R., *Robot Manipulators: Mathematics, Programming, and Control (Artificial Intelligence)*, MIT Press, Cambridge, 1981.

[482] Rosheim, M., *Robot Wrist Actuators*, John Wiley & Sons, New York, NY, 1989.

[483] Merlet, J.-P., *Parallel Robots*, Springer, Dordrecht, The Netherlands, 2006.

[484] Rosheim, M., *Robot Evolution: The Development of Anthrobotics*, John Wiley & Sons, New York, NY, 1994.

[485] *Grasping in Robotics*, G. Carbone (Ed.), Springer-Verlag, London, United Kingdom, 2013.

Robot Behavior

[486] Walter, W.G., "An Imitation of Life", *Scientific American*, Vol. 182, No. 5, May 1950, pp. 42–45.

[487] Watson, J., *Behaviorism*, Kegan Paul, Trench, Trubner & Co., Ltd., London, UK, Jan. 1924.

[488] Grupen, R., *The Developmental Organization of Robot Behavior*, The MIT Press, Cambridge, MA, Mar. 2023.

[489] Bonabeau, E., Dorigo, M., and Theraulaz, G., *Swarm Intelligence: From Natural to Artificial Systems*, Oxford University Press, New York, NY, 1999.

[490] Brooks, R., "Intelligence without Representation", *Artificial Intelligence*, Vol. 47, No. 1–3, Jan. 1991, pp. 139–159.

[491] Billard, A., Mirrazavi, S., and Figueroa, N., *Learning for Adaptive and Reactive Robot Control: A Dynamical Systems Approach*, The MIT Press, Cambridge, MA, Feb. 2022.

[492] Walter, W.G., "An Electro-Mechanical Animal", *Dialectica*, Vol. 4, No. 3, Sept. 1950, pp. 206–213.

[493] Thorndike, E., *Animal Intelligence: Experimental Studies*, The Macmillan Co., New York, NY, 1911.

[494] Sutherland, W., Mugglin, M., and Sutherland, I., "An Electro-Mechanical Model of Simple Animals", *Computers and Automation for February*, Vol. 7, No. 2, Feb. 1958, pp. 6–32.

[495] Talbott, J., Anderson, T., and Donath, M., "Scarecrow: An Implementation of Behavioral Control on a Mobile Robot", *Proceedings of the SPIE Mobile Robots IV*, Boston, MA, Mar. 1990.

[496] Firby, R.J., *The RAP Language Manual, Version 1*, Artificial Intelligence Lab, University of Chicago, Chicago, IL, Mar. 1995.

[497] Birk, A., "Behavior-based Robotics, its Scope and its Prospects", *Proceedings of the 24th Annual Conference of the IEEE Industrial Electronics Society, IECON '98, Aachan,* Germany, IEEE Press, Sept. 1998.

[498] Agre, P. and Chapman, D., "What are Plans For?", A.I. Memo 1050a, MIT Artificial Intelligence Laboratory, Revised Oct. 1989 (and in New Architectures for Autonomous Agents: Task-Level Decomposition and Emergent Functionality, P. Maes (Ed.), MIT Press, Cambridge, MA, 1990.

[499] Arkin, R., *Behavior-based Robotics*, The MIT Press, Cambridge, MA, 1998.

[500] Brooks, R., *A Robot that Walks; Emergent Behaviors from Carefully Evolved Network*, A.I. Memo 1091, Massachusetts Institute of Technology Artificial Intelligence Laboratory, Feb. 1989.

[501] Braitenberg, V., *Vehicles: Experiments in Synthetic Psychology*, MIT Press, Cambridge, MA, 1986.

[502] Jones, J., Flynn, A., and Seiger, B., *Mobile Robots: Inspiration to Implementation*, AK Peters Press/CRC Press, Wellesley, MA, 1997.

[503] Arkin, R., "Governing Lethal Behavior: Embedding Ethics in a Hybrid Deliberative/Reactive Robot Architecture", Technical Report GIT-GVU-07-11, Georgia Institute of Technology, 2007.

[504] Lashley, K., "The Problem of Serial Order in Behavior", in *Cerebral Mechanisms in Behavior*, L. Jeffress (Ed.), The Hixon Symposium, Wiley, 1951, pp. 112–146.

[505] Skinner, B.F., *Science and Human Behavior*, The Free Press, The MacMillan Company, New York, NY, 1953.

[506] Skinner, B.F., *About Behaviorism*, Vintage Books, New York, NY, Feb. 1976.

Mobile Robots

[507] Durrant-Whyte, H. and Bailey, T., "Simultaneous Localisation and Mapping (SLAM): Part I The Essential Algorithms, *IEEE Robotics Automation Magazine*. Vol. 13, No. 2, pp. 99–110, 2006.

[508] Leonard, J. and Durrant-Whyte, H., *Directed Sonar Sensing for Mobile Robot Navigation*, Kluwer Academic Publishers, Boston, MA, 1992.

[509] Shneier, M., Shackleford, W., Hong, T., and Chang, T., "Performance Evaluation of a Terrain Traversability Learning Algorithm in the DARPA LAGR Program", *Proceedings of the Performance Metrics for Intelligent Systems (PerMIS) Workshop*, Gaithersburg, MD, Aug. 2006.

[510] Correll, N., Hayes, B., Heckman, C., and Roncone, A., *Introduction to Autonomous Robots: Mechanisms, Sensors, Actuators, and Algorithms*, MIT Press, Cambridge, MA, Dec. 2022.

[511] *Artificial Intelligence and Mobile Robots: Case Studies of Successful Robot Systems*, D. Kortenkamp, P. Bonasso, and R. Murphy (Eds.), AAAI Press, MIT Press, Cambridge, MA, Mar. 1998.

[512] Russell, A., *Odour Detection by Mobile Robots*, World Scientific Publishing Co, Singapore, Jul. 1999.

[513] *Sensing and Control for Autonomous Vehicles: Applications to Land, Water and Air Vehicles*, T. Fossen, K. Pettersen and H. Nijmeijer (Eds.), Springer International Publishing, Switzerland, 2017.

[514] Khatib, O., "Real-time Obstacle Avoidance for Manipulators and Mobile Robots", *The International Journal of Robotics Research*, Vol. 5, No. 1, Spring 1986, pp. 90–98.

[515] Fox, D., Burgard, W., Dellaert, F., and Thrun, S., "Monte Carlo Localization: Efficient Position Estimation for Mobile Robots", *Proceedings of the Sixteenth National Conference on Artificial Intelligence*, Orlando, FL, Jul. 1999.

[516] Borenstein, J. and Koren, Y., "The Vector Field Histogram - Fast Obstacle Avoidance for Mobile Robots, *Proceedings of the IEEE Journal of Robotics and Automation*, Vol. 7, No. 3, Jun. 1991, pp. 278–288.

[517] Holland, J., *Designing Autonomous Mobile Robots*, Elsevier, Burlington, MA, 2004.

[518] Murphy, R., *Disaster Robotics*, The MIT Press, Cambridge, MA, 2014.

[519] Borenstein, J., Everett, B., and Feng, L., *Navigating Mobile Robots: Systems and Techniques*, A.K. Peters, Ltd., Natick, MA, 1996.

[520] Dudek, G. and Jenkin, M., *Computational Principles of Mobile Robotics*, Cambridge University Press, Cambridge, UK, 2000.
[521] Everett, H., *Sensors for Mobile Robots: Theory and Applications*, A. K. Peters, Ltd., Natick, MA, 1995.
[522] Siegwart, R. and Nourbakhsh, I., *Introduction to Autonomous Mobile Robots*, MIT Press, 2004.
[523] Stentz, A., "Optimal and Efficient Path Planning for Partially-Known Environments", In *Proceedings IEEE International Conference on Robotics and Automation*, San Diego, CA, May 1994.
[524] Koren, Y. and Borenstein, J., "Potential Field Methods and Their Inherent Limitations for Mobile Robot Navigation", *Proceedings of the IEEE Conference on Robotics and Automation*, Sacramento, CA, Apr. 1991, pp. 1398–1404.
[525] Minguez, J. and Montano, L., "Nearness Diagram (ND) Navigation: Collision Avoidance in Troublesome Scenarios", *IEEE Transactions on Robotics and Automation*, Vol. 20, No. 1, Feb. 2004, pp. 45–59.
[526] *Autonomous Robot Vehicles*, I.J. Cox and G.T. Wilfong (Eds.), Springer-Verlag, New York, NY, 1990.
[527] Correll, N., *Introduction to Autonomous Robots*, CreateSpace Independent Publishing, Scotts Valley, CA, 2014.
[528] Gervet, T., Chintala, S., Batra, D., Malik, J., and Chaplot, D.S., "Navigating to Objects in the Real World", *Science Robotics*, Vol. 8, No. 79, Jun. 2023.
[529] Thrun, S., Burgard, W., and Fox, D., *Probabilistic Robots*, MIT Press, Cambridge, 2005.
[530] Reif, J. and Sharir, M., "Motion Planning in the Presence of Moving Obstacles", *Proceedings of the 26th IEEE Symposium on Foundations of Computer Science*, IEEE, New York, NY, 1985, pp. 144–154 (Also Published in the Journal of ACM, Vol. 41, No. 4, Jul. 1994, pp. 764–790).
[531] Marani, G. and Yu, J., *Introduction to Autonomous Manipulation: Case Study with an Underwater Robot*, Springer Verlag, London, England, 2014.
[532] Antonelli, G., *Underwater Robots: Motion and Force Control of Vehicle-Manipulator Systems*, Springer-Verlag, Berlin, Heidelberg, 2006.
[533] Rosheim, M., *Leonardo's Lost Robots*, Springer, New York, NY, 2006.
[534] Bekey, G., *Autonomous Robots: From Biological Inspiration to Implementation and Control*, MIT Press, Cambridge, MA, 2005.
[535] Choset, H., Lynch, K., Hutchinson, S., Kantor, G., Burgard, W., Kavraki, L., and Thrun, S., *Principles of Robot Motion: Theory, Algorithms, and Implementations*, The MIT Press, Cambridge, MA, 2005.
[536] Kala, R., *Autonomous Mobile Robots: Planning, Navigation and Simulation*, Academic Press, Cambridge, MA, Sept. 2023.
[537] Matari, M., *The Robotics Primer*, MIT Press, Cambridge, MA, 2007.
[538] Bailey, T. and Durrant-Whyte, H., "Simultaneous Localisation and Mapping (SLAM): Part II State of the Art", *IEEE Robotics & Automation Magazine*, Vol. 13, No. 3, Sept. 2006, pp. 108–117.

IV Artificial Intelligence

Artificial Intelligence

[539] Holland, J., *In Natural and Artificial Systems*, A Bradford Book, MIT Press, Cambridge, MA, 1975.
[540] Heidegger, M., *Being and Time*, Blackwell Publishers Ltd, Oxford, UK, 1962.
[541] McCarthy, J., Minsky, M., Rochester, N., and Shannon, C., "A Proposal for the Dartmouth Summer Research Project on Artificial Intelligence", *AI Magazine*, Vol. 27, No. 4, Aug. 1955, p. 12.
[542] Licklider, J.C.R., "Televistas: Looking Ahead through Side Windows", In *Public Television: A Program for Action*, Carnegie Commission on Educational Television (Ed.), New York, Bantam Books, 1967, pp. 201–225.
[543] *The Handbook of Artificial Intelligence, Vol. 1V*, A. Barr, P. Cohen, and E. Feigenbaum (Eds.), Addison-Wesley Publishing Co., Inc, Reading, MA, 1989.
[544] *Encyclopedia of Artificial Intelligence*, Second Edition, S. Shapiro (Ed.), Vol. 1, John Wiley & Sons, Inc., New York, NY, Jan. 1992.
[545] Luger, G. and Stubblefield, W., "Artificial Intelligence", MacMillan Encyclopedia of Computer Science, New York, NY, May 1991.
[546] Finlay, J. and Dix, A., *An Introduction to Artificial Intelligence*, CRC Press, Boca Raton, FL, 1996.
[547] Winston, P.H., *Artificial Intelligence*, Third Edition, Addison-Wesley Publishing Co, Reading, MA, 1992.
[548] Vilone, G. and Longo, L., "Explainable Artificial Intelligence: A Systematic Review", arXiv:2006.00093v4, Oct. 2020.
[549] Dreyfus, H., *Alchemy and AI*, Rand Corporation, Santa Monica, CA, Dec. 1965.
[550] Langley, P. and Laird, J., "Artificial Intelligence and Intelligent Systems", *AI Magazine*, Vol. 27, 2000, pp. 33-44
[551] Dreyfus, H., *What Computers Still Can't Do: A Critique of Artificial Reason*, MIT Press, Cambridge, MA, 1972.
[552] Dreyfus, H., *What Computers Still Can't Do*, MIT Press, Cambridge, MA, 1993.
[553] Michie, D., *On Machine Intelligence*, Ellis Horwood Ltd., West Sussex, England, 1986 199 pages.
[554] Michie, D., "Machines and the Theory of Intelligence", *Nature*, Vol. 241, Feb. 1973.
[555] Phillips, J.P., et al., "Four Principles of Explainable Artificial Intelligence", Report No. NISTIR 8312, National Institute of Standards and Technology, Sept. 2021.

[556] Voulgaris, Z. and Bulut, Y., *AI for Data Science: Artificial Intelligence Frameworks and Functionality for Deep Learning, Optimization, and Beyond*, Technics Publications, Basking Ridge, NJ, 2018.

[557] Corea, F., *An Introduction to Data: Everything You Need to Know about AI*, Big Data, and Data Science, Springer Nature, Cham, Switzerland, 2019.

[558] Kaplan, J., *Artificial Intelligence: What Everyone Needs to Know*, Oxford University Press, New York, NY, 2016.

[559] McCorduck, P., *Machines who Think: A personal Inquiry into the History and Prospects of Artificial Intelligence*, A K Peters, Ltd, Natick, MA, 2004.

[560] Miller, T., "Explanation in Artificial Intelligence: Insights from the Social Sciences", *Journal of Artificial Intelligence*, Vol. 267, Feb. 2019, pp. 1–38.

[561] Boden, M., *Artificial Intelligence and Natural Man*, Basic Books, Inc., New York, NY, 1977.

[562] *Machine Intelligence 1*, N. Collins and D. Michie (Eds.), Oliver & Boyd, Edinburgh, 1967.

[563] *The Cambridge Handbook of Artificial Intelligence*, K. Frankish and W. Ramsey (Eds.), Cambridge University Press, Cambridge, UK, 2014.

[564] "Artificial Intelligence and Life in 2030: One Hundred Year Study on Artificial Intelligence", Report of the 2015 Study Panel, Stanford University, Sept. 2016.

[565] Nilsson, N., *The Quest for Artificial Intelligence: A History of Ideas and Achievement's*, Cambridge University Press, Cambridge, UK, Oct. 2009.

[566] Marcus, G., "The Next Decade in AI: Four Steps towards Robust Artificial Intelligence", Not Published.

[567] Schwartz, J., *The Limits of Artificial Intelligence*, Palala Press, Warsaw, Mar. 2018.

[568] Nilsson, N., *Artificial Intelligence: A New Synthesis*, Morgan Kaufmann Publishers, Inc., Burlington, MA,1998.

[569] Mitchell, M., *Artificial Intelligence: A Guide for Thinking Humans*, Farrar, Straus, and Giroux, New York, NY, 2019.

[570] Luger, G. and Stubblefield, W., *Artificial Intelligence: Structures and Strategies for Problem Solving*, Addison Wesley Longman, Inc., Reading, MA, 1997.

[571] Ertel, W., *Introduction to Artificial Intelligence*, Springer-Verlag, London, 2011.

[572] McDermott, D., "Artificial Intelligence Meets Natural Stupidity", In *Mind Design*, Haugeland, J. (Ed.), MIT Press, Cambridge, MA, 1981.

[573] Bernard, D., Dorais, G., Fry, C., Gamble Jr., E., Kanefsky, B., Kurien, J., Millar, W., Muscettola, N., Nayak, P., Pell, B., Rajan, K., Rouquette, N., Smith, B., and Williams, B., "Design of the Remote Agent Experiment for Spacecraft Autonomy", *IEEE Aerospace Conference*, Big Sky, Montana, Mar. 1998.

[574] McClelland, J. and Cleeremans, A., "Connectionist Models", In *Oxford Companion to Consciousness*, T. Byrne, A. Cleeremans, and P. Wilken (Eds.), Oxford University Press, New York, NY, 2009.

[575] Negnevitsky, M., *Artificial Intelligence: A Guide to Intelligent Systems*, Addison Wesley, Essex, England, 2002.

[576] Haugeland, J., *Artificial Intelligence: The Very Idea*, A Bradford Book, MIT Press, Cambridge, MA, 1985.

[577] Copeland, B.J. and Proudfoot, D., "Artificial Intelligence: History, Foundations, and Philosophical Issues", *Part of the Handbook of the Philosophy of Science*, D. Gabbay, P. Thagard, and J. Woods (Eds.), Elsevier B.V., 2007.

[578] Zerilli, J., *A Citizen's Guide to Artificial Intelligence*, The MIT Press, Cambridge, MA, Feb. 2021.

[579] Boden, M., *Artificial Intelligence: A Very Short Introduction*, Oxford University Press, Oxford, UK, Dec. 2018.

[580] Richardson, K., *An Anthropology of Robotics and AI: Annihilation Anxiety and Machines*, Routledge, Taylor & Francis Group, New York, NY, 2015.

[581] Johnson, J. and Picton, P., *Vol. 2 Designing Intelligent Machines: Concepts in Artificial Intelligence*, Butterworth Heinemann, Oxford, England, UK, the Open University, England, UK, 1995.

[582] Gracilla, N., "Two Approaches to Artificial Intelligence: An Analysis of Heideggerian and Dreyfusian Critiques", *Episteme*, Vol. 4, No. 1, May 1993, p. 4.

[583] Cleeremans, A. and McClelland, J., "Learning the Structure of Event Sequences", *Journal of Experimental Psychology: General*, Vol. 120, No. 3, 1991, pp. 235–253.

[584] *The Philosophy of Artificial Intelligence*, M. Boden (Ed.), Oxford University Press, Oxford, UK, Jul. 1990.

[585] Cleeremans, A., Servan-Schriber, D., and McClelland, J., "Finite State Automata and Simple Recurrent Networks", *Neural Computation*, Vol. 1, Sept. 1989, pp. 372–381.

[586] Servan-Schriber, D., Cleeremans, A., and McClelland, J., "Graded State Machines: The Representation of Temporal Contingencies in Simple Recurrent Networks", *Machine Learning*, Vol. 7, 1991, pp. 57–89.

[587] Dreyfus, H., "Misrepresenting Human Intelligence", *Thought: Fordham University Quarterly*, Vol. 61, No. 4, Dec. 1986, pp. 430–441.

[588] Dreyfus, H. and Dreyfus, S., *Mind over Machine*, The Free Press, New York, NY, 1986.

[589] Dreyfus, H., "Why Heideggerian AI Failed and How Fixing it Would Require Making it More Heideggerian", In *The Mechanical Mind in History*, P. Husbands, O. Holland, and M. Wheeler (Eds.), The MIT Press, Cambridge, MA, 2008.

[590] Henderson, H., *Artificial Intelligence: Mirrors for the Mind*, Chelsea House Publishers, New York, NY, 2007.

[591] Jackson, P., Introduction to Artificial Intelligence, Dover Publications, New York, NY, 1974.

[592] Joshi, P., *Artificial Intelligence with Python*, Packt Publishing, Birmingham, UK, 2017.

[593] Luger, G., *Artificial Intelligence: Structures and Strategies for Complex Problem Solving*, Pearson Education, Boston, MA, 2009.

[594] Poole, D. and Mackworth, A., *Artificial Intelligence: Foundations of Computational Agent*, Cambridge University Press, Cambridge, UK, 2010.

[595] Coppin, B., *Artificial Intelligence Illuminated*, Jones and Bartlett Publishers, Sudbury, MA, 2004.

[596] Voulgaris, Z. and Bulut, Y., *AI for Data Science*, Technics Publications, Basking Ridge, NJ, 2018.
[597] Dreyfus, H. and Dreyfus, S., "Making a Mind Versus Modeling the Brain: Artificial Intelligence Back at a Branchpoint", In *Understanding the Artificial: On the Future Shape of Artificial Intelligence*, M. Negrotti (Ed.), Springer, London, UK, 1991, pp. 33–54.
[598] Cawsey, A., *The Essence of Artificial Intelligence*, Prentice Hall, Hertfordshire, UK, 1997.
[599] *Artificial Intelligence and Molecular Biology*, L. Hunter (Ed.), AAAI Press, American Association for Artificial Intelligence, Menlo Park, CA, 1993.
[600] *Human-Like Machine Intelligence*, S. Muggleton and N. Chater (Eds.), MIT Press, Cambridge, MA, Jul. 2021.
[601] Genesereth, M. and Nilsson, N., *Logical Foundations of Artificial Intelligence*, Morgan Kaufmann Publishers, Inc., Los altos, CA, 1987.
[602] Goertzel, B., *AGI Revolution: An Inside View of the Rise of Artificial General Intelligence*, Humanity+ Press, San Jose, CA, 2016.
[603] Rich, E. and Knight, K., *Artificial Intelligence*, Second Edition, McGraw-Hill, New York, NY, 1991.
[604] Russell, S., *Human Compatible: Artificial Intelligence and the Problem of Control*, Viking, Random House LLC, London, UK, 2019.
[605] McDermott, D., "Artificial Intelligence meets Natural Stupidity", *ACM SIGART Bulletin*, Vol. 57, Issue 57, Apr. 1976, pp. 4–9.
[606] McDermott, D., "Artificial Intelligence and Consciousness", In *The Cambridge Handbook of Consciousness*, P. Zelazo, M. Moscovitch, and E. Thompson (Eds.), Cambridge University Press, Cambridge, UK, 2007, pp. 117–150.
[607] Dreyfus, H., *What Computers Can't Do: Of Artificial Reason*, Harper & Row Publishers, New York, NY, 1972.
[608] *The Robot's Dilemma: The Frame Problem in Artificial Intelligence*, K. Ford and Z. Pylyshyn (Eds.), Praeger Publishers, Westport, CT, 1996.
[609] Schank, R., "What is AI, Anyway?", *AI Magazine*, Vol. 8., No. 4, 1987, p. 59.
[610] Ford, M., *Architects of Intelligence: The Truth about AI from the People Building It*, Packt Publishing, Ltd., Birmingham, UK, 2018.
[611] Raynor, W., *The International Dictionary of Artificial Intelligence*, Glenlake Publishing Company, Ltd., New York, NY, 1999.
[612] Hawkins, J. and Blakeslee, S., *On Intelligence*, Owl Books, Henry Holt and Company, LLC, New York, NY, Aug. 2005.
[613] Charniak, E. and McDermott, D., *Introduction to Artificial Intelligence*, Addison-Wesley Publishing Co, Reading, MA, 1985.
[614] Crevier, D., *AI: The Tumultuous History of the Search for Artificial Intelligence*, Basic Books, New York, NY, Dec. 1993.
[615] Norvig, P. and Russell, S., *Artificial Intelligence: A Modern Approach*, Prentice Hall, 1995.
[616] Murphy, R., *Introduction to AI Robotics*, MIT Press, Cambridge, MA, 2000.
[617] Broussard, M., *Artificial Unintellligence: How Computers Misunderstand the World*, The MIT Press, Cambridge, MA, 2018.
[618] *The Routledge Social Science Handbook of AI*, A. Elliott (Ed.), Routledge, New York, NY, 2022.
[619] *The Handbook of Artificial Intelligence*, Vol. 1, A. Barr and E. Feigenbaum (Eds.), William Kaufmann, Inc, Los Altos, CA, 1981.
[620] Blackwell, A., "The Two Kinds of Artificial Intelligence, or How not to Confuse Objects and Subjects", *Interdisciplinary Science Reviews*, Vol. 48, No. 1, Informa, London, UK, Jan. 2023.
[621] Agre, P., "Toward a Critical Technical Practice: Lessons Learned in Trying to Reform AI", In *Bridging the Great Divide: Social Science, Technical Systems, and Cooperative Work*, G. Bowker, L. Gasser, L. Star, B. Turner (Eds.), L. Erlbaum Associates, Hillsdale, NJ, Apr. 1997.
[622] *Multiagent Systems: A Modern Approach to Distributed Modern Approach to Artificial Intelligence*, G. Weiss (Ed.), The MIT Press, Cambridge, MA, 1999.
[623] Li, P., Yang, J., Islam, M., Ren, S., "Making AI Less "Thirsty", Uncovering and Addressing the Secret Water Footprint of AI Models", arXiv:2304.03271v3, Oct. 2023.
[624] Kimmerly, W., *Enterprise Transformation to Artificial Intelligence and the Metaverse: Strategies for the Technology Revolution*, Mercury Learning and Information, Boston, MA, 2024.
[625] Russell, S., *Human-Compatible: Artificial Intelligence and the Problem of Control*, Penguin Books, London, UK, Nov. 2020.
[626] Korteling, J.E., van de Boer-Visschedijk, Blankendaal, R., Boonekamp, R., and Eikelboom, A., "Human-versus Artificial Intelligence", *Frontiers in Artificial Intelligence*, Vol. 4, Article 622364, Lausanne, Switzerland, March 2021.
[627] Kowalski, R., "Artificial Intelligence and Human Thinking", Twenty-Second International Joint Conference on Artificial Intelligence, IJCAI, Catalonia, Spain, AAAI Press, Menlo Park, CA, July 2011.
[628] Bonasso, R.P., Firby, R.J., Gat, E., Kortenkamp, D., and Miller, D.P., Slack, M., "Experiences with an Architecture for Intelligent, Reactive Agents", *Journal of Experimental & Theoretical Artificial Intelligence*, Vol. 9, No. 2-3, Jan. 1995, pp. 187–202.
[629] Bach, J., *Principles of Synthetic Intelligence: Building Blocks for an Architecture of Motivated Cognition*, PhD Submission at Universität Osnabrück, Osnabrück, Germany, Mar. 2007.

AI Modeling

[630] Kermack, W. and McKendrick, A., "A Contribution to the Mathematical Theory of Epidemics", *Proceedings of the Royal Society, Series A*, Vol. 115, No. 772, Aug. 1927, pp. 700–721.

[631] Pólya, G., *Mathematics and Plausible Reasoning, Volume II Patterns of Plausible Reasoning*, Princeton University Press, Princeton, NJ, 1954.

[632] Bellman, R., "The Theory of Dynamic Programming", *Bulletin of the American Mathematical Society*, Vol. 60, No. 6, Sept. 1954, pp. 503–515.

[633] Dempster, A.P., "A Generalization of Bayesian Inference", Technical Report No. 20, Department of Statistics, Harvard University, Nov. 1967.

[634] Luger, G., Bower, T., and Wishart, J., "A Model of the Development of the Early Infant Object Concept", *Perception*, Vol. 12, 1983, pp. 21–34.

[635] Tamkin, A., Brundage, M., Clark, J., and Ganguli, D., "Understanding the Capabilities, Limitations, and Societal Impacts of Large Language Models", arXiv:2102.02503[cs.CL], Cornell University, Feb. 2021.

[636] Mitchell, B. and Mancoridis, S., "Using Heuristic Search Techniques to Extract Design Abstractions from Source Code", *Proceedings of the 4th Annual Conference on Genetic and Evolutionary Computation (GECCO '02)*, New York, NY, Jul. 2002, pp. 1375–1382.

[637] Bellman, R., *Applied Dynamic Programming*, Rand Corporation Report, No. R-352, May 1962.

[638] Hofstadter, D. and Mitchell, M., "The Copycat Project: A Model of Mental Fluidity and Analogy-Making", In *Fluid Concepts and Creative Analogies*, K.J. Holyoak and J.A. Barnden (Eds.), Basic Books, New York, NY, 1995.

[639] Pearl, J., "An Introduction to Causal Inference", *The International Journal of Biostatistics*, Vol. 6, No. 2, 2010, p. 7.

[640] Breese, J., Heckerman, D., and Kadie, C., "Empirical Analysis of Predictive Algorithms for Collaborative Filtering", *Proceedings of the 14th Conference on Uncertainty in Artificial Intelligence*, Madison, WI, Morgan Kaufmann Publishers, Burlington, MA, Jul. 1998, pp. 43–52.

[641] Anderson, J., *How can the Human Mind Occur in the Physical Universe*, Oxford University Press, Sept. 2007.

[642] Mitchell, M., *An Introduction to Genetic Algorithms*, A Bradford Book, MIT Press, Cambridge, MA, 1996.

[643] *Genetic Algorithms: Applications*, Second Edition, L. Chambers (Ed.), Chapman & Hall/CRC, Boca Raton, FL, 2001.

[644] *The Practical Handbook of Genetic Algorithms Applications*, L. Chambers (Ed.), CRC Press, Boca Raton, FL, 2001.

[645] Davis, L.D., *Handbook of Genetic Algorithms*, Thomson Publishing Group, Washington DC, Jan. 1991.

[646] Smolensky, P., "Tensor Product Variable Binding and the Representation of Symbolic Structures in Connectionist Systems", *Artificial Intelligence*, Vol. 46, 1990, pp. 159–216.

[647] Anderson, J. and Labiere, C., *The Atomic Components of Thought*, Lawrence Erlbaum Associates, Inc., Mahwah, NJ, 1998.

[648] Bellman, R., "A Markovian Decision Press", Report P-1066, Rand Corporation, Apr. 1957.

[649] Fogel, D., "Using Evolutionary Programming to Create Networks that are Capable of Playing Tic-Tac-Toe", *IEEE International Conference on Neural Networks*, Vol. 2, 1993, pp. 875–880.

[650] Amarel, S., "On Representations of Problems of Reasoning about Actions", In *Readings in Artificial Intelligence*, Morgan Kaufmann, Burlington, MA, 1981, pp. 2–22.

[651] Smolensky, P., "On the Proper Treatment of Connectionism", *Behavioral and Brain Sciences*, Vol. 11, No. 1, Mar. 1988, pp. 1–23.

[652] Fogel, D., "Foundations of the Evolutionary Computation", *Proc. SPIE*, Vol. *6228*, May 2006, p. 622801.

[653] Sternberg, S., "Memory-Scanning: Mental Processes Revealed by Reaction-Time Experiments", *American Scientist*, Vol. 57, No. 4, 1969, pp. 421–457.

[654] Bryson, A. and Denham, W., "A Steepest-Ascent Method for Solving Optimum Programming Problems", *Journal of Applied Mechanics, Transactions of the ASME*, Vol. 29, No. 2, Jun. 1962, pp. 247–257.

[655] Charniak, E., "Bayesian Networks without Tears", *AI Magazine*, Vol. 12, No. 4, Winter 1991, p. 50.

[656] Evans, T., "A Heuristic Program to Solve Geometric-Analogy Problems", *Proceedings of the Spring Joint Computer Conference (AFIPS'64)*, Washington DC, Association for Computing Machinery (ACM), Apr. 1964, pp. 327–338.

[657] John, B. and Kieras, D., "The GOMS Family of Analysis Techniques: Tools for Design and Evaluation", CMU Tech Report CMU-CS-94-181/CMU-HCII-94-106, Carnegie Mellon University, Aug. 1994.

[658] Spivak, D., "Database Queries and Constraints via Lifting Problem", *Mathematical Structures in Computer Science*, Vol. 24, No. 6, Oct. 2013, p. e240602.

[659] Hochreiter, S. and Schmidhuber, J., "Long Short-Term Memory", *Neural Computation*, Vol. 9, No. 8, 1997, pp. 1735–1780.

[660] *Graphical Models: Foundations of Neural Computation*, M. Jordan and T. Sejnowski (Eds.), The MIT Press, Cambridge, MA, 2001.

[661] Anderson, J., "Automaticity and the ACT Theory", *American Journal of Psychology*, Vol. 105, No. 2, Summer 1992, pp. 165–180.

[662] Anderson, J., Bothell, D., and Byrne, M., "An Integrated Theory of the Mind", *Psychological Review*, Vol. 111, No. 4, 2004, p. 1036.

[663] Shafer, G., *A Mathematical Theory of Evidence*, Princeton University Press, Princeton, London, 1976.

[664] Zadeh, L., "On the Validity of Dempster's Rule of Combination of Evidence", Electronics Research Laboratory, memorandum No. UCB/ERL M79/24, University of California Berkeley, Mar. 1979.

[665] Johnson-Laird, P.N., "Mental Models in Cognitive Science", *Cognitive Science*, Vol. 4, 1980, pp. 71–115.

[666] Pólya, G., *Mathematics and Plausible Reasoning, Volume I Induction and Analogy in Mathematics*, Princeton University Press, Princeton, NJ, 1954.
[667] Klopf, A.H., "Brain Function and Adaptive Systems - A Heterostatic Theory", Air Force Cambridge Research Laboratories, Report #AFCRL-72-0164, Special Report No. 133, Air Force Systems Command, USAF, Mar. 1972
[668] Anderson, J., Matessa, M., and Lebiere, C., "ACT-R: A Theory of Higher Level Cognition and Its Relation to Visual Attention", *Human-Computer Interaction*, Vol. 12, 1997, pp. 439–462.
[669] Kleene, S.C., "Representation of Events in Nerve Nets and Finite Automata", Rand Report # RM-704, Rand Corporation, Santa Monica, CA, Dec. 1951.
[670] Kieras, D., "GOMS Models for Task Analysis", In *Task Analysis for Human-Computer Interaction,* D. Diaper and N. Stanton (Eds.), Lawrence Erlbaum Associates Publishers, Mahwah, NJ, 1998.
[671] Forbus, K., "Qualitative Process Theory", Technical Report 789, MIT Artificial Intelligence Laboratory, Massachusetts Institute of Technology, Cambridge, MA, 1984.
[672] Laird, J., "Extending the Soar Cognitive Architecture", *Proceedings of the First Artificial General Intelligence (AGI) Conference*, Memphis, TN, Mar. 2008.
[673] Siciliano, B., Sciavicco, L., Villani, L., and Oriolo, G., *Robotics: Modeling, Planning and Control*, Springer Publishing, New York, NY, 2009.
[674] Shafer, G., "Perspectives on the Theory and Practice of Belief Functions", *International Journal of Approximate Reasoning*, Vol. 4, No. 5–6, 1990, pp. 323–362.
[675] Pearl, J., "Reasoning with Belief Functions: An Analysis of Compatibility", International Journal of Approximate Reasoning, Vol. 4, No. 5–6, 1990, pp. 363–389.
[676] Pearl, J., *Causality: Models, Reasoning, and Inference*, Cambridge University Press, New York, NY, 2000.
[677] Bellman, R., *Dynamic Programming*, Princeton University Press, Princeton, NJ, 1972.
[678] MacKay, D., "Bayesian Non-linear Modeling for the Prediction Competition", In *Maximum Entropy and Bayesian Methods*, G. Heidbreder (Ed.), Kluwer Academic Publishers, Santa Barbara, CA, 1993.
[679] *Artificial Intelligence Theory, Models, and Applications*, P. Kaliraj and T. Devi (Eds.), CRC Press, Taylor & Francis Group, Boca Raton, FL, 2022.
[680] Sarker, I., "AI-Based Modeling: Techniques, Applications and Research Issues towards Automation, Intelligent and Smart Systems", *SN Computer Science*, Vol. 3, No. 2, 2022, p. 158.
[681] Sloman, A., *The Computer Revolution Philosophy: Philosophy Science and Models of the Mind*, The Harvester Press, Sussex, UK, Jan. 1978.
[682] Dick, S., "Of Models and Machines: Implementing Bounded Rationality", *Isis*, Vol. 106, No. 3, 2015, pp. 623–634.
[683] Anderson, J., "ACT: A Simple Theory of Complex Cognition", *American Psychologist*, Vol. 51, No. 4, Apr. 1996, pp. 355–365.
[684] Sniedovich, M., "Dijkstra's Algorithm Revisited: the Dynamic Programming Connexion", *Control and Cybernetics*, Vol. 35, No. 3, 2006, pp. 599–620.
[685] Sniedovich, M., and Lew, A., "Dynamic Programming: An Overview", *Control and Cybernetics*, Vol. 35, No. 3, 2006 pp. 5134-533
[686] Rosenbloom, P., Laird, J., Newell, A., and McCarl, R., "A Preliminary Analysis of the Soar Architecture as a Basis for General Intelligence", *Artificial Intelligence*, Vol. 47, 1991, pp, 289–325.
[687] *Mental Models,* D. Gentner and A. Stevens (Eds.), Psychology Press, Lawrence Erlbaum Associates, Inc., New York, NY, 1983.
[688] Jones, N., Ross, H., Lynam, T., Perez, P., and Leitch, A., "Mental Models: An Interdisciplinary Synthesis of Theory and Methods", *Ecology and Society*, Vol. 16, No. 1, 2011, p. 46.

Cognitive Maps and Mental Models

[689] Jones, N., Ross, H., Lynam, T., Perez, P., Leitch, A., "Mental Models: International Synthesis of Theory and Methods", *Ecology and Society*, Vol. 16, No. 1, Article 46, 2011.
[690] Jones, N., Ross, H., Lynam, T., Perez, P., and Leitch, A., "Mental Models: An Interdisciplinary Synthesis of Theory and Methods", *Ecology and Society*, Vol 16, No. 1, Mar 2011.
[691] Johnson-Laird, P.N., *Mental Models*, Harvard University Press, Cambridge, MA, Nov. 1983.
[692] Rescorla, M., "Cognitive Maps and the Theory of Thought", *The British Journal for the Philosophy of Science*, Vol. 60, No. 2, University of Chicago Press, Chicago, IL, Jun. 2009.
[693] Cummins, R., "Mental Representation", In *A Companion to Epistemology, Blackwell Companions to Philosophy*, J. Dancy, E. Sosa, and M. Steup (Eds.), Blackwell Publishing, Hoboken, NJ, Feb. 1994.
[694] Andrews, R., Lilly, J.M., Srivastava, D., and Feigh, K., "The Role of Shared Mental Models in Human-AI teams: a Theoretical View", *Theoretical Issues in Ergonomics Science*, Vol. 24, No. 2, Taylor & Francis, Oxfordshire, England, 2023, pp. 129–175.
[695] Johnson-Laird, P.N., "Mental Models in Cognitive Science", *Cognitive Science*, Vol. 4, John Wiley & Sons, Hoboken, NJ, 1980, pp. 71–115.
[696] Johnson-Laird, P.N., "History of Mental Models", In *Psychology of Reasoning*, Psychology Press, Routledge, Taylor & Francis, London, UK, 2024.

World Model

[697] von Uexkll, J., *A Stroll Through the Worlds of Animals and Men*, International Universities Press, Inc., New York, NY, 1934.

[698] Corkill, D., "Blackboard Systems", *AI Expert*, Vol. 6, No. 9, Jan. 1991.

[699] Ha, D. and Schmidhuber, J., "World Models", arXiv:1803.10122v4, May 2018.

[700] Taniguchi, T., Murata, S., Suzuki, M., Ognibene, D., Lanillos, P., Ugur, E., Jamone, L., Nakamura, T., Ciria, A., Lara, B., and Pezzulo, G., "World Models and Predictive Coding for Cognitive and Developmental Robotics: Frontiers and Challenges", *Advanced Robotics*, Vol. 37, No. 13, 2023, pp. 780–806.

[701] Nii, H.P., "Blackboard Systems, Part One: The Blackboard Model of Problem Solving and the Evolution of Blackboard Architectures", *AI Magazine*, Vol. 7, No. 2, Summer 1986, pp. 38–53.

[702] Nii, H.P., "Blackboard Systems, Part Two: Blackboard Application Systems, Blackboard Systems from a Knowledge Engineering Perspective", *AI Magazine*, Vol. 7, No. 3, Aug. 1986, pp. 82–106.

[703] Fox, C., Evans, M., Pearson, M., and Prescott, T., "Towards Hierarchical Blackboard Mapping on a Whiskered Robot", *Robotics and Autonomous Systems*, Vol. 60, No. 11, Nov. 12, pp. 1356–1366.

[704] Nii, H.P., "Blackboard Systems at the Architecture Level", *Expert Systems with Applications*, Vol. 7, 1994, pp. 43–54.

[705] Ha, D. and Schmidhuber, J., "World Models", arXiv:1803.10122v4, May 2018.

[706] Freeman, C.D., Metz, L., and Ha, D., "Learning to Predict without Looking Ahead: World Models Without Forward Prediction", 33rd Conference on Neural Information Processing Systems (NeurIPS2019), Vancouver, Canada, 2019.

[707] Kim, K., Sano, M., De Freitas, J., Haber, N., and Yamins, D., "Active World Model Learning with Progress Curiosity", Proceedings of the 37th International Conference on Machine Learning (PMLR 119), Vol. 119, Virtual, Jul. 2020.

[708] Friston, K., Moran, R., Nagai, Y., Taniguchi, T., Gomi, H., and Tenenbaum, J., "World Model Learning and Inference", *Neural Networks*, Vol. 144, Elsevier, Amsterdam, Netherlands, 2021, pp. 573–590.

[709] Liu, H., Yan, W., Zaharia, M., and Abeel, P., "World Model on Million-Length Video and Language with Blockwise Ring Attention", arXiv:2402.08268v3, Jul. 2024.

AI Software

[710] Brooks, F., *The Mythical Man-Month: Essays on Software Engineering*, Addison -Wesley Publishing Co., London, UK, 1972.

[711] Norvig, P., *Paradigms of Artificial Intelligence Programming: Case Studies in Common Lisp*, Morgan Kaufmann, Burlington, MA, Oct. 1991.

[712] Bobrow, D., Falkenhainer, B., Farquhar, A., Fikes, R., Forbus, K., Gruber, T., Iwasaki, Y., and Kuipers, B., "A Computational Modeling Language", AAAI Technical Report-96-01, Association for the Advancement of Artificial Intelligence, MIT Press, Cambridge, MA, 1996.

[713] Graham, P., *ANSI Common LISP*, Prentice Hall, Upper Saddle River, NJ, 1996.

[714] Nilsson, N., "Triangle Tables: A Proposal for a Robot Programming Language", SRI Technical Note 347, Feb. 1985.

[715] Barlow, H. and Tenenbaum, J., "Interpreting Line Drawings as Three-Dimensional Surfaces", *Proceedings of the First National Conference on Artificial Intelligence (AAAI-80)*, Palo Alto, CA, Aug. 1980.

[716] Longuet-Higgins, H.C., "A Computer Algorithm for Reconstructing a Scene from Two Projections", *Nature*, Vol. 293, Sept. 1981, pp. 133–135.

[717] Colmerauer, A., "An Introduction to Prolog III", *Communications of the ACM*, Vol. 33, No. 7, Jul. 1990, pp. 69–90.

[718] Bobrow, D., Darley, D.L., Murphy, D., Solomon, C., and Teitelman, W., "The BBN-LISP System", Bolt, Beranek and Newman Scientific Report No. 1, Prepared for Cambridge Research Laboratories, USAF, Feb. 1966.

[719] Kowalski, R., "Predicate Logic as Programming Language", *Information Processing 74*, North-Holland Publishing Company, 1974, pp. 569–574.

[720] de Bruijn, N.G., "The Mathematical Language AUTOMATH, its Usage, and Some of its Extensions", *Conference Proceedings of the Symposium on Automatic Demonstration*, Versailles, France, Dec. 1968, pp. 29–61.

[721] Bratko, I., *Prolog Programming for Artificial Intelligence*, Addison-Wesley Publishing Company, Boston, MA, 1986.

[722] Hughes, C. and Hughes, T., *Robot Programming: A Guide to Controlling Autonomous Robots*, Que Publishing, Pearson, London, May 2016.

[723] De Jong, K., Fogel, D., and Schwefel, H.-P., "A History of Evolutionary Computation", In *Handbook of Evolutionary Computation*, IOP Publishing Ltd. & Oxford University Press, Oxford, 1997.

[724] Dempster, A., Laird, N., and Rubin, D., "Maximum Likelihood from Incomplete Data via the *EM* Algorithm", *Journal of the Royal Statistical Society, Series B (Methodological)*, Vol. 39, No. 1, 1977, pp. 1–38.

[725] Vaughan, R., Gerkey, B., and Howard, A., "On Device Abstractions for Portable, Reusable Robot Code", *IEEE/RSJ International Conference on Intelligent Robots and Systems (IROS 2003)*, Las Vegas, NV, Oct. 2003.

[726] Edmonds, J., *How to Think about Algorithms*, Cambridge University Press, Cambridge, UK, 2008.

[727] Zingaro, D., *Algorithmic Thinking: A Problem-based Introduction*, No Starch Press, Inc., San Francisco, CA, 2021.

[728] Viterbi, A., "Error Bounds for Convolutional Codes and an Asymptotically Optimum Decoding Algorithm", *IEEE Transactions on Information Theory*, Vol. IT-13, No. 2, Apr. 1967, pp. 260–269.

[729] Iverson, K., *A Programming Language*, John Wiley & Sons, New York, NY, 1962.
[730] Aho, A. and Ullman, J., *Foundations of Computer Science*, Computer Science Press, 1992.
[731] Christian, B. and Griffiths, T., *Algorithms to Live By: The Computer Science of Human Decisions*, William Collins, An imprint of Harper Collins Publishers, London, Great Britain, 2016.
[732] Scott, D. and Strachey, C., "Toward a Mathematical Semantics for Computer Languages", *Proceedings of the Symposium on Computers and Automata*, Fox, J. (Ed.), *Polytechnic Press*, New York City, 1971.
[733] Scott, M., "The Interface between Distributed Operating System and High-Level Programming Language", Report TR 182, *Proceedings of the 1986 International Conference on Parallel Processing, St.* Charles, IL, Sept. 1986.
[734] Lozano-Prez, T., "Robot Programming", *Proceedings of the IEEE*, Vol. 71, No. 7, Jul. 1983, pp. 821–841.
[735] Chaitin, G.J., *Algorithmic Information Theory*, Cambridge University Press, Cambridge, MA, 1987.
[736] Jones, N., *Computability and Complexity: From a Programming Perspective*, The MIT Press, Cambridge, MA, 1997.
[737] Hewitt, C. and Baker Jr., H., "Actors and Continuous Functions", MIT/LCS/TR-194, Laboratory of Computer Science, Massachusetts Institute of Technology, Cambridge, MA, Dec. 1977.
[738] Sentz, K. and Ferson, S., "Combination of Evidence in Dempster-Shafer Theory", Sandia Report, No. SAND2002-0835, Albuquerque, NM, Apr. 2002.
[739] Tononi, G., Boly, M., Massimini, M., and Koch, C., "Integrated Information Theory: from Consciousness to its Physical Substrate", *Nature Reviews Neuroscience*, Vol. 17, Jul. 2016, pp. 450–461.
[740] Bobrow, D., "Natural Language Input for a Computer Problem Solving System", Artificial Intelligence Project, Memo 66, Memorandum MAC-M-148, Project MAC, MIT, Cambridge, MA, Mar. 1964.
[741] Bobrow, D. and Raphael, B., "A Comparison of List-Processing Computer Languages", Memorandum RM-3842-PR, The Rand Corporation, Oct. 1963.
[742] Buhmann, M.D., "Radial Basis Functions", *Acta Numerica*, Vol. 9, 2000, pp. 1–38
[743] Johnson, J., Roberts, T., Verplank, W., Smith, D., Irby, C., Beard, M., and Mackey, K., "The Xerox 'Star': A Retrospective", *Computer*, Vol. 29, Sept. 1989, pp. 11–26.
[744] Smith, D., "Pygmalion: A Creative Programming Environment", Stanford Artificial Intelligence Laboratory, Memo AIM-260, Report No. STAN-CS-75-499, Palo Alto, CA, Jun. 1975.
[745] Kay, A., "The Early History of SmallTalk", *ACM SIGPLAN Notices*, Vol. 28, No. 3, Mar. 1993, pp. 69–95.
[746] Krishnamurthi, S., *Programming Languages: Application and Interpretation*, Creative Commons Attribution-ShareAlike, Los Angeles, CA, 2003.
[747] Singh, S., *The Code Book: The Science of Secrecy from Ancient Egypt to Quantum Cryptography*, Anchor Books, New York, 1999.
[748] Robinson, J.A., "Computational Logic: Memories of the Past and Challenges of the Future", In *Computational Logic - CL 2000*, J. Loyld, et al. (Eds.), Springer-Verlag, Berlin, Heidelberg, 2000
[749] Newell, A., Tonge, F., Feigenbaum, E., Green, B., and Mealy, G., *Information Processing Language-V Manual*, Prentice-Hall, Inc., Englewood Cliffs, NJ, 1961.
[750] Petzold, C., *CODE: The Hidden Language of Computer Hardware and Software*, Microsoft Press, Redmond, WA, 2000.
[751] McCarthy, J., Abrahams, P., Edwards, D., Hart, T., and Levin, M., *LISP 1.5 Programmer's Manual*, The MIT Press, Cambridge, MA, 1985.
[752] Friedman, D., Wand, M., and Haynes, C., *Essentials of Programming Languages*, The MIT Press, Cambridge, MA, 2001.
[753] Riguzzi, F., *Foundations of Probabilistic Logic Programming: Languages, Semantics, Inference and Learning*, River Publishers, Delft, the Netherlands, 2018.
[754] Lee, K., *Foundations of Programming Languages*, Springer International Publishing Switzerland, Cham, Switzerland, 2014.
[755] Graham, P., *On Lisp*, Prentice Hall, Hoboken, NJ, 1993.
[756] Sebesta, R., *Concepts of Programming Languages*, Pearson Education Inc., Upper Saddle River, NJ, 2004.
[757] Strachey, C., "Fundamental Concepts in Programming", *Higher-Order and Symbolic Computation*, Vol. 13, Kluwer Academic Publishers, Dordrecht, Netherlands, 2000, pp. 11–49.
[758] Jackson, D., *Software Abstraction: Logic, Language, and Analysis*, The MIT Press, Cambridge, MA, 2006.
[759] Grand, G., Wong, L., Bowers, M., Olusson, T., Liu, M., Tenenbaum, J., and Andreas, J., "LILO: learning Interpretable Libraries by Compressing and Documenting Code", arXiv:2310.19791v4, Mar 2024.

Decision Making

[760] Post, E., "A Variant of a Recursively Unsolvable Problem", *Bulletin of the American Mathematical Society*, Vol. 52, 1946, pp. 264–268.
[761] Mitchell, T., "Decision Tree Learning", In *Machine Learning*, McGraw-Hill, New York, NY, 1997.
[762] Yudkowsky, E., *Timeless Decision Theory*, The Singularity Institute, San Francisco, CA, 2010.
[763] Tversky, A. and Kahneman, D., "Rational Choices and the Framing of Decisions", Office Naval Research Report AD-A168-687, May 1986.
[764] Baron, J., *Thinking and Deciding*, Cambridge University Press, Cambridge, UK, 2000.
[765] Rokach, L. and Maimon, O., "Decision Trees", In *Data Mining and Knowledge Discovery Handbook*, O. Maimon and L. Rokach (Eds.), Springer, Boston, MA, 2005.
[766] Quinlan, J.R., "Induction of Decision Trees", *Machine Learning*, Vol 1, 1986, pp. 81–106.

[767] Cheng, H.-F., Wang, R., Zhang, Z., O'Connell, F., Gray, T., Harper, M., and Zhu, H., "Explaining Decision-Making Algorithms thru UI: Strategies to Help Non-Expert Stakeholders", *Proceedings of the 2019 CHI Conference on Human Factors in Computing Systems (CHI 2019), SIGCHI*, Glasgow, Scotland, UK, May 2019.

[768] McDermott, J., "R1: A Rule-Based Configurer of Computer Systems", *Artificial Intelligence*, Vol. 19, 1982, pp. 39–88

[769] Slagle, J., "A Heuristic Program that Solved Symbolic Integration", *Journal of the ACM*, Vol. 10, No. 4, Oct. 1963, pp. 507–520.

[770] Forgy, C., "Rete: A Fast Algorithm for the Many Pattern/Many Object Pattern Match Problem", *Artificial Intelligence*, Vol. 19, 1982, pp. 17–37.

[771] Ho, T.K., "Random Decision Forests", *Proceedings of 3rd International Conference on Document Analysis and Recognition*, Vol. 1, Montreal, Quebec, CA, 1995, pp. 278–282.

[772] Schwarting, W., Alonso-Mora, J., and Rus, D., "Planning and Decision-Making for Autonomous Vehicles", *Annual Review of Control, Robotics, and Autonomous Systems*, 2018, https://doi.org/10.1146/annurev-control-060117-105157

[773] Post, E., "Formal Reductions of the General Combinatorial Decision Problem", *American Journal of Mathematics*, Vol. 65, No. 2, Apr. 1943, pp. 197–215.

Search

[774] Werbos, P., "New Tools for Prediction and Analysis in the Behavioral Sciences" (Thesis), Harvard University, Aug. 1974.

[775] Pearl, J., *Heuristics: Intelligent Search Strategies for Computer Problem Solving*, Addison-Wesley Publishing Company, Inc., Reading, MA, 1984.

[776] Lenat, D.B., "Theory Formation by Heuristic Search", *Artificial Intelligence*, Vol. 21, 1983, pp. 31–59

[777] Rumelhart, D., Durbin, R., Golden, R., and Chauvin, Y., "Backpropagation: The Basic Theory", In *Mathematical Perspectives on Neural Networks*, P. Smolensky, M. Mozer, and D. Rumelhart (Eds.), Lawrence Erlbaum Associates, Publishers, Hillsdale, NJ, 1996.

[778] Rumelhart, D. and McClelland, J., "Interactive Processing Through Spreading Activation", In *Interactive Processes in Reading*, A. Lesgold and C. Perfetti (Eds.), Lawrence Erlbaum Associates, Publishers, Hillsdale NJ, 1981.

[779] McClelland, J. and Rumelhart, D., "An Interactive Activation Model of Context Effects in Letter Perception: Part 1. An Account of Basic Findings", *Psychological Review*, Vol. 88, No. 5, Sept. 1981, pp. 60–94.

[780] Werbos, P., "Backpropagation through Time: What is Does and How to Do it", *Proceedings of the IEEE*, Vol. 78, No. 10, Oct. 1990, pp. 1550–1560.

[781] Boner, B. and Geffner, H., "Planning as Heuristic Search", *Artificial Intelligence*, Vol. 129, 2001, p. 5–33.

[782] Lenat, D.B., "Theory Formation by Heuristic Search by Heuristic Search: The Nature of Heuristics II: Background and Examples", *Artificial Intelligence*, Vol. 21, No. 1–2, Mar. 1983, pp. 31–59.

Reasoning

[783] Gelernter, H., "Realization of a Geometry - Theorem Proving Machines", *Proceedings of the International Federation for Information Processing (IFIP) Congress*, New York City, Oct. 1959.

[784] Wos, L., Overbeek, R., Lusk, E., and Boyle, J., *Automated Reasoning*, 2nd Edition, McGraw-Hill Inc., New York, NY, 1992.

[785] Cover, T. and Hart, P., "Nearest Neighbor Pattern Classification", *IEEE Transaction on Information Theory*, Vol. IT-13, No. 1, Jan. 1967, pp. 21–27.

[786] Kalman, J., *Automated Reasoning with OTTER*, Rinton Press, Paramus, NJ, Feb. 2001.

[787] Brachman, R. and Levesque, H., *Knowledge Representation and Reasoning*, Morgan Kauffman Publishers, San Francisco, CA, 2004.

[788] Reiter, R., "A Logic for Default Reasoning", *Artificial Intelligence*, Vol. 13, 1980, pp. 81–132.

[789] Harrison, J., *Handbook of Practical Logic and Automated Reasoning*, Cambridge University Press, Cambridge, UK, 2009.

[790] Parfit, D., *Reasons and Persons*, Clarendon Press, Oxford, UK, 1984.

[791] Dezert, J. and Smarandache, F., "On the Tweety Penguin Triangle Problem", In *Advances and Applications of DSmT for Information Fusion, Vol. I*, F. Smarandache and J. Dezert (Eds.), American Research Press, Rehoboth, DE, 2004.

[792] Reiter, R., "Nonmonotonic Reasoning", *Annual Review of Computer Science*, Vol. 2, 1987, pp. 147–186.

[793] Doyle, J., "Analysis by Propagation of Constraints in Elementary Geometry Problem Solving", MIT AI Lab, AI Working Paper 108, Jun. 1976.

[794] Brewka, G., Niemelä, I., and Truszczyski, M., "Chapter 1: Nonmonotonic Reasoning", In *Handbook of Knowledge Representation*, F. Harmelen, V. Lifschitz, and B. Porter (Eds.), Vol. 3, Elsevier, Amsterdam, the Netherlands, 2008, pp. 239–284.

[795] Wos, L., "A Spectrum of Applications of Automated Reasoning", arXiv preprint cs/0205078, Jan. 2002.

[796] Barber, D., *Bayesian Reasoning and Machine Learning*, Cambridge University Press, Cambridge, UK, 2012.

[797] Pal, S. and Mandal, D.P., "Fuzzy Logic and Approximate Reasoning: An Overview", *Journal of the Institution and Telecommunication Engineers*, Vol. 37, No. 5 & 6, 1991, pp. 548–560.

[798] Pearl, J., *Probabilistic Reasoning in Intelligent Systems: Networks of Plausible Inference*, Morgan Kaufmann Publishers, Inc., San Mateo, CA, 1988.

[799] *Handbook of Automated Reasoning*, A. Robinson and A. Voronkov (Eds.), The MIT Press, Cambridge, MA, Jun. 2001.

[800] Li, M. and Vitnyi, P., *An Introduction to Kolmogorov Complexity and Its Applications*, Springer-Verlag, Berlin, NJ, 1997.
[801] McDermott, D., "A Critique of Pure Reason", *Computer Intelligence*, Vol. 3, No. 3, 1987, pp. 151–160.
[802] Shahaf, D. and Amir, E., "Towards a Theory of AI Completeness", Logical Formalizations of Commonsense Reasoning, Technical Report SS-07-05, AAAI Spring Symposium, Stanford, CA, Mar. 2007.
[803] Wittocx, J., Marin, M., and Denecker, M., "Approximate Reasoning in First-Order Logic Theories", *Proceedings from the Eleventh International Conference on Principles of Knowledge Representation and Reasoning (KR '08)*, AAAI Press, Sydney, Australia, Sept. 2008, pp. 103–111.
[804] Voronkov, A., "Automated Reasoning: Past Story and New Trends", *Proceeding of International Joint Conference on Artificial Intelligence (IJCAI)*, Acapulco, Mexico, 2003.
[805] Johnson-Laird, P., "Deductive Reasoning", *Annual Reviews Psychology*, Vol. 50, 1999, pp. 109–135
[806] Johnson-Laird, P., "Models and Deduction", *Trends in Cognitive Science*, Vol. 5, No. 10, Oct. 2001, pp. 434–442.
[807] Johnson-Laird, P. and Byrne, R., "Précis of Deduction", *Behavioral and Brain Sciences*, Vol. 16, 1993, pp. 323–380.
[808] *The Oxford Handbook of Thinking and Reasoning*, K. Holyoak and R. Morrison (Eds.), Oxford University Press, Oxford, UK, May 2013.
[809] Shortliffe, E. and Buchannan, B., "A Model of Inexact Reasoning in Medicine", *Mathematical Biosciences*, Vol. 23, American Elsevier Publishing Company, Inc., 1975, pp. 351–379.
[810] Roth, D., "A Connectionist Framework for Reasoning: Reasoning with Examples", Rule-Based Reasoning & Connectionism, Proceedings of AAAI 96, AAAI, 1996, pp. 1256–1261.
[811] Bishop, J.M., "Artificial Intelligence is Stupid and Casual Reasoning will Not Fix It", *Frontiers in Psychology*, Vol. 11, Article 513474, A. Tolmie (Ed.), Frontiers Media, Lausanne, Switzerland, Jan. 2021.
[812] Yao, S., Zhao, J., Yu, D., Du, N., Shafran, I., Narasimhan, K., and Cao, Y., "REACT: Synergizing Reasoning and Acting in Language Models", The Eleventh International Conference on Learning Representation (ICLR 2023), IEEE Information Theory Society, Kigali, Rwanda, May 2023.
[813] Norman, D. and Rumelhart, D., *Explorations in Cognition*, W.H. Freeman & Company, Jan. 1975.

Knowledge

[814] Lewis, C., *Mind and the World-Order: Outline of a Theory of Knowledge*, Charles Scribner's Sons, New York, NY, 1929.
[815] Russell, B., *Theory of Knowledge*, Routledge, New York, NY, 1992 (First published in 1984).
[816] Ha, D. and Schmidhuber, J., "World Models", *32nd Annual Conference on Neural Information Processing Systems (NIPS 2018)*, Montreal, Canada, Mar. 2018.
[817] Chisholm, R., *Theory of Knowledge*, Prentice-Hall International, Inc., Englewood Cliffs, NJ, 1989.
[818] Cornforth, M., *The Theory of Knowledge*, International Publishers Co., Inc., New York, NY, 1955.
[819] Tanwar, P., Prasad, T., and Aswal, M., "Comparative Study of Three Declarative Knowledge Representation Techniques", *International Science and Engineering*, Vol. 2, No. 7, 2010, pp. 2274–2281.
[820] Goldman, A., *Epistemology and Cognition*, Harvard University Press, Cambridge, MA, 1956.
[821] Rescher, N., *Epistemology: An Introduction to the Theory of Knowledge*, State University of New York Press, Albany, NY, 2003.
[822] Lenat, D. and Guha, R., *Building Large Knowledge-Based Systems: Representation and Inference in the Cyc Project*, Addison-Wesley, Boston, MA, Jan. 1990
[823] Freeman, C.D., Metz, L., and Ha, D., "Learning to Predict Without Looking Ahead: World Models Without Forward Prediction", *33rd Annual Conference on Neural Information Processing Systems (NIPS 2019)*, Vancouver, Canada, Dec. 2019.
[824] Levesque, H. and Brachman, R., "A Fundamental Tradeoff in Knowledge Representation and Reasoning", *Proceedings of the CSCSI-84, Canadian Society for Computational Studies of Intelligence*, London, Ontario, May 1984, pp. 141–152.
[825] Copi, I., *Symbolic Knowledge*, Macmillan Publishing Co., Inc, New York, NY, 1979.
[826] Lemos, N., *An Introduction to the Theory of Knowledge*, Cambridge University Press, Cambridge, UK, 2007.
[827] Koller, D. and Friedman, N., *Probabilistic Graphical Models: Principles and Techniques*, The MIT Press, Cambridge, MA, 2009.
[828] Levesque, H. and Brachman, R., "Expressiveness and Tractability in Knowledge Representation and Reasoning", *An International Journal on Computer Intelligence*, Vol. 3, No. 1, Feb. 1987, pp. 78–93.
[829] Audi, R., *Epistemology: A Contemporary Introduction to the Theory of Knowledge*, Routledge, New York, NY, 1998.
[830] Nii, H.P., "Blackboard Systems", *The AI Magazine*, Vol. 7, No. 3 & 4, Summer 1986, pp. 82–106.
[831] Moore, R., "Reasoning about Knowledge and Action", *Proceedings of the Fifth International Joint Conference on Artificial Intelligence (IJCAI)*, Cambridge, MA, Aug. 1977.

Expert Systems

[832] Gevarter, W., "An Overview of Expert Systems", NBSIR 82-2505, Prepared for NASA HQ, National Bureau of Standards, US Department of Commerce, May 1982.
[833] Buchanan, B. and Smith, R., "Fundamentals of Expert Systems", *Annual Review of Computer Science*, Vol. 3, No. 1, Nov. 2003, pp. 23–58.

[834] Giarratano, J. and Riley, G., *Expert Systems: Principles and Programming*, Addison-Wesley, Boston, MA, 1991.
[835] Davis, R. and King, J., "The Origin of Rule-Based Systems in AI", In *Rule Based Expert Systems: Mycin Experiments of the Stanford Heuristic Programming Project*, B. Buchannan and E. Shortliffe (Eds.), Addison Wesley, Reading, MA, 1984.
[836] Lindsay, R., Buchanan, B., Feigenbaum, E., and Lederberg, J., "DENDRAL: a case study of the first expert system for scientific hypothesis formation", *Artificial Intelligence*, Vol. 61, Elsevier, Amsterdam, Netherlands, 1993, pp. 209–261.
[837] McDermott, J., "R1: An Expert in the Computer Systems Domain", *Proceedings of the First National Conference on Artificial Intelligence*, AAAI Press, Stanford, CA, Aug. 1980.
[838] Holland, C. R., "Trends in Expert Systems", *Proceedings of IEEE Southeastcon '90, Vol. 1 of 3*, New Orleans, LA, Apr. 1990.
[839] Rothenberg, J., Paul, J., Kameny, I.M., Kipps, J.R., and Swenson, M., "Evaluating Expert System Tools: A Framework and Methodology", Report No. R-3542-DARPA, The Rand Corporation, Santa Monica, CA, Jul. 1987.
[840] *Expert Systems: The Technology of Knowledge Management and Decision Making for the 21st Century*, Leondes, C. (Ed.), Academic Press, London, UK, 2002.

Data Mining

[841] Tukey, J., "The Future of Data Analysis", *The Annals of Mathematical Statistics*, Vol. 33, No. 1, Mar. 1962, pp. 1–67.
[842] Berry, M. and Linoff, G., *Data Mining Techniques: For Marketing, Sales, and Customer Relationship Management*, Wiley Publishing, Inc., Indianapolis, IN, 2004.
[843] Frawley, W., Piatetsky-Shapiro, G., and Matheus, C., "Knowledge Discovery in Databases: An Overview", *AI Magazine*, Vol. 13, No. 3, 1992, pp. 57–70.
[844] Friedman, J., "Data Mining and Statistics: What's the Connection?", *29th Symposium on the Interface: Computing Science and Statistics*, Houston, TX, May 1997.
[845] Sutton, C., "Classification and Regression Trees, Bagging, and Boosting", In *Handbook of Statistics*, Vol. 24, C. Rao, E. Wegman, J. Solka (Eds.) Elsevier B.V., Amsterdam, the Netherlands, Jun. 2005, pp. 303–329.
[846] Han, J., Kamber, M., and Pei, J., *Data Mining: Concepts and Techniques*, Morgan Kaufmann, Waltham, MA, 2012.
[847] Bramer, M., *Principles of Data Mining*, Springer-Verlag, London, UK, 2007.
[848] Hand, D., Mannila, H., and Smyth, P., *Principles of Data Mining*, The MIT Press, Cambridge, MA, 2001.
[849] Lehman, E., Leighton, F.T., and Meyer, A., *Mathematics for Computer Science*, Samurai Media Ltd, Surrey, UK, 2010.

Planning

[850] Fikes, R. and Nilsson, N., "STRIPS: A New Approach to the Application of Theorem Proving to Problem Solving", *Proceedings of the 2nd International Joint Conference on Artificial Intelligence (IJCAI)*, London, UK, Sept. 1971.
[851] Feigenbaum, E., McCorduck, P., and Nii, P., *The Rise of the Expert Company: How Visionary Businesses are Using Intelligent Computers to Achieve Higher Productivity and Profits*, MacMillan London, London, UK, Nov. 1988.
[852] Polit, S., "R1 and Beyond: AI Technology Transfer at DEC", *AI Magazine*, Vol. 5, No. 4, 1984, p. 76.
[853] Latombe, J.-C., Lazanas, A., and Shekhar, S., "Robot Motion Planning in Control and Sensing", Stanford Research Report No. STAN-CS-89-1292, Nov., 1989.
[854] Hillyer, B., "A Knowledge-Based Expert Systems Primer and Catalog", Columbia University Computer Science Technical Report, CUCS-195-85, Nov., 1985.
[855] Latombe, J.-C., *Robot Motion Planning*, Kluwer Academic Publishers, New York, NY, 1991.
[856] LaValle, S., *Planning Algorithms*, Cambridge University Press, New York, NY, May 2006.
[857] *Handbook of Constraint Programming*, F. Rossi, P. van Beck, and T. Walsh (Eds.), Elsevier Science, Oxford, UK, Aug. 2006.
[858] Huang, Y.-C., Selman, B., and Kautz, H., "Learning Declarative Control Rules for Constraint-Based Planning", *Proceedings of the Seventeenth International Conference on Machine Learning (ICML '00)*, San Francisco, CA, Morgan Kaufmann Publishers, Jun. 2000, pp. 425–422.
[859] Anderson, S., Karumanchi, S., and Iagnemma, K., "Constraint-Based Planning and Control: for Safe, Semi-Autonomous Operation of Vehicles", *2012 Intelligent Vehicles Symposium, IEEE ITS Society, Alcalá de Henares*, Spain, Jun. 2012.
[860] Loula, J., Allen, K., Silver, T., and Tenenbaum, J., "Learning Constraint-Based Planning Models from Demonstrations", *2020 IEEE/RSJ International Conference on Intelligent Robots and Systems (IROS)*, Las Vegas, NV, Oct. 2020.
[861] Wilkins, D., *Practical Planning: Extending the Classical AI Planning Paradigm*, Morgan Kaufmann Publishers, Inc., San Mateo, CA, 1988.
[862] Agre, P. and Chapman, D., "Pengi: An Implementation of a Theory of Activity", *Six National Conference on Artificial Intelligence, AAAI-87 Proceedings*, AAAI, Seattle, WA, Jul. 1987, pp. 268–272.
[863] Agre, P. and Chapman, D., "What are Plans for?", A.I. Memo 1050, MIT AI Lab, Sept. 1988.
[864] Kavraki, L., Svestka, P., Latombe, J.-C., and Overmars, M., "Probabilistic Roadmaps for Path Planning in High-Dimensional Configuration Spaces", Technical Report UU-CS-1994-32, Utrecht University, Netherlands, Aug. 1994.
[865] Buchanan, B. and Duda, R., "Principles of Rule-Based Expert Systems", *Advances in Computers*, Vol. 22, 1983, pp. 163–216.

[866] Lucas, P. and van der Gaag, L., *Principles of Expert Systems*, Addison-Wesley Longman, Boston, MA, 1991.
[867] Szczerba, R., Galkowski, P., Glickstein, I., and Ternullo, N., "Robust Algorithm for Real-time Route Planning", *IEEE Transactions on Aerospace and Electronic Systems*, Vol. 36, No. 3, Jul. 2000, pp. 869–878.
[868] Kriegel, S., Rink, C., Bodenmller, T., and Suppa, M., "Efficient Next-Best-Scan Planning for Autonomous 3D Surface Reconstruction of Unknown Objects", *Journal of Real-Time Image Processing*, Vol. 10, 2013, pp. 611–631.
[869] Sacerdoti, E., "A Survey of Expert Systems", Software Development '89, San Francisco, CA, 1989.
[870] Durkin, J., *Expert Systems: Design and Development*, Macmillan Publishing Company, New York, NY, 1994.
[871] Schank, R. and Abelson, R., "Scripts, Plans, and Knowledge", Proceedings of the Fourth International Joint Conference on Artificial Intelligence (IJCAI-75), Tbilisi, Georgia, USSR, Sep. 1975, pp. 151–157.
[872] Charniak, E. and Goldman, R., "A Bayesian Model of Plan Recognition", *Artificial Intelligence*, Vol. 64, Elsevier Science Publishers B.V., Amsterdam, Netherlands, 1993, pp. 53–79.

Pattern Recognition

[873] Duda, R., Hart, P., and Stork, D., *Pattern Classification*, Wiley & Sons, New York, NY, 2001.
[874] Fukushima, K., "Visual Feature Extraction by a Multilayered Network of Analog Threshold Elements", *IEEE Transactions on Systems, Man, and Cybernetics*, Vol. SSC-5, No. 4, Oct. 1969, pp. 322–333.
[875] Lenat, D., Witbrock, M., Baxter, D., Blackstone, E., Deaton, C., Schneider, D., Scott, J. and Shepard, B., "Harnessing Cyc to Answer Clinical Researchers' Ad Hoc Queries", *AI Magazine*, Vol. 31, 2010, pp. 13–32.
[876] Wittkuhn, L. and Schuck, N., "Faster than Thought: Detecting Sub-Second Activation Sequences with Sequential fMRI Pattern Analysis", *Nature Communications*, Vol. 12, 2021, p. 1795.
[877] Bishop, C., *Neural Network for Pattern Recognition*, Clarendon Press, Oxford, UK, 1995.
[878] Watanabe, S., *Pattern Recognition: Human and Mechanical*, John Wiley & Sons, New York, NY, 1985.
[879] Uttley, A., "Temporal and Spatial Patterns in a Conditional Probability Machine", *Automata Studies (AM-34)*, Vol. 34, 1956, p. 277.
[880] Theodoridos, S. and Koutroumbas, K, *Pattern Recognition*, Academic Press, Imprint of Elsevier, Oxford, London, UK, 2009.
[881] Zhou, B., Lapedriza, A., Khosla, A., Oliva, A., and Torralba, A., "Places: A 10 Million Image Database for Scene Recognition", *IEEE Transactions on Pattern Analysis and Machine Intelligence*, Vol. 40, No. 6, Jun. 2018, pp. 1452–1464.
[882] Fukushima, K., "Neocognition: Self-organizing Neural Network Model for a Mechanism of Pattern Recognition Unaffected by Shift in Position", *Biological Cybernetics*, Vol. 36, 1980, pp. 193–202.
[883] Duda, R. and Hart, P., *Pattern Classification and Scene Analysis*, John Wiley & Sons, Inc., New York, NY, 1973.
[884] Fukushima, K., "Neocognitron: A Hierarchical Neural Network Capable of Visual Pattern Recognition", *Neural Networks*, Vol. 1, 1988, pp. 119–130.
[885] Bishop, C., *Pattern Recognition and Machine Learning*, Springer, New York, NY, Aug. 2006.
[886] Ghahramani, Z., "An Introduction to Hidden Markov Models and Bayesian Networks", *International Journal of Pattern Recognition and Artificial Intelligence*, Vol. 15, No. 1, World Scientific, Singapore, Singapore, 2001, pp. 9–42.

Computer Vision

[887] Uhr, L., Vossler, C., and Uleman, J., "Pattern Recognition over Distortions, by Human Subjects and by a Computer Simulation of a Model for Human Perception", *Journal of Experimental Psychology*, Vol. 63, No. 3, 1962, p. 227.
[888] Bledsoe, W. and Chan, H., "A Man-Machine Facial Recognition System: Some Preliminary Results," Technical Report PRI 19A, Panoramic Research, Inc., Palo Alto, CA, 1965.
[889] Lowe, D., "Object Recognition from Local Scale-Invariant Features", *Proceedings of the Seventh IEEE International Conference on Computer Vision*, Kerkyra, Greece, Vol. 2, 1999.
[890] Hubel, D. and Wiesel, T., "Brain Mechanisms of Vision", *Scientific American*, Vol. 241, No. 3, Sept. 1979, pp. 150–163.
[891] Hubel, D. and Wiesel, T., "Receptive Fields, Binocular Interaction and Functional Architecture in the Cat's Visual Cortex", *Journal of Physiology*, Vol. 160, 1962, pp. 106–154.
[892] Block, N., *The Border Between Seeing and Thinking*, Oxford University Press, Oxford, UK, 2023.
[893] Barlow, H.B., "Summation and Inhibition in the Frog's Retina", *The Journal of Physiology*, Vol. 119, Jan. 1953, pp. 69–88.
[894] Guzman, A., "Decomposition of a Visual Scene into Three-Dimensional Bodies", *Proceedings of the December 9-11, 1968, Fall Joint Computer Conference, Part I*, American Federation of Information Processing Societies, San Francisco, CA, Dec. 1968, pp. 291–304.
[895] Szeliski, R., *Computer Vision: Algorithms and Applications*, Springer -Verlag, Berlin, Heidelberg, Nov. 2010.
[896] Marr, D. and Poggio, T., "A Theory of Human Stereo Vision", A.I. Memo No. 451, Massachusetts Institute of Technology Artificial Intelligence Laboratory, Nov. 1977.

[897] Mitchell, M., *Analogly-Making as Perception,* Bradford Books, MIT, Cambridge, MA, May 1993.

[898] Bajcsy, R., "Computer Identification of Textured Visual Scenes", Report No. STAN-CS-72-321, Stanford University, Oct. 1972.

[899] Canny, J., "Finding Edges and Lines in Images", MIT Technical Report 720, MIT Artificial Intelligence Lab, Jun. 1983.

[900] Ponce, J., Bajcsy, R., Metaxas, D., Bedford, T., Forsyth, D., Herbert, M., Ikeuchi, K., Kak, A., Shapiro, L., Sclaroff, S., Pentland, A., and Stockman, G., "Object Representation for Object Recognition", CMU Paper 453, Robotics Institute, Jan. 1994.

[901] Mumford, D., "Pattern Theory: the Mathematics of Perception", *International Congress of Mathematics,* Vol. 3, Beijing, China, 2002.

[902] Zhang, J., Marszalek, M., Lazebnik, S., and Schmid, C., "Local Features and Kernels for Classification of Texture and Object Categories: A Comprehensive Study", *International Journal of Computer Vision,* Vol. 73, 2007, pp. 213–238.

[903] Hildreth, E. and Ullman, S., "The Measurement of Visual Motion", A.I. Memo No. 699, Massachusetts Institute of Technology, Dec. 1982.

[904] Stephens, M. and Harris, C., "3D Wire-frame Integration from Image Sequences", *Image and Vision Computing,* Vol. 7, No. 1, Feb. 1989, pp. 24–30.

[905] Rosenfeld, A. and Kak, A., *Digital Signal Processing,* Academic Press, Inc., New York, NY, 1982.

[906] Lucas, B. and Kanade, T., "An Iterative Registration Technique with an Application to Stereo Vision", *Proceedings of the 7th International Conference on Artificial Intelligence (IJCAI),* Vancouver, British Columbia, Aug. 1981, pp. 674–679.

[907] Gonzalez, R. and Woods, R., *Digital Image Processing,* Prentice Hall, Upper Saddle River, NJ, 1992.

[908] Ballard, D. and Brown, C., *Computer Vision,* Prentice-Hall, Englewood Cliffs, NJ, May 1982.

[909] *Artificial Intelligence: Handbook of Perception and Cognition,* M. Boden (Ed.), Academic Press, Inc., San Diego, CA, 1996.

[910] Fischler, M. and Bolles, R., "Random Sample Consensus: A Paradigm for Model Fitting with Applications to Image Analysis and Automated Cartography", *Communications of the ACM,* Vol. 24, No. 6, Jun. 1981, pp. 381–395.

[911] Harris, C. and Stephens, M., "A Combined Corner and Edge Detector", *Proceedings of the Alvey Vision Conference (AVC),* Southampton, UK, 1988.

[912] Viola, P. and Jones, M., "Robust Real-Time Face Detection", *International Journal of Computer Vision,* Vol. 57, No. 2, 2004, pp. 137–154.

[913] Viola, P. and Jones, M., "Rapid Object Detection using a Boosted Cascade of Simple Features", *Conference on Computer Vision and Pattern Recognition (CVPR 2001),* IEEE Computer Society, Kauai, HI, 2001.

[914] Bajcsy, R., "Active Perception", *Proceedings of the IEEE,* Vol. 76, No. 8, Aug. 1988, pp. 966–1005.

[915] Forbus, K., Gentner, D., Markman, A., and Ferguson, R., "Analogy just Looks like High Level Perception: Why a Domain-general approach to Analogical Mapping is Right", *Journal of Experimental & Theoretical Artificial Intelligence,* Vol. 10, No, 2, 1998, pp. 231–257.

[916] Horn, B., *Robot Vision,* MIT Press, Cambridge, MA, Jan. 1986.

[917] Roberts, L., "Machine Perception of Three-Dimensional Solids", In *Computer Methods in Image Analysis,* J. Aggarwal, R. Duda, and A. Rosenfeld (Eds.), IEEE Press, New York, NY, 1963.

[918] Hubel, D. and Wiesel, T., "Receptive Fields and Functional Architecture of Monkey Striate Cortex", *Journal of Physiology,* Vol. 195, 1968, pp. 215–243.

[919] Krizhevsky, A., Sutskever, I., and Hinton, G., "ImageNet Classification with Deep Convolutional Neural Networks", *Twenty-Sixth Conference on Neural Information Processing Systems (NIPS-2012),* Lake Tahoe, NV, Dec. 2012.

[920] Fischler, M. and Elschlager, R., "The Representation and Matching of Pictorial Structure", *IEEE Transactions on Computers*, New York, NY, Jan. 1973, pp. 67–92.

[921] Huffman, D., "Impossible Objects as Nonsense Sentences", in *Approaches for Picture Analysis,* 2012, pp. 295–323

[922] Turk, M. and Pentland, A., "Eigenfaces for Recognition", *Journal of Cognitive Neuroscience,* Vol. 3, No. 1, MIT Press, Cambridge, MA, Jan. 1991, pp. 71–86

[923] Kanade, T., "A Theory of Origami World", *Artificial Intelligence,* Vol. 13, North-Holland Publishing Co., Elsevier, Amsterdam, Netherlands, 1980, pp. 279–311

[924] Leung, T., Burl, M., Perona, P., "Finding Faces in Cluttered Scenes Using Random Labeled Graph Matching", Proceedings of IEEE International Conference on Computer Vision (ICCV), Cambridge, MA, June 1995, pp. 637–644

[925] Malik, J., "Interpreting Line Drawings of Curved Objects", *International Journal of Computer Vision,* Vol. 1, Kluwer Academic Publishers, Dordrecht, Netherlands, 1987, pp. 73–103

[926] Marr, D., *Understanding Complex Information-Processing Systems,* W.H. Freeman and Company, London, UK, 1982, pp. 69–83

[927] Lal, M., Kumar, K., Arain, R., Maitlo, A., Ruk, S., Shaikh, H., "Study of Face Recognition Techniques: A Survey", International Journal of Advanced Computer Science and Applications (IJACSA), Vol. 9, No. 6, Cleckheaton, United Kingdom, 2018, pp. 42–49

[928] Bierderman, I., "Recognition-by-Components: A Theory of Human Image Understanding", Psychological Review, Vol. 94, No. 2, American Psychological Association, Inc., 1987, pp. 115–147

[929] Sirovich, L. and Kirby, M., "Low-dimensional Procedure for the Characterization of Human Faces", Journal of the Optical Society of America, A, Optics and Image Science, Vol. 4, No. 3, Mar. 1987, pp. 519–524

[930] Weber, M., Unsupervised Learning of Models for Object Recognition, PhD Submission at California Institute of Technology, May 2000

[931] Girshick, R., Donahue, J., Darrell, T., Malik, J., "Rich Feature Hierarchies for Accurate Object Detection and Semantic Segmentation, Tech Report (v5)", arXiv:1311.2524v5, Oct. 2014

[932] Forsyth, D. and Ponce, J., Computer Vision: A Modern Approach, Prentice Hall, Hoboken, NJ, Mar. 2002

[933] Bishop, R., Computer Vision: Algorithms and Applications, Springer-Verlag, London, UK, 2011

[934] Torralba, A., Isola, P., Freeman, W., Foundations of Computer Vision, The MIT Press, Cambridge, MA, Apr. 2024

[935] Prince, S., Understanding Deep Learning, The MIT Press, Cambridge, MA, Dec. 2023

[936] Peters, J., Foundations of Computer Vision: Computational Geometry, Visual Image Structures and Object Shape Detection, Springer International Publishing, Cham, Switzerland, 2017

[937] Reynolds, D., "Gaussian Mixture Models", Encyclopedia of Biometrics, Vol. 741, A. Li and A. Jain (Eds.), Springer Reference, New York, NY, Jul. 2009, pp. 659–663

[938] Shapiro, L., and Stockman, G., Computer Vision, Prentice Hall, Hoboken, NJ, Feb. 2001

[939] Trucco, E. and Verri, A., Introductory Techniques for 3-D Computer Vision, Prentice Hall, Upper Saddle River, NJ, 1998

[940] Fischer, R., "The Use of Multiple Measurements in Taxonomic Problems", Annals of Eugenics, Vol. 7, Blackwell Publishing Ltd., Sep. 1936, pp. 179–188

[941] Cortes, C. and Vapnik, V., "Support-Vector Networks", Machine Learning, Vol. 20, Kluwer Academic Publishers, Boston, MA, 1995, pp. 273–297

[942] Fahlman, S. and Lebiere, C., "The Cascade-Correlation Learning Approach", Advances in Neural Information Processing Systems 2, Morgan Kaufmann Publishers, San Francisco, CA, Jun. 1990, pp. 524–532

[943] Lowe, D., "Distinctive Features from Scale-Invariant Keypoints", International Journal of Computer Vision, Vol. 60, Springer Science + Business Media, Berlin, Germany, Nov. 2004, pp. 91–110

[944] Schwarz, M., Schultz, H., Behnke, S., "RGB-D Object Recognition and Pose Estimation based on Pre-trained Neural Network Features", IEEE International Conference on Robotics and Automation (ICRA), Seattle, WA, May 2015

[945] Gao, K., Gao, Y., He, H., Lu, D., Xu, L., Li, J., "NeRF: Neural Radiance Field in 3D Vision, Introduction and Review", arXiv:2210.00379v5, Nov. 2023

[946] Hu, Y., Chun, B., Lin, J., Wang, Yu., Wang, Yi., Mehlman, C., Lipson, H., "Human-Robot Facial Coexpression", Science Robotics, Vol. 9, American Association for the Advancement of Science (AAAS), Mar. 2024

[947] Radford, A., Kim, J.-W., Hallacy, C., Ramesh, A., Goh, G., Agarwal, S., Sastry, G., Askell, A., Mishkin, P., Clark, J., Krueger, G., Sutskever, I., "Learning Transferable Visual Models from Natural Language Supervision", Preprint arXiv:2103.00020v1, Feb. 2021

[948] Krizhevsky, A., Sutskever, I., Hinton, G., "ImageNet Classification with Deep Convolutional Neural Networks", NeurIPS Proceedings, Advances in Neural Information processing Systems 25 (NIPS 2012) and in Communications of the ACM, Vol. 60, No. 6.

[949] Helmholtz, H., *Handbuch der Physiologischen Optik*, Leopold Voss, Leipzig, Germany, 1867.

[950] Lettvin, J., Maturana, H., McCulloch, W., and Pitts, W., "What the Frog's Eye Tells the Frog's Brain", *Proceedings of the IRE*, Vol. 47, Issue 11, Nov. 1959, pp. 1940–1951.

[951] Riesenhuber, M. and Poggio, T., "Hierarchical Models of Object Recognition in Cortex", *Nature Neuroscience*, Vol. 2, No. 11, Nature America Inc., Nov. 1999.

[952] McClelland, J. and Rumelhart, D., *An Interactive Activation Model of the Effect of Context in Perception, Part I*, 1980, Technical Report AD-A086 946, California University San Diego, May 1980.

[953] Roberts, L., Machine Perception of Three-Dimensional Solids, PhD Submission at Massachusetts Institute of Technology, Jun. 1963.

[954] Fei-Fei, L., Fergus, R., and Perona, P., "A Bayesian Approach to Unsupervised One-shot Learning of Object Categories", *Proceedings of the Ninth IEEE International Conference on Computer Vision (ICCV 2003)*, Nice, France, Oct. 2003.

[955] Ha, D. and Schmidhuber, J., "World Models", arXiv:1803.10122v4, May 2018.

[956] Marr, D., *Vision: A Computational Investigation into the Human Representation and Processing of Visual Information*, W.H. Freeman and Company, London, UK, 1982.

[957] Guzmán, A., "Decomposition of a Visual Scene into Three-Dimensional Bodies", *Proceedings of the AFIPS '68 Fall Joint Computer Conference, San Francisco, CA*, Dec. 1968, pp. 291–304.

[958] Duda, R. and Hart, P., *Pattern Classification and Scene Analysis*, Wiley-Interscience, John Wiley & Sons, New York, NY, 1973.

[959] Guzmán-Arenas, A., Computer Recognition of Three-Dimensional Objects in a Visual Scene, PhD Submission at Massachusetts Institute of Technology, Dec. 1968.

[960] Kanade, T., *Computer Recognition of Human Faces*, Birkhauser Verlag, Bagel, und Stuttgart, 1977.

[961] Kanade, T., Picture Processing System by Computer Complex and Recognition of Human Faces, PhD Submission at Kyoto University, Nov. 1973.

[962] Mackworth, A.K., "Interpreting Pictures of Polyhedral Scenes", *Artificial Intelligence*, Vol. 4, North-Holland Publishing Co., Elsevier, Amsterdam, Netherlands, 1973, pp. 121–137.

[963] Duda, R., Hart, P., and Stork, D., *Pattern Classification*, John Wiley & Sons, Inc., New York, NY, Nov. 2000.

[964] Fergus, R., Perona, P., and Zisserman, A., "Object Class Recognition by Unsupervised Scale-Invariant Learning", *Proceedings of the 2003 IEEE Computer Society on Computer Vision and Pattern Recognition (CVPR)*, Madison, WI, 2003.

Learning

[965] "New Navy Device Learns by Doing", *New York Times*, Jul. 1958.

[966] Goodfellow, I., Pouget-Abadie, J., Mirza, M., Xu, B., Warde-Farley, D., Ozair, S., Courville, A., and Bengio, Y., "Generative Adversarial Nets", In *Advances in Neural Information Processing Systems 28 (NIPS 2014), Montreal, Canada*, Z. Ghahramani, M. Welling, C. Cortes, N. Lawrence, and K.Q. Weinberger (Eds.), MIT Press, Cambridge, MA, Dec. 2014.

[967] Rumelhart, D., Hinton, G., and Williams, R., "Learning Representations by Back-propagating Errors", *Nature*, Vol. 323, Oct. 1986, pp. 533–536.

[968] Memisevic, R. and Hinton, G., "Unsupervised Learning of Image Transformation", *2007 IEEE Conference on Computer Vision and Pattern Recognition (CVPR 2007)*, Minneapolis, MN, Jun. 2007.

[969] Song, Y., Lukasiewicz, T., Xu, Z., and Bogacz, R., "Can the Brain Do Backpropagation? - Exact Implementation of Backpropagation in Predictive Coding Networks", *Proceedings of the Advances in Neural Information Processing System, NeurIPS 2020, Online Conference*, Dec. 2020.

[970] Badrinarayanan, V., Kendall, A., and Cipolla, R., "SegNet: A Deep Convolutional Encoder-Decoder Architecture for Image Segmentation", *IEEE Transactions on Pattern Analysis and Machine Intelligence (PAMI)*, Vol. 39, No. 12, Dec. 2017, pp. 2481–2495.

[971] Ben-Iwhiwhu, E., Dick, J., Ketz, N., Pilly, P., and Soltoggio, A., "Context meta-reinforcement Learning via Neuromodulation", *Neural Networks*, Vol. 152, 2022, pp. 70–79.

[972] Matsuo, Y., LeCun, Y., Sahani, M., Precup, D., Silver, D., Sugiyama, M., Uchibe, E., and Morimoto, J., "Deep Learning, Reinforcement Learning, and World Models", *Neural Networks*, Vol. 152, 2022, pp. 267–275.

[973] Haykin, S., *Neural Networks: A Comprehensive Foundation*, Pearson Prentice Hall, Upper Saddle River, NJ, 1994.

[974] Boser, B., Guyon, I., and Vapnik, V., "A Training Algorithm for Optimal Margin Classifiers", *Proceedings of the 5th Annual Workshop on Computational Learning Theory (COLT '92)*, Pittsburgh, PA, Jul. 1992, pp. 144–152.

[975] Sejnowski, T., *The Deep Learning Revolution*, The MIT Press, Cambridge, MA, 2018.

[976] Burkov, A., *The Hundred-Page Machine Learning Book*, Andriya Burkov Self Published, Quebec, Jan. 2019.

[977] Rumelhart, D., Smolensky, P., McClelland, J.L., and Hinton, G., "Schemata and Sequential Thought Process in PDP Models", In *Readings in Cognitive Science*, Morgan Kaufmann, Burlington, MA, 1988, pp. 224–249.

[978] Mitchell, T., *Machine Learning*, McGraw-Hill Education, New York, NY, Mar. 1997.

[979] Cristianini, N. and Shawe-Taylor, J., *An Introduction to Support Vector Machines and other kernel-based Learning Methods*, Cambridge University Press, Cambridge, UK, 2000.

[980] Rumelhart, D. and Norman, D., "Analogical Processes in Learning", In *Cognitive Skills and Their Acquisition*, J.R. Anderson (Ed.), Psychology Press, Sussex, 1981.

[981] Kaelbling, L., Littman, M., and Moore, A., "Reinforcement Learning: A Survey", *Journal of Artificial Intelligence Research*, Vol. 4, May 1996, pp. 237–285.

[982] Wiering, M. and Schmidhuber, J., "Fast Online Q(λ)", *Machine Learning*, Vol. 33, 1998, pp. 105–115.

[983] Sutton, R. and Barto, A., *Reinforcement Learning: An Introduction*, A Bradford Book, MIT Press, Cambridge, MA, 2014.

[984] Kober, J., Bagnell, J.A., and Peters, J., "Reinforcement Learning in Robotics: A Survey", *International Journal of Robotics Research*, Vo. 32, No. 11, Sept. 2013, pp. 1238–1274.

[985] Skansi, S., *Introduction to Deep Learning: From Logical Calculus to Artificial Intelligence*, Springer International Publishing AG, Switzerland, 2018.

[986] Alpaydin, E., *Machine Learning*, The MIT Press, Cambridge, MA, Oct. 2016.

[987] Friedberg, R., "A Learning Machine: Part I", *IBM Journal*, Vol. 2, No. 1, Jan. 1958, pp. 2–13.

[988] Rumelhart, D., McClelland, J.L., and PDP Research Group, "On Learning the Past Tenses on English Verbs", *In Parallel Distributed Processing: Explorations in the Microstructure of Cognition*, Vol. 2, Psychological and Biological Models, Bradford Books, MIT Press, Cambridge, MA, 1986.

[989] Hastie, T., Tibshirani, R., and Friedman, J., *The Element of Statistical Learning: Data Mining, Inference, and Prediction*, Springer, New York, NY, 2009.

[990] Shyam, R. and Singh, R., "A Taxonomy of Machine Learning Techniques", *Journal of Advancements in Robotics (JoARB), STM Journals*, Vol. 8, No. 3, 2021, pp. 18–25.

[991] Campaigne, H., "Some Experiments in Machine Learning", *Proceedings of the Western Joint Computer Conference*, San Francisco, Mar. 1959.

[992] Friedberg, R., Dunham, B., and North, J., "A Learning Machine: Part II", *IBM Journal*, Vol. 3, No. 3, Jul. 1959, pp. 282–287.

[993] Tesauro, G., "Practical Issues in Temporal Difference Learning", *Machine Learning*, Vol. 8, May 1992, pp. 259–266.

[994] Ng, A. and Jordan, M., "On Discriminative vs. Generative Classifiers: A Comparison of logistic regression and Naïve Bayes", *Proceedings of the*

Advances in Neural Information Processing System, NeurIPS 2001, T. Dietterich, S. Becker, and Z. Ghahramani (Eds.), MIT Press, Cambridge, MA, 2001.

[995] Carbonell, J., Michalski, R., and Mitchell, T., "An Overview of Machine Learning", In *Machine Learning: An Artificial Intelligence Approach*, R. Michalski, J. Carbonell, and T. Mitchell (Eds.), Tioga Publishing Co., Palo Alto, CA, 1983.

[996] Smola, A. and Vishwanathan, S.V., *Introduction to Machine Learning*, Cambridge University Press, Cambridge, UK, 2008.

[997] Hecht-Nielsen, R., "Kolmogorov's Mapping Neural Network Existence Theorem", *Proceedings of the IEEE First International Conference on Neural Networks*, San Diego, CA, Jun. 1987, pp. 11–13.

[998] Smolensky, P., "Information Processing in Dynamical Systems: Foundations of Harmony Theory", In *Parallel Distributed Processing*, D.E. Rumelhart and J.L. McClelland, (Eds.), Vol. 1, MIT Press, Cambridge, MA, 1986, pp. 194–281.

[999] LeCun, Y., Bengio, Y., and Hinton, G., "Deep Learning", *Nature*, Vol. 521, May 2015, pp. 436–444.

[1000] Werbos, P., "Backpropagation: Past and Future", *IEEE International Conference on Neural Networks*, Jul. 1988.

[1001] Barlow, H.B., "Unsupervised Learning", *Neural Computation*, Vol. 1, No. 3, 1989, pp. 295–311.

[1002] Murphy, K., *Probabilistic Machine Learning: An Introduction*, The MIT Press, Cambridge, MA, Mar. 2022.

[1003] Dawid, A. and LeCun, Y., "Introduction to Latent Variable Energy-Based Models: A Path Towards Autonomous Machine Intelligence", arXiv:2306.02572v1. Jun. 2023.

[1004] Carbonell, J., Michalski, R., and Mitchell, T., "An Overview of Machine Learning", In *Machine Learning: Symbolic Computation*, R. Michalski, J. Carbonell, and T. Mitchell (Eds.), Springer, Berlin, Heidelberg, 1983.

[1005] Langley, P., *Elements of Machine Learning*, Morgan Kaufmann Publishers, Inc., San Francisco, CA, 1996.

[1006] Cortes, C. and Vapnik, V., "Support-Vector Networks", *Machine Learning*, Vol. 20, 1995, pp. 273–297.

[1007] Mitchell, T., "Machine Learning and Data Mining", *Communications of the ACM*, Vol. 42, No. 11, Nov. 1999, pp. 30–36.

[1008] Dietterich, T. and Michalski, R., "A Comparative Review of Selected Methods for Learning from Examples", In *Machine Learning: An Artificial Intelligence Approach*, R. Michalski, J. Carbonell, and T. Mitchell (Eds.), Tioga Publishing Co., Palo Alto, CA, 1983, pp. 41–81.

[1009] Millidge, B., Salvatori, T., Song, Y., Lukasiewicz, T., and Bogacz, R., "Universal Hopfield Networks: A General Framework for Single-Shot Associative Memory Models", *39th International Conference on Machine Learning*, Baltimore, MD, PMLR 162, 2022.

[1010] Aizenberg, I., Aizenberg, N., and Vandewalle, J., *Multi-Valued and Universal Binary Neurons*, Springer, New York, NY, 2000.

[1011] James, G., Witten, D., Hastie, T., and Tibshirani, R., *An Introduction to Statistical Learning: with Applications in R*, Springer, New York, NY, Jul. 2021.

[1012] Deisenroth, M., Faisal, A.A., and Ong, C.S., *Mathematics for Machine Learning*, Cambridge University Press, Cambridge, UK, 2020.

[1013] Zhang, A., Lipton, Z., Li, M., and Smola, A., *Dive into Deep Learning*, Cambridge University Press, Cambridge, UK, Sept. 2020.

[1014] *The Handbook of Brain Theory and Neural Networks*, M. Arbib (Ed.), A Bradford Book, The MIT Press, Cambridge, MA, 2003.

[1015] Schapire, R., "Strength of Weak Learnability", *Machine Learning*, Vol. 5, Jun. 1990, pp. 197–227.

[1016] Principe, J., Euliano, N., and Lefebvre, W.C., "Chapter VI - Hebbian Learning and Principal Component Analysis", In *Neural and Adaptive Systems: Fundamentals through Simulation*, Wiley, New York, NY, Dec. 1999.

[1017] Medler, D., "A Brief History of Connectionism", *Neural Computing Surveys*, Vol. 1, No. 2, 1998, pp. 18–73.

[1018] *Elements of Dimensionality Reduction and Manifold Learning*, B. Ghojogh, M. Crowley, F. Karray, and A. Ghodsi (Eds.), Springer Nature, Cham, Switzerland, 2023.

[1019] Parisi, G., Kemker, R., Part, J., Kanan, C., and Wermter, S., "Continual Lifelong Learning with Neural Networks: A Review", *Neural Networks*, Vol. 113, 2019, pp. 54–71.

[1020] Additional References on Machine Learning continues in Vol. 2.

Q-Learning

[1021] Mnih, V., Kavukcuoglu, K., Silver, D., Rusu, A., Veness, J., Bellemare, M., Graves, A., Riedmiller, M., Fidjeland, A., Ostrovski, G., Petersen, S., Beattie, C., Sadik, A., Autonoglou, I., King, H., Kumaran, D., Wierstra, D., Legg, S., and Hassabis, D., "Human-level Control Through Deep Reinforcement Learning", *Nature*, Vol. 518, Feb. 2015, pp. 529–541.

[1022] Watkins, C. and Dayan, P., "Q-Learning", *Machine Learning*, Vol. 8, 1992, pp. 279–292.

[1023] van Hasselt, H., Guez, A., and Silver, D., "Deep Reinforcement Learning with Double Q-learning", *Proceedings of Thirteenth AAAI Conference on Artificial Intelligence, (AAAI'16), Association for the Advancement of Artificial Intelligence*, Phoenix, AZ, 2016.

[1024] Rummery, G.A. and Niranjan, M., *On-line Q-Learning using Connectionist Systems*, University of Cambridge, Cambridge, 1994.

[1025] van Hasselt, H., "Double Q-learning", *Advances in Neural Information Processing Systems 23 (NIPS 2010), Twenty-Fourth Conference on Neural Information Processing Systems*, Vancouver, Canada, Dec. 2010.

[1026] van Hasselt, H., Guez, A., and Silver, D., "Deep Reinforcement Learning with Double Q-learning", arXiv:1509.06461v3, Dec. 2015.

Linguistics

[1027] Dudley, H.W., *Signaling System*, US Patent No. 3,470,323, Sept. 1969

[1028] Saussure, F.D., Cours de linguistique gnrale: tudes et Documents Payot, Payot, 106 Boulevard Saint-Germain, Paris, 1971.

[1029] Winograd, T., "Understanding Natural Language", *Cognitive Psychology*, Vol. 3, 1972, pp. 1–191.

[1030] Chowdhury, G., "Natural Language Processing", Annual Review of Information Science and Technology, No. 37, 2003, pp. 51–89.

[1031] Radford, A., Narasimhan, K., Salimans, T., and Sutskever, I., "Improving Language Understanding by Generative Pre-training", arXiv:2012.11747v3, Sept 2021.

[1032] *Natural Language Processing*, Second Edition, N. Indurkhya and F. Damerau (Eds.), CRC Press, Taylor & Francis Group, Boca Raton, FL, 2010.

[1033] Barr, A., "Natural Language Understanding", *AI Magazine*, Vol. 1, No. 1, 1980, pp. 5-10

[1034] Taniguchi, T., Mochihashi, D., Nagai, T., Uchida, S., Inoue, N., Kobayashi, I., Nakamura, T., Hagiwara, Y., Iwahashi, N. and T. Inamura, T., "Survey on frontiers of language and robotics", *Advanced Robotics, Journal of the Robotics Society of Japan*, Vol. 33, No. 15–16, 2019, pp. 700–730.

[1035] Woods, W.A., "Transition Network Grammars for Natural Language Analysis", *Communications of the ACM*, Vol. 13, No. 10, Oct. 1970, pp. 591–606.

[1036] Gazdar, G., Klein, E., Pullum, G., and Sag, I., *Generalized Phrase Structure Grammar*, Harvard University Press, Cambridge, MA, Jan. 1985.

[1037] Cherry, E.C., "Some Experiments on the Recognition of Speech, with One and with Two Ears", *The Journal of the Acoustical Society of America*, Vol. 25, No. 5, Sept. 1953, pp. 975–979.

[1038] Thomason, J., Zhang, S., Mooney, R., and Stone, P., "Learning to Interpret Natural Language, Commands, through Human-Robot Dialog", *Proceedings of the Twenty-Fourth International Joint Conference on Artificial Intelligence (IJCAI 2015)*, Buenos Aires, Argentina, Jul. 2015, pp. 1923–1929.

[1039] Jurafsky, D. and Martin, J., *Speech and Language Processing: An Introduction to Natural Language Processing, Computational Linguistics, and Speech Recognition*, Prentice Hall, Hoboken, NJ, May 2008.

[1040] Brown, J., Frishkoff, G., and Eskenazi, M., "Automatic Question Generation for Vocabulary Assessment", *Proceedings of Human Language Technology Conference on Empirical Methods in Natural Language Processing (HLT/EMNLP)*, Vancouver, Canada, Association for Computational Linguistics, Oct. 2005, pp. 819–826.

[1041] Schank, R., "Conceptual Dependency: A Theory of Natural Language Understanding", *Cognitive Psychology*, Vol. 3, 1972, pp. 552–631.

[1042] Papadimitriou, S., Kitagawa, H., Gibbons, P., and Faloutsos, C., "LOCI: Fast Outlier Detection Using the Local Correlation Integral", Proceedings of the 19th International Conference on Data Engineering, Bangalore, India, 2003, pp. 315–326.

[1043] Shyu, M.-L., Chen, S.-C., Sarinnapakorn, K., and Chang, L., "A Novel Anomaly Detection Scheme Based on Principal Component Classifier", IEEE Foundations and New Directions of Data Mining Workshop, in conjunction with the Third IEEE International Conference on Data Mining (ICDM '03), Melbourne, FL, Nov. 2003.

[1044] Eskin, E., "Anomaly Detection over Noisy Data using Learned Probability Distributions", Proceedings of the Seventeenth International Conference On Machine Learning (ICML '00), Stanford, CA, June 2000, pp. 255–262.

[1045] Tang, J., Fu, A., Chun, Z., and Cheung, D., "Capabilities of Outlier Detection Schemes in Large Datasets, Framework and Methodologies", *Knowledge and Information Systems*, Vol. 11, Springer-Verlag, Berlin, Germany, Jan. 2007, pp. 45–84.

[1046] Song, Y., Wen, Z., Lin, C.-Y., and Davis, R., "One-Class Conditional Random Fields for Sequential Anomaly Detection", 23rd International Joint Conference on Artificial Intelligence (IJCAI 13), Beijing, China, Association for Computing Machinery (ACM), Aug. 2013, pp. 1685–1691.

[1047] Warrender, C., Forrest, S., and Pearlmutter, B., "Detecting Intrusions Using System Calls: Alternative Data Models", Proceedings of the 1999 IEEE Symposium on Security and Privacy", Oakland, CA, May 1999, pp. 133–145.

[1048] Lee, W. and Xiang, D., "Information-Theoretic Measures for Anomaly Detection", Proceedings of the 2001 IEEE Symposium on Security and Privacy, Oakland, CA, May 2000, pp. 130–143.

[1049] Joshi, M., Agarwal, R., and Kumar, V., "Mining Needles in a Haystack: Classifying Rare Classes via Two-Phase Rule Induction", Special Interest Group on Management of Data (SIGMOD), Association for Computing Machinery (ACM), Santa Barbara, CA, May 2001.

[1050] Chandola, V., Banerjee, A., and Kumar, V., "Outlier Detection: A Survey", ACM Computing Surveys, Association for Computing Machinery (ACM), Jan. 2009.

[1051] Hawkins, S., He, H., Williams, G., and Baxter, R., "Outlier Detection Using Replicator Neural Networks", In *Data Warehousing and Knowledge Discovery (DaWaK 2002)*, Y. Kambayashi, W. Winiwarter, and M. Arikawa (Eds.), Springer-Verlag, Berlin, Germany, Jan. 2002, pp. 170–180.

[1052] Knorr, E. and Ng, R., "Algorithms for Mining Distance-based Outliers in Large Datasets", Proceedings of the 24th International Conference on Very Large Data Bases, New York City, NY, Morgan Kaufmann, Aug. 1998, pp. 392–403.

[1053] Eskin, E., Arnold, A., Prerau, M., Portnoy, L., and Stolfo, S., "A Geometric Framework for Unsupervised Anomaly Detection: Detecting Intrusions in Unlabeled Data", In *Applications of Data Mining in Computer Security*, D. Barbara and S. Jajodia (Eds.), Kluwer Academic Publishers, Feb. 2002, pp. 77–101.

[1054] D'Oro, P., Nasca, E., Masci, J., and Matteucci, M., "Group Anomaly Detection via Graph Autoencoder", 33rd Annual Conference on Neural Information Processing Systems (NeurIPS Workshop), Vancouver, BC, Canada, Dec. 2019.

Anomaly Detection

[1055] Leuschen, M., Cavallaro, J., and Walker, I., "Testing on the Curve: Nonlinear Analytical Redundancy for Fault Detection", 9th Topical Meeting on Robotics and Remote Systems, American Nuclear Society, Seattle, WA, March 2001.

[1056] Lazarevic, A., Ertoz, L., Kumar, V., Ozgur, A., and Srivasta, J., "A Comparative Study of Anomaly Detection Schemes in Network Intrusion Detection", Proceedings of the 2003 SIAM International Conference on Data Mining (SDM), Society for Industrial and Applied Mathematics (SIAM), 2003, pp. 25–36.

[1057] Joshi, M. and Kumar, V., "CREDOS: Classification Using Ripple Down Structure (A Case for Rare Classes)", Proceedings of the Fourth SIAM International Conference on Data Mining (SDM), Society for Industrial and Applied Mathematics (SIAM), Lake Buena Vista, FL, Apr. 2004.

[1058] Yu, D., Sheikholeslami, G., and Zhang, A., "FindOut: Finding Outliers in Very Large Datasets", *Knowledge and Information Systems: An International Journal*, Vol. 4, No. 4, Springer, Sept. 2002, pp. 387–412.

[1059] Pol, A., Berger, V., Cerminara, G., Germain, C., and Pierini, M., "Anomaly Detection with Conditional Variational Autoencoders", Eighteenth International Conference on Machine Learning and Applications (ICMLA 2019), Boca Raton, FL, Dec. 2019.

[1060] Breunig, M., Kriegel, H.-P., Ng, R., and Sander, J., "LOF: Identifying Density-Based Local Outliers", ACM SIGMOD Record, Special Interest Group on Management of Data (SIGMOID), Vol.29, No. 2, May 2000, pp. 93–104.

[1061] Ramaswamy, S., Rastogi, R., and Shim, K., "Efficient Algorithms for Mining Outliers from Large Data Sets", *ACM SIGMOD Record, Special Interest Group on Management of Data (SIGMOID)*, Vol.29, No. 2, May 2000, pp. 427–438.

[1062] Yamanishi, K., Takeuchi, J.-I., and Williams, G., "On-line Unsupervised Outlier Detection Using Finite Mixtures with Discounting Learning Algorithms", Proceedings of the 6th ACM SIGKDD Conference on Knowledge Discovery and Data Mining, Association for Computing Machinery, Boston, MA, Aug. 2000, pp. 320–324.

V Integrating AI and Robots

Situational Awareness

[1063] Endsley, M., Bolt, B., and Jones, D., *Designing for Situational Awareness: An Approach to User-Centered Design*, CRC Press, Taylor & Francis Group, Boca Raton, FL, 2003.

[1064] Wickens, C. and McCarley, J., *Applied Attention Theory*, CRC Press, Taylor & Francis Group, Boca Raton, FL, 2008.

[1065] Endsley, M., "Theoretical Underpinnings of Situation Awareness: A Critical Review", In *Situation Awareness Analysis and Measurement*, M. Endsley and D. Garland (Eds.), Lawrence Erlbaum Associates, Mahwah, NJ, 2000.

[1066] Endsley, M., "Situation Awareness Misconceptions and Misunderstandings", *Journal of Cognitive Engineering and Decision Making*, Vol. 9, No. 1, Mar. 2015, pp. 4–32.

Hardware: Computing & Memory

[1067] Menabrea, L., "Sketch of the Analytical Engine invented by Charles Babbage, Esq.", originally published in French in 1842 in the Bibliothèque Universelle de Genève, No. 82, and Translation originally published in 1843 in the Scientific Memoirs, Vol. 3, pp. 666–731.

[1068] Hartree, D., *Calculating Instruments and Machines*, The University of Illinois Press, Urbana, IL, 1949.

[1069] Uttley, A.M., "The Design of Conditional Probability Computers", Information and Control, Vol. 2, 1959, pp. 1–24.

[1070] Shain, E., Nov. 1964, *Computer Method and Apparatus*, US 3,155,821 Patent.

[1071] *Heterogeneous Computing Architectures: Challenges and Vision*, O. Terzo, K. Djemame, A. Scionti, and C. Pezuela (Eds.), CRC Press, Taylor & Francis Group, Boca Raton, FL, 2020.

[1072] Widrow, B. and Hoff, M., "Adaptive Switching Circuits", In *Neurocomputing: Foundations of Research*, J. Anderson and E. Rosenfeld (Eds.), MIT Press, Cambridge, MA, Jan. 1988, pp. 123–134

[1073] Randell, B., "From Analytical Engine to Electronic Digital Computer: The Contributions of Ludgate, Torres, and Bush", *Annals of the History of Computing*, Vol. 4, No. 4, Oct. 1982, pp. 327–341.

[1074] Willshaw, D., Buneman, O., and Longuet-Higgins, H., "Non-Holographic Memory", *Nature*, Vol. 222, Jun. 1969, pp. 960–962.

[1075] Baddeley, A., *Working Memory, Thought, and Action*, Oxford University Press, Oxford, UK, 2007.

[1076] Oh, K.-S. and Jung, K., "GPU Implementation of Neural Networks", *Pattern Recognition*, Vol. 37, 2004, pp. 1311–1314.

[1077] *FPGA Implementations of Neural Networks*, A. Omondi and J. Rajapakse (Eds.), Springer, Dordrecht, the Netherlands, 2006.

[1078] Jouppi, N.P., et al., "In-datacenter Performance Analysis of a Tensor Processing Unit", *44th International Symposium on Computer Architecture (ISCA)*, Toronto, Canada, Jun. 2017.

[1079] Gazzaniga, M., Ivry, R., and Mangun, G., *Cognitive Neuroscience: The Biology of the Mind*, W. W. Norton & Company, New York, NY, 1998.

[1080] Stern, N., *From Uniac to Univac: An Appraisal of the Eckert-Mauchly Computers*, Digital Press, Bedford, MA, 1981.

[1081] Merolla, P., et al., "A Million Spiking-Neuron Integrated Circuit with a Scalable Communication Network", *Science*, Vol. 345, No. 6197, Aug. 2014, pp. 668–673.

[1082] Baddeley, A., "Working Memory", *Science*, Vol. 255, Jan. 1992, pp. 255–259.

[1083] Eckert, J. and Mauchly, J., US 3,120,606A Patent, *Electronic Numerical Integrator and Computer*, Feb. 1964.

[1084] Luccioni, A., Jernite, Y., and Strubell, E., "Power Hungry Processing: Watts: Driving the Cost of AI Deployment", arXiv:2311.16863v1, Nov. 2023

[1085] *Perspectives on the Computer Revolution*, Z. Pylyshyn and L. Bannon (Eds.), Ablex Publishing Corp., Norwood, NJ, 1989.

[1086] *Models of Working Memory: Mechanisms of Active Maintenance and Executive Control*, A. Miyake and P. Shah (Eds.), Cambridge University Press, Cambridge, UK, 1999.

[1087] Baddeley, A. and Sala, S., "Working Memory and Executive Control", *Philosophical Transactions of the Royal Society of London. Series B: Biological Sciences*, Vol. 351, 1996, pp. 1397–1404.

[1088] Widrow, B., "An Adaptive 'Adaline' Neuron Using Chemical 'Memistors'", Technical Report No. 1553-2, Solid-State Electronics Laboratory, Stanford, University, Palo Alto, CA, Oct. 1960.

[1089] Ceruzzi, P., *Computing: A Concise History*, The MIT Press, Cambridge, MA, 2012.

[1090] Stokes, J., *Inside the Machine: An Illustrated Introduction to Microprocessors and Computer*, No Starch Press, San Francisco, CA, 2007.

[1091] Peddie, J., *The History of the GPU - Steps to Invention*, Springer Nature, Cham, Switzerland, 2022.

Cognitive Architectures

[1092] Anderson, J. and Bower, G., *Human Associative Memory*, Lawrence Erlbaum Assoc., New York, NY, 1973.

[1093] Vernon, D., Hofsten, C., and Fadiga, L., *A Roadmap for Cognitive Development in Humanoid Robots*, Springer-Verlag, Berlin, Heidelberg, 2010.

[1094] *Logic-Based Artificial Intelligence*, J. Minker (Ed.), Kluwer Academic Publishers, Amsterdam, the Netherlands, 2000.

[1095] Hassanien, A.E. and Emary, E., *Swarm Intelligence: Principles, Advances, and Applications*, CRC Press, Boca Raton, FL, 2016.

[1096] Miyazawa, K., Horii, T., Aoki, T., and Nagai, T., "Integrated Cognitive Architecture for Robot Learning of Action and Language", *Frontiers in Robotics and AI*, Vol. 6, 2019, p. 131.

[1097] Anderson, J., "A Spreading Activation Theory of Memory", *Journal of Verbal Learning and Verbal Behavior*, Vol. 22, 1983, pp. 261–295.

[1098] Pearl, J., "Bayesian Networks: A Model of Self-Activated Memory for Evidential Reasoning", *7th Conference of the Cognitive Science Society*, University of California, Irvine, CA, Aug. 1985.

[1099] Hawkins, J. and Blakeslee, S., *On Intelligence*, Times Books, New York City, NY, Apr. 2007.

[1100] Fodor, J. and Pylyshyn, Z., "Connectionism and Cognitive Architecture: A Critical Analysis", *Cognition*, Vol. 28, 1988, pp. 3–71.

[1101] Kurup, U., Bignoli, P., Scally, J.R., and Cassimatis, N., "An Architectural Framework for Complex Cognition", *Cognitive Systems Research*, Vol. 12, 2011, pp. 281–292.

[1102] Langley, P., Laird, J., and Rogers, S., "Cognitive Architectures: Research Issues and Challenges", *Cognitive Systems Research*, Vol. 10, No. 2, 2006, pp. 141–160.

[1103] Cassimatis, N., "A Cognitive Substrate for Achieving Human-Level Intelligence", *AI Magazine*, Vol. 27, No. 2, Summer 2006, pp. 45–56.

[1104] *Handbook of Swarm Intelligence: Concepts, Principles and Applications*, B. Panigrahi, Y. Shi, and M.-H. Lim (Eds.), Springer-Verlag, Berlin, Germany, 2011.

[1105] Goertzel, B., "Artificial General Intelligence: Concept, State of the Art, and Future Prospects", *Journal of Artificial General Intelligence*, Vol. 5, No. 1, 2014, pp. 1–46.

[1106] *Artificial General Intelligence*, B. Goertzel and C. Pennachin (Eds.), Springer-Verlag, Berlin, Heidelberg, 2007.

Random Trees

[1107] Karger, D., Lehman, E., Leighton, T., Levine, M., Lewin, D., and Panigrahy, R., "Consistent Hashing and Random Trees: Distributed Caching Protocols for Relieving Hot Spots on the World Wide Web", *Proceedings of the 29th Annual Symposium on Theory of Computing (STOC '97), Association of Computing Machinery*, El Paso, TX, May 1997.

[1108] Biau, G., "Analysis of a Random Forests Model", *Journal of Machine Learning Research*, Vol. 13, 2012, pp. 1063–1095.

[1109] Drmota, M., *Random Trees: An Interplay between Combinatorics and Probability*, Springer Wien, New York, NY, 2009.

[1110] Geurts, P., Ernst, D., and Wehenkel, L., "Extremely Randomized Trees", *Machine Learning*, Vol. 63, Mar. 2006, pp. 3–42.

[1111] Strobl, C., Malley, J., and Tutz, G., "An Introduction to Recursive Partitioning: Rationale, application, and characteristics of Classification and Regression Trees, Bagging, and Random Forests", *Psychological Methods*, Vol. 14, No. 4, 2009, pp. 323–348.

[1112] Seidel, R. and Aragon, C., "Randomized Search Tree", *Algorithmica*, Vol. 16, Oct. 1996, pp. 464–497.

Miscellaneous

[1113] Goldberg, E., *Statistical Machine*, US Patent 1,838,389, Patented Dec. 29, 1931

[1114] Cornfold, F.M., *The Republic of Plato*, Translated with Introduction, Oxford University Press, London, UK, 1941.

[1115] *Faster than Thought*, B.V. Bowden (Ed.), Sir Isaac Pitman & Sons Ltd, London, UK, 1953.

[1116] Nyquist, H., "Certain Topics in Telegraph Transmission Theory", *Transactions of the American Institute of Electrical Engineers*, Vol. 47, No. 2, Apr. 1928, pp. 617–644.

[1117] Bayes, T., "An Essay towards Solving a Problem in the Doctrine of Chances", *Philosophical Transactions*, Vol. 53, Dec. 1763, pp. 370–418.

[1118] Gabor, D., "Theory of Communication", *Journal of Institution of Electrical Engineers*, Vol. 93, No. 3, 1946, pp. 429–457.

[1119] Wooldridge, M. and Jennings, N., "Intelligent Agents: Theory and Practice", *The Knowledge Engineering Review*, Vol. 10, No. 2, 1995, pp. 115–152.

[1120] Crossman, F., et al., *Computer-Aided Materials Selection during Structural Design*, National Materials Advisory Board, Commission on Engineering and Technical Systems, National Research Council, National Academy Press, Washington, DC, 1995.

[1121] Norman, D. and Bobrow, D., "On Data-limited and Resource-limited Processes", *Cognitive Psychology*, Vol. 7, 1975, pp. 44–64.

[1122] Breazeal, C., *Designing Sociable Robots*, MIT Press, Cambridge, MA, 2002.

[1123] Berners-Lee, T., *Weaving the Web: The Original Design and Ultimate Destiny of the World Wide Web by its Inventor*, Harper Business, New York, NY, 1999.

[1124] Brady, M., "Artificial Intelligence and Robotics", A.I. Memo 756, MIT AI Lab, Feb. 1984.

[1125] Pierson, H. and Gashler, M., "Deep Learning in Robotics: A Review of Recent Research", *Advanced Robotics*, Vol. 31, No. 2, Jul. 2017, pp. 821–835.

[1126] Gordon, M.J.C. and Melham, T.F., *Introduction to HOL: A Theorem Proving Environment for Higher Order Logic*, Cambridge University Press, Cambridge, UK, 1993.

[1127] Blackburn, P., Gardent, C., and Meyer-Viol, W., "Talking about Trees", *Sixth Conference of the European Chapter of the Association for Computational Linguistics*, Utrecht, the Netherlands, Association for Computational Linguistics, Apr. 1993.

[1128] Marcus, M., Hindle, D., and Fleck, M., "D-Theory: Talking about Talking about Trees", *Proceedings of the 21st Annual Meeting on Association for Computational Linguistics*, Cambridge, MA, Linguistics, Association for Computational Linguistics, Jun. 1983, pp. 129–136.

[1129] Kirkpatrick, S., Gelatt, C.D., and Vecchi, M.P., "Optimization by Simulated Annealing", *Science*, Vol. 220, No. 4598, May 1983, pp. 671–680.

[1130] Stigler, S., "Gauss and the Invention of Least Squares", *The Annals of Statistics*, Vol. 9, No. 3, 1981, pp. 465–474.

[1131] Koza, J., *Genetic Programming: On the Programming of Computers by Means of Natural Selection, A Bradford Book*, MIT Press, Cambridge, MA, 1992.

[1132] Carnap, R., "On Inductive Logic", *Philosophy of Science*, Vol. 12, No. 2, Apr. 1945, pp. 72–97.

[1133] Smith, A., *An Inquiry into the Nature and Causes of the Wealth of Nations, Books I, II, III, IV and V*, MetaLibri Digital Library, First Edition 1776 Lausanne, Switzerland.

[1134] Cormen, T., Leiserson, C., Rivest, R., and Stein, C., *Introduction to Algorithms*, The MIT Press, Cambridge, MA, 1990.

[1135] Rembold, U. and Dillmann, R., "Artificial Intelligence in Robotics", *Robot Control (SYROCO '85), Proceedings of the 1st IFAC Symposium, International Federation of Automatic Control*, Barcelona, Spain, 1985.

[1136] *The Handbook of Brain Theory and Neural Networks*, M. Arbib (Ed.), The MIT Press, Cambridge, MA, Sept. 1998.

[1137] Masterman, M., *Language, Cohesion and Form*, Cambridge University Press, Cambridge, England, 2005.

[1138] Browne, C., Powley, E., Whitehouse, D., Lucas, S., Cowling, P., Rohlfshagen, P., Tavener, S., Perez, D., Samothrakis, S., and Colton, S., "A Survey of Monte Carlo Tree Search Methods", *IEEE Transactions on Computational Intelligence and AI in Games*, Vol. 4, No. 1, Mar. 2012, pp. 1–43.

[1139] *Cognitive Robotics*, A. Cangelosi and M. Asada (Eds.), The MIT Press, Cambridge, MA, 2022.

[1140] *Cognitive Robotics*, H. Samani (Ed.), CRC Press, Boca Raton, FL, 2015.

[1141] Barsalou, L., "Grounded Cognition", *Annual Review of Psychology*, Vol. 59, Aug. 2007, pp. 617–645.

[1142] Winograd, T., *Language as a Cognitive Process*, Addison-Wesley Publishing, Reading, MA, 1982.

[1143] Rajan, K. and Saffiotti, A., "Towards a Science of Integrated AI and Robotics", *Artificial Intelligence*, Vol. 247, 2017, pp. 1–9.

[1144] Abelson, H. and Sussman, G.J., *Structure and Interpretation of Computer Programs*, The MIT Press, Cambridge, MA, 1996.

[1145] Jones, K.S., "A Statistical Interpretation of Term Specificity and its Application in Retrieval", *Journal of Documentation*, Vol. 28, No. 1, 1972, pp. 11–21.

[1146] Manning, C., Raghavan, P., and Schtze, H., *An Introduction to Information Retrieval*, Cambridge University Press, Cambridge, UK, 2009.

[1147] *Readings in Model-based Diagnosis*, W. Hamscher, L. Console, and J. de Kleer (Eds.), Morgan Kaufmann, Elsevier, San Francisco, CA, 1992.

[1148] *Readings in Nonmonotonic Reasoning*, L. Ginsberg (Ed.), Morgan Kaufmann, Elsevier, San Francisco, CA, 1987.

[1149] Reiter, R., "A Theory of Diagnosis from First Principle", *Artificial Intelligence*, Vol. 32, 1987, pp. 57–95.

[1150] Billinghurst, M., Clark, A., and Lee, G., "A Survey of Augmented Reality", *Foundations and Trends in Human-Computer Interaction*, Vol. 8, No. 2–3, 2014, pp. 73–272.

[1151] Martello, S. and Toth, P., *Knapsack Problems: Algorithms and Computer Implementations*, John Wiley & Sons, West Sussex, England, 1990.

[1152] Tambe, M., *Security and Game Theory: Algorithms, Deployed Systems, Lessons Learned*, Cambridge University Press, New York, NY, Dec. 2011.

[1153] Weise, T., *Global Optimization Algorithms: Theory & Applications*, Self-Published, GNU Free Documentation License, 2006–2009.

[1154] Knuth, D., *The Art of Computer Programming: Vol. 1 Fundamental Algorithms*, Addison Wesley Publishing Co., Reading, MA, 1968.

[1155] Shanahan, M., "The Ramification Problem in the Event Calculus", *Proceedings of the Sixteenth International Joint Conference on Artificial Intelligence (IJCAI)*, Stockholm, Sweden, Jul. - Aug. 1999.

[1156] Van Wyk, C., "Mathematics and Engineering in Computer Science", NBSIR 75-780, Final Report for Computer Science and Technology, National Bureau of Standards, Aug. 1975.

[1157] McClelland, J., Rumelhart, D., and Hinton, G., "Chapter 1: The Appeal of Parallel Distributed Processing", *Readings in Cognitive Science: A Perspective from Psychology and Artificial Intelligence*, A. Collins and E. Smith (Eds.), Morgan Kaufmann, Burlington, MA, 1988, pp. 52–72.

[1158] Snderhauf, N., Brock, O., Scheirer, W., Hadsell, R., Fox, D., Leitner, J., Upcroft, B., Abbeel, P., Burgard, W., Milford, M., and Corke, P., "The Limits and Potentials of Deep Learning for Robotics", *The International Journal of Robotics Research*, Vol. 37, No. 4–5, 2018, pp. 405–420.

[1159] Cangelosi, A. and Schlesinger, M., *Developmental Robotics: From Babies to Robots*, MIT Press, Cambridge, MA, 2015.

[1160] Gagniuc, P., *Markov Chains: From Theory to Implementation and Experimentation*, John Wiley & Sons, Inc., Hoboken, NJ, Aug. 2017.

[1161] Colledanchise, M. and Ögren, P., *Behavior Trees in Robotics and AI: An Introduction*, Chapman & Hall, CRC Press, Oxfordshire, England, Jul. 2018.

[1162] Garrabrant, S., Benson-Tilsen, T., Critch, A., Soares, N., and Taylor, J., "Logical Induction", arXiv preprint arXiv:1609.03543v5, Dec. 2020.

[1163] Murphy, R., *Robotics through Science Fiction: Artificial Intelligence Explained through Six Robot Short Stories*, The MIT Press, Cambridge, MA, Dec. 2018.

[1164] *Soft Robotics: Transferring Theory to Application*, A. Verl, A. Albu-Schffer, O. Brock, and A. Raatz (Eds.), Springer, Berlin, Heidelberg, Germany, 2015.

[1165] Lehmann, F., "Semantic Networks", *Computers Mathematics Application*, Vol. 23, No. 2–5, 1992, pp. 1–50.

[1166] Sowa, J., *Principles of Semantic Networks: Explorations in the Representation of Knowledge*, Elsevier Science, Amsterdam, 2014.

[1167] Robert, C., Chopin, N., and Rousseau, J., "Harold Jeffrey's Theory of Probability Revisited", *Statistical Science*, Vol. 24, No. 2, pp. 141–172, 2009.

[1168] Laird, J., Kinkade, K., Mohan, S., and Xu, J., "Cognitive Robotics Using the Soar Cognitive Architecture", AAAI Technical Report WS-12-06, Association for the Advancement of Artificial Intelligence, 2012.

[1169] Laird, J., *The Soar Cognitive Architecture*, MIT Press, Cambridge, MA, Apr. 2012.

[1170] Fisher, M., *An Introduction to Practical Formal Methods Using Temporal Logic*, A John Wiley & Sons, Ltd., Sussex, UK, 2011.

[1171] Kurzweil, R., *The Age of Intelligent Machines*, The MIT Press, Cambridge, MA, 1992.

[1172] Nilsson, N., *The Quest for Artificial Intelligence: A History of Ideas and Achievements*, Cambridge University Press, New York, NY, 2010.

[1173] Hecht-Nielsen, R., *Neurocomputing*, Addison-Wesley Publishing Co., Reading, MA, 1990.

[1174] Hecht-Nielsen, R., "Neurocomputer Applications", In *Neural Computers*, R. Eckmiller and C. Malsburg (Eds.), Vol. 41, Springer, Berlin, Heidelberg, 1989.

[1175] More, L.T., "Dartmouth Conference on Artificial Intelligence: Report on the 5th and 6th Weeks", IR-00057, Technical Research Group, Jul. 1956.

[1176] Solomonoff, R., "The Universal Distribution and Machine Learning", Manno-Lugano, Switzerland, https://world.std.com/~rjs/pubs.html.

[1177] Solomonoff, R., "Lecture 1: Algorithmic Probability", https://world.std.com/~rjs/.

[1178] Solomonoff, R., "Lecture 2: Applications of Algorithmic Probability", https://world.std.com/~rjs/.

[1179] Gao, J.B., and Harris, C., "Some Remarks on Kalman Filters for the Multisensor Fusion", *Information Fusion*, Vol. 3, 2002, pp. 191–201.

[1180] Davis, M. and Travers, M., "A Brief Overview of the Narrative Intelligence Reading Group", In *Narrative Intelligence*, M. Mateas and P. Sengers (Eds.), John Benjamins Company, Amsterdam, 2003, pp. 27–38.

[1181] Grosz, B., "Collaborative Systems (AAAI-94 Presidential Address)", *AI Magazine*, Vol. 17, No. 2, Summer 1996, p. 67.

[1182] Grosz, B. and Sidner, C., "Plans for Disclosure", In *Intentions in Communication*, P.R. Cohen, J. Morgan, and M.E. Pollack (Eds.), MIT Press, Cambridge, MA, 1990.

[1183] Holland, J., *Adaptation in Natural and Artificial Systems*, A Bradford Book, MIT Press, Cambridge, MA, 1992.
[1184] van Etten, W., Introduction to Random Signals and Noise, John Wiley & Sons, Ltd., West Sussex, England, 2005.
[1185] Gardner, H., *Frames of Mind: The Theory of Multiple Intelligences*, Basic Book, New York, NY, 1983.
[1186] Solomonoff, R. J., A *Preliminary Report on a General Theory of Indictive Inference*, Report V-131, Zator Company, Cambridge, MA, Feb. 1960.
[1187] Bhaumik, D., Khalifa, A., Green, M., and Togelius, J., "Tree Search versus Optimization for Map Generation", *Proceedings of the Sixteenth AAAI Conference on Artificial Intelligence and Interactive Digital Entertainment (AIIDE-20), Worchester Polytechnic Institute*, Worcester, MA, Oct. 2020, pp. 24–30.
[1188] Algoet, P. and Cover, T., "A Sandwich Proof of the Shannon-McMillan-Breiman Theorem", *The Annals of Probability*, Vol. 16, No. 2, 1988, pp. 899–909.
[1189] Smolensky, P., "Connectionist Modeling: Neural Computation/Mental Connections", In *Mind Design II*, J. Haugeland (Ed.), The MIT Press, Cambridge, MA, 1997.
[1190] Kruse, R., Borgelt, C., Braune, C., Mostaghim, S., and Steinbrecher, M., *Computational Intelligence: A Methodological Introduction*, Second Edition, Springer-Verlag, London, UK, 2016.
[1191] Eberhart, R. and Shi, Y., *Computational Intelligence: Concepts to Implementation*, Elsevier, Morgan Kaufmann Publishers, Burlington, MA, 2007.
[1192] Kruse, R., Moewes, C., Borgelt, C., Steinbrecher, M., Klawonn, F., and Held, P., *Computational Intelligence: A Methodological Introduction*, Springer-Verlag, London, UK, 2013.
[1193] Engelbrecht, A., *Computational Intelligence: An Introduction*, John Wiley & Sons, Ltd., West Sussex, England, 2007.
[1194] Salton, G. and McGill, M., *Introduction to Modern Information Retrieval*, McGraw-Hill Book Co., New York, NY, 1983.
[1195] Pierce, J., *An Introduction to Information Theory: Symbols*, Signals & Noise, Dover Publications, Inc., Mineola, NY, 1961.
[1196] Smith, B.C., *Origin of Objects*, The MIT Press, Cambridge, MA, 1996.
[1197] Holland, J.H., *Adaptation in Natural and Artificial Systems: An Introductory Analysis with Applications to Biology, Control, and Artificial Intelligence*, MIT Press, Cambridge, MA, 1992.
[1198] Amari, S.-I., "Information Geometry on Hierarchy of Probability Distributions", *IEEE Transactions on Information Theory*, Vol. 47, No. 5, Jul. 2001, pp. 1701–1711.
[1199] Solomonoff, R.J., "A Formal Theory of Inductive Reference, Part 1", *Information and Control*, Vol. 7, No. 1, Mar. 1964, pp. 1–22.
[1200] Bobrow, D., "Natural Language Input for a Computer Problem System", AI Project Memo 66, Memorandum MAC-M-148, Massachusetts Institute of Technology, Cambridge, MA, Mar. 1964.
[1201] Winograd, T. and Flores, F., *Understanding Computers and Cognition: A New Foundation for Design*, Addison-Wesley Professional, Boston, MA, 1987.
[1202] Dantzig, G., Fulkerson, R., and Johnson, S., "Solution of a Large Scale Traveling Salesman Problem", Report P-510, The Rand Corporation, Santa Monica, CA, Apr. 1954.
[1203] Shubik, M., *Strategy and Market Structure*, John Wiley & Sons, New York, NY, 1959.
[1204] Shubik, M., "Game theory, Behavior, and the Paradox of the Prisoner's Dilemma: Three Solutions", *The Journal of Conflict Resolution*, Vol. 14, No. 2, Jun. 1970, pp. 181–193
[1205] Halevy, A., Norvig, P., and Pereira, F., "The Unreasonable Effectiveness of Data", *IEEE Intelligent System, IEEE Computer Society*, Vol. 24, Mar./Apr. 2009, pp. 8–12.
[1206] Gigerenzer, G., Todd, P., and ABC Research Group, *Simple Heuristics that Make us Smart*, Oxford University Press, Inc., New York, NY, 1999.
[1207] Rasmussen, J., "Skills, Rules, and Knowledge; Signals, Signs, and Symbols, and Other Distinctions in Human Performance Models", *Transactions on Systems, Man, and Cybernetics*, Vol. smc-13, No. 3, May 1983, pp. 257–266.
[1208] Von Foerster, H., "From Stimulus to Symbol: The Economy of Biological Computation", In *Sign, Image and Symbol*, G. Kepes (Ed.), George Braziller, Inc., New York, NY, 1966.
[1209] Changizi, M., Zhang, Q., Ye, H., and Shimojo, S., "The Structures of Letters and Symbols Throughout Human History are Selected to Match those Found in Objects in Nature", *American Naturalist*, Vol. 167, No. 5, May 2006, pp. E117–E139.
[1210] Zhao, H., She, Q., Zhu, C., Yang, Y., and Xu, K., "Online 3D Bin Packing with Constrained Deep Reinforcement Learning", arXiv abs/2006.14978, 2021.
[1211] Hunt, E., "Mimicking Thought", University of Washington Technical Report No. 68-1-03, Feb. 1968.
[1212] Ashby, W.R., *Design for a Brain: The Origin of Adaptive Behaviour*, John Wiley & Sons, Inc., Chapman & Hall, Ltd., London, Great Britain, 1960.
[1213] Gilmore, P. and Gomory, R., "Sequencing a One-state Variable Machine: A Solvable Case of the Traveling Salesman Problem", *Operations Research*, Vol. 12, No. 5, Sept.-Oct. 1964, pp. 655–679.
[1214] Anderson, J. and Bower, G., "A Propositional Theory of Recognition Memory", *Memory & Cognition*, Vol. 2, No. 3, 1974, pp. 406–412.
[1215] Sutherland, I.E., "Sketchpad: A Man-Machine Graphical Communication System", *Proceedings of the AFIPS*, Vol. 23, American Federation of Information Processing Societies, Las Vegas, NV, Nov. 1963, pp. 323–328.
[1216] Friedman, J., Hastie, T., and Tibshirani, R., "Additive Logistic Regression: A Statistical View of Boosting", *The Annals of Statistics*, Vol. 28, No. 2, 2000, pp. 337–407.

[1217] Burges, C., "A Tutorial on Support Vector Machines for Pattern Recognition", *Data Mining and Knowledge Discovery*, Vol. 2, 1998, pp. 121–167.

[1218] *Handbook of Computational Social Choice*, F. Brandt, V. Conitzer, U. Endriss, J. Lang, and A. Procaccia (Eds.), Cambridge University Press, New York, NY, 2016.

[1219] Floridi, L., *What is the Philosophy of Information?*, Oxford University Press, Oxford, UK, 2011.

[1220] Kirk, D. and Hwu, W., *Programming Massively Parallel Processors: A Hands-on Approach*, Morgan Kaufmann, Elsevier, Waltham, MA, 2010.

[1221] Yakir, B., *Introduction to Statistical Thinking*, CreateSpace Independent Publishing Platform, Amazon, Charleston, SC, Sept. 2014.

[1222] Bruner, J., "The Narrative Construction of Reality", *Critical Inquiry*, Vol. 18, No. 1, 1991, pp. 1–21.

[1223] *Handbook of the History of Logic*, D. Gabbay and J. Woods (Eds.), Vol. 5, North-Holland, Elsevier, Amsterdam, the Netherlands, 2009.

[1224] Dennett, D., "Granny's Campaign for Safe Science", In *Meaning in Mind: Fodor and His Critics*, B. Loewer and G. Rey (Eds.), Blackwell Books, Oxford, UK, 1990, pp. 87–95.

[1225] Waltz, D., "Scenes with Shadows", In *The Psychology of Computer Vision*, McGraw-Hill, New York, NY 1975.

[1226] Nagel, E., *The Structure of Science: Problems in the Logic of Scientific Explanation*, Harcourt, Brace & World, Inc., New York and Burlingame, 1961.

[1227] Hinton, G. and Sojnowski, T., "Optimal Perceptual Inference", *Proceedings of the IEEE Conference on Computer Vision and Pattern Recognition*, Washington DC, Jun. 1983.

[1228] *Traveling Salesman Problem, Theory and Applications*, D. Davendra (Ed.), INTECH, Republic of Croatia, 2019.

[1229] Bregler, C., Covell, M., and Slaney, M., "Video Rewrite: Driving Visual Speech with Audio", *ACM SIGGRAPH 97*, Addison Wesley Publishing Co., New York, NY, Aug. 1997

[1230] Ivakhnenko, A.G., *Cybernetic Predicting Devices*, U.S. Dept of Commerce, Joint Publication Research Service, Washington, DC, Sept. 1966.

[1231] Green, C., "Application of Theorem Proving to Problem Solving", *Proceedings of the 1st International Joint Conference on Artificial Intelligence (IJCAI '69)*, Morgan Kaufmann Publishers, Washington, DC, May 1969, pp. 219–239.

[1232] Baum, L. and Petrie, T., "Statistical Inference for Probabilistic Functions of Finite State Markov Chains", *The Annals of Mathematical Statistics*, Vol. 37, No. 6, Dec. 1966, pp. 1554–1563

[1233] Pearl, J. and Russel, S., "Bayesian Networks", In *Handbook of Brain Theory and Neural Networks*, M. Arbib (Ed.), MIT Press, Cambridge, MA, Nov. 2000, pp. 157–160.

[1234] Bryson, A. and Ho, Y.-C., *Applied Optimal Control: Optimization, Estimation, and Control*, Taylor & Francis Group, New York, NY, 1975.

[1235] Michie, D., "Experiments on the Mechanization of Game-learning: Part I. Characterization of the Model and its Parameters", *The Computer Journal, The British Computer, Science*, Vol. 6, No. 3, Nov. 1963, pp. 232–236.

[1236] Winston, P., *The Psychology of Computer Vision*, McGraw-Hill, New York, NY, Jan. 1975.

[1237] Arora, S. and Barak, B., *Computational Complexity: A Modern Approach*, Cambridge University Press, New York, NY, Apr. 2009.

[1238] Schneider, J., Broschat, S., and Dahmen, J., *Algorithmic Problem Solving with Python*, Washington State University, Creative Commons Attribution-ShareALike License, Pullman, WA, Feb. 2019.

[1239] Berkeley, I., "The Curious Case of Connectionism", *Open Philosophy*, Vol. 2, 2019, pp. 190–205.

[1240] Pareto, V., *Manual of Political Economy*, Translated by A. Schwier, MacMillan Press Ltd., Augustus M. Kelley Publishers, London, UK, 1971 (*Manuale di Economia Politica*, Piccola Biblioteca Scientifica - 13, Societa Editrice Libraria, Milano, Italy, 1906).

[1241] Erickson, J., *Algorithms*, Creative Commons Attribution 4.0 International License, Dec. 2018.

[1242] Papadkis, N. and Plexousakis, D., "Actions with Durations and Constraints: The Ramification Problem in Temporal Databases", *Proceedings of the 14th IEEE International Conference on Tools with Artificial Intelligence (ICTAI 2002)*, Washington DC, USA, Nov. 2002, pp. 83–90.

[1243] *Robotic Intelligence*, P. Sheu (Ed.), World Scientific Encyclopedia with Semantic Computing and Robotic Intelligence, Vol. 2, World Scientific Publishing Co. Pte. Ltd., Singapore, May 2019.

[1244] Kalman, R.E., "A New Approach to Linear Filtering and Prediction Problems", *Transactions of the ASME Journal of Basic Engineering*, Vol. 82, 1960, pp. 35–45.

[1245] Kim, J., *Philosophy of Mind*, Third Edition, Westview Press, Routledge Publishers, 2010.

[1246] Raina, R., Madhavan, A., and Ng, A., "Large-scale Deep Unsupervised Learning using Graphics Processors", *Proceedings of the 26th Annual International Conference on Machine Learning (ICML '09)*, Montreal, Quebec, Canada, Jun. 2009, pp. 873–88.

[1247] Carnap, R., *Statistical and Inductive Probability: Inductive Logic and Science*, Galois Institute of Mathematics and Art, Brooklyn, NY 1955.

[1248] Rich, C. and Waters, R., "The Programmer's Apprentice Project: A Research Overview", A.I. Memo No. 1004, Massachusetts Institute of Technology, Nov. 1987.

[1249] Spring, A., "A History of Laser Scanning, Part 1: Space and Defense Applications", *Photogrammetric Engineering & Remote Sensing*, Vol. 86, No. 7, American Society for Photogrammetry and Remote Sensing, Baton Rouge, LA, July 2020, pp. 419–491.

[1250] Hartley, R.V.L., "Transmission of Information", International Congress of Telegraphy and Telephony, Lake Como, IT, Sept. 1927 and *The Bell System Technical Journal*, Vol. 7, No. 3, Jul. 1928, pp. 535–563.

[1251] Huffman, D., "A Method for the Construction of Minimum-Redundancy Codes", *Proceedings of the I.R.E.*, Sept. 1952, pp. 1098–1101.
[1252] Korzybski, A., *Science and Sanity: An Introduction to Non-Aristotelian Systems and General Semantics, Institute of General Semantics*, Brooklyn, NY, 1933.
[1253] Kuhns, T., *The Structure of Scientific Revolutions*, The University of Chicago, Chicago, IL, 1962.
[1254] Drexler, K.E., *Engines of Creation*, Anchor Books, Random House, Inc., New York, NY, 1986.
[1255] Huang, Y., "Levels of AI Agents", arXiv:2405.06643, Mar. 2024.
[1256] Maass, W., "Liquid State Machines: Motivation, Theory, and Applications", In *Computability in Context: Computation and Logic in the Real World*, S.B. Cooper and A. Sorbi (Eds.), Imperial College Press, Feb. 2011, pp. 275–296.
[1257] Shmailov, M., *Intellectual Pursuits of Nicolas Rashevsky: The Queer Duck of Biology*, Birkhäuser, Springer International Publishing, Switzerland, 2016.
[1258] Eagle, C. and Nance, K., The Ghidra Book: *The Definitive Guide*, No Starch Press, San Francisco, CA, 2020.
[1259] Bogoni, Luca and Bajcsy, R., "Characterization of Functionality in a Dynamic Environment", University of Pennsylvania Technical Report No. MS-CIS-93-76, Sept 1993.
[1260] Whitley, D. and Watson, J.P., "Complexity Theory and the No Free Lunch Theorem", In *Search Methodologies: Introductory Tutorials in Optimization and Decision Support Techniques*, E. Burke and G. Kendall (Eds.), Springer Science + Business Media, New York, NY, 2005.
[1261] Wolpert, D. and Macready, W., "No Free Lunch Theorems for Optimization", *IEEE Transactions on Evolutionary Computation*, Vol. 1, No. 1, Apr. 1997.
[1262] Jumper, J., Evans, R., Pritzel, A., Green, T., Figurnov, M., Ronneberger, O., Tunyasuvunakool, K., Bates, R., Žídek, A., Potapenko, A., Bridgland, A., Meyer, C., Kohl, S., Ballard, A., Cowie, A., Romera-Paredes, B., Nikolov, S., Jain, R., Adler, J., Back, T., Petersen, S., Reiman, D., Clancy, E., Zielinski, M., Steinegger, M., Pacholska, M., Berghammer, T., Bodenstein, S., Silver, D., Vinyals, O., Senior, A., Kavukcuoglu, K., Kohli, P., and Hassabis, D., "Highly Accurate Protein Structure Prediction with AlphaFold", *Nature*, Vol. 596, Nature Portfolio, Springer Nature, Berlin, Germany, Aug. 2021, pp. 583–589.
[1263] Willshaw, D. and Dayan, P., "Optimal Plasticity from Matrix Memories: What Goes Up Must Come Down", *Neural Computation*, Vol. 2, MIT Press, Cambridge, MA, 1990, pp. 85–93.
[1264] Holland, J., "Genetic Algorithms", *Scientific American*, Vol. 267, No. 1, Nature America Inc., New York, NY, July 1992, pp. 66–73.
[1265] Dumoncel, J.-C., "Between the Time of Physics and the Time of Metaphysics, the Time of Tense Logic?", In *Einstein vs. Bergson: An Enduring Quarrel on Time*, A. Campo and S. Gozzano (Eds.), De Gruyter, Boston, MA, 2022, pp. 331–348.
[1266] Hawkes, E., An, B., Benbernou, N., Tanaka, H., Kim, S., Demaine, E., and Rus, D., "Programmable Matter by Folding", *Proceedings of the National Academy of Science (PNAS)*, Vol. 107, No. 28, Washington, D.C., Jul. 2010, pp. 12441–12445.

VI Artificial Intelligence: Today

[1267] Piaget, J., *The Language and Thought of the Child*, Kegan Paul, Trench, Trubner & Co. Ltd, New York, NY, 1926.
[1268] Schopenhauer, A., *The World as Will and Idea*, translated from German, Die Welt Als Willie und Vorstellung (WWV), and initially published in 1818. Seventh Edition, Vol. 1, Kegan Paul, 5 Keynes, J. M., *A Treatise of Probability*, MacMillan and Co., Ltd., London, UK, 1921.
[1269] *Logic and Philosophy of Time: Themes from Prior*, P. Hasle, P. Blackburn, and P. Ohrstrom (Eds.), Vol. 1, Aalborg University Press, Aalborg, Denmark, 2017.
[1270] Schneider, S., *The Language of Thought: A New Philosophical Direction*, MIT Press, Cambridge, MA, 2011.
[1271] Heidegger, M., *Logic: The Question of Truth*, Translated by T. Sheehan, Indiana University Press, Bloomington, IN, 1976.
[1272] Black, S., Gao, L., Golding, L., He, H., Leahy, C., McDonell, K., Phang, J., Pieler, M., Prashanth, S., Purohit, S., Reynolds, L., Tow, J., Wang, B., and Weinbach, S., "GPT-NeoX-20B: An Open-Source Autoregressive Language Model", Proceedings of the ACL Workshop on Challenges & Perspectives in Creating Large Language Models, Dublin, Ireland, May 2022.
[1273] Plaut, D., "Connectionist Modeling", In *Encyclopedia of Psychology*, A. Kasdin (Ed.), American Psychological Association, Washington, DC, Jan. 1999.
[1274] Grossberg, S., "Adaptive Resonance Theory: How a Brain Learns to Consciously Attend, Learn, and Recognize a Changing World", *Neural Networks*, Vol. 37, 2013, pp. 1–47.
[1275] Carpenter, G. and Grossberg, S., "Adaptive Resonance Theory", In *Encyclopedia of Machine Learning and Data Mining*, C. Sammut and G. Webb (Eds.), Springer Science + Business Media, New York, NY, 2016.
[1276] Clark, A., *Being There: Putting Brain, Body, and World Together Again*, A Bradford Book, MIT Press, Cambridge, MA, 1998.
[1277] Scarselli, F., Gori, M., Tsoi, A.-C., Hagenbuchner, M., and Monfardini, G., "The Graph Neural Network Model", *IEEE Transactions on Neural Networks*, Vol. 20, No. 1, Jan. 2009, pp. 61–80.

[1278] Corpina, F., "The Ancient Master Argument and Some Examples of Tense Logic", *Argumenta*, Vol. 1, No. 2, May 2016, pp. 245–258.
[1279] Davis, M., *Computability and Unsolvability*, McGraw-Hill Book Company, New York, NY, 1958.
[1280] *Connectionist-Symbolic Integration: From Unified to Hybrid Approaches*, R. Sun and F. Alexandre (Eds.), Psychology Press, Lawrence Erlbaum, Mahwah, NJ, 1997.
[1281] *Parallel Models of Associative Models*, G. Hinton and J. Anderson (Eds.), Lawrence Erlbaum Assoc., Hillsdale, NJ, 1989.
[1282] Bechtel, W. and Abrahamsen, A., *Connectionism and the Mind: Parallel Processing, Dynamics, and Evolution in Networks*, Published Work A-80, Blackwell Publishers, Malden, MA, 1991.
[1283] Pierce, B., *Types and Programming Languages*, The MIT Press, Cambridge, MA, 2002.
[1284] Olson, M. and Ramirez, J., *An Introduction to the Theories of Learning*, Pearson, London, UK, 2005.
[1285] Kahneman, D., Thinking, *Fast and Slow*, Farrar, Straus and Giroux, New York, NY, Apr. 2013.
[1286] Rochester, N., Holland, J., Haibt, L., and Duda, W., *Tests on a Cell Assembly Theory of the Action of the Brain, Using a Large Digital Computer*, Neurocomputing: Foundations of Research, MIT Press, Cambridge, MA, Jan. 1988, pp. 65–79.
[1287] Schank, R. and Abelson, R., "Scripts, Plans, and Knowledge", *Proceedings of the 4th International Joint Conference on Artificial Intelligence (IJCAI'75)*, Sept. 1975, pp. 151–157.
[1288] Feldman, J.A. and Ballard, D.H., *Connectionist Models and Their Properties*, Cognitive Science, Vol. 6, 1982, pp. 205–254.
[1289] Scheider, W., "Connectionism: Is it a Paradigm Shift for Psychology?", *Behavior Methods Research, Instruments, & Computers*, Vol. 19, No. 2, 1987, pp. 73–83.
[1290] Smith, L. and Thelen, E., *A Dynamic Systems Approach to the Development of Cognition and Action*, The MIT Press, Cambridge, MA, Jul. 1994.
[1291] Nakano, K., "Association - A Model of Associative Memory", *IEEE Transactions on Systems, Man, and Cybernetics*, Vol. SMC-2, No. 3, Jul. 1972, pp. 380–388.
[1292] Amarel, S., "On Representations of Problems of Reasoning about Actions", In *Machine Intelligence 3*, D. Mitchie (Ed.), Edinburgh University Press, Edinburgh, Scotland, UK, 1968, pp. 131–171.
[1293] Fodor, J. and Pylyshyn, Z., "Connectionism and Cognitive Analysis: A Critical Review", *Cognition*, Vol. 28, 1988, pp. 3–71.
[1294] van Gerven, M., "Computational Foundations of Natural Intelligence", *Frontiers in Computational Neuroscience*, Vol. 11, Dec. 2017, p. 299674.
[1295] Russell, S., "Artificial Intelligence and the Problem of Control", In *Perspectives on Digital Humanism*, H. Werthner, E. Prem, E. Lee, and C. Ghezzi (Eds.), Springer, Cham, Switzerland, 2022.
[1296] Netsky, M., "What is a Brain, and Who Said So?", *British Medical Journal*, Vol. 293, Dec. 1986, pp. 1670–1672.
[1297] Boden, M., "Creativity and Artificial Intelligence", *Artificial Intelligence*, Vol. 103, 1998, pp. 347–356.
[1298] Holyoak, K., Gentner, D., and Kokinov, B., "Introduction: The Place of Analogy in Cognition", In *The Analogical Mind: Perspectives from Cognitive Science*, D. Gentner, K.K. Holyoak, and B. Kokinov (Eds.), MIT Press, Cambridge, MA, 2001, pp. 1–19.
[1299] Bechtel, W. and Abrahamsen, A., *Connectionism and the Mind: An Introduction to Parallel Processing in Networks*, Basil Blackwell Inc., Cambridge, MA, 1991.
[1300] Pinker, S. and Prince, A., "On Language and Connectionism: Analysis of a Parallel Distributed Processing Model of Language Acquisition", *Cognition*, Vol. 28, No. 1–2, Mar. 1988, pp. 73–193.
[1301] Thomas, M. and McClelland, J., "Connectionist Models of Cognition", In *The Cambridge Handbook of Computational Psychology*, R. Sun (Ed.), Cambridge University Press, 2008, pp. 23–58.
[1302] Schank, R. and Abelson, R., Scripts, Plans, *Goals and Understanding: An Inquiry into Human Knowledge Structures*, Lawrence Erlbaum Associates, Publishers, Hillsdale, NJ, 1977.
[1303] Valiant, L.G., "A Theory of the Learnable", *Communications of the ACM*, Vol. 27, No. 11, New York, NY, Nov. 1984, pp. 1134–1142.
[1304] Davis, M., "What is Computation?", In *Mathematics Today - Twelve Informal Essays*, 1978, pp. 241–267.
[1305] Winsberg, E., *Science in the Age of Computer Simulation*, University of Chicago Press, Chicago, IL, Oct. 2010.
[1306] Johnson-Laird, P.N., "Mental Models", In *Foundations of Cognitive Science*, M. Posner (Ed.), MIT Press, Cambridge, MA, 1989, pp. 469–499 (Also Published by Harvard University Press, Cambridge, MA, Nov. 1983).
[1307] Trenholme, R., "Analog Simulation", *Philosophy of Science*, Vol. 61, No. 1, 1994, pp. 115–131.
[1308] Hergenhahn, B. and Olson, M., *An Introduction to the Theories of Learning*, Prentice Hall, Upper Saddle River, NJ, 1976.
[1309] *Connectionist Models in Cognitive Psychology*, G. Houghton (Ed.), Psychology Press, Hove, East Sussex, UK, 2005.
[1310] McMurray, B., "Connectionist Modeling for ...er...linguists", In *University of Rochester Working Papers in the Language Sciences (WPLS: UR)*, K. Crosswhite and J. McDonough (Eds.), Vol. Spring, No. 1, Rochester, NY, 2000.
[1311] Pfeifer, R. and Scheier, C., *Understanding Intelligence*, The MIT Press, Cambridge, MA, 2001.
[1312] Fodor, J., *The Language of Thought*, Thomas Y. Crowell Company, Inc., New York, NY, 1975.
[1313] Piaget, J., *Six Psychological Studies*, Random House, Inc., New York, NY, 1967.
[1314] Marcus, G., *The Algebraic Mind: Integrating Connectionism and Cognitive Science*, A Bradford Book, The MIT Press, Cambridge, MA, 2001.

[1315] Dreyfus, H., "From Micro-Worlds to Knowledge Representation: AI at an Impasse", *Mind Design*, J. Haugel (Ed.), MIT Press, Cambridge, MA, 1981, pp. 161–204.
[1316] Dorner, D., "The Logic of Failure", *Philosophical Transactions of the Royal Society B Biological Sciences*, Vol. 327, No. 1241, May 1990, pp. 463–473.

Information Science

[1317] *Historical Studies in Information Science*, T. Hahn and M. Buckland (Eds.), American Society for Information Science, Information Today, Inc., Medford, NJ, 1998.
[1318] Luenberger, D., *Information Science*, Princeton University Press, Princeton, NJ, 2006.
[1319] Luenberger, D. and Ye, Y., *Linear and Nonlinear Programming*, Springer International Publishing, Cham, Switzerland, 1973.

Computer Science

[1320] Brookshear, J.G. and Brylow, D., *Computer Science: An Overview*, Pearson Education Limited, Essex, England, 2015.
[1321] Deisenroth, M., Faisal, A.A., and Ong, C.S., *Mathematics for Machine Learning*, Cambridge University Press, Cambridge, UK, 2020.
[1322] VanderPlas, J., *Python Data Science Handbook: Essential Tools for Working with Data*, O'Reilly Media, Inc., Sebastopol, CA, 2017.
[1323] Sweigart, A., *Automating the Boring Stuff with Python: Practical Programming for Total Beginners*, No Starch Press, Inc., San Francisco, CA, 2015.
[1324] Forouzan, B., *Foundations of Computer Science*, Cengage Learning EMEA, Hampshire, United Kingdom, 2018.
[1325] Aho, A. and Ullman, J., *Foundations of Computer Science*, Computer Science Press, Jan. 1992.
[1326] Rasmussen, J., *Information Processing and Human-Machine Interaction: An Approach to Cognitive Engineering*, Series Volume 12, North Holland, Elsevier Science Publishing Co., New York, NY, 1986.

Cognitive Science

[1327] Minds, *Brains, and Computers: The Foundations of Cognitive Science*, R. Cummins and D. Cummins (Eds.), Blackwell Publishing, Ltd., Malden, MA, 2000.
[1328] Stephens, R., *Beginning Software Engineering*, John Wiley & Sons, Inc., Hoboken, NJ, 2023.
[1329] Stephens, R., *Essential Algorithms: A Practical Approach to Computer Algorithms*, John Wiley & Sons, Inc., Hoboken, NJ, 2013.
[1330] Johnson-Laird, P.N., *Mental Models: Towards a Cognitive Science of Language, Inference, and Consciousness*, Harvard University Press, Cambridge, MA, 1983.
[1331] Barr, M. and Wells, C., *Category Theory for Computer Science*, Prentice Hall, Hoboken, NJ, Jul. 1990.
[1332] Pierce, B., *Basic Category Theory for Computer Scientists*, The MIT Press, Cambridge, MA, Aug. 1991.
[1333] Tversky, A. and Kahneman, D., "Judgment under Uncertainty: Heuristics and Biases", *Science*, Vol. 185, No. 4157, Sept. 1974, pp. 1124–1131.
[1334] Schning, U., *Logic for Computer Scientists*, Birkhuser, Boston, MA, 2008.
[1335] Piccinini, G. and Scarantino, A., "Computation vs Information Processing: Why their Difference Matters to Cognitive Science", *Studies in History and Philosophy of Science*, Vol. 41, 2010, pp. 237–246.
[1336] Grossberg, S., *Studies of Mind and Brain: Neural Principles of Learning, Perception, Development, Cognition, and Motor Control*, Springer Dordrecht, Holland, Jun. 1982.
[1337] Van Gelder, T., "The Dynamical Hypothesis in Cognitive Science", *The Behavioral and Brain Sciences*, Vol. 21, 1998, pp. 615–665.
[1338] *The MIT Encyclopedia of the Cognitive Sciences*, R. Wilson and F. Keil (Eds.), The MIT Press, Cambridge, MA, 1999.
[1339] Thagard, P., *Mind: Introduction to Cognitive Science*, MIT Press, Cambridge, MA, 2005.
[1340] Gleick, J., *The Information: A History, A Theory, A Flood*, Pantheon Books, New York, NY, 2011.
[1341] Tversky, A., Preference, *Belief, and Similarity: Selected Writings*, E. Shafir (Ed.), A Bradford Book, The MIT Press, Cambridge, MA, 2004.
[1342] Grossberg, S., "Linking Mind to Brain: The Mathematics of Biological Intelligence", *Notice of the AMS*, Vol. 47, No. 11, Dec. 2000, pp. 1361–1372.

Computer Architecture

[1343] Tanenbaum, A. and Austin, T., *Structured Computer Organization*, Prentice Hall, Inc., Upper Saddle River, NJ, 1999.
[1344] *OpenSPARC Internals: OpenSPARC T1/T2 Chip Multithreaded throughput Computing*, D. Weaver (Ed.), Sun Microsystems, Inc., Santa Clara, CA, Oct. 2008.
[1345] Appel, A., *Modern Compiler Implementation in ML*, Cambridge University Press, Cambridge, UK, 1998.
[1346] Bryant, R. and O'Hallaron, D., *Computer Systems: A Programmer's Perspective*, Pearson, 2016.
[1347] Hennessy, J. and Patterson, D., *Computer Organization and Design: The Hardware/Software Interface*, Morgan Kaufmann, Burlington, MA, 2008.
[1348] Akenine-Mller, T., Haines, E., Hoffman, N., Pesce, A., Iwanicki, M., and Hillaire, S., *Real-time Rendering*, CRC Press, Taylor & Francis Group, Boca Raton, FL, 2018.
[1349] Mano, M.M., *Computer Science Architecture*, Pearson Education, London, UK, Oct. 1992.
[1350] Gonzlez, A., Latorre, F., and Magklis, G., *Processor Microarchitecture: An Implementation Perspective*, Morgan & Claypool, Kentfield, CA, 2011.

[1351] Neufert, E., *Architect's Data*, Blackwell Science Ltd., Oxford, UK, 1980.

[1352] Hughes, J., Van Dam, A., McGuire, M., Sklar, D., Foley, J., Feiner, S., and Akeley, K., *Computer Graphics: Principles and Practice*, Addison-Wesley, Upper Saddle River, NJ, 2014.

[1353] Shen, J.P. and Lipasti, M.H., *Modern Processor Design: Fundamentals of Superscalar Processors*, Waveland Press, Inc., Long Grove, IL, 2005.

[1354] Hennessy, J. and Patterson, D., *Computer Architecture: A Quantitative Approach*, Morgan Kaufmann, Elsevier, Waltham, MA, 2012.

[1355] Harris, D. and Harris, S., *Digital Design and Computer Architecture*, Morgan Kaufmann, Elsevier, Waltham, MA, 2013.

Computational Logic

[1356] Nilsson, U. and Matuszynski, J., Logic, *Programming and Prolog*, John Wiley & Sons, Ltd., 1990.

[1357] Sipser, M., *Introduction to the Theory of Computation*, Thomson Learning, Inc., Boston, MA, 2006.

[1358] Hefferon, J., *Theory of Computation: Making Connections*, Independently Published, Jul. 2023.

[1359] Hao, W., *Computation, Logic, Philosophy: A Collection of Essays*, Science Press, Beijing, China, 1990.

[1360] Vince, J., *Foundation Mathematics for Computer Science: A Visual Approach*, Springer International Publishing, Cham, Switzerland, 2015.

[1361] Sipser, M., *Introduction to the Theory of Computation*, Cengage Learning, Boston, MA, Jun. 2012.

[1362] Page, R. and Gamboa, R., *Essential Logic for Computer Science*, The MIT Press, Cambridge, MA, Jan. 2019.

[1363] Barak, B., *Introduction to Theoretical Computer Science*, https://github.com/boazbk/tcs, a Creative Commons "Attribution-NonCommercial-NoDerivatives 4.0 International, Mar. 2023.

[1364] Sterling, L. and Shapiro, E., *The Art of Prolog*, MIT Press, Cambridge, MA, 1986.

[1365] Schning, U., *Logic for Computer Scientist*, Wissenschaftsverlag, Mannheim, Germany, 1987.

[1366] Reeves, S. and Clarke, M., *Logic for Computer Science*, Addison-Wesley, Boston, MA, Jan. 1990.

[1367] de Bruijn, N.G., "A Survey of the Project Automath", *Studies in Logic and the Foundations of Mathematics*, Vol. 133, 1994, pp. 141–161.

[1368] Wasilewska, A., *Logics for Computer Science: Classical and Non-Classical*, Springer Nature Switzerland, Cham, Switzerland, 2019.

[1369] Boyer, R. and Moore, J.S., *A Computational Logic*, Academic Press, Inc., New York, NY, 1979.

[1370] Flach, P., *Simply Logical: Intelligent Reasoning by Example*, John Wiley & Sons, Ltd., Hoboken, NJ, 1994.

Algorithms

[1371] Kreher, D. and Stinson, D., *Combinatorial Algorithms: Generation, Enumeration, and Search*, CRC Press LLC, Boca Raton, FL, 1999.

[1372] Nijenhuis, A. and Wilf, H., *Combinatorial Algorithms: For Computers and Calculators*, Academic Press Inc., New York, NY, 1978.

[1373] *Deep Learning Algorithms*, Z. Gacovski (Ed.), Arcler Press, Burlington, Ontario, Canada, 2022.

[1374] *Deep Learning: Algorithms and Applications*, W. Pedrycz and S.-M. Chen (Eds.), Springer Nature Switzerland AG, Cham, Switzerland, 2020.

[1375] Smolyakov, V., *Machine Learning Algorithm in Depth*, Manning Publications Co., Shelter Island, NY, 2023.

Algorithmic Logic

[1376] *Algorithmic Game Theory*, N. Nisan, T. Roughgarden, E. Tardos, and V. Vazirani (Eds.), Cambridge University Press, New York, NY, 2007.

[1377] Wooldridge, M., *An Introduction to MultiAgent Systems*, John Wiley & Sons, Ltd., Hoboken, NJ, Aug. 2002.

[1378] Shoham, Y. and Leyton-Brown, K., *Multiagent Systems: Algorithmic, Game-Theoretic, and Logical Foundations*, Cambridge University Press, Cambridge, UK, 2008.

[1379] Banachowski, L., Kreczmar, A., Mirkowska, G., Rastowa, H., and Salwicki, A., "An Introduction to Algorithmic Logic: Metamathematical Investigations in the Theory of Programs", *Banach Center Publications*, Vol. 2, 1977, pp. 7–99.

[1380] Mirkowska, G. and Salwicki, A., Algorithmic Logic, D. Reidel Publishing Company, PWN-Polish Scientific Publishers, Warszawa, Poland, 1987.

[1381] Skiena, S., *The Algorithm Design Manual*, Springer-Verlag London Limited, London, UK, 2008.

Mathematics

[1382] Samartskii, A.A. and Mikhailov, A.P., *Principles of Mathematical Modeling: Ideas, Methods, Examples*, CRC Press, Taylor & Francis Group, Boca Raton, FL, 2002.

[1383] Shen, S., *Introduction to Modern Mathematics Modeling – Developing Mathematical Models Using Data*, Wiley-Interscience, A John Wiley & Sons, Inc. Publication, 2016.

[1384] Bender, E., *An Introduction to Mathematical Modeling*, Dover Publications, Mineola, NY, Mar. 2000.

[1385] Bernacchi, G., *TENSORS made easy with SOLVED PROBLEMS*, Lulu.com, Self-Published, 2017.

[1386] Frieze, A. and Karoński, M., *Introduction to Random Graphs*, Cambridge University Press, Cambridge, United Kingdom, Nov. 2015.

[1387] Sidorov, Y., Fedoryuk, M., and Shabunin, M., *Lectures on the Theory of Functions of a Complex Variable*, Translated by E. Yankovsky, Mir Publishers, Moscow, Russia, 1985.

[1388] Mcgrayne, S., *The Theory that Would Not Die: How Bayes' Rule Cracked the Enigma Code, Hunted Down Russian Submarines, & Emerged Triumphant from Two Centuries of Controversy*, Yale University Press, New Haven CT, 2011.

[1389] Cybenko, G., "Approximation by Superpositions of a Sigmoidal Function", *Mathematics of Control, Signals, and Systems*, Vol. 2, Springer-Verlag, New York, NY, 1989, pp. 303–314.

[1390] Parr, T. and Howard, J., "The Matrix Calculus You Need for Deep Learning", arXiv:1802.01528v3, Jul. 2018.

[1391] Netz, R., *The Shaping of Deduction in Greek Mathematics: A Study in Cognitive History*, Cambridge University Press, Cambridge, UK, 2004.

[1392] Enderton, H., *A Mathematical Introduction to Logic*, Academic Press, A Hartcourt Science and Technology Co., London, UK, 1972.

[1393] Koller, D. and Friedman, N., *Probabilistic Graphical: Principles and Techniques*, The MIT Press, Cambridge, MA, 2009.

[1394] Evans, T., "A Heuristic Program to Solve Geometric-Analogy Problems", Proceedings of the Spring Joint Computer Conference, 1964, pp. 327–338.

[1395] Davis, M., "What Is a Computation?", In *Mathematics Today – Twelve Informal Essays*, L. Steen (Ed.), Random House Inc., Springer-Verlag, New York, NY, 1978, pp. 241–267.

Neural Networks

[1396] Kelley, H., "Gradient Theory of Optimal Flight Paths", *Aerospace Research Central (ARC) Journal*, Vol. 30, Oct. 1960, pp. 947–954.

[1397] Kulkarni, T., Whitney, W., Kohli, P., and Tenebaum, J., "Deep Convolutional Inverse Graphics Network", *Advances in Neural Information Processing Systems*, 2015.

[1398] Maass, W., Natschläger, and Markram, H., "Real-time Computing without Stable States: A New Framework for Neural Computation based on Perturbations", *Neural Computation*, Vol. 14, No. 11, 2002, pp. 2531–2560.

[1399] Huang, G.-B., Zhu, Q.-Y., and Siew, C.-K., "Extreme Learning Machine: Theory and Applications", *Neurocomputing*, Vol. 70, No. 1–3, 2006, pp. 489–501.

[1400] Ranzato, M., Poultney, C., Chopra, S. and LeCun, Y., "Efficient Learning of Sparse Representations with an Energy-based Model", *Proceedings of NIPS*, 2007.

[1401] Nielsen, M., *Neural Networks and Deep Learning*, Determination Press, San Francisco, CA, 2015.

[1402] Hayes, B., "First Links in the Markov Chain", *American Scientist*, Vol. 101, No. 2, 2013, p. 252.

[1403] Sabour, S., Frosst, N., and Hinton, G., "Dynamic Routing Between Capsules", *Advances in Neural Information Processing Systems*, 2017, pp. 3856–3866.

[1404] Kohonen, T., "Self-organized Formation of Topologically Correct Feature Maps", *Biological Cybernetics*, Vol. 43, No. 1, 1982, pp. 59–69.

[1405] Jaderberg, M., Simonyan, K., Zisserman, A., and Kavukcuoglu, K., "Spatial Transformer Networks", *Advances in Neural Information Processing Systems*, 2015, pp. 2017–2025.

[1406] Muselli, M., "Switching Neural Networks: A New Connectionist Model for Classification", *16th Italian Workshop on Neural Networks (WIRN 2005)/ International Workshop on Natural and Artificial Immune Systems, NAIS 2005: Neural Nets*, Vietri Sul Mare, Italy, Jun. 2005, pp. 23–30.

[1407] Dayhoff, J., *Neural Network Architectures: An Introduction*, International Thomson Computer Press, Devon, 1996.

[1408] Broomhead, D. and Lowe, D., "Radial Basis Functions, Multi-variable Functional Interpolation and Adaptive Networks", No. RSRE-MEMO-4148, Royal Signals and Radar Establishment Malvern, United Kingdom, 1988.

[1409] Elman, J., "Finding Structure in Time", *Cognitive Science*, Vol 14, No. 2, 1990, pp. 179–211.

[1410] Chung, J., Gulcehre, C., Cho, K., and Bengio, Y., "Empirical Evaluation of Gated Recurrent Neural Networks on Sequence Modeling", arXiv preprint arXiv:1412.3555, 2014.

[1411] Velikovi, P., "Everything is Connected: Graph Neural Networks", In *AI Methodologies in Structural Biology*, A. Bender, C. de Graaf, and N. O'Boyle (Eds.), Vol. 29, Elsevier, Amsterdam, Apr. 2023.

[1412] Graves, A., Wayne, G., Reynolds, M., Harley, T., Danihelka, I., Grabska-Barwinska, A., Colmenarejo, S., Grefenstette, E., Ramalho, T., Agapiou, J., Badia, A., Hermann, K., Zwols, Y., Ostrovski, G., Cain, A., King, H., Summerfield, C., Blunsom, P., Kaukcuoglu, K., and Hassabis, D., "Hybrid Computing Using a Neural Network with Dynamic External Memory", *Nature*, Vol. 538, 2016, pp. 471–476.

[1413] Burger, J., *Brain Theory from A Circuits and Systems Perspectives: How Electrical Science Explains Neuro-circuits, Neuro-systems, and Qubits*, Springer, New York, NY, 2013.

[1414] Gashler, M. and Martinez, T., "Temporal Nonlinear Dimensionality Reduction", *Proceedings of the International Joint Conference on Neural Networks*, San Jose, CA, Jul. 2011.

[1415] Vincent, P., Larohelle, H., Bengio, Y., and Manzagol, P.-A., "Extracting and Composing Robust Features with Denoising Autoencoders", *Proceedings of the 25th International Conference on Machine Learning, ACM*, 2008.

[1416] Haykin, S., *Neural Networks and Learning Machines*, Pearson, Prentice Hall, Upper Saddle River, NJ, 1999.

[1417] Lewis, F.L., Jagannathan, S., and Yesildirek, A., *Neural Network Control of Robot Manipulators and Nonlinear Systems*, Taylor & Francis, London, UK, 1999.

[1418] Haykin, S., *Neural Network: A Comprehensive Foundation*, Pearson Prentice Hall, Hoboken, NJ, 1999.

[1419] Neofytou, A., Chatzikostanis, G., Magkanaris, I., Smaragdos, G., Strydis, C., and Soudris, D., "GPU Implementation of Neural-Network Simulations based on Adaptive-Exponential Models", *2019 IEEE 19th International Conference on Bioinformatics and Bioengineering (BIBE)*, Athens, Greece, Oct. 2019.

[1420] Chen, R., Rubanova, Y., Bettencourt, J., and Duvenaud, D., "Neural Ordinary Differential Equations", arXiv:1806.07366v5, Dec. 2019.

[1421] Vaswani, A., Shazeer, N., Parmar, N., Uszkoreit, J., Jones, L., Gomez, A., Kaiser, L., and Polosukhin, I., "Attention is All You Need", *31st Conference on Neural Information Processing Systems (NIPS 2017)*, Long Beach, CA, Dec. 2017.

[1422] Zeiler, M., Krishnan, D., Taylor, G., and Fergus, R., "Deconvolutional Networks", *Computer Vision and Pattern Recognition (CVPR), 2010 IEEE Conference on.* IEEE, 2010.

[1423] Jaeger, H. and Haas, H., "Harnessing Nonlinearity: Predicting Chaotic Systems and Saving Energy in Wireless Communication", *Science*, Vol. 304, No. 5667, 2004, pp. 78–80.

[1424] Tenorio, M. and Lee, W.-T., "Self Organizing Neural Networks for the Identification Problem", *Advances in Neural Information Processing Systems 1 (NIPS 1988)*, D. Touretzky (Ed.), Denver, CO, Morgan Kaufmann, 1988.

[1425] Graves, A., Wayne, G. and Danihelka, I., "Neural Turing Machines", arXiv preprint arXiv:1410.5401, 2014.

[1426] Hinton, G. and Sejnowski, T., "Learning and Relearning in Boltzmann Machines", *Parallel Distributed Processing: Explorations in the Microstructure of Cognition*, Vol. 1, 1986, pp. 282–317.

[1427] LeCun, Y, Lamblin, P., Popovici, D., and Larochelle, H., "Gradient-based Learning Applied to Document Recognition", *Proceedings of the IEEE*, Vol. 86, No. 11, 1998, pp. 2278–2324.

[1428] Schuster, M., and Paliwal, K., "Bidirectional Recurrent Neural Networks", *IEEE Transactions on Signal Processing*, Vol 45, No. 11, 1997, pp. 2673–2681.

[1429] Bourlard, H., "Auto-Association by Multilayer Perceptrons and Singular Value Decomposition", *Biological Cybernetics*, Vol. 59, No. 4–5, 1988, pp. 291–294.

[1430] Kingma, D. and Welling, M., "Auto-Encoding Variational Bayes", arXiv preprint arXiv:1312.6114, 2013.

[1431] Amari, S.-I., "Learning Patterns and Pattern Sequences by Self-Organizing Nets of Threshold Elements", *IEEE Transactions on Computers*, Vol. C-21, No. 11, Nov. 1972, pp. 1197–1206.

[1432] Pearlmutter, B., "Gradient Calculations for Dynamic Recurrent Neural Networks: A Survey", IEEE Transactions on Neural Networks, Vol. 6, No. 5, Sept. 1995, pp. 1212–1228.

[1433] Anderson, J. and Rosenfeld, E., *Talking Nets: An Oral History of Neural Networks*, A Bradford Book, MIT Press, Cambridge, MA, 1998.

[1434] Haykin, S., *Neural Networks: A Comprehensive Foundation*, Pearson, Prentice Hall, Delhi, India, 2001.

[1435] Aggarwal, C., *Neural Networks and Deep Learning: A Textbook*, Springer International Publishing, Cham, Switzerland, 2018.

[1436] He, K., Zhang, X., Ren, S., and Sun, J., "Deep Residual Learning for Image Recognition", arXiv preprint arXiv:1512.03385, 2015.

[1437] McCulloch, W. and Pitts, W., "A Logical Calculus of the Ideas Immanent in Nervous Systems", *Bulletin of Mathematical Biophysics*, Vol. 5, 1943, pp. 115–133.

[1438] Fukushima, K., "Neocognitron: A Self-organizing Neural Network Model for a Mechanism of Pattern Recognition Unaffected by Shift in Position", *Biological Cybernetics, Vol. 36, Springer-Verlag*, 1980, pp. 193–202.

[1439] Ballard, D., "Modular Learning in Neural Networks", Proceedings of the Sixth National Conference on Artificial Intelligence (AAAI-87), Seattle, WA, Association for the Advancement of Artificial Intelligence, Jul. 1987, pp. 279–284.

[1440] Sutskever, I., Vinyals, O., and Le, Q., "Sequence to Sequence Learning with Neural Networks, 27th International Conference on Neural Information Processing (NIPS '14), Vol. 2, Montreal, Canada, MIT Press, Dec. 2014, pp. 3104–3112.

[1441] Ciresan, D., Meier, U., and Schmidhuber, J., "Multi-column Deep Neural Networks for Image Classification", Proceedings of the IEEE Conference on Computer Vision and Pattern Recognition (CVPR), Providence, RI, Feb. 2012, pp. 3642–3649.

[1442] Santoro, A., Raposo, D., Barrett, D., Malinowski, M., Pascanu, R., Battaglia, P., and Lillicrap, T., "A Simple Neural Network Module for Relational Reasoning", arXiv:1706.01427v1, Jun. 2017.

[1443] Hopfield, J., "Neural Networks and Physical Systems with Emergent Collective Computational Abilities", Proceedings of the National Academy of Science, Vol. 79, Biophysics, National Academies of Science, Washington, D.C. April 1982, pp. 2554–2558.

[1444] Schmidhuber, J., "Deep Learning in Neural Networks: An Overview", Technical Report IDSIA-03-14, and also arXiv:1404.7828v4, Oct 2014.

[1445] Simard, P., Steinkraus, D., and Platt, J., "Best Practices for Convolutional Neural Networks Applied to Visual Document Analysis", Seventh International Conference on Document Analysis and Recognition (ICDAR 2003), IEEE, Edinburgh, Scotland, Aug. 2003.

[1446] Ciresan, D. Meier, U., Masci, J., and Schmidhuber, J., "Multi-column Deep Neural Network for Traffic Sign Classification", *Neural Networks*, Vol. 32, Elsevier, Aug. 2012, pp. 333–338.

[1447] Ciresan, D. Meier, U., Masci, J., Gambardella, L., and Schmidhuber, J., "High-Performance Neural Network for Visual Object Classification", arXiv:1102.0183v1, Feb 2011.

[1448] Hinton, G. and Salakhutdinov, R., "Reducing the Dimensionality of Data with Neural Networks", Science, Vol. 313, American Association for the Advancement of Science, July 2006, pp. 504–507.

[1449] Srivastava, N., Hinton, G., Krizhevsky, A., Sutskever, I., and Salakhutdinov, R., "Dropout: A Simple Way to Prevent Neural Networks from Overfitting", *Journal of Machine Learning Research*, Vol. 15, MIT Press, Cambridge, MA, 2014, pp. 1929–1958.

[1450] Veličković, P., "Everything Is Connected: Graph Neural Networks", arXiv:2301.08210v1, Jan. 2023.

[1451] Lin, L.-J., "Programming Robots Using Reinforcement Learning and Teaching", Ninth Conference on Artificial Intelligence (AAAI-91), Association for the Advancement of Artificial Intelligence, Anaheim, CA, Jul. 1991, pp. 781–786.

[1452] Kriz, M., "Nonlinear Principal Component Analysis Using Autoassociative Neural Networks", *American Institute of Chemical Engineers (AIChE)*, Vol. 37, No. 2, Wiley & Sons, Feb. 1991, pp. 233–243.

[1453] Lippmann, R., "An Introduction to Computing with Neural Networks", *IEEE Acoustics, Speech, and Signal Processing (ASSP) Magazine*, Vol. 3, No. 4, Piscataway, NJ, Apr. 1987, pp. 4–22.

[1454] Hornik, K., Stinchcombe, M., and White, H., "Multilayer Feedforward Networks are Universal Approximators", *Neural Networks*, Vol. 2, Pergamon Press, 1989, pp. 359–366.

Neuro-Symbolic Artificial Intelligence

[1455] Goodman, R., Higgins, C., Miller, J., and Smyth, P., "Rule-Based Neural Networks for Classification and Probability Estimation", *Neural Computation*, Vol. 4, No. 6, Nov. 1992, pp. 781–804.

[1456] "Connectionist Models of Cognition and Perception", *Proceedings of the Seventh Neural Computation and Psychology Workshop*, J. Bullinaria and W. Lowe (Eds.), World Scientific, 2002.

[1457] Hitzler, P., Eberhart, A., Ebrahimi, M., Sarker, M.K., and Zhou, L., "Neuro-symbolic Approaches to Artificial Intelligence", *National Science Review*, Vol. 9, Mar. 2022, p. nwaco35.

[1458] Hilario, M., "An Overview of Strategies for Neurosymbolic Integration", books.google.com, Jan. 1996.

[1459] Greenspan, H., Goodman, R., and Chellappa, R., "Texture Analysis via Unsupervised and Supervised Learning", *Seattle International Joint Conference on Neural Networks (IJCNN-91)*, Vol. 1, IEEE, Piscataway, NJ, Jul. 1991, pp. 639–644.

[1460] Towell, G. and Shavlik, J., "Interpretation of Artificial Neural Networks: Mapping Knowledge-based Neural Networks into Rules", *Advances in Neural Information Processing Systems (NIPS2)*, Denver, CO, Morgan Kaufmann, Dec. 1991, pp. 977–984.

[1461] Svato, M., Sourek, G., and Zelezný, F., "Revisiting Neural-Symbolic Learning Cycle", *14th International Workshop on Neural-Symbolic Learning and Reasoning (NeSy 19), 28th International Joint Conference on Artificial Intelligence, Neural-Symbolic Learning and Reasoning Association*, Macao, China, Aug. 2019.

[1462] Besold, T., et al., "Neural-Symbolic Learning and Reasoning: A Survey and Interpretation", In *Neuro-Symbolic Artificial Intelligence: The State of the Art*, P. Hitzler and M. Sarker (Eds.), Vol. 342, IOS Press Ebook, Amsterdam, Dec. 2021, pp. 1–51.

[1463] Hatzilygeroudis, I. and Prentzas, J., "Neuro-symbolic Approaches for Knowledge Representation in Expert Systems", *International Journal of Hybrid Intelligent Systems*, Vol. 1, No. 3–4, 2004, pp. 111–126.

[1464] Garcez, A., Gori, M., Lamb, L., Serafini, L., Spranger, M., and Tran, S., "Neural-Symbolic Computing: An Effective Methodology for Principled Integration of Machine Learning and Reasoning", *Journal of Applied Logics*, Vol. 6, No. 4, 2019, pp. 612–631.

[1465] Hassabis, D., Kumaran, D., Summerfield, C., and Botvinick, M., "Neuroscience-Inspired Artificial Intelligence", *Neuron*, Vol. 95, Jul. 2017, pp. 245–258.

[1466] Moreno, M., Civitarese, D., Brandao, R., and Cerqueira, R., "Effective Integration of Symbolic and Connectionist Approaches through a Hybrid Representation", *2019 International Workshop on Neuro-Symbolic Learning and Reasoning (NeSy 2019)*, Macao, China, Aug. 2019.

[1467] Churchland, P., *Neurophilosophy: Towards a Unified Science of the Mind-Brain*, The MIT Press, Cambridge, MA, Sept. 1989.

[1468] Fu, L.-M. and Fu, L.-C., "Mapping Rule-Based Systems into Neural Architecture", *Knowledge-Based Systems*, Vol. 3, No. 1, Mar. 1990, pp. 48–56.

[1469] Cummins, R., *Meaning and Mental Representation*, Bradford Book, MIT Press, Cambridge, MA, 1989.

[1470] Marblestone, A., Wayne, G., and Kording, K., "Towards an Integration of Deep Learning and Neuroscience", *Frontiers in Computational Neuroscience*, Vol. 10, Sept. 2016, p. 94.

[1471] Lake, B., Ullman, T., Tenenbaum, J., and Gershman, S., "Building Machines that Learn and Think like People", *Proceedings of the 17th International Conference on Autonomous Agents and MultiAgent Systems (AAMAS)*, Stockholm, Sweden, Jul. 2018 (Also Published in Behavioral and Brain Sciences 2017, Cambridge University Press).

[1472] Ilkou, E. and Koutraki, M., "Symbolic Vs Sub-symbolic AI Methods: Friends or Enemies?", *Proceedings of the CIKM 2020 Workshop*, Galway, Ireland, Oct. 2020.

[1473] Moreno, M., Civitarese, D., Brandao, R., and Cerqueira, R., "Effective Integration of Symbolic Approaches through a Hybrid Representation", arXiv:1912:08740, Dec. 2019.

[1474] Hilario, M., Lallement, Y., and Alexandre, F., "Neurosymbolic Integration: Unified versus Hybrid", *European Symposium on Artificial Neural Networks (ESANN 1995)*, Brussels, Belgium, Apr. 1995, pp. 181–186.

[1475] Sun, R., "Hybrid Connectionist-Symbolic Modules: A Report from the IJCAI-95 Workshop on Connectionist-Symbolic Integration", *AI Magazine*, Vol. 17, No. 2, 1996, p. 99.

[1476] Moreno, M., Civitarese, D., Brandao, R., and Cerqueira, R., "Effective Integration of Symbolic and Connectionist Approaches through a Hybrid Representation", arXiv:1912.08740 [cs.AI], 2019.

[1477] Garcez, A., Besold, T., Raedt, L., Foldiak, P., Hitzler, P., Icard, T., Kuhnberger, K.-U., Lamb, L., Miikkulainen, R., and Silver, D., "Neural-Symbolic Learning and Reasoning: Contributions and Challenges", *Knowledge Representation and Reasoning: Integrating Symbolic and Neural Approaches, 2015 AAAI Spring Symposium*, Palo Alto, CA, Mar. 2015.

[1478] Cassimatis, N., "Integrating Cognitive Models based on Different Computational Methods", *Proceedings of the Annual Meeting of the Cognitive Science Society*, Vol. 27, No. 27, 2005, pp. 402–407.

[1479] Calegari, R., Ciatto, G., and Omicini, A., "On the Integration of Symbolic and Sub-symbolic techniques for XAI: A Survey", *Intelligenza Artificiale*, Vol. 14, No. 1, Sept. 2020, pp. 7–32.

[1480] Roth, D., "A Connectionist Framework for Reasoning: Reasoning with Examples", *Proceedings of the Thirteenth National Conference on Artificial Intelligence (AAAI-1996), Association for the Advancement of Artificial Intelligence*, Portland, OR, Aug. 1996.

[1481] Smolensky, P., Legendre, G., and Miyata, Y., "Integrating Connectionist and Symbolic Computation for the Theory of Language", *Current Science*, Vol. 64, No. 6, Mar. 1993, pp. 381–391.

[1482] Davidson, D., *Essays on Actions and Events*, Clarendon Press, Oxford, UK, 2001.

[1483] Besold, T., Garcez, A., Bader, S., Bowman, H., Domingos, P., Hitzler, P., Kuehnberger, K.-U., Lamb, L., Lowd, D., Lima, P.M.V., Penning, L., Pinkas, G., Poon, H., and Zaverucha, G., "Neural-Symbolic Learning and Reasoning: A Survey and Interpretation", arXiv:1711.03902v1, Nov. 2017.

[1484] Bader, S., Hitzler, P., and Hölldobler, S., "The Integration of Connectionism and First-Order Knowledge Representation and Reasoning as a Challenge for Artificial Intelligence", arXiv:cs/0408069v1, Aug. 2004.

[1485] Shastri, L., "The Relevance of Connectionism to AI: A Representation and Reasoning Perspective", MS-CIS-89-05, LINC Lab 140, Univ. of Pennsylvania, Philadelphia, PA, Sept. 1989.

Neuro-Symbolic AI

[1486] Walker, S., "A Brief History of Connectionism and its Psychological Implications", *AI and Society*, Vol. 4, No. 1, Springer Science + Business Media, Berlin, Germany, 1990, pp. 17–38.

[1487] Feldman, J. and Ballard, D., "Connectionist Models and Their Properties", *Cognitive Science*, Vol. 6, Cognitive Science Society, Ablex Publishing, Norwood, NJ, 1982, pp. 205–254.

[1488] Medler, D., "A Brief History of Connectionism", *Neural Computing Surveys*, Vol. 1, No. 2, Lawrence Erlbaum Associates, Inc., 1998, pp. 18–73.

[1489] Smolensky, P., "Information Processing in Dynamical Systems: Foundations of Harmony Theory", In *Parallel Distributed Processing, Vol. 1: Explorations in the Microstructure of Cognition: Foundations*, D. Rumelhart and J. McClelland (Eds.), The MIT Press, Cambridge, MA, Jan. 1986.

[1490] Thórisson, K., Benko, H., Abramov, D., Arnold, A., Maskey, S., and Vaseekaran, A., "Constructionist Design Methodology for Interactive Intelligences", *A.I. Magazine*, Vol. 25, No. 4, Association for the Advancement of Artificial Intelligence, Winter 2004.

[1491] Sheth, A., Roy, K., and Gaur, M., "Neurosymbolic AI – Why, What, and How", arXiv:2305.00813v1, May 2023.

[1492] Hassan, M., Guan, H., Melliou, A., Wang, Y., Sun, Q., Zeng, S., Liang, W., Zhang, Y., Zhang, Z., Hu, Q., Liu, Y., Shi, S., An, L., Ma, S., Gul, I., Rahee, M., You, Z., Zhang, C., Pandey, V., Han, Y., Zhang, Y., Xu, M., Huang, Q., Tan, J., Xing, Q., Qin, P., and Yu, D., "Neuro-Symbolic Learning: Principles and Applications in Ophthalmology", arXiv:2208.00374v1, Jul 2022.

[1493] *Neuro-symbolic Artificial Intelligence: the State of the Art*, P. Hitzler and K. Sarker (Eds.), IOS Press, Amsterdam, Netherlands, 2022.

[1494] Chalmers, D., "Subsymbolic Computation and the Chinese Room", In *The Symbolic and Connectionist Paradigms: Closing the Gap*, J. Dinsmore (Ed.), Lawrence Erlbaum, 1992.

[1495] Goel, A., "Looking Back, Looking Ahead: Symbolic versus Connectionist AI", *AI Magazine*, Vol. 42, Association for the Advancement of Artificial Intelligence, 2021, pp. 83–85.

[1496] Ilkou, E. and Koutraki, M., "Symbolic vs Sub-Symbolic AI Methods: Friends or Enemies?", Proceedings of the 29th ACM International Conference of Information and Knowledge Management (CIKM 2020), Galway, Ireland, Association for Computing Machinery, New York, NY, Oct. 2020.

Combined Traditional and Deep Learning Computer Vision

[1497] Kelley, T., "Symbolic and Sub-symbolic Representations in Computational Models of Human Cognition: What Can be Learned from Biology?", *Theory & Psychology*, Vol. 13, No. 6, 2003, pp. 847–860.

[1498] Stanovich, K. and West, R., "Individual Differences in Reasoning: Implications for the Rationality Debate?", *Behavioral and Brain Sciences*, Vol. 23, 2000, pp. 645–726.

[1499] Yosinski, J., Clune, J., Nguyen, A., Fuchs, T., and Lipson, H., "Understanding with Neural Networks through Deep Visualization", *Deep Learning Workshop, 31st International Conference on Machine Learning*, Lille, France, 2015.

[1500] Han, K., Wang, Y., Chen, H., Chen, X., Guo, J., Liu, Z., Tang, Y., Xiao, A., Xu, C., Xu, Y., Yang, Z., Zhang, Y., and Tao, D., "A Survey on Vision Transformer", *IEEE Transactions on Pattern Analysis and Machine Intelligence (PAMI)*, Vol. 45, No. 1, Jan. 2023, pp. 87–110.

[1501] Mahony, N., Campbell, S., Carvalho, A., Harapanahalli, S., Hernandez, G., Krpalkova, L., Riordan, D., and Walsh, J., "Deep Learning vs. Traditional Computer Vision", arXiv:1910.13796, Oct. 2019.

[1502] Eslami, S.M.A., et al., "Neural Scene Representation and Rendering", *Science Magazine*, Vol. 360, Jun. 2018, pp. 1204–1210.

[1503] Vig, J., "Visualization Attention in Transformer-based Language Representation Models", arXiv:1904.02679v2, Apr. 2019.

[1504] Yeh, C., Chen, Y., Wu, A., Chen, C., Vigas, F., and Wattenberg, M., "AttentionViz: A Global View of Transformer Attention", arXiv:12305.03210v2, Aug. 2023.

[1505] Orlandi, N., "Perception without Computation?", In *The Routledge Handbook of the Computational Mind*, M. Sprevak and M. Colombo (Eds.), Routledge Publishers, London, Aug. 2018, pp. 410–423.

[1506] Szegedy, C., Vanhoucke, V., Ioffe, S., Shlens, J., and Wojna, Z., "Rethinking the Inception Architecture for Computer Vision", *2016 IEEE Conference on Computer Vision and Pattern Recognition (CVPR 2016)*, Las Vegas, NV, Jun. 2016.

[1507] Szeliski, R., "Computer Vision: Algorithms and Applications", 2021, https://szeliski.org/Book.

[1508] Voulodimos, A., Doulamis, N., Doulamis, A., and Protopapadakis, E., "Deep Learning for Computer Vision: A Brief Review", *Computational Intelligence and Neuroscience*, Vol. 2018, 2018, p. 7068349.

[1509] Henderson, H., *Artificial Intelligence: Mirrors for the Mind*, Chelsea House Publishers, New Year, NY, 2007.

[1510] Simard, P., Steinkraus, D., and Platt, J., "Best Practices for Convolutional Neural Networks Applied to Visual Document Analysis", *Seventh International Conference on Document Analysis and Recognition (ICDAR 2003)*, IEEE, Edinburgh, Scotland, Aug. 2003.

[1511] Riesenhuber, M. and Poggio, T., "Hierarchical Models of Object Recognition in Cortex", *Nature Neuroscience*, Vol. 2, No. 11, Nov. 1999, pp. 1019–1025.

[1512] Schwarz, M., Schultz, H., and Behnke, S., "RGB-D Object Recognition and Pose Estimation based on Pre-trained Neural Network Features", *IEEE International Conference on Robotics and Automation (ICRA)*, Seattle, WA, May 2015.

[1513] Jones, N., Ross, H., Lynam, T., Perez, P., and Leitch, A., "Mental Models: International Synthesis of Theory and Methods", *Ecology and Society*, Vol. 16, No. 1, 2011, p. 46.

[1514] Bernard, D., Dorais, G., Gamble, E., Kanefsky, B., Kurien, J., Millar, W., Muscettola, N., Nayak, P., Rouquette, N., Rajan, K., Smith, B., Taylor, W., and Tung, Y.-W., "Remote Agent Experimentation Validation", *Deep Space 1 Technology Validation Symposium*, Pasadena, CA, Feb. 2000.

Mathematical Modeling

[1515] Samartskii, A.A. and Mikhailov, A.P., *Principles of Mathematical Modeling: Ideas, Methods, Examples*, CRC Press, Taylor & Francis Group, Boca Raton, FL, 2002.

[1516] Shen, S., *Introduction to Modern Mathematics Modeling – Developing Mathematical Models Using Data*, Wiley-Interscience, A John Wiley & Sons, Inc. Publication, New York, NY, 2016.

[1517] Bender, E., An Introduction to Mathematical Modeling, Dover Publications, Mineola, NY, Mar. 2000.

[1518] Hoare, G.T.Q., "1936: Post, Turing and 'A Kind of Miracle' in Mathematical Logic", *The Mathematical Gazette*, Vol. 88, No. 511, Mar. 2004, pp. 2–15.

[1519] Petzold, C., *The Annotated Turing: A Guided Tour through Alan Turing's Historic Paper on Computability and the Turing Machine*, Wiley Publishing, Inc., Indianapolis, IN, 2008.

[1520] Copeland, J. and Fan, Z., "Turing and Von Neumann: From Logic to the Computer", *Philosophies*, Vol. 8, No. 22, 2023, p. 22.

[1521] *Multiagent Systems: A Modern Approach to Distributed Modern Approach to Artificial Intelligence*, G. Weiss (Ed.), The MIT Press, Cambridge, MA, 1999.

[1522] Shannon, C., "Communication in the Presence of Noise", *Proceedings of the IRE*, Vol. 37, No. 1, Jan. 1949, pp. 10–21.

[1523] Shannon, C., "A Mathematical Theory of Communication", *The Bell System Technical Journal*, Vol. 27, Jul/Oct. 1948, pp. 379–423 and pp. 623–656.

[1524] Hartley, R.V.L., "Transmission of Information", *International Congress of Telegraphy and Telephony*, Lake Como, IT, Sept. 1927 and *The Bell System Technical Journal*, Vol. 7, No. 3, Jul. 1928, pp. 535–563.

[1525] Nyquist, H., "Certain Topics in Telegraph Transmission Theory", *Winter Convention of the A.I.E.E.*, New York, NY, Feb. 1928 and *Transactions of the A.I.E.E.*, New York, NY, Feb. 1928, pp. 617–644.

[1526] *The Essential Turing: Seminal Writings in Computing, Logic, Philosophy, Artificial Intelligence, and Artificial Life: Plus the Secrets of Enigma*, B.J. Copeland (Ed.), Oxford University Press, Oxford, UK, 2004.

[1527] Forouzan, B., *Foundations of Computer Science*, Cengage Learning EMEA, Hampshire, United Kingdom, 2018.

[1528] Zadeh, L., "Is Probability Theory Sufficient for Dealing with Uncertainty in AI: A Negative View", *Machine Intelligence and Pattern Recognition*, Vol. 4, 1986, pp. 103–116.

[1529] Zadeh, L., "The Role of Fuzzy Logic in the Management of Uncertainty in Expert Systems", Fuzzy Sets and Systems, Vol.11, No. 1–3, 1983, pp. 199–227.

[1530] Roth, D., "A Connectionist Framework for Reasoning: Reasoning with Examples", Rule-Based Reasoning & Connectionism, Proceedings of AAAI 96, AAAI, 1996, pp. 1256–1261.

[1531] Koffka, K., Principles of Gestalt Psychology, Kegan Paul, Trench, Trubner & Co., Ltd., London, England, 1935.

[1532] Shagrir, O., The Nature of Physical Computation, Oxford University Press, Oxford, UK, 2022.

[1533] Ciresan, D., Meier, U., and Schmidhuber, J., "Multi-Column Deep Neural Networks for Image Classification", Proceedings of the IEEE Conference on Computer Vision and Pattern Recognition (CVPR), Providence, RI, Feb. 2012, pp. 3642–3649.

[1534] Francois, C., "Systemics and Cybernetics in a Historical Perspective", *Systems Research and Behavioral Science System Research*, Vol. 16, 1999, pp. 203–219.

[1535] Sheth, A., Roy, K., and Gaur, M., "Neurosymbolic AI – Why, What, and How", arXiv: 2305.00813v1, May 2023.

[1536] Sowa, J.F., "Architectures for Intelligent Systems", IBM System Journal, Vol. 41, No. 3, 2002, pp. 331–349.

[1537] Xiong, W., Droppo, J., Huang, X., Seide, F., Seltzer, M., Stolcke, A., Yu, D., and Zweig, G., "Towards Human Parity in Conversational Speech Recognition", IEEE/ACM Transactions on Audio, Speech, and Language Processing, Vol. 25, No. 12, Dec. 2017, pp. 2410–2423.

[1538] Thórisson, K., Benko, H., Abramov, D., Arnold, A., Maskey, S., and Vaseekaran, A., "Constructionist Design Methodology for Interactive Intelligences", A.I. Magazine, Vol. 25, No. 4, Winter 2004, p. 77.

[1539] Shannon, C., "A Mathematical Theory of Communication", The Bell System Technical Journal, Vol. 27, Jul./Oct. 1948, pp. 379–423 and pp. 623–656.

[1540] Newell, A. and Simon, H., "Computer Science as Empirical Inquiry: Symbols and Search", *Communications of the ACM*, Vol. 19, No. 3, Mar. 1976, pp. 113–126.

[1541] Nilsson, N., "The Physical Symbol System Hypothesis: Status and prospects", In 50 Years of Artificial Intelligence, M. Lungarella, F. Iida, J. Bongard, and R. Pfeifer (Eds.), Lecture Notes in Computer Science, Vol. 4850, Springer, Berlin, Heidelberg, 2007.

[1542] Zadeh, L., "A New Direction in AI: Toward a Computational Theory of Perception", *AI Magazine*, Vol. 22, No. 1, 2001, p. 73.

[1543] Gao, K., Gao, Y., He, H., Lu, D., Xu, L., and Li, J., "NeRF: Neural Radiance Field in 3D Vision, Introduction and Review", arXiv: 2210.00379v5, Nov. 2023.

[1544] Ciresan, D. Meier, U., Masci, J., and Schmidhuber, J., "Multi-Column Deep Neural Network for Traffic Sign Classification", *Neural Networks*, Vol. 32, Aug. 2012, pp. 333–338.

[1545] Zhuge, M., Wang, W., Kirsch, L., Faccio, F., Khizbullin, D., and Schmidhuber, J., "Language Agents as Optimizable Graphs", arXiv: 2402.16823v2, Feb 2024.

[1546] Johnson-Laird, P.N., *Mental Models*, Harvard University Press, Cambridge, MA, Nov. 1983.

[1547] Ciresan, D., Meier, U., Masci, J., Gambardella, L., and Schmidhuber, J., "High-Performance Neural Network for Visual Object Classification", arXiv: 1102.0183v1, Feb 2011.

[1548] Hassan, M., et al., "Neuro-Symbolic Learning: Principles and Applications in Ophthalmology", arXiv: 2208.00374v1, Jul 2022.

[1549] *The Logical Legacy of Nikolai Vasiliev and Modern Logic*, V. Markin and D. Zaitsev (Eds.), Vol. 387, Studies in Epistemology, Logic, Methodology and Philosophy of Science, Springer International Publishing, Cham, Switzerland, 2017.

[1550] Korzybski, A., *Science and Sanity: An Introduction to Non-Aristotelian Systems and General Semantics*, Institute of General Semantics, Brooklyn, NY, 1933.

[1551] Netz, R., *The Shaping of Deduction in Greek Mathematics: A Study in Cognitive History*, Cambridge University Press, Cambridge, UK, 2004.

[1552] Kuhns, T., *The Structure of Scientific Revolutions*, The University of Chicago, Chicago, IL, 1962.

[1553] Locke, J., *An Essay Concerning Human Understanding*, Thomas Davison, Whitefriars, London, UK, 1825.

[1554] Vinge, V., "The Coming Technological Singularity: How to Survive in the Post-Human Era", Proceedings of the VISION-21 Symposium, NASA CP-10129, Cleveland, OH, Mar. 1993.

[1555] Van Vogt, A.E., The World of Null-A, Ace Books, Inc., Simon & Schuster, Inc., New York, NY, 1945.

[1556] Newell, A. and Simon, H., *Human Problem Solving*, Prentice-Hall, Inc., Englewood Cliffs, NJ, 1972.

[1557] Bronstein, M., Bruna, J., LeCun, Y., Szlam, A., and Vandergheynst, P., "Geometric Deep Learning: Going Beyond Euclidean Data", IEEE Signal Processing Magazine, Vol. 34, Jul. 2017, pp. 18–42.

[1558] Bahdanau, D., Cho, K.-H., and Bengio, Y., "Neural Machine Translation by Jointly Learning to Align and Translate", 3rd International Conference on Learning Representation (ICLR 2015), San Diego, CA, May 2015.

[1559] Beziau, J.-Y., "Is Modern Logic Non-Aristotelian?", In The Logical Legacy of Nikolai Vasiliev and Modern Logic, V. Markin and D. Zaitsev (Eds.), Springer-Verlag, Cham, Switzerland, 2017.

[1560] Hu, Y., Chun, B., Lin, J., Wang, Y., Wang, Y., Mehlman, C., and Lipson, H., "Human-Robot Facial Coexpression", *Science Robotics*, Vol. 9, Mar. 2024, p. eadi4724.

[1561] Hillis, W.D., *The Connection Machines*, The MIT Press, Cambridge, MA, 1985.

[1562] Drexler, K.E., Engines of Creation, Anchor Books, Random House, Inc., New York, NY, 1986.

[1563] Bartlett, F., Thinking: An Experimental and Social Study, Basic Books, Inc., New York, NY, 1958.

[1564] Thorndike, E. and the Staff of the Division of Psychology of the Institute of Educational Research of Teachers College, Columbia University, The Fundamentals of Learning, Bureau of Publications, Teachers College, Columbia University, New York, NY, 1932.

Data Mining

[1565] Bramer, M., *Principles of Data Mining*, Springer-Verlag, London, UK, 2007.

[1566] Hand, D., Mannila, H., and Smyth, P., Principles of Data Mining, The MIT Press, Cambridge, MA, 2001.
[1567] Stefferud, E., "The Logic Theory Machine: A Model Heuristic Program", Memorandum RM-3731-CC, The Rand Corporation, Santa Monica, CA, Jun. 1963.
[1568] Villani, M. and McBurney, P., "The Topos of Transformer Networks", arXiv: 2403.18415v1, Mar. 2024.
[1569] Blackwell, A., "The Two Kinds of Artificial Intelligence, or How not to Confuse Objects and Subjects", Interdisciplinary Science Reviews, Vol. 48, No. 1, Jan. 2023, pp. 5–14.
[1570] Agre, P., "Toward a Critical Technical Practice: Lessons Learned in Trying to Reform AI", In Bridging the Great Divide: Social Science, Technical Systems, and Cooperative Work, G. Bowker, L. Gasser, L. Star, and B. Turner (Eds.), L. Erlbaum Associates, Hillsdale, NJ, Apr. 1997.
[1571] Ghahramani, Z., "An Introduction to Hidden Markov Models and Bayesian Networks", International Journal of Pattern Recognition and Artificial Intelligence, Vol. 15, No. 1, 2001, pp. 9–42.
[1572] Goguen, J., "Towards a Social, Ethical Theory of Information", In Social Science Research, Technical Systems and Cooperative Work: Beyond the Great Divide, G. Bowker, L. Gasser, L. Star, and B. Turner (Eds.), L. Erlbaum Associates, Hillsdale, NJ, Apr. 1997.
[1573] Radford, A., Wu, J., Child, R., Luan, D., Amodei, D., and Sutskever, I., "Language Models are Unsupervised Multitask Learners", OpenAI Blog, Vol. 1, No. 8, 2019, p. 9.
[1574] Radford, A., Narasimhan, K., Salimans, T., and Sutskever, I., "Improving Language Understanding by Generative Pre-training", Work in Progress, 2018.
[1575] Radford, A., Kim, J.-W., Hallacy, C., Ramesh, A., Goh, G., Agarwal, S., Sastry, G., Askell, A., Mishkin, P., Clark, J., Krueger, G., and Sutskever, I., "Learning Transferable Visual Models from Natural Language Supervision", Preprint arXiv: 2103.00020v1, Feb. 2021.
[1576] Krizhevsky, A., Sutskever, I., and Hinton, G., "ImageNet Classification with Deep Convolutional Neural Networks", *NeurIPS Proceedings, Advances in Neural Information Processing Systems 25 (NIPS 2012) and in Communications of the ACM*, Vol. 60, No. 6, May 2017, pp. 84–90.
[1577] Graham, P., *On Lisp*, Prentice Hall, Hoboken, NJ, 1993.
[1578] *Deep Learning Algorithms*, Z. Gacovski (Ed.), Arcler Press, Burlington, ON, Canada, 2022.
[1579] *Deep Learning: Algorithms and Applications*, W. Pedrycz and S.- M. Chen (Eds.), Springer Nature Switzerland AG, Cham, Switzerland, 2020.
[1580] Sebesta, R., *Concepts of Programming Languages*, Pearson Education Inc., Upper Saddle River, NJ, 2004.
[1581] Zhou, Z.-H., Yu, Y., and Qian, C., *Evolutionary Learning: Advances in Theories and Algorithms*, Springer Nature Singapore Pte Ltd., Singapore, Singapore, 2019.
[1582] Thurner, S., Hanel, R., and Klimek, P., *Introduction to the Theory of Complex Systems*, Oxford University Press, Oxford, UK, 2018.
[1583] Raff, E., *Inside Deep Learning: Math, Algorithms, Models*, Manning Publications Co., Shelter Island, NY, 2020.
[1584] Prosise, J., *Applied Machine Learning and AI for Engineers: Solve Business Problems That Can't Be Solved Algorithmically*, O'Reilly Media, Sebastopol, CA, 2023.
[1585] Smolyakov, V., *Machine Learning Algorithm in Depth*, Manning Publications Co., Shelter Island, NY, 2023.
[1586] Hinton, G. and Salakhutdinov, R., "Reducing the Dimensionality of Data with Neural Networks", *Science*, Vol. 313, July 2006, pp. 504–507.
[1587] Srivastava, N., Hinton, G., Krizhevsky, A., Sutskever, I., and Salakhutdinov, R., "Dropout: A Simple Way to Prevent Neural Networks from Overfitting", *Journal of Machine Learning Research*, Vol. 15, 2014, pp. 1929–1958.
[1588] Ioffe, S. and Szegedy, C., "Batch Normalization: Accelerating Deep Network Training by Reducing Internal Covariate Shift", *Proceedings of the 32nd International Conference on Machine Learning*, Vol. 37, JMLR.org, Lille, France, Jul. 2015.
[1589] He, K., Zhang, X., Ren, S., and Sun, J., "Deep Residual Learning for Image Recognition", *2016 IEEE Conference on Computer Vision and Pattern Recognition (CVPR)*, Las Vegas, NV, *Computer Vision Foundation*, Jun. 2016.
[1590] Singhal, K., et al., "Large Language Models Encode Clinical Knowledge", *Nature*, Vol. 620, Aug. 2023, pp. 172–180.
[1591] Bernard, D., Dorais, G., Fry, C., Gamble, E., Kanefsky, B., Kurien, J., Millar, W., Muscettola, N., Nayak, P.P., Pell, B., Rajan, K., Rouquette, N., Smith, B., and Williams, B., "Design of the Remote Agent Experiment for Spacecraft Autonomy", *Proceedings of the* 1998 IEEE Aerospace Conference, Piscataway, NJ, Mar. 1998.
[1592] Sowa, J.F., "Architectures for Intelligent Systems", IBM Systems Journal, Vol. 41, No. 3, 2002, pp. 331–349.
[1593] Titiloye, O. and Crispin, A., "Quantum Annealing of the Graph Coloring Problem", Discrete Optimization, Vol. 8, 2011, pp. 376–384.
[1594] Kasneci, E., et al., "ChatGPT for Good? On Opportunities and Challenges of Large Language Models for Education", Learning and Individual Differences, Vol. 103, 2023, p. 102274.
[1595] Nixon, M. and Aguado, A., Feature Extraction and Image Processing, Newnes, Oxford, UK, 2002.
[1596] Rein, D., Hou, B., Stickland, A., Petty, J., Pang, R., Dirani, J., Michael, J., and Bowman, S., "GPQA: A Graduate-level Google-Proof Q & A Benchmark, arXiv: 2311.12022v1, Nov. 2023.
[1597] Li, P., Yang, J., Islam, M., and Ren, S., "Making AI Less "Thirsty", Uncovering and Addressing the Secret Water Footprint of AI Models", arXiv: 2304.03271v3, Oct. 2023.
[1598] Xiong, W., Droppo, J., Huang, X., Seide, F., Seltzer, M., Stolcke, A., Yu, D., and Zweig, G., "Achieving Human Parity in Conversations Speech Recognition", arXiv: 1610.05256v2, Feb. 2017.

[1599] Neuro-Symbolic Artificial Intelligence: The State of the Art, P. Hitzler and K. Sarker (Eds.), IOS Press, Amsterdam, Netherlands, 2022.
[1600] Kimmerly, W., Enterprise Transformation to Artificial Intelligence and the Metaverse: Strategies for the Technology Revolution, Mercury Learning and Information, Boston, MA, 2024.
[1601] Eagle, C. and Nance, K., The Ghidra Book: The Definitive Guide, No Starch Press, San Francisco, CA, 2020.
[1602] Kreher, D. and Stinson, D., Combinatorial Algorithms: Generation, Enumeration, and Search, CRC Press LLC, Boca Raton, FL, 1999.
[1603] Nijenhuis, A. and Wilf, H., Combinatorial Algorithms: For Computers and Calculators, Academic Press Inc., New York, NY, 1978.
[1604] Gloeckle, F., Idrissi, B., Roziere, B., Lopex-Paz, D., and Synnaeve, G. "Better & Faster Large Language Models via Multi-Token Prediction", arXiv: 2404.19737v1. Apr. 2024.
[1605] Rasmussen, J., *Information Processing and Human-Machine Interaction: An Approach to Cognitive Engineering*, Vol. 12, North Holland, Elsevier Science Publishing Co., New York, NY, 1986.
[1606] Chameleon Team, "Chameleon: Mixed-Modal Early-Fusion Foundation Models", arXiv: 2405.09818v1, May 2024.
[1607] Grand, G., Wong, L., Bowers, M., Olusson, T., Liu, M., Tenenbaum, J., and Andreas, J., "LILO: Learning Interpretable Libraries by Compressing and Documenting Code", arXiv: 2310.19791v4, Mar 2024.
[1608] Wong, L., Mao, J., Sharma, P., Siegel, Z., Feng, J., Korneev, N., Tenenbaum, J., and Andreas, J., "Learning Adaptive Planning Representations with Natural Language", arXiv: 2312.08566v1, Dec 2023.
[1609] Peng, A., Sucholutsky, I., Li, B., Sumers, T., Griffiths, T., Andreas, J., and Shah, J., "Learning with Language-Guided State Abstractions", *Proceedings of the Twelfth International Conference on Learning Representations, ICLR 2024*, Vienna, Austria, May 2024.
[1610] Raschka, S., *Build a Large Language Model (from Scratch)*, Manning Publication Co., Shelter Island, NY, 2023.

Computer Science

[1611] Brookshear, J.G. and Brylow, D., *Computer Science: An Overview*, Pearson Education Limited, Essex, England, 2015.
[1612] Parr, T. and Howard, J., "The Matrix Calculus You Need for Deep Learning", arXiv: 1802.01528v3, Jul. 2018.
[1613] Lester, B., Al-Rfou, R., and Constant, N., "The Power of Scale for Parameter-Efficient Prompt Tuning", arXiv: 2104.08691v2, Sep. 2021.
[1614] Kurakin, A., Ponomareva, N., Syed, U., MacDermed, L., and Terzis, A., "Harnessing Large-Language Models to Generate Private Synthetic Text", arXiv: 2306.01684v2, Jan. 2024.
[1615] Niu, X., Bai, B., Deng, L., and Han, W., "Beyond Scaling Laws: Understanding Transformer Performance with Associative Memory", arXiv: 2405.08707v1, May 2024.
[1616] Lu, K., Grover, A., Abbeel, P., and Mordatch, I., "Pretrained Transformers as Universal Computation Engines", arXiv: 2103.05247v2, Jun 2021.
[1617] Hu, E., Li, Y., Shen, Y., Wang, S., Wallis, P., Wang, L., Allen-Zhu, Z., and Chen, W., "LoRA: Low-Rank Adaption of Large Language Models", arXiv: 2106.09685v2, Oct 2021.
[1618] Naveed, H., Khan, A., Qiu, S., Saqib, M., Anwar, S., Usman, M., Akhtar, N., Barnes, N., and Mian, A., "A Comprehensive Overview of Large Language Models", arXiv: 2307.06435v9, Apr. 2024.
[1619] Bogoni, L. and Bajcsy, R., "Characterization of Functionality in a Dynamic Environment", University of Pennsylvania Technical Report No. MS-CIS-93-76, Sept 1993.
[1620] Lewis, P., Perez, E., Piktus, A., Petroni, F., Karpukhin, V., Goyal, N., Küttler, H., Lewis, M., Yih, W.-T., Rocktäschel, T., Riedel, S., and Kiela, D., "Retrieval-Augmented Generation for Knowledge-Intensive NLP Tasks", Thirty-Eight Annual *Conference on Neural Information Processing Systems (NeurIPS 2024)*, Vancouver, Canada, Dec. 2024.
[1621] James, G., Witten, D., Hastie, T., Tibshirani, R., and Taylor, J., An Introduction to Statistical Learning with Applications in Python, Springer Nature Switzerland AG, Cham, Switzerland, Jul. 2023.
[1622] James, G., Witten, D., Hastie, T., and Tibshirani, R., An Introduction to Statistical Learning with Applications in R, Springer Nature Switzerland AG, Cham, Switzerland, Jun. 2023.
[1623] Chollet, F., Deep Learning with Python, Manning Publications Co., Shelter Island, NY, 2018
[1624] Turner, R., "An Introduction to Transformers", arXiv: 2304.10557v5, Feb. 2024.
[1625] Whitley, D. and Watson, J.P., "Complexity Theory and the No Free Lunch Theorem", In Search Methodologies: Introductory Tutorials in Optimization and *Decision Support Techniques*, E. Burke and G. Kendall (Eds.), Springer Science + Business Media, New York, NY, 2005.
[1626] Wolpert, D. and Macready, W., "No Free Lunch Theorems for Optimization", IEEE Transactions on Evolutionary Computation, Vol. 1, No. 1, Apr. 1997, pp. 67–82.
[1627] The Aftermath of Syllogism: Aristotelian Logical Argument from Avicenna to Hagel, M. Sgarbi and M. Cosci (Eds.), Bloomsbury Academic, London, UK, 2018.

Index

Note: **Bold** page numbers refer to tables and *italic* page numbers refer to figures.

Printed in the United States
by Baker & Taylor Publisher Services